VACUUM TECHNOLOGY

VACUUM TECHNOLOGY

*Third, Updated
and Enlarged
Edition*

A. ROTH [†]

*Former Deputy Director, and
Head of Department of Vacuum Technology,
Soreq Nuclear Research Centre,
Israel*

ELSEVIER
AMSTERDAM • LAUSANNE • NEW YORK • OXFORD • SHANNON • TOKYO

ELSEVIER SCIENCE B.V.
Sara Burgerhartstraat 25
P.O. Box 211, 1000 AE Amsterdam, The Netherlands

Library of Congress Cataloging in Publication Data

Roth, Alexander.
Vacuum technology.

Includes bibliographical references and index.
I. Vacuum technology.   I. Title.
TJ940.R63   621.5'5   74-30919
ISBN 0-444-88010-0 (Elsevier)

First edition 1976
Second impression 1978
Third impression 1979
Second, revised edition 1982
Second impression 1983
Third impression 1986
Fourth impression 1989
Third, updated and enlarged edition 1990
Second impression 1994
Third impression 1996
Fourth impression 1998

ISBN 0 444 88010 0

Printed in The Netherlands

Professor Alexander Roth
(1921–1989)

In Memory of Professor Alexander Roth

Prior to the publication of this updated and enlarged edition of Vacuum Technology, Professor Alexander Roth died suddenly in September 1989. Professor Roth had completed this third revision early in 1989, and passed away during the galley proof stage.

The book Vacuum Technology has first been published in 1976. The vast success of this book is evidenced by the fact that the first edition was reprinted four times and the second edition three times. It is no exaggeration to say that his work is a classic, present in each laboratory working with vacuum techniques, and is used in many courses on this subject.

During his professional life Professor Roth published over 40 papers, as well as other books on vacuum technology. He was a member of the American Vacuum Society since 1963 and later was elected as one of its Senior Members. He founded the Israeli Society for Vacuum Technology in 1968 and was its first president. Professor Roth was also one of the individual founder members of the International Organization for Vacuum Science and Technology, IUVSTA, and was its treasurer for the period 1971–74. For almost two decades he was the head of the Vacuum & Technology Department at the Soreq Research Centre, Yavne, Israel, and later acted as the director of the Centre.

Professor Roth was a unique combination of a scientist, engineer and an excellent educator. His untimely death was a shock for many, and the void he leaves behind will be felt for a long time.

The Publisher

Preface

"The Aristotelians and Cartesians have said that there is no vacuum at all in the tube of Torricellius since glass has small pores, which the beams of light, the effluvia of the loadstone, and other very thin fluids may go through." *Third Paper of Leibnitz to Clarke.*

Although the phenomena are not precisely those assumed by the early investigators, the "horror vacuum" exists, at least here on the earth, where the surrounding exerts its "opposition" to our attempts at isolating a part of the space and completely evacuating its gas content. This is—in fact—just a particular case of the general tendency to oppose any disturbance of the equilibrium. In outer space, where the equilibrium pressure is in the vacuum range, no efforts are required to evacuate our vessel; on the contrary, many difficulties are encountered in keeping it at atmospheric pressure.

The "horror vacuum" appears to the vacuum technologist as the combined effect of conductance, leak rate, desorption, permeation, vapour pressure, diffusion, etc., and Vacuum Technology comprises to date the knowledge developed to avoid or to use these phenomena in order to achieve, measure and maintain lower and lower pressures.

This book is a result of the Postgraduate Course given by the author at the Faculty of Engineering of the Tel-Aviv University. It attempts to cover vacuum technology from the low vacuum up to the ultra-high vacuum as it is known today.

The material presented in the book was selected with a double aim:
—to give enough detailed explanations on principles and phenomena that the author considered as fundamental for any worker in the field, and
—to give the reader the necessary connections to the literature on those topics which are not treated in detail. The book includes numerical examples, tabulated and plotted data, as well as a collection of nomograms which may provide the answers to most of the questions which occur to workers in this field.

It is hoped that the book will provide a useful background for graduates and undergraduates in Universities and Technical Colleges, and serve as a "handbook" for scientists and engineers having to cope with problems of vacuum technology in R & D work or in industry.

To the many authors cited as reference throughout the book is due the credit for their publications, and to them I acknowledge my indebtedness. I wish to express my thanks to the authors and to the publishers for their kind permission to reproduce material from their publications.

I am very grateful to my daughter, Miss Michaela Roth, B.Sc., for preparing and drawing the figures for this book, and to Mrs. Rebeca Kuznetz for typing the manuscript.

I also wish to express my indebtedness to the Management and Staff of North-Holland Publishing Company, for their cooperation.

Alexander ROTH

Preface to the Third Edition

The first edition of this book (1976) was the result of a Postgraduate Course given by the author. The first, and then the second edition of the book (1982) were, and are still used in many Vacuum Technology courses, including some of those of the American Vacuum Society. Besides its role in educational activities, the book serves as one of the "handbooks" for those working in this field, or in fields connected to "Vacuum Technology".

Almost a decade has passed since the writing of the second edition. The activities of the Vacuum Technology were much extended as a backing to space technologies, surface investigations, solar energy, accelerators and storage rings, microelectronics, thermonuclear fusion, lasers, superconductors, biotechnologies.

As a result of these requirements the turbomolecular, the nonevaporable getter- and the cryopumps received important roles.

Extension of pumping technologies brought also some extension of pressure measuring techniques, especially regarding the calibration techniques and international comparison of standards.

Material technology succeeded to reduce outgassing rates by appropriate discharge and bombardment treatments – even on very large systems.

The progress of Vacuum Technology in the last decade was presented in the many national conferences held in various countries, in the last four international congresses of IUVSTA (Cannes – 1980, Madrid – 1983, Baltimore – 1986, Köln – 1989) and in the hundreds of papers published in the journals dedicated to High- and Ultrahigh vacuum and related subjects.

This third edition includes about 350 new papers added to the previous list of References.

The third edition is revised and updated with:
- Thermomolecular pumping;
- Throughput;
- Transmission probability;
- Electronic circuit simulation;
- Sorption on charcoal;
- Desorption from porous materials;
- Desorption from stainless steel, Al alloys (outgassing rates);
- Ion bombardment (glow discharge) cleaning;

- Claw-type pumps;
- Turbomolecular pumps – improvements;
- Cryosorption;
- NEG (nonevaporable getter) linear pumps;
- Standards for measurement of pumping speed (Recommended practice, test domes);
- Spinning rotor gauges;
- Quartz friction gauges;
- Increase of sensitivity of thermocouple gauges;
- Lubrication in vacuum;
- Calibration of diffusion leaks;
- Improvements in leak detection.

The author wants to express his indebtedness to all those authors and acknowledge the credit for their publication.

I also wish to express my thanks to the Staff of Elsevier Publishing Co. for their kind help and cooperation in the publishing of the third edition.

A. ROTH
August 1989

Contents

Commonly used symbols

A – area (also depth of grooves on sealing surfaces)
a – distance (also radius)
a, b – sides of rectangle
B – circumference, perimeter
C – conductance
C_r – compression ratio
c_p – specific heat (constant pressure)
c_v – specific heat (constant volume)
D – diameter
D_{12} – coefficient of diffusion
E – energy
e – charge of the electron
F – force
f – molecular sticking coefficient
h – height (also thickness)
i, I – current
J – mechanical equivalent of heat (also factor defining molecular-viscous flow)
K – heat conductivity (also correction factor molecular flow)
k – Boltzmann's constant
L – length
L_T, L_0 – latent heat of evaporation
m – mass (of molecule)
M – molecular weight
n – molecular density
N – total number (of molecules)
N_A – number of molecules per mole
P – pressure
$\overline{P}$ – average pressure
P_u – ultimate pressure
P_v – vapour pressure
P_r – probability factor
q – gas flow (molecules per second)

q_L – specific leak rate
q_D – specific outgassing rate
q_P – specific permeation rate
Q – throughput
r, R – radius
R – sealing factor
R_0 – gas constant (per mole)
R_e – Reynold's number
S – pumping speed
S_p – pumping speed at pump inlet
t – time
t_c – temperature ($^\circ$C)
T – temperature ($^\circ$K)
V – volume
v – velocity
w – width (of seals)
W – mass or specific mass (per sec, per cm^2)
Y – correction factor (viscous flow)
α (alpha) – accommodation coefficient
γ (gamma) – ratio c_p/c_v (also surface tension)
δ (delta) – molecular-viscous flow ratio
ε (epsilon) – slip coefficient
η (eta) – viscosity
λ (lambda) – mean free path
Λ (lambda) – free molecular heat conductivity
ξ (xi) – molecular diameter
ρ (rho) – density, mass per unit volume
τ (tau) – period (time)
ϕ (phi) – molecular incidence rate
ψ (psi) – correction factor (viscous flow)

CHAPTER 1

Introduction

1.1. The vacuum

Although the Latin word *vacuum* means "empty", the object of vacuum techniques is far from being spaces without matter. At the lowest pressures which can be obtained by modern pumping methods there are still hundreds of molecules in each cm³ of evacuated space.

According to the definition of the American Vacuum Society (1958) the term "vacuum" refers to a given space filled with gas at pressures below atmospheric, i.e. having a density of molecules less than about 2.5×10^{19} molecules/cm³.

The general term "vacuum" includes nowadays about 19 orders of magnitude of pressures (or densities) below that corresponding to the standard atmosphere. The lower limit of the range is continuously decreasing, as the vacuum technology improves its pumping and measuring techniques.

1.1.1. Artificial vacuum

Here on the earth vacuum is achieved by pumping on a vessel, the degree of vacuum increasing as the pressure exerted by the residual gas decreases below atmospheric. Measuring a system's absolute pressure is the traditional way to classify the degree of vacuum. Thus, we speak of low, medium, high and ultra-high vacuum corresponding to regions of lower and lower pressures (fig. 1.1).

At first approach the limits of these various ranges may look as arbitrary, since for each range there are specific kinds of pumps and measuring instruments. In fact, each of these various vacuum ranges corresponds to a different physical situation. In order to describe these situations it is useful to utilize the concepts of *molecular density*, *mean free path*, and *the time constant to form a monolayer*, concepts which are related to the pressure, as well as to the kind of gas and its temperature.

1

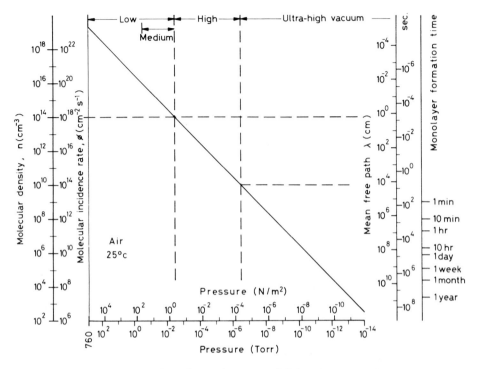

Fig. 1.1 Relationship of several concepts defining the degree of vacuum.

These terms will be mathematically analyzed in further chapters. For the sake of this introduction, they can be defined as:

– *Molecular density* is the average number of molecules per unit volume.

– *Mean free path* is the average distance that a molecule travels in a gas between two successive collisions with other molecules of that gas.

– *Time to form a monolayer* is the time required for a freshly cleaved surface to be covered by a layer of the gas of one molecule thickness. This time is given by the ratio between the number of molecules required to form a compact monolayer (about 8×10^{14} molec/cm^2) and the molecular incidence rate (at which molecules strike a surface).

Tables 1.1 and 1.2 list values of these terms, and fig. 1.1 shows their relationship.

By analysing the ranges shown on fig. 1.1 and the values of the terms listed in

Table 1.1.

Values of molecular density n, molecular incidence rate ϕ, mean free path λ, and time to form a monolayer τ, as a function of pressure P, for air at 25°C.

P Torr	n molec/cm^3	ϕ molec/cm^2·sec	λ cm	τ sec
760	2.46×10^{19}	2.88×10^{23}	6.7×10^{-6}	2.9×10^{-9}
1	3.25×10^{16}	3.78×10^{20}	5.1×10^{-3}	2.2×10^{-6}
10^{-3}	3.25×10^{13}	3.78×10^{17}	5.1	2.2×10^{-3}
10^{-6}	3.25×10^{10}	3.78×10^{14}	5.1×10^{3}	2.2
10^{-9}	3.25×10^{7}	3.78×10^{11}	5.1×10^{6}	2.2×10^{3}
10^{-12}	3.25×10^{4}	3.78×10^{8}	5.1×10^{9}	2.2×10^{6}
10^{-15}	3.25×10	3.78×10^{5}	5.1×10^{12}	2.2×10^{9}

Table 1.2.

Values* of ϕ, λ and τ for various gases at 25°C and 10^{-3} Torr.

Gas	ϕ molec/cm^2·sec	λ cm	τ sec
H_2	14.4×10^{17}	9.3	1×10^{-3}
He	10.4×10^{17}	14.7	2.3×10^{-3}
N_2	3.85×10^{17}	5.0	2.1×10^{-3}
O_2	3.60×10^{17}	5.4	2.4×10^{-3}
A	3.22×10^{17}	5.3	2.6×10^{-3}
Air	3.78×10^{17}	5.1	2.2×10^{-3}
H_2O	4.80×10^{17}	3.4	1.1×10^{-3}
CO_2	3.07×10^{17}	3.3	1.7×10^{-3}

*Notations as in table 1.1.

tables 1.1 and 1.2, it results that the physical situations characterizing the various *vacuum ranges are :*

Low (and medium) vacuum – the number of molecules of the gas phase is large compared to that covering the surfaces, thus in this range the pumping is directed toward rarefying the existing gas phase. The range extends from atmospheric pressure to about 10^{-2} Torr.

High vacuum – the gas molecules in the system are located principally on surfaces, and the mean free path equals or is greater than the pertinent dimensions of the enclosure. Therefore the pumping consists in evacuating or capturing the molecules leaving the surfaces and individually reaching (molecular flow) the pump.

This is the range where particles can travel in the vacuum enclosure without colliding with other particles. The range extends from about 10^{-3} to 10^{-7} Torr.

Ultra-high vacuum – the time to form a monolayer is equal or longer than the usual time for laboratory measurements, thus "clean" surfaces can be prepared and their properties can be determined before the adsorbed gas layer is formed. This vacuum range extends from about 10^{-7} to 10^{-16} Torr (lower limit decreasing with the progress of the technology). Hobson (1973) calculated that a pressure of 10^{-33} Torr can be achieved (theoretically) by cryopumping.

The classification of Kaminsky and Lafferty (1980) proposes to divide the lowest pressure range in: very high vacuum 10^{-4}–10^{-7} Pa (about 10^{-6}–10^{-9} Torr); ultra-high vacuum 10^{-7}–10^{-10} Pa (about 10^{-9}–10^{-12} Torr); and extreme ultrahigh vacuum less than 10^{-10} Pa (10^{-12} Torr).

Composition of the gas – While the total pressure in a vacuum chamber decreases, the composition of the gas phase changes as well. In the *low vacuum* range the composition of the gas mainly resembles that of the atmosphere (table 1.3). In the *high vacuum* range the composition changes continuously, toward one which contains 70–90 percent water vapour. The water molecules come from the sur-

Table 1.3.
Gas compositions.

Component	Atmosphere[1]		Ultra-high vacuum	
	Percent by volume	Partial pressure Torr	Partial pressure Torr	
			(2)	(3)
N_2	78.08	5.95×10^2	2×10^{-11}	—
O_2	20.95	1.59×10^2	—	3×10^{-13}
Ar	0.93	7.05	6×10^{-12}	—
CO_2	0.033	2.5×10^{-1}	6.5×10^{-11}	6×10^{-12}
Ne	1.8×10^{-3}	1.4×10^{-2}	5.2×10^{-11}	—
He	5.24×10^{-4}	4×10^{-3}	3.6×10^{-1}	—
Kr	1.1×10^{-4}	8.4×10^{-4}	—	—
H_2	5.0×10^{-5}	3.8×10^{-4}	1.79×10^{-9}	2×10^{-11}
Xe	8.7×10^{-6}	6.6×10^{-5}	—	—
H_2O	1.57	1.19×10^1	1.25×10^{-10}	9×10^{-13}
CH_4	2×10^{-4}	1.5×10^{-3}	7.1×10^{-11}	3×10^{-13}
O_3	7×10^{-6}	5.3×10^{-5}	—	—
N_2O	5×10^{-5}	3.8×10^{-4}	—	—
CO	—	—	1.4×10^{-10}	9×10^{-12}

(1) Norton (1962) p. 11, (2) Dennis and Heppel (1968) p. 105, (3) Singleton (1966) p. 355.

faces. As pumping is continued and heating is applied, the carbon monoxide content increases. In the *ultra-high vacuum* range hydrogen is the dominant component (table 1.3), coming mostly from the bulk of the materials (permeation).

1.1.2. *Natural vacuum*

Vacuum on earth Nature uses "low vacuum techniques" in some of the functions of life of animals, but no natural high vacuum is known on earth. Some of these "applications" are very vital, as our own respiration, others like the vacuum action of mosquitos are rather bothersome.

Human beings are pumping to about 740 Torr during their respiration, and may achieve pressures as low as 300 Torr by suction. The octopus is able to achieve pressures of about 100 Torr (Champeix, 1965).

Vacuum in space As the pressure of 760 Torr at sea level is a result of the "atmospheric column", the pressure decreases with the altitude. Up to 100 km altitude (troposphere and stratosphere) the pressure decreases quite regularly by a factor of 10 for each increase in altitude of 15 km, which results in a pressure of 10^{-3} Torr at about 90 km altitude. At higher altitudes *high vacuum exists*.

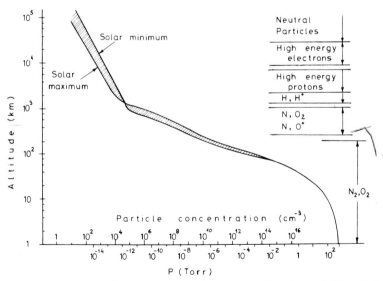

Fig. 1.2 Characteristics of the high altitude atmosphere surrounding the earth. Plotted after data from Nicolet (1960), Dushman and Lafferty (1962), Holkeboer *et al.* (1967), Rittehouse and Singletary (1968), Champion (1969).

The ionosphere (100–400 km) contains a large number of ionized atoms, and its pressure decreases only by a factor of 10 every 100–200 km. This decrease results in a pressure of about 10^{-10} Torr at an altitude of 1000 km. According to figs. 1.2 and 1.1 above 400 km, *ultra-high vacuum* conditions exist. Above this altitude the pressure decreases at an even slower rate, thus at 10 000 km a pressure of about 10^{-13} Torr exists.

Since the average spacecraft travels at a velocity considerably in excess of that of the average gas molecule, the pressures measured on spacecrafts are actually determined by the spacecraft velocity and gas particle concentration. Thus the diagram (fig. 1.2) of the high altitude atmosphere is expressed in concentration (density) units.

The gas molecule concentration (density) is estimated to fall in the shaded area of fig. 1.2, since the density varies with the time of day and the amount of solar activity. At an altitude below 200 km, the atmosphere is essentially air. Between 200–1000 km the gas is principally atomic nitrogen and oxygen, which may be largely ionized at periods of solar maxima. There is some evidence of an appreciable amount of helium at about 700–1000 km altitude. Above an altitude of 1500 km, the gas consists of neutral atomic hydrogen, protons and electrons.

1.2. Fields of application and importance

1.2.1. *Applications of vacuum techniques*

The large variety of applications of vacuum can be classified either according to the physical situation achieved by vacuum technology (table 1.4) or according to the fields where the application belongs.

Obviously each of the applications of vacuum technology utilizes one or more physical situations obtained by rarefying the gas. Some of them achieve products or facilities in which the vacuum exists during all their life (lamps, tubes, accelerators, etc.), others only use vacuum technology as a step in the production, the final product being used in atmospheric conditions (vacuum coating, drying, impregnation, etc.).

According to the physical situations created by vacuum the various applications may be resumed (table 1.4) as follows:

The pressure difference achieved by evacuating a vessel can realize forces on the walls up to 1 kg/cm². These forces are used for *holding* or *lifting* solids, for the *transport* of solids or liquids, and for *forming* (shaping) objects.

Plastic or rubber cups applied on surfaces so that the air be excluded from the cup, can *hold* small objects. The same principle is used to fasten tools on work

Table 1.4.
Applications of vacuum techniques.

Physical situation	Objective	Applications
Low pressure	Achieve pressure difference	Holding, Lifting Transport (pneumatic, cleaners, filtering) Forming
Low molecular density	Remove active atmospheric constituents	Lamps (incandescent, fluorescent, electric discharge tubes) Melting, sintering Packaging Encapsulation, Leak detection
	Remove occluded or dissolved gas	Drying, dehydration, concentration, Freeze drying, Degassing Lyophylisation, Impregnation
	Decrease energy transfer	Thermal insulation Electrical insulation Vacuum microbalance Space simulation
Large mean free path	Avoid collisions	Electron tubes, cathode ray tubes, television tubes, photocells, photo-multipliers, X-ray tubes Accelerators, storage rings, mass spectrometers, isotope separators, Electron microscopes, Electron beam welding, heating Coating (evaporation, sputtering), Molecular distillation
Long monolayer formation time	Clean surfaces	Friction, adhesion, emission studies, Materials testing for space

tables (chucks). Here the middle part of a large rubber membrane forming the base of the tool is mechanically pulled away, to form a vacuum enclosure with its periphery sitting on the table.

By using sniffers which are evacuated after being placed with their mouth on the object to be *lifted*, very small objects can be precisely lifted and transferred (e.g. filaments in the mass production of lamps). Relatively large (flat) objects can be lifted (plates, cars) if the mouth of the lifting cup is large, 5–7 tons can be lifted with a mouth of 1 m².

The vacuum cleaner is the simplest example of a widely used vacuum *transport* system. Vacuum cleaners are usually able to achieve pressures of 600 Torr, thus to suck objects of tens of grams/cm². Vacuum transport systems for grains and powders are based on features similar to vacuum cleaners.

The pneumatic transport systems connecting post offices in Paris or London are examples of very large vacuum transport facilities. That of Paris has a length of about 300 km, of double, 60 or 80 mm bore tubes; they are using a pressure of 450 Torr for the transport from post offices toward pumping stations, and an over-pressure of 0.8 atm for transport in the opposite direction. The transport cylinders containing the letters move at speeds of 8–10 m/s.

It is interesting to mention that pneumatic trains working on this principle were in function at Dublin (Ireland) and Saint–Germain (France) in the 1840–1860 years.

Vacuum is commonly used in laboratory and chemical industry to accelerate *filtering* speed. The pressure difference obtained by evacuation is used in the vacuum *forming* (molding) of plastics.

The necessity of *removing the chemically active constituents* of the atmosphere (oxygen, water vapour) by vacuum pumping appeared together with the invention of the incandescent lamps. In order to avoid oxidation of the filament heated at very high temperature, it must be an inert atmosphere. This atmosphere is constituted either by a high vacuum (about 10^{-6} Torr), or by an inert gas filled into the lamp after its evacuation at a high vacuum.

The possibility of evacuating large chambers at a high vacuum level is used in *vacuum metallurgy* to protect active metals from oxidation during melting, casting, sintering, etc. (Bunshah, 1958; Winkler and Bakish, 1971; Coates *et al.*, 1977).

Vacuum packaging of food, or materials sensitive to reactions with atmospheric components is used at a large scale in modern industry, the level of evacuation being usually in the low vacuum range. *Vacuum encapsulation* of sensitive devices (transistors, capacitors, etc.) is often carried out at high vacuum levels. The *leak testing* techniques using high sensitivity detectors can control the tightness of the encapsulation.

Vacuum technology is used to *remove humidity* from food, chemicals, pharmaceutic products, concrete, etc., and occluded (dissolved) gas from oils, plastics, etc. The fabrication of fruit juice, and concentrated milk, are examples of large-scale productions based on vacuum concentration. This process does not require extensive heating in order to evaporate the water or solvents contained in the products.

By using the *vacuum drying* process in conjunction with cooling, the products are first frozen, the water being then removed by sublimation. This is the basic feature of *freeze drying*. In the products of freeze drying the final water content is very low, chemical changes are minimal, volatile constituents are essentially kept in the product (e.g. instant coffee), coagulation is avoided (blood plasma) and storage properties are excellent (Pirani and Yarwood, 1961; De Luca, 1977).

Vacuum impregnation process consists in removing the occluded humidity or gases, and filling their place by another material. Although the commonly-known impregnation processes are those used to improve the dielectric properties of insulations (motor windings, capacitors, cables), vacuum impregnation techniques are also used to increase strength, or decrease combustibility of textiles, paper, wood, etc. (Holland–Merten, 1953).

High vacuum is a *thermal and electrical insulant*. This property is used in the Dewar flasks for the storage of liquid air, nitrogen, helium, etc., as well as in the "thermos flasks" used to keep cool drink or food. Both are double-walled flasks, the space between the walls being evacuated at high vacuum.

The *electrical insulation* properties of high vacuum are used in vacuum switches, as well as in high voltage devices (accelerators, tubes). Vacuum considerations play a central role in fusion devices for energy production such as Tokamaks and laser-fusion systems. These aspects are discussed by Lewin and Tenney (1974), Prévot (1974), Pustovoit (1974), Clausing (1976), Cohen (1976), Borghi and Ferrario (1977), Cullingford and Beal (1977), Heiland (1977), Abel (1978), Glaros *et al.* (1979), Gomay *et al.* (1979), Schwenterly *et al.* (1980), Wilson and Watts (1980), Murakami (1983), Cecchi and Knize (1984), Fuller and Haines (1984), Parris *et al.* (1984), Bennett (1987), Sedgley *et al.* (1987), Brooks *et al.* (1988), Sedgley *et al.* (1988), Yatsu *et al.* (1988).

As the energy transfer in outer space is similar to that which occurs in ultra-high vacuum, *space simulation* became one of the sophisticated applications of vacuum technology. Space simulator chambers extend to volumes of more than 1000 m³, and some of them are evacuated to the lowest pressures which can be achieved today, (e.g. Forth and Frank, 1977). A recent application is the Space Shuttle, a vacuum laboratory travelling in space (at 200–500 km altitude) and using the high vacuum existing outside; the Spacelab and its problems are discussed by Hamacher (1977), Kleber (1977), Mark (1977), Oran and Naumann (1977, 1978), Outlaw and Brock (1977), Seibert (1977), Falland (1981), Patrick (1981), Dauphin (1982), Curien and Rolfo (1983), Kleber (1983), Breckenridge and Russel (1986), Cohen (1986), Debe (1986, 1987), Czanderna and Thomas (1987), Schäfer and Häfner (1987).

Vacuum microbalance techniques use high and ultra-high vacuum to avoid any "background" resulting from the surrounding gas.

The large mean free paths existing in high vacuum is used to *avoid collisions* between molecules, electrons, ions, in electron tubes, photocells, cathode ray tubes, X-ray tubes, accelerators, mass spectrometers, electron microscopes, etc. The techniques used to achieve high and ultrahigh vacuum in large accelerators and storage rings are discussed by Cummings *et al.* (1971), Fischer (1974, 1977), Bostic *et al.* (1975), Halama and Aggus (1975), Aggus *et al.* (1977), Hartwig and Kouptsidis (1977a, c), Rees (1977), Trickett (1977), Bennet *et al.* (1978), Schuchman *et al.* (1979), Briggs *et al.* (1980), Bartelson (1983), Halama (1983), Reinhard (1983), Schuchman (1983), Ishimaru (1984), Bourgeois *et al.* (1987), Cruz *et al.* (1987), Halama and Hseuh (1987), Benvenuti and Francia (1988).

This same property is used in *vacuum coating* plants where the coating material evaporated from its source reaches the substrate being coated, by travelling in straight lines, without collisions. In this way thin films are deposited for a large number of optical, research, or ornamental uses (Holland, 1956; Maissel and Glang, 1970; Reale, 1976; Eckertova, 1977; Poulsen, 1977; Holland, 1978).

Molecular distillation is another field where high vacuum is used in order to obtain very pure fractions by evaporating and condensing the molecules without any collisions to other gas molecules (Burrows, 1960, 1973; Watt, 1963).

Ultra-high vacuum permits to study the real properties of surfaces (friction, adhesion, emission, etc.) since at these low pressures the *times of formation of a monolayer* are sufficiently long (hours, fig. 1.1).

1.2.2. *Importance of vacuum technology*

The list of applications of vacuum technology includes a large number of items which became symbols of the progress. From this point of view the importance of vacuum technology is evident.

The size of the field can be shown by the number of persons (scientists, engineers, technicians and workers) involved in the world in the various aspects of vacuum technology. This number was in 1965 over one million, receiving a total of salaries of about three milliard dollars. At that time it was evaluated that more than four milliard lamps and one milliard electron tubes were produced per year.

The number of persons active in the progress of vacuum science and technology can be evaluated to tens of thousands, according to the number of members of IUVSTA. I.U.V.S.T.A. is the International Union for Vacuum Science, Techniques and Applications, which includes (in 1987), 24 National Vacuum Societies.

The history and activities of IUVSTA are related by Lafferty (1987).

The number of commercial firms producing general and specialized vacuum equipment ranges to about 100 (companies ranging from hundreds to thousands of persons).

1.3. Main stages in the history of vacuum techniques

It can be considered that the history of vacuum techniques begins in 1643, when Torricelli discovered the vacuum which is produced at the top of a column of mercury when a long tube sealed at one end is filled with mercury and inverted in a trough containing Hg.

A review of the early applications of vacuum is presented by Madey (1984).

The pioneer period of vacuum techniques continues up to the invention of the electric lamp. In this period important theoretical and experimental scientific progress is achieved in the fundamentals of gas laws (Boyle–Mariotte, Charles, Gay–Lussac, Bernoulli, Avogadro, Maxwell, Boltzmann, etc.). The first progress in the practical use of vacuum was connected to the mechanical effects which can be achieved by using the pressure difference between vacuum and atmosphere. The classic experiment of Guericke (1654) showing that the two hemispheres of an 119 cm "evacuated" ball cannot be separated by pulling with 2×8 horses, demonstrated the atmospheric forces.

The application of this knowledge to drive railway cars (Dublin) was used only a few years, but the pneumatic–vacuum transport systems begun in 1850–1860 in London and Paris are still in use (slightly modernized!).

The development of the incandescent lamp (Edison, 1879) was also a consequence of the pumping system invented during the previous years (Toepler, Sprengel, see table 1.5). The McLeod gauge* (1874) gave for the first time the possibility of measuring low pressures. The incandescent lamp has shown the usefulness of low molecular densities (removal of the active atmospheric constituents), the cathode ray tube of Crookes (1879) was the first application of the increased mean free path, while the Dewar flask (1893) constitutes the first appliaction of vacuum thermal insulation.

The invention of the vacuum diodes (1902) and triode (1907), and of the tungsten filament (1909), begins the development of the electron tubes, and brought that of the incandescent lamps to a maturity (Langmuir, 1915). The "quality" of the vacuum used in the production of the incandescent lamps was revealed to be insufficient in the new field of electron tubes, and this brought about research and development work on pumping and measurement (Dunkel, 1975).

The Pirani gauge (1906), Gaede's (1915) and Langmuir's (1916) diffusion pumps, and the hot cathode ionization gauge 1916), opened the possibilities of high vacuum technology. The development of high vacuum technology continued up to the second world war, in the years 1935–1936 receiving three new items : the gas ballast pumps, the oil diffusion pump, and the Penning cold cathode ioniza-

---

* Marland (1973), Thomas and Leyniers (1974).

Table 1.5.
Stages in the history of vacuum techniques.

Year	Author	Work (Discovery)
1643	Evangelista Torricelli	Vacuum in the 760 mm mercury column
1650	Blaise Pascal	Variation of Hg column with altitude
1654	Otto von Guericke	Vacuum piston pumps, Magdeburg hemispheres
1662	Robert Boyle	Pressure–volume law of ideal gases
1679	Edme Mariotte	
1775	A.L. Lavoisier	Atmospheric air : a mixture of nitrogen and oxygen
1783	Daniel Bernoulli	Kinetic theory of gases
1802	J.A. Charles	Volume temperature law of gases
	J. Gay–Lussac	
1810	Medhurst	Propose first vacuum post lines
1811	Amedeo Avogadro	Constant molecular density of gases
1843	Clegg and Samuda	First vacuum railways (Dublin)
1850	Geissler and Toepler	Mercury column vacuum pump
1859	J.K. Maxwell	Gas molecule velocity laws
1865	Sprengel	Mercury drop vacuum pump
1874	H. McLeod	Compression vacuum gauge
1879	T.A. Edison	Carbon filament, incandescent lamp
1879	W. Crookes	Cathode ray tube
1881	J. Van der Waals	Equation of state of real gases
1893	James Dewar	Vacuum insulated flask
1895	Wilhelm Roentgen	X-rays
1902	A. Fleming	Vacuum diode
1904	Arthur Wehnelt	Oxide-coated cathode
1905	Wolfgang Gaede	Rotary vacuum pump
1906	Marcello Pirani	Thermal conductivity vacuum gauge
1907	Lee de Forest	Vacuum triode
1909	W.D. Coolidge	Powder metallurgy of tungsten, Tungsten filament lamp
1909	M. Knudsen	Molecular flow of gases
1913	W. Gaede	Molecular vacuum pump
1915	W.D. Coolidge	X-ray tube
1915	W. Gaede	Diffusion pump
1915	Irving Langmuir	Gas filled incandescent lamp
1915	Saul Dushman	The kenotron
1916	Irving Langmuir	Condensation diffusion pump
1916	O.E. Buckley	Hot cathode ionization gauge
1923	F. Holweck	Molecular pump
1935	W. Gaede	Gas-ballast pump
1936	Kenneth Hickman	Oil diffusion pump
1937	F.M. Penning	Cold cathode ionization gauge
1950	R.T. Bayard and D. Alpert	Ultra-high vacuum gauge
1953	H.J. Schwartz, R.G. Herb	Ion pumps

tion gauge, items which together with the Pirani gauge remained up to now the usual components of most vacuum systems.

After 1940 vacuum technology had a very large development in the direction of equipment for nuclear research (cyclotron, isotope separation, etc.), vacuum metallurgy, vacuum coating, freeze drying, etc.

Up to 1950 the usual vacuum range extended to 10^{-6}–10^{-7} Torr. Perhaps lower pressures were obtained before also, but no possibility existed for measuring lower pressures. The Bayard–Alpert gauge (1950) opened the way to measure lower pressures, in the range called later ultra-high vacuum. The ion-pumps produced after 1953 permitted to obtain very low pressures, and the so-called "clean vacuum" (Redhead, 1976).

In the past decade, the space research gave a new quantitative jump to vacuum techniques, by the numerous vacuum problems which had to be solved for space missions.

Hobson (1984) predicts that in the future, "vacuum technology" will continue to contribute to: Space Science, Surface Science, Semiconductor Technology, etc. The main vacuum application will be in very large accelerators and colliders using superconductivity and cryopumping as well as in research of new materials for fusion technology.

1.4. Literature sources

Vacuum Technology is based on a very extensive literature of *books*, *journals* and *conference transactions* dealing with the various aspects of the subject.

1.4.1. *Books*

The first book on vacuum was published in Latin 300 years ago : Ottonis de Guericke : *Experimenta Nova Magdeburgica de Vacuo Spatio* (J. Jansson, Amsterdam) 1672.

It was republished in German (Von Guericke, 1968). Guericke's activity is remembered by Jahrreis (1987).

Books on vacuum science and technology appeared after 1920; their list in chronological order is : Dushman (1922,) Newman (1925), Goetz 1926, Kaye (1927), Mönch (1937), Strong (1938), Yarwood (1943), Inanananda (1947), Bachman (1948), Tjagunov (1948), Dushman (1949), Guthrie and Wakerling (1949), Martin and Hill (1949), Jaeckel (1950), Korolev (1950), Wagner (1950), Davy (1951), Dunoyer (1951), Leblanc (1951), Reimann (1952), Holland–Merten (1953), Millner and Szalkay (1953), Zobac (1954), Groszkowski (1955), Heinze (1955), Roth (1955), Steyskal (1955), Laporte (1957), Champeix (1958), Morand

(1958), Danilin (1959), Kuznetzov (1959), Mönch (1959), Delafosse and Mongodin (1961), Pirani and Yarwood (1961), Boutry (1962), Buch (1962), Dushman and Lafferty (1962), Eschbach (1962), Pupp (1962/64), Turnbull *et al.* (1962), Barrington (1963), Guthrie (1963), Spinks (1963), Steinherz (1963), Beck (1964/66), Brunner and Batzer (1965), Champeix (1965), Grigorov and Kanev (1965), Lewin (1965), Rosebury (1965), Van Atta (1965), Wutz (1965), Castaner (1966), Diels and Jaeckel (1966), Holkeboer *et al.* (1967), Ward and Bunn (1967), Yarwood (1967), Dennis and Heppel (1968), Green (1968), Paty (1968), Henry (1971), Lapalle (1972), American Vacuum Soc. (1973a), Holland, Steckelmacher and Yarwood (1974), Duval (1975), Roth (1976), Göllnitz *et al.* (1978), Sigmond (1979). Weissler and Carlson (1979), O'Hanlon (1980), Harris (1979), Haefer (1981), Roth (1982), Wutz, Adam and Walcher (1982), Ferrario (1983), Stuart (1983).

Most of these books are in English, German or French; a few of those appeared in Russian, and those published in Bulgarian, Czech, Hungarian, Polish, Rumanian, Norvegian, Italian and Spanish are also included in this list.

The field of *ultra-high vacuum* is covered by books dealing exclusively with this specific subject : Roberts and Vanderslice (1963), Trendelenburg (1963), Redhead, Hobson and Kornelsen (1968), Robinson (1968), Juilet and Nieuwenhuyze (1969), Weston (1985).

Various aspects of vacuum technology were treated by specialized books, going into the very details of these subjects : *measurements* are treated by Leck (1964), Berman (1985), *pumping* by Power (1966), *sealing* by Roth (1966), and *leak detection* by Marr (1968), Maurice (1971), Wilson and Beavis (1976).

The technology of *materials* used in vacuum techniques was treated by Espe and Knoll (1936), Knoll (1959), Mönch (1959), Kohl (1960), Von Ardenne (1962), Rosebury (1965), Espe (1966/68), Kohl (1967), Katz (1974).

As regarding the *applications* of vacuum technology, the field of *thin films* is covered by the books of Holland (1956), Auwarter (1957), Berry *et al.* (1968), Maissel and Glang (1970), Reale (1976), Eckertova (1977), Stuart (1983), and that of *vacuum metallurgy* by Bunshah (1958), Cable (1960), Belk (1963), Winkler and Bakish (1971), *vacuum insulation* by Latham (1981), and *micro-weighing* in vacuum by Czanderna and Wolsky (1980).

Weber (1968) and Hurrle *et al.* (1973) published dictionaries of high vacuum and Brombacher (1961) a bibliography on vacuum literature.

A vacuum technology vocabulary was published by ISO/DIS-1981 and a dictionary of vacuum terms by Kaminsky and Lafferty (1980).

1.4.2. *Journals*

The journals exclusively dedicated to vacuum techniques and their applications are :

a. Vacuum (England), monthly, founded 1950.
b. Vakuum-Technik (Germany), monthly, founded 1952.
c. Le Vide (France), bimonthly, founded 1945.
d. Journal of Vacuum Science and Technology (USA), bimonthly, founded 1965.
e. Vuoto (Italy), trimestrial, founded 1968.
f. Nederlands Tijdschrift voor Vacuumtechniek (The Netherlands).
g. Shinkum (Vacuum) (Japan), founded 1958.
h. Vacuum Science and Technology (China), founded 1980.

Besides these journals, a large number of papers on vacuum techniques and applications are published in: British J. Applied Phys., Experimentelle Technik der Physik, Japan J. Applied Physics (U.S.A.), J. Scientific Instruments, Materials Evaluation, Nuclear Instruments and Methods, Review of Scientific Instruments, Research and Development, R.C.A. Review, Solid State Technology, and Thin Solid Films.

The *bibliography* is covered by:

a. Index Bibliographique du Vide (Vol. 1–4, 1966–1969), Internat. Union for Vacuum Science, Bruxelles.
b. Vacuum and Surface Physics Index (founded 1965), Max Planck Inst., Garching bei München.
c. NASA – Scientific and Technical Aerospace Reports.

1.4.3. *Conference transactions*

The International Congresses of Vacuum Science, Techniques and Applications are held every three years, and their Transactions (Proceedings) were published as follows:

a. 1st International Congress, held Namur (Belgium) 10–13 June, 1958, 2 volumes, (Pergamon Press, Oxford) 824p.
b. 2nd Internat. Congress, held Washington 16–19 Oct., 1961, together with the 8th Symposium of the American Vacuum Society, 2 volumes, (Pergamon Press, Oxford) 1351p.
c. 3rd Internat. Congress, held Stuttgart, 28 June–2 July, 1965, 4 volumes, (Pergamon Press, Oxford) 929p.
d. 4th Internat. Congress, held Manchester, 17–20 Apr., 1968, 2 volumes, (Instit. Physics London) 827p.
e. 5th Internat. Congress, held Boston, Mass. 11–15 Oct., 1971, 1 volume, (American Vacuum Society, New York) 1018p.
f. 6th Internat. Congress, held Kyoto (Japan), 25–29 March, 1974, Vol. 1 (Japanese Journal of Applied Physics, Suppl. 2), 865p.

g. 7th Internat. Congress, held Vienna (Austria), 12–16 Sept., 1977, Vol. 1
    (R. Dobrozemsky *et al.*, Vienna), 378p.
h. 8th Internat. Congress, held Cannes (France), 22–26 Sept., 1980, 4 volumes
    (Le Vide, Paris), 3570 p.
i. 9th Internat. Congress, held Madrid (Spain), 26–30 Sept., 1983, 2 volumes
    (Spanish Vacuum Society).
j. 10th Internat. Congress, held Baltimore (USA), 27–31 Oct. 1986, 4 volumes
    (American Vacuum Society, New York).
k. 11th Internat. Congress, Köln (W. Germany), 26–29 Sept. 1989.

Most of the National Vacuum Societies publish the Proceedings of their
Annual Symposia as well as their Conferences on Specific Subjects in the
Journals mentioned in §1.4.2.

1.4.4. *Visual and audio-visual aids*

The cooperative work of specialists from various countries produced a set of
326 slides (mainly in colour) covering the subjects of vacuum technology (fundamentals, gauges, pumps, gas analyzers, vacuum deposition, leak detection). This
set of slides (and explanatory booklets to each part) is available from the
International Vacuum Union (IUVSTA, 1978). A video tape course on Vacuum
Technology (11 cassettes, 10 hours) covering fundamentals, pumps, gauges, is
also available (Anderson, 1978).

CHAPTER 2

Rarefied gas theory for vacuum technology

2.1. Physical states of matter

A collection of molecules can occur either in the solid, liquid or gaseous state depending on the strength of the intermolecular forces, and the average kinetic energy per molecule (temperature).

The state in which molecules are most independent of each other is called an *ideal* or *perfect gas*. This is a theoretical concept which corresponds to the assumptions that : (a) the molecules are minute spheres; (b) their volume is very small compared with that actually occupied by the gas; (c) the molecules do not exert forces upon each other; (d) they travel along rectilinear paths in a perfectly random fashion; (e) the molecules make perfectly elastic collisions.

Some real gases, such as hydrogen, nitrogen, oxygen, argon, helium, krypton, neon, xenon, approximate closely at atmospheric pressures the behavior assumed for ideal gases. At lower pressures (vacuum) many more gases approach the ideal gases.

Real gases, unlike ideal ones, have intermolecular forces. At pressures and temperatures where the molecules of the gas are brought close to each other they will begin to form new structures, which will have properties very different from those of the gas. When these new structures begin to form, the gas is said to be *liquefying*.

Figure 2.1 shows a plot of pressure versus volume for different temperatures of a real gas (e.g. carbon dioxide). Curves A and B, for which the temperatures are high, are hyperbolas conforming to Boyle's law, describing a behavior assumed for ideal gases. At temperature T_3, curve C is no longer completely hyperbolic.

A small bump has formed at point P. At still lower temperatures, curves D and E show complete departure from the hyperbola of ideal gases; a flat plateau appears. When the system has the pressure and volume associated with points

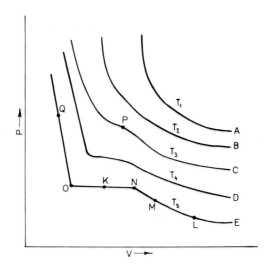

Fig. 2.1 Variation of pressure with volume in a real gas at various temperatures. $T_1 > T_2 > T_3 > T_4 > T_5$.

L or M, the material (CO_2) is in the gaseous state. Along the plateau N–O the temperature and pressure of the system are both constant while the volume changes. At N the material is in gaseous state, while at O it is a liquid. At K a fraction of the system is liquid. It is important to note that each curve (fig. 2.1) has only one plateau, that is, there is *only one pressure* for a given temperature, at which the gas will liquefy. At temperatures higher than that of curve C, *there is no pressure* at which the gas can be liquefied. Point P on curve C is called the *critical point*. Table 2.1 lists the values of the critical temperature and pressure for various gases.

From fig. 2.1 it is clear that at higher pressures the liquefaction process takes place at higher temperatures. The temperature at which a gas liquefies is called the *boiling point* and depends on the pressure of the system. For example, to boil at 20°C, water requires that the pressure of the surrounding gas be 17.54 Torr (table 2.2), mercury requires 1.2×10^{-3} Torr, while CO_2 requires 56.5 atm.

The pressure exerted by the molecules on the surrounding atmosphere and liquid is called the *vapour pressure*. The vapour pressure depends on the temperature of the material.

The *boiling point* of a liquid is that temperature at which the vapour pressure of the liquid is equal to the surrounding pressure. The *normal boiling point* corresponds to a vapour pressure of one atmosphere. If the vapour pressure is plotted vs. the temperature, curves as that in fig. 2.2 are obtained. At the right side

Table 2.1.
Triple points and critical points of various substances.

Substance	Formula	Triple point °C	Normal boiling point °C	Critical point	
				Temp. °C	Pressure Atm.
Water	H_2O	0.01	100.00	374.0	217.7
Carbon dioxide	CO_2	−56.6	−78.4*	31.1	73.0
Carbon monoxide	CO	−205.0	−191.5	−139.0	35.0
Oxygen	O_2	−218.8	−182.9	−118.8	49.7
Nitrogen	N_2	−210.0	−195.8	−147.1	33.5
Argon	A	−189.3	−185.8	−122.4	48.0
Neon	Ne	−248.7	−246.0	−228.5	26.9
Hydrogen	H_2	−259.2	−252.8	−239.9	12.8
Helium[4]	He^4	—	−268.9	−267.9	2.26
Helium[3]	He^3	—	−269.9	−269.8	1.17
Krypton	Kr	−157.2	−153.4	−63.8	54.2
Carbon tetrachloride	$C Cl_4$	−23.0	76.7	263.0	45.0
Freon 11	$CF Cl_2$	−111.0	23.8	198.0	43.2
Freon 12	CF_2Cl_2	−158.0	−29.8	112.0	40.1
Freon 22	CHF_2Cl	−160.0	−40.8	96.0	48.7
Methane	CH_4	−184.0	−161.5	−82.1	45.8

*Sublimation.

(below) of these curves vapour exists, while at the left side (above) of the curve liquid exists. Table 2.2 lists the vapour pressures of water and mercury at various temperatures.

If (fig. 2.1) the liquid is compressed at pressures above point Q, a second plateau appears (fig. 2.3); it is here that the liquid undergoes a change of phase, into a solid.

The temperature corresponding to the liquid–solid phase change at atmospheric pressure is called the *freezing* or *melting* point.

The solid–liquid transition point (freezing at various pressures) varies according to curves as those shown on fig. 2.4, and their slope is negative, or positive, depending if the substance expands on freezing (e.g. water) or contracts on freezing (e.g. mercury). The known experiment of the ice cut by the wire under load shows that as the pressure is increased the "freezing point" is lowered.

At all points on the vapour pressure curves (fig. 2.2) the liquid and vapour coexist in equilibrium, and at all points on the "freezing point" curves (fig. 2.4)

Table 2.2.
Vapour pressure of water (ice), and mercury (Torr).

Temperature (°C)	Water	Mercury	Temperature (°C)	Water	Mercury
−183	1.4×10^{-22}	3.48×10^{-32}	30	31.82	2.8×10^{-3}
−150	7.4×10^{-15}	—	40	55.32	6.1×10^{-3}
−140	2.9×10^{-10}	—	50	92.51	1.27×10^{-2}
−130	6.98×10^{-9}	—	60	149.3	2.52×10^{-2}
−120	1.13×10^{-7}	—	70	233.7	4.82×10^{-2}
−110	1.25×10^{-6}	—	80	355.1	8.88×10^{-2}
−100	1.1×10^{-5}	2.39×10^{-11}	90	525.7	1.58×10^{-1}
−90	7.45×10^{-5}	—	100	760	2.72×10^{-1}
−80	4.1×10^{-4}	2.38×10^{-9}	150	3570.4	2.80
−70	1.98×10^{-3}	1.68×10^{-8}	200	11 650	17.28
−60	8.1×10^{-3}	9.89×10^{-8}	250	29 817	74.37
−50	2.9×10^{-2}	4.94×10^{-7}	300	64 432	246.8
−40	9.7×10^{-2}	2.51×10^{-6} (−38.9°C)	400	—	1574.0
−30	2.9×10^{-1}	4.78×10^{-6}	500	—	7691
−20	7.8×10^{-1}	1.81×10^{-5}	600	—	22.8 atm
−10	1.95	6.06×10^{-5}	700	—	52.5 atm
0	4.58	1.85×10^{-4}	800	—	103.3 atm
10	9.2	4.9×10^{-4}	900	—	180.9 atm
20	17.54	1.2×10^{-3}	1000	—	290.5 atm

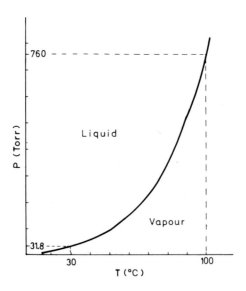

Fig. 2.2 Vapour pressure curve (water).

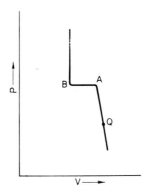

Fig. 2.3 Liquid-to-solid phase change, represented by the plateau AB. Curve AQ is identical with curve QO from fig. 2.1.

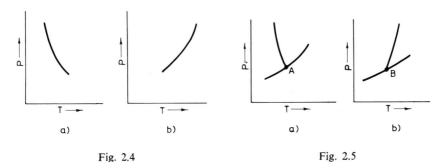

Fig. 2.4   Fig. 2.5

Fig. 2.4 Liquid-to-solid transition curves: (a) for substances which expand upon freezing; (b) for substances which contract upon freezing.

Fig. 2.5 Boiling, freezing and sublimation curves: (a) for substances which expand upon freezing, (b) for substances which contract upon freezing. Points A and B refer to the respective triple points.

the solid and liquid coexist. At the intersection of these two curves (fig 2.5) all three phases coexist. The point is called the *triple point* of the substance. Values of the pressure and temperature corresponding to the triple point of various substances are listed in table 2.1.

At pressures and temperatures below the triple point substances are changing from the solid to the vapour phase without passing through the liquid phase. This process is known as *sublimation*, and the line representing pressure – temperatures at which a solid and its vapour coexist – is the *sublimation curve* (fig. 2.5).

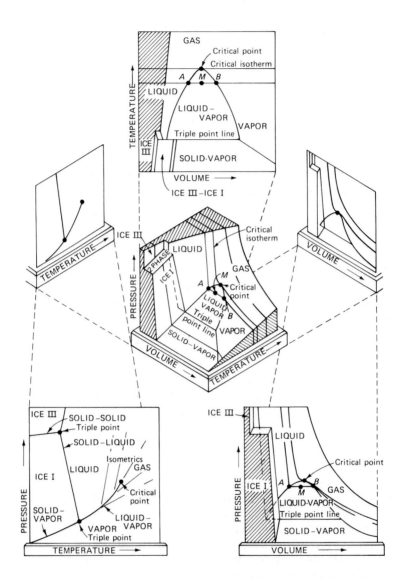

Fig. 2.6 The $P—V—T$ surface for water and its projections. Reprinted from Himmelblau (1974), by permission of Prentice-Hall Inc., Englewood Cliffs, New Jersey.

Any equation of state which describes the changes in a thermodynamic system must be a function of three variables: pressure, volume and temperature, and such an equation can be represented by a three-dimensional $P—V—T$ surface (fig. 2.6). The equations of this surface are discussed in §2.2.3 and 2.2.5.

2.2. Perfect and real gas laws

2.2.1. *Boyle's law*

By an ideal or perfect gas we mean one which obeys Boyle's law at all temperatures. The relationship established by Boyle (1662) and Mariotte (1679) is valid for gases over those ranges of pressures and temperatures for which the forces between the molecules of the gas can be considered negligible. Referring to table 2.1, at temperatures higher than the critical point any gas behaves as a perfect gas. The hyperbolas A and B on fig. 2.1 represent the Boyle's law

$$PV = \text{const} \tag{2.1}$$

Considering two different points on a hyperbola the relationship between them is written

$$P_1 V_1 = P_2 V_2 \tag{2.2}$$

describing an isothermal compression.

If the apparatus shown in fig. 2.7a is considered, it can be seen that for any position of the mercury column, the pressure P of the enclosed gas is equal to the atmospheric pressure P_0 minus the gauge pressure caused by the column of mercury of height h. The product of the pressure P and volume V remains a constant.

Equation (2.2) can be used for determining volumes by measuring pressures or vice versa. The volume can be determined either by expanding the gas into the unknown volume (and into supplementary, known volumes) and measuring the fractional decrease of pressure (Lee and Peavey, 1976; Norström *et al.*, 1977; Fortucci and Meyer, 1979), or by displacing the gas at a constant pressure and measuring the amount displaced (Kendall, 1974). If the volumes are known (calibrated), the principle described by eq. (2.2) can be used for determining the pressure of the gas.

This principle was used by McLeod (1874) in his high vacuum gauge, which has remained until now the reference gauge in calibrating other vacuum gauges.

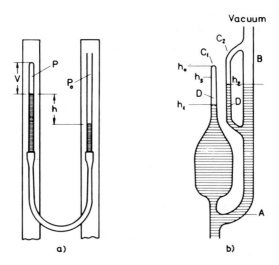

Fig. 2.7 (a) Boyle's law apparatus; (b) McLeod gauge.

The essential elements of a *McLeod gauge* are shown in fig. 2.7b, and consist of a glass bulb with a capillary tube extension on the top, a side arm connecting to the vacuum system, and some means of raising and lowering the mercury level within the gauge (see §6.3.5). By lowering the level of the mercury below A (fig. 2.7b), the bulb of volume V is evacuated through side arm B, and the gas in the bulb reaches the same pressure P as that in the system (vacuum). By raising the level of the mercury the bulb is cut off from the side arm and the sample of gas trapped in the bulb is compressed into the capillary C_1. The capillary C_2 is in parallel with a section of the side arm B and has the same bore D as C_1. Since the surface tension or capillary effect in C_1 and C_2 is the same, the difference in level of the mercury is due only to the pressure difference resulting from compression of the gas sample from the large volume V into the small volume of C_1 above the mercury level h_1. The pressure of the compressed gas in the closed capillary is proportional to $P + (h_2 - h_1)$.

According to Boyle's law

$$[P+(h_2-h_1)] \, A \, (h_o-h_1)=PV \tag{2.3}$$

and

$$P=A(h_2-h_1) \, (h_o-h_1)/[V-A(h_o-h_1)] \tag{2.4}$$

where P is the pressure (vacuum) of the system to be measured; V is the volume of of the bulb; A is the cross sectional area of the capillaries C_1, C_2; h_o is the effective height of the closed end of C_1; h_2 is the mercury level in capillary C_2.

A McLeod gauge may conveniently be read by bringing the mercury level up to the point where $h_2 = h_o$ (i.e. the level in the open capillary opposite the end of the closed capillary) or the mercury level can be set at some standard level h_s in the closed capillary. In the *first method* with $h_2 = h_o$, and $h_2 - h_1 = (\Delta h)_1$, the pressure is

$$P = A(\Delta h)_1^2 / [V - A(\Delta h)_1] \approx (A/V) (\Delta h)_1^2 \qquad (2.5)$$

By using a volume $V = 1300$ cm^3, and capillaries of 0.63 mm bore ($A = 0.32$ mm^2), for a difference $(\Delta h)_1 = 1$ mm between the levels in the open and closed capillaries, the pressure which can be determined is $P = 0.32/(1.3 \times 10^6 - 0.32) = 2.4 \times 10^{-7}$ Torr. In the *second method* with $h_1 = h_s$, $h_o - h_s$ is a constant, thus the pressure is

$$P = A (h_o - h_s) (h_2 - h_s) / [V - A(h_o - h_s)] \approx (A/V) (h_o - h_s) (\Delta h)_2 \qquad (2.6)$$

where $(\Delta h)_2 = h_2 - h_s$.

Thus the first method results in a pressure reading proportional to the *square* of the reading (2.5) whereas the second method leads to eq. (2.6) in which the pressure is proportional to the first power of the reading (*linear* scale).

Details on the McLeod gauge are discussed in §6.3.5.

2.2.2. Charles' law

Charles and Gay-Lussac (1802) observed that at constant volume the pressure of the gas increases linearly with its temperature, and that at constant pressure, the same phenomena happen with the volume. Their experimental results were described by the relations

$$P_t = P_o(1 + \beta t_c) \qquad (2.7)$$

and

$$V_t = V_o(1 + \beta t_c) \qquad (2.8)$$

where t_c is the temperature in degrees Celsius,

$$\beta = 1/273;$$

P_o and V_o pressure and volume at 0°C. The extrapolation of these expressions to $P_t = V_t = 0$ has shown that this would theoretically happen at $t_c = -273°C$. On this basis the *absolute temperature scale* was established, where the zero point of the scale was set at $t_c = -273.16°C$ exactly, so that the temperature T in °K (degrees Kelvin) is given by

$$T = t_c + 273$$

By introducing this last relation in eqs. (2.7) and (2.8) it results that

$$P_T = (T/273)P_o \tag{2.9}$$

and

$$V_T = (T/273)V_o \tag{2.10}$$

2.2.3. The general gas law

Consider an ideal gas which at a given instant is specified by the thermodynamic quantities P_o, V_o, T_o. The change of these initial quantities, to a set of final co-ordinates P_2, V_2, T_2, can be interpreted as first expanding the gas at constant pressure to coordinates P_1, V_1, T_1 and then expanding it at constant temperature to P_2, V_2, T_2 (fig. 2.8). In the isobaric expansion AB, Charles' law is valid, where $P_1 = P_o$, while in the isothermal expansion BC, Boyle's law can be used, where $T_1 = T_2$. By combining these two laws it results

$$V_1 = V_o(T_1/T_o) = V_o(T_2/T_o) \tag{2.11}$$

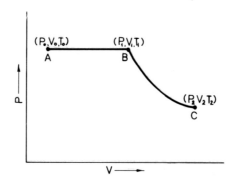

Fig. 2.8 Isobaric expansion (AB) followed by an isothermal expansion (BC).

$$V_1 = V_2(P_2/P_1) = V_2(P_2/P_o) \tag{2.12}$$

thus

$$P_oV_o / T_o = P_2V_2 / T_2 \tag{2.13}$$

which shows that for a given gas PV/T is a constant value.

It was found that by relating PV/T to the concept of *mole*, its value is the same for any perfect gas. A *mole is the weight in grams equal numerically to the molecular weight of a substance.* One mole of oxygen (O_2) is 32 g, one mole of H_2 is 2.016 g, thus one mole of H_2O is 18.016 g.

Avogadro demonstrated that at the same temperature and pressure the mass of a standard volume of gas is proportional to its molecular weight. Experiments have shown that under standard conditions of temperature ($0°C$ or $273.16°K$) and pressure (normal atmospheric pressure defined as 760 Torr), one *mole* (or gram molecular weight) of any gas occupies a volume of 22 415 cm^3 (22.4 liter). Values of molecular weights of some gases are listed in table 2.3.

Table 2.3.
Molecular weight M of gases.

Gas	Formula	M(g/mole)
Helium	He	4.003
Neon	Ne	20.18
Argon	A	39.944
Krypton	Kr	83.70
Xenon	Xe	131.30
Hydrogen	H_2	2.016
Nitrogen	N_2	28.02
Air	—	28.98 (average)
Oxygen	O_2	32.00
Chlorine	Cl_2	70.91
Ammonia	NH_3	17.03
Carbon monoxide	CO	28.01
Carbon dioxide	CO_2	44.01
Methane	CH_4	16.01
Acetylene	C_2H_2	26.04

For one *mole* of gas, the expression

$$PV / T = R_o \tag{2.14}$$

was written, where R_o is an universal constant*.

The numerical value of R_o depends upon the units of pressure, volume and temperature used. If the pressure is measured in Torr, the volume in liters and the temperature in degrees Kelvin, then under standard conditions where $P=760$ Torr, $V=22.415$ liters and $T=273.16°K$, the value of R_o is

$$R_o = PV/T = 760 \times 22.415/273.16 = 62.364 \text{ Torr} \cdot \text{lit}/°K \cdot \text{mole}.$$

By expressing pressure and volume in CGS units, $1 \text{ atm} = 1.0133 \times 10^6$ dynes/cm², and

$$R_o = (1.0133 \times 10^6)(2.24 \times 10^4)/273.16 = 8.314 \times 10^7 \text{ ergs}/°K \cdot \text{mole}.$$

since $1 \text{ cal} = 4.186 \times 10^7$ ergs, $R_o \approx 2$ cal/$°K \cdot$mole. Table 2.4 lists numerical values of R_o for various systems of units.

Table 2.4.
Numerical values of R_0 for various systems of units.

P	V	T	R_0
dyne/cm²	cm³	$°K$	8.314×10^7 erg/$°K$ mole
Newton/m²	m³	$°K$	8.314 Joule/$°K$ mole
Torr	cm³	$°K$	6.236×10^4 Torr·cm³/$°K$ mole
Torr	liter	$°K$	62.364 Torr·liter/$°K$ mole
atm	cm³	$°K$	82.057 atm·cm³/$°K$ mole

For a gas sample of mass W, of a gas having a molecular weight M, the general gas law is written

$$PV = (W/M)R_o T \tag{2.15}$$

2.2.4. Molecular density

Avogadro concluded that equal volumes of all gases under the same conditions of temperature and pressure contain equal numbers of molecules.

The number of molecules in one mole is defined as Avogadro's number, N_A. By X-ray techniques that accurately determine the interatomic spacing of solid crystals, the mass of the hydrogen atom is known to be 1.67×10^{-24} g. The mole-

*The symbol R refers to H.V. Regnault (1810–1878), successor of Gay-Lussac.

cular weight of H_2 (mole of hydrogen) being 2.016 g, it results that

$$N_A = 2.016/(2 \times 1.67 \times 10^{-24}) = 6.023 \times 10^{23} \text{ molec/mole.}$$

The Avogadro number also results from the precise measurement of the Faraday, $F = 96\ 488$ coulomb, defined as the electrical charge necessary to deposit a mole of a substance in electrolysis. The charge of an electron being $e = 1.602 \times 10^{-19}$ coulomb, it follows that

$$N_A = F/e = 96\ 488/(1.602 \times 10^{-19}) = 6.023 \times 10^{23} \text{ molec/mole.}$$

In eq. (2.15) W/M denotes the number of moles, thus

$$(W/M)\ (N_A/V) = n \tag{2.16}$$

is the *number of molecules per unit volume*. From eqs. (2.15) and (2.16), it results that

$$n = (N_A/R_0)\ (P/T) \tag{2.17}$$

thus if P is expressed in Torr, and R_0 in Torr·cm^3/°K, the number of molecules per cm^3 is given by

$$n = [6.023 \times 10^{23}/(6.236 \times 10^4)]\ (P/T) = 9.656 \times 10^{18}\ (P/T) \tag{2.18}$$

At normal pressure ($P = 760$ Torr) and temperature ($T = 273.16$°K), eq. (2.18) gives $n = 2.687 \times 10^{19}$ molec/cm^3, which is known as the *Loschmidt number*.

From the molecular weight M and the Avogadro number N_A, it results that *the mass of a molecule*, m (in grams), is

$$m = (M/N_A) = 1.66035 \times 10^{-24}M \tag{2.19}$$

The *distances between molecules* can be visualized by using a model in which all the molecules are steady and at the same distance from each other. In this case, the distance L (cm) between molecules is given by using eq. (2.18), and is

$$L = n^{-1/3} \approx 4.6 \times 10^{-7} \left[\frac{T}{P} \right]^{1/3} \text{ (cm)} \tag{2.20}$$

which gives at $T = 273$°K

$L \approx 3.3 \times 10^{-7}$ cm for $P = 760$ Torr,
$L \approx 3.0 \times 10^{-6}$ cm for $P = 1$ Torr,
$L \approx 3.0 \times 10^{-3}$ cm for $P = 10^{-9}$ Torr.

It should be mentioned that these distances are (much) smaller than the mean free path (see §2.4.1), but are (very) large compared to the molecular diameters (see table 2.9). Eq. (2.17) can also be written

$$P = n(R_o/N_A)T = nkT \tag{2.21}$$

by using $k = R_o/N_A$, known as the *Boltzmann constant*.
The value of the Boltzmann constant is

$$k = 8.314 \times 10^7/(6.023 \times 10^{23}) = 1.3805 \times 10^{-16} \text{ erg}/^\circ\text{K}$$

At very low pressures ($P \leqslant 10^{-10}$ Pa $\approx 7 \times 10^{-13}$ Torr) the walls of the vessel/pipe have a high sticking coefficient f (only a part of the incident gas molecules are reemitted). Da and Da (1987) calculated that in this case eq. (2.21) has to be written as

$$P = \bar{n}(2 - f)kT/2 \tag{2.21a}$$

where $\bar{n}$ is the average density of the gas molecules in the chamber at a given time.

The *molecular weight of gas mixtures* is established by using eq. (2.15). The partial pressure of the various gases being $P_1, P_2, P_3 \ldots P_n$, their masses $W_1, W_2 \ldots W_n$, and their molecular weights $M_1, M_2 \ldots M_n$, eq. (2.15) becomes

$$PV = (\Sigma P_n)V = [\Sigma(W_n/M_n)]R_oT \tag{2.22}$$

If the average molecular weight of the mixture is $\overline{M}$, then

$$PV = (W/\overline{M})R_oT = (\Sigma W_n/\overline{M})R_oT \tag{2.23}$$

thus

$$\overline{M} = (\Sigma W_n)/\Sigma(W_n/M_n) \tag{2.24}$$

2.2.5. Equation of state of real gases

The general gas law (eqs. 2.14, 2.15) is valid for the region above the critical point (fig. 2.6) where the matter is in a state of gas. The $P-V$ curve of real gases, (fig. 2.1) shows a flat plateau, corresponding to the liquid–gas transition. Near the critical point the behavior of real gases can be described very satisfactorily by a modified form of eq. (2.14), deduced by *Van der Waals*,

$$[P + (A/V^2)](V - b) = R_oT \tag{2.25}$$

In this equation, the term A/V^2 takes account of the fact that the attractive forces between molecules will bring them closer together and will thus have the same effect as an additional pressure governed by the constant A. This "pressure" must be the stronger, the closer the molecules are together, hence A is divided by V^2.

The correction b reduces the total volume, b representing that part of it which is occupied by the molecules themselves. The volume which is excluded was established to be for each molecule four times that of the molecule itself, thus

$$b = 4N_A \ (4/3) \ (\pi\xi^3/8) \tag{2.26}$$

where N_A is Avogadro's number, and ξ is the *molecular diameter*.

The plot of eq. (2.25) appears in fig. 2.9. Exclusive of the region inside the dashed curves (region of liquid–vapour equilibrium), fig. 2.9 agrees with fig. 2.1 (experimental data); point P corresponds to the critical point on fig. 2.1. The dashed portions of the curves which show the pressure and volume both decreasing simultaneously, such as RS, are physically untenable. However in the region where Van der Waals' equation fails to agree with experimental results, the plateau can be inserted so that the areas of the two lobes I and II are equal (fig. 2.9). With this understanding, Van der Waals' equation can be used as a fair approximation of the behavior of real gases. Attempts have been made to explain the portions ST and UR of the curves, by asserting that they refer to the states of supercooled gases and superheated liquids.

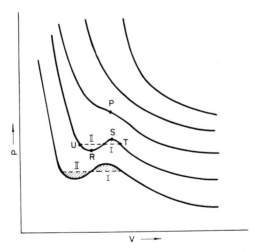

Fig. 2.9 Isotherms corresponding to Van der Waals' equation of state.

The values A and b in eq. (2.25) were determined by writing that at the critical point the three roots where the curve cuts the horizontal (fig. 2.9) are equal.

Eq. (2.25) can be written

$$V^3 - [b+(R_oT)/P]\ V^2+(A/P)\ V - Ab/P = 0 \tag{2.27}$$

while the cubic equation with three roots at the critical volume V_c, is

$$(V - V_c)^3 = V^3 - 3V_c V^2 + 3V_c^2 V - V_c^3 = 0 \tag{2.28}$$

Comparing the coefficients of eq. (2.28) with those of eq. (2.27), it results that the constants are

$$b = \tfrac{1}{8} R_o T_c/P_c \text{ and } A = 27b^2 P_c \tag{2.29}$$

The values of the constants calculated for various gases are listed in table 2.5.

Table 2.5.
Critical constants T_c, P_c, Van der Waals' constants A, b, molecular diameters ξ, and mean free paths λ, computed from eqs. (2.26), (2.29), (2.56).

Gas	Formula	T_c	P_c	A	b	ξ	$\lambda*$
		°C	atm	$\left[\dfrac{cm^3}{mole}\right]^2 \cdot atm$	$cm^3/mole$	cm	cm
Helium	He	−267.9	2.26	3.412×10^4	23.70	2.61×10^{-8}	9.26×10^{-3}
Neon	Ne	−228.5	25.9	2.107×10^5	17.09	2.38×10^{-8}	1.03×10^{-2}
Argon	A	−122.0	48.0	1.345×10^6	32.19	2.94×10^{-8}	7.34×10^{-3}
Krypton	Kr	−63.0	54.2	2.318×10^6	39.78	3.16×10^{-8}	6.38×10^{-3}
Xenon	Xe	16.6	58.2	4.194×10^6	51.05	3.43×10^{-8}	5.40×10^{-3}
Hydrogen	H_2	−239.9	12.8	2.450×10^5	26.61	2.76×10^{-8}	8.33×10^{-3}
Nitrogen	N_2	−147.1	33.5	1.390×10^6	39.13	3.14×10^{-8}	6.44×10^{-3}
Air	—	—	—	$1.33\ \times10^6$	36.6	—	—
Oxygen	O_2	−118.8	49.7	1.360×10^6	31.83	2.93×10^{-8}	7.40×10^{-3}
Mercury	Hg	>1500	>200	$8\ \ \times10^6$	17.0	2.38×10^{-8}	1.12×10^{-2}
Ammonia	NH_3	132.4	111.5	$4.17\ \times10^6$	37.07	3.09×10^{-8}	6.68×10^{-3}
Carbon monoxide	CO	−139.0	35.0	1.485×10^6	39.9	3.16×10^{-8}	6.36×10^{-3}
Carbon dioxide	CO_2	31.1	73.0	$3.59\ \times10^6$	42.67	3.22×10^{-8}	6.13×10^{-3}
Acetylene	C_2H_2	36.0	62.0	$4.39\ \times10^6$	51.40	3.44×10^{-8}	5.38×10^{-3}

*Mean free path for $P=1$ Torr, $T=273°K$.

2.3. Motion of molecules in rarefied gases

2.3.1. *Kinetic energy of molecules*

The kinetic theory of gases rests upon the fundamental assumptions that the matter is made up of molecules, and that the molecules of a gas are in constant motion, intimately related to the temperature of the gas.

During their motion the molecules suffer collisions between themselves, and also impinge on the walls of the confining vessel. The momentum transfer from the molecules to the walls of the vessel results in the pressure P, which appears in previous chapters. Thus the pressure can be related to the kinetic energy of the molecules.

Consider the collision of just one particle (molecule) of mass m, travelling with the velocity v in the x direction of a box of length L (in the x direction), with walls of area A perpendicular to the x direction. The time between successive collisions with the wall A is $\Delta t = 2L/v_x$. The change of momentum $\Delta(mv)$ of the particle in each collision is

$$\Delta(mv) = mv_x - m(-v_x) = 2mv_x$$

Newton's second law defines the force F as the rate of change of momentum with respect to time, thus

$$F = \Delta(mv)/\Delta t = 2mv_x/(2L/v_x) = mv_x^2/L \tag{2.30}$$

The average pressure due to this particle is

$$P_x = F/A = mv_x^2/(AL) = mv_x^2/V \tag{2.31}$$

where V is the volume of the box. The pressure due to N molecules is

$$P_x = (N/V)\, m\overline{v_x^2} = nm\overline{v_x^2} \tag{2.32}$$

where $\overline{v_x^2}$ is the average value of v_x^2 for n molecules. Similarly $P_y = nm\overline{v_y^2}$, and $P_z = nm\overline{v_z^2}$. Since this motion is random there is no difference in the average motion in the various directions, so that

$$\overline{v_x^2} = \overline{v_y^2} = \overline{v_z^2}$$

According to the Pythagorean theorem

$$\overline{v^2} = \overline{v_x^2} + \overline{v_y^2} + \overline{v_z^2} = 3\overline{v_x^2} = 3\overline{v_y^2} = 3\overline{v_z^2}$$

thus the pressure measured in any direction is

$$P = P_x = P_y = P_z = \tfrac{1}{3} nm\overline{v^2} \tag{2.33}$$

Comparing the expression of pressure given in eq. (2.33) with that given by eq. (2.21), it results

$$P = \tfrac{1}{3} nm\overline{v^2} = nkT \tag{2.34}$$

The average kinetic energy of a molecule being $\overline{E} = \tfrac{1}{2} m\overline{v^2}$, one concludes that

$$\overline{E} = \tfrac{3}{2} kT \tag{2.35}$$

i.e. the *average kinetic energy of the molecules is the same for all gases, and is proportional to the absolute temperature.*

2.3.2. *Molecular velocities*

The constant occurrence of collisions produces a wide distribution of velocities. In a collision of two molecules with velocities v_1 and v_2, the total kinetic energy is preserved, thus the quantity $m(v_1^2 + v_2^2)/2$ is the same before and after the collision, even if v_1 and v_2 must change.

Maxwell and *Boltzmann* expressed the distribution of the velocities by the law

$$\frac{1}{n}\frac{dn}{dv} = f_v = \frac{4}{\pi^{1/2}}\left(\frac{m}{2kT}\right)^{3/2} v^2 \exp\left(\frac{-mv^2}{2kT}\right) \tag{2.36}$$

where f_v is the fractional number of molecules in the velocity range between v and $v+dv$, per unit of velocity range.

The value of f_v is zero for $v = 0$, and $v = \infty$, and has its maximum at a value

$$v_p = (2kT/m)^{1/2} = (2R_0T/M)^{1/2} \tag{2.37}$$

given by differentiating f_v with respect to v and setting the result equal to zero

$$\frac{df_v}{dv} = \frac{4}{\pi^{1/2}}\left(\frac{m}{2kT}\right)^{3/2}\left(2v - \frac{mv^3}{kT}\right)\exp\left(\frac{-mv^2}{2kT}\right) = 0$$

Figure 2.10 shows eq. (2.36), plotted versus v/v_p. The value v_p is known as

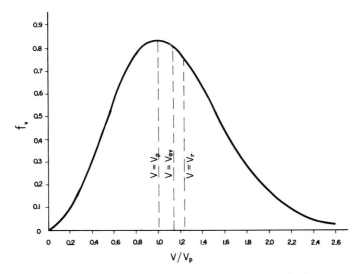

Fig. 2.10 Maxwell–Boltzmann molecular velocity distribution curve.

the *most probable velocity*, i.e. more molecules have this velocity than any other value of the velocity. The v_p value is different from the *arithmetic average value* v_{av}, which results from

$$v_{av} = \int_0^\infty v f_v \, dv \Big/ \int_0^\infty f_v \, dv = (2/\pi^{1/2})\,(2kT/m)^{1/2} = 1.128\,v_p \tag{2.38}$$

The *mean square velocity* $\overline{v^2}$ is obtained from

$$\overline{v^2} = \int_0^\infty v^2 f_v \, dv \Big/ \int_0^\infty f_v \, dv = 3kT/m \tag{2.39}$$

and is the same as obtained in eq. (2.34).

The *root-mean-square* velocity is therefore

$$v_r = [\overline{v^2}]^{1/2} = (3kT/m)^{1/2} = 1.225\,v_p \tag{2.40}$$

Which of these velocities is of interest as representing the average behavior of a gas depends upon the process under consideration. When the molecules directly influence the process by their velocity (e.g. flow of gases), the arithmetic average is used, while when the kinetic energy of the molecule influences the process the root-mean-square should be used.

Based on eqs. (2.19) and (2.21)

$$k/m = R_o/M \tag{2.41}$$

thus

$$v_{av} = 1.45 \times 10^4 (T/M)^{1/2} \text{ cm/s} \tag{2.42}$$

it results that the average air molecule ($M=29$) at $T=300°K$ has a velocity of about 4.6×10^4 cm/sec.

2.3.3. Molecular incidence rate

In a similar way to eq. (2.36) the distribution function fv_x of the velocities of molecules in the x direction was written as:

$$\frac{1}{n} \frac{dn_x}{dv_x} = fv_x = \left(\frac{m}{2\pi kT} \right)^{1/2} \exp \left(\frac{-mv_x^2}{2kT} \right) \tag{2.43}$$

The *number ϕ of molecules striking an element of surface* (perpendicular to the x direction), per unit time is given by

$$\phi = \int_0^\infty v_x dn_x \tag{2.44}$$

By introducing dn_x from eq. (2.43) into (2.44) and integrating, it results that

$$\phi = [n/(2\pi^{1/2})] \ (2kT/m)^{1/2} \text{ molec/cm}^2 \cdot \text{s} \tag{2.45}$$

and by using eqs. (2.38), (2.42), (2.17), it also results*

$$\phi = \tfrac{1}{4} nv_{av} = 3.513 \times 10^{22} \ [P/(MT)^{1/2}] \text{ molec/cm}^2 \cdot \text{s} \tag{2.46}$$

where P is in Torr.

Table 1.1 lists some values of ϕ.

If a hole of area A is cut in the thin wall of the vessel beyond which the gas density is zero, the rate at which molecules of gas leave the vessel is

---

* Equation (2.46) is valid only when the wall scattering follows exactly the cosine law (Edwards, 1978b).

$$q = \phi A = \tfrac{1}{4}\, n v_{av}\, A = 3.64 \times 10^3\ (T/M)^{1/2}\ nA\ \text{molec/sec} \tag{2.47}$$

The *volume of gas* at the pressure in the vessel *escaping* each second is obtained by dividing the flow q by the density n, thus

$$dv/dt = q/n = 3.64 \times 10^3\ (T/M)^{1/2}\ A\ \text{cm}^3/\text{s} \tag{2.48}$$

which for air at $20°C$ would be

$$dv/dt = 11.6\ A\ \text{liter/s} \tag{2.49}$$

where A is in cm^2.

The mass W of gas escaping can be found by combining eqs. (2.46) and (2.19), thus

$$W = 5.83 \times 10^{-2}\ P(M/T)^{1/2}\ \text{g/s·cm}^2 \tag{2.50}$$

2.4. Pressure and mean free path

2.4.1. *Mean free path*

During their motion the molecules suffer collisions between themselves. The distance traversed by a molecule between successive collisions is its *free path*. Since the magnitude of this distance is a function of the velocities of the molecules, the conception of *mean free path* λ is used. This is defined as the *average distance traversed by all the molecules between successive collisions with each other, or as the average of the distances traversed between successive collisions by the same molecule, in a given time.*

A molecule having a diameter ξ and a velocity v moves at a distance $v\delta t$ in the time δt. The molecule suffers a collision with another molecule if anywhere its center is within the distance ξ of the center of another molecule, therefore sweeps out without collision a cylinder of diameter 2ξ. This cylinder has a volume

$$\delta V = (\pi/4)\ (2\xi)^2\ v\delta t \tag{2.51}$$

Since there are n molecules/cm^3, the volume associated with one molecule is on the average $1/n$ cm^3. When the volume δV is equal to $1/n$, it must contain on the average one other molecule, thus a collision has occurred. If $\tau = \delta t$ is the average time between collisions,

$$1/n = \pi \xi^2 v \tau \tag{2.52}$$

and the mean free path λ is

$$\lambda = v\tau = 1/(\pi n \xi^2) \tag{2.53}$$

If we consider the more realistic case, in which not only the reference molecule is in motion but also the others, then eq. (2.53) must be written

$$\lambda = [1/(\pi n \xi^2)] \, (v/v_r) \tag{2.54}$$

where v is the absolute velocity, while v_r is the relative velocity of the molecules. *Clausius* established that $v/v_r = \frac{3}{4}$, thus

$$\lambda = 3/(4\pi n \xi^2) \tag{2.55}$$

Finally if the Maxwell–Boltzmann distribution of velocities is also considered, it results that

$$\lambda = 1/[2^{1/2} \pi n \xi^2] = kT/[2^{1/2} \pi \xi^2 P] \tag{2.56}$$

by using eq. (2.21).
This gives the relation

$$\lambda = 2.33 \times 10^{-20} \, T/(\xi^2 P) \text{ (cm)} \tag{2.57}$$

where T is in $^\circ$K, ξ in cm and P in Torr. For air, at ambient temperature the simple formula

$$\lambda = 5 \times 10^{-3}/P \tag{2.58}$$

can be used, with P–Torr and λ–cm. It can be seen that at $P = 10^{-6}$ Torr $\lambda = 50$ m, thus much larger than the dimensions of a vacuum enclosure, therefore at such pressures the molecules collide only with the walls of the vessel.

Figure 1.1 shows the values of λ for air, while table 1.2 lists some values for other gases. The values of ξ are listed in table 2.5, as calculated from eq. (2.26), and in table 2.9 as calculated from eq. (2.71).

Eq. (2.53) shows no influence of the temperature on the mean free path; it was established considering pure mechanical (elastic) collisions between molecules.

Sutherland established experimentally that at constant n, the mean free path is influenced by the temperature, and the dependence was described by the relation

$$\lambda = 1/[2^{1/2}\pi n\xi^2 \ (1+c/T)] \qquad (2.59)$$

where the constant c (Sutherland's constant) is a measure of the strength of the attractive forces between the molecules. From this equation it was deduced that

$$\lambda_T = \lambda_\infty/[1+(c/T)] \qquad (2.60)$$

where λ_T is the mean free path at the temperature T, λ_∞ is the mean free path at very high temperatures ($T=\infty$), and c is a constant. (See Table 2.6.)

Table 2.6.
Values of Sutherland's constant c.

Gas	λ_∞ (cm) at 1 Torr	c (°K)
Hydrogen	1.056×10^{-2}	76
Nitrogen	$6.1 \ \times 10^{-3}$	112
Oxygen	6.87×10^{-3}	132
Water vapour	$9.5 \ \times 10^{-3}$	600
Helium	$1.6 \ \times 10^{-2}$	79
Neon	1.12×10^{-2}	56
Argon	$7 \ \ \times 10^{-3}$	169
Krypton	$5.9 \ \times 10^{-3}$	142

Similarly to eq. (2.56), *the mean free path $\lambda_{1,2}$ of a gas mixture* of two gases 1 and 2, (i.e. λ of gas 1 colliding with molecules of gas 2) was written

$$\lambda_{1,2} = 4kT/[\pi(\xi_1+\xi_2)^2 \ P_2 \ (1+M_1/M_2)^{1/2}] \qquad (2.61)$$

where ξ_1, ξ_2 are the molecular diameters of the gases 1 and 2 respectively, M_1 and M_2 their molecular weights, and

$$P_2 = n_2kT$$

the partial pressure of gas 2, in the mixture. It can be seen that for $\xi_1=\xi_2$ and $M_1=M_2$ this equation leads to

$$\lambda_2 = kT/[2^{1/2}\pi\xi^2 P_2] \qquad (2.62)$$

which is identical with eq. (2.56).

By introducing the value given by eq. (2.62), in eq. (2.61), and the fact that $\xi_1/\xi_2 = (\lambda_2/\lambda_1)^{1/2}$, it results that the mean free path of a gas mixture is given by

$$\lambda_{1,2} = 4 \times 2^{1/2}\lambda_2 \left[1 + (\lambda_2/\lambda_1)^{1/2}\right]^{-2} \left[1 + (M_1/M_2)\right]^{-1/2} \tag{2.63}$$

If the partial pressures of the two components are $P_1 \ll P_2$, the molecules of gas 1 will have much more collisions with those of gas 2, than with those of gas 1. Thus in this case $\lambda_1 \approx \lambda_{1,2}$. By adding to air ($M_2 = 28.7$; $\lambda_2 = 5 \times 10^{-3}$ cm) a very small quantity of He ($M_1 = 4$; $\lambda_1 = 14.5 \times 10^{-3}$) it results that

$$\lambda_{He} \approx \lambda_{1,2} = 4 \times 2^{1/2} (1.59)^{-2} \, 1.14^{-1/2} \, \lambda_2 \approx 2.1 \, \lambda_{air}$$

Therefore the mean free path of helium molecules is twice that of the other molecules, thus they diffuse fast in the mixture; leak detection takes advantage of these phenomena.

The *mean free path of electrons* λ_e (very small mass and diameter), according to eq. (2.63) is

$$\lambda_e = 4\sqrt{2} \, \lambda \tag{2.64}$$

while for the *mean free path of ions* λ_i the relationship

$$\lambda_i = \sqrt{2} \, \lambda \tag{2.65}$$

was established.

The mean free path of gas molecules emitted from the external surfaces of spacecrafts is discussed by Walters (1982).

2.4.2. *Pressure units*

Pressure is the most widely quoted parameter in vacuum technology, and this brought into use a large number of pressure units, which appear in various texts.

Pressure in a gas, defined in terms of gas impingement on a surface (see §2.3.1), is the time rate of change of the normal component of momentum of the impinging gas molecules per unit area of surface. Thus, the *pressure exerted by a gas on a real surface is defined as the force applied per unit area*. The various pressure units belonging to the *coherent unit systems* are based on this definition.

As pressure can be measured by the height of a liquid column, the various *non-coherent units* of the pressure are related to these columns.

In a *coherent unit system*, the unit of force [F] is expressed as

$$[F] = [l] \, [m] \, [t]^{-2}$$

thus the pressure $p = Fl^{-2}$ is expressed as

$$[p] = [l]^{-1} \, [m] \, [t]^{-2}$$

where $[l]$, $[m]$, $[t]$ are the base units of length, mass and time, respectively (SI, CGS, MTS systems).

In the CGS system the dyn/cm^2 is the pressure unit (table 2.7). This unit is called *microbar*. The microbar is also called "barye" (in the French literature). The name of "*vac*" was proposed for the *millibar*.

The MKS system uses the Newton per square meter (N/m^2), which is called *Pascal* (Pa) in the French literature. The *Gaede*, $Gd = 10^{-3} N/m^2$.

The British system uses the pound per square inch (psi, or lb/in^2), while the MTS (meter, ton, second) system has a pressure unit called "*pieze*" (1 pz = 10^3Pa).

The *technical atmosphere* (at) is the name given to the kg (force)/cm^2.

The base unit of pressure recommended by the International Organization for Standardization (ISO) is *pascal (Pa)*, but the use of *bar* is also accepted.

These various units are summarized in table 2.7, their conversion to each other and to the various non-coherent units is given in table 2.8 (see also Moss, 1987).

The *non-coherent pressure units used are:* the physical atmosphere, the millimeter and micron of mercury and the Torr, the millimeter and centimeter of water, and the inch of mercury.

Table 2.7.
Coherent pressure units.

System	Unit of area	Unit of force	Unit of pressure
CGS	cm^2	dyne 1 dyn = 1 g·cm/s^2	dyne/cm^2 1 bar = 10^6 dyn/cm^2 = 10^5 N/m^2
SI (MKS)	m^2	Newton (N) 1 N = 1 kg·m/s^2	N/m^2 1 Pascal (Pa) = 1 N/m^2
Technical	m^2	kgf 1 kgf = 9.81 N	kgf/m^2 1 at = 1 kgf/cm^2 = 9.81 × 10^4 N/m^2
British	ft^2 in^2	1 lb — ≈ 4.448 N	lb/ft^2 ≈ 47.88 N/m^2 lb/in^2 ≈ 6894.7 N/m^2
MTS	m^2	Sthene (sn) 1 sn = 1t·m/s^2 = 10^3N	pieze (pz) 1 pz = 10^3N/m^2

Table 2.8.
Conversion factors (n) for pressure units ($1X = nY$).

X \ Y	dyne/cm²	microns of Hg	N/m² (Pascal)	mm of water or kg/m²	millibar	cm of water
dyne/cm² or microbar	1	7.5×10^{-1}	1×10^{-1}	1.01×10^{-2}	1×10^{-3}	1.01×10^{-3}
microns of Hg	1.33	1	1.33×10^{-1}	1.35×10^{-2}	1.33×10^{-3}	1.35×10^{-3}
N/m² (Newton per m²)	10	7.5	1	1.01×10^{-1}	1×10^{-2}	1.01×10^{-2}
mm of water or kg/m²	98	73	9.8	1	9.8×10^{-2}	1×10^{-1}
millibar (mb)	1×10^3	750	100	10.1	1	1.01
cm of water or Ger (Guericke)	980	730	98	10	9.8×10^{-1}	1
Torr	1.33×10^3	1×10^3	133.3	13.59	1.33	1.35
in. of Hg	3.3×10^4	2.54×10^4	3386	340	33	34
lb/in² (p.s.i.)	6.8×10^4	5.17×10^4	6894.7	700	68	70
Techn. atmosphere (at) kg/cm²	9.8×10^5	7.3×10^5	9.81×10^4	1×10^4	980	1×10^3
Bar	1×10^6	7.5×10^5	1×10^5	1.01×10^4	1×10^3	1.01×10^3
Physical atmosphere (atm)	1.01×10^6	7.6×10^5	1.01×10^5	1.03×10^4	1.04×10^3	1.03×10^3

Table 2.8. (contd.)
Conversion factors (n) for pressure units $(1X = nY)$

Torr	in. of Hg	lb/in² (psi)	at (kg/cm²)	Bar	atm
7.5×10^{-4}	2.95×10^{-5}	1.45×10^{-5}	1.01×10^{-6}	1×10^{-6}	9.8×10^{-7}
1×10^{-3}	3.93×10^{-5}	1.93×10^{-5}	1.33×10^{-6}	1.33×10^{-6}	1.31×10^{-6}
7.5×10^{-3}	2.95×10^{-4}	1.45×10^{-4}	1.10×10^{-5}	1×10^{-5}	9.8×10^{-6}
7.3×10^{-2}	2.89×10^{-3}	1.42×10^{-3}	1×10^{-4}	9.8×10^{-5}	9.6×10^{-5}
7.5×10^{-1}	2.95×10^{-2}	1.45×10^{-2}	1.01×10^{-3}	1×10^{-3}	9.8×10^{-4}
7.3×10^{-1}	2.89×10^{-2}	1.42×10^{-2}	1×10^{-3}	9.8×10^{-4}	9.6×10^{-4}
1	3.93×10^{-2}	1.93×10^{-2}	1.35×10^{-3}	1.33×10^{-3}	1.31×10^{-3}
25.4	1	4.9×10^{-1}	3.4×10^{-2}	3.3×10^{-2}	3.3×10^{-2}
51.7	2.03	1	7×10^{-2}	6.8×10^{-2}	6.8×10^{-2}
735	28.9	14.2	1	9.8×10^{-1}	9.6×10^{-1}
750	29.5	14.5	1.01	1	9.8×10^{-1}
760	29.92	14.7	1.03	1.01	1

The *physical or normal atmosphere* (*atm*) was defined as the pressure exerted by a mercury column of 760 mm when the specific gravity of mercury is 13.595 g/cm³ (at 0°C).

$$1 \ atm = 76 \ cm \times 13.595 \ g/cm^3 \times 980.665 \ cm/s^2$$
$$= 1.013 \times 10^6 \ dyn/cm^2 = 1.013 \times 10^5 \ N/m^2$$

The *Torr* (Torricelli) is defined as the 760th part of the normal atmosphere thus

$$1 \ Torr = 1.333 \times 10^3 \ dyn/cm^2 = 133.32 \ N/m^2$$

Practically 1 Torr = 1 mm Hg, theoretically 1 mm Hg = 1.00000014 Torr.

The *m Torr* (milli Torr) is equal to the *micron of Hg* (μ). The inch of mercury (in Hg) = 3.386 × 10³N/m².

The mm $H_2O \approx 1$ kgf/m²; the cm of water was called *Guericke (Ger)*.

2.5. Transport phenomena in viscous state

2.5.1. *Viscosity of a gas*

A gas streaming through a narrow-bore tube experiences a resistance to flow, so that the velocity in the direction of the flow decreases uniformly (parabolic distribution) from the axis until it reaches zero on the walls. In the same way the gas between two plates (fig. 2.11), one at rest and the other pulled in the plane, has a drift velocity zero at the contact with the steady plate, and a maximum velocity at the contact with the moving plate. Each layer of gas parallel to the direction of flow exerts a tangential force on the adjacent layer, tending to decrease the velocity of the faster-moving and to increase that of the slower-moving layers. The property of the fluid by virtue of which it exhibits this phenomenon is known as *internal viscosity*. Newton assumed that *the internal viscous forces are directly proportional to the velocity gradient in the fluid.*

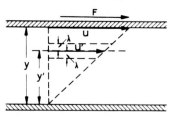

Fig. 2.11 Drift velocity distribution due to viscosity.

Considering the gas between two parallel plates (fig. 2.11) separated by a distance y, the upper plate being in motion with a velocity u. The gas will be steady at the level of the lower plate, whereas its drift velocity will be u at the contact with the upper plate. The drift velocity of the gas u' at some intermediate level y' will be

$$u' = u(y'/y) \tag{2.66}$$

The *coefficient of viscosity* η is defined as *the tangential force per unit area for unit rate of decrease of velocity with distance.* Imagining the gas divided into layers parallel to the surface, each having a depth λ, the mean free path (layers in which the molecule has no collisions), the tangential force between adjacent layers of area A is written

$$F = \eta A(u/y) \tag{2.67}$$

where η is the coefficient of viscosity.

According to the kinetic theory, the tangential force per unit area is measured by the rate at which momentum is transferred between adjacent layers. Molecules from a distance λ above move down into the layer u' with a momentum

$$(mu')_+ = m(y' + \lambda)u/y$$

while those from a distance λ below move up with a momentum

$$(mu')_- = m(y' - \lambda)u/y$$

The number of molecules that cross unit area per unit time in any direction in a gas at rest is equal to $\frac{1}{6} n v_{av}$. Hence the net rate of transfer of momentum across the area A is equal to

$$F = \tfrac{1}{6} A \, n v_{av} \left[(mu')_+ - (mu')_- \right] = \tfrac{1}{3} A n v_{av} \, m\lambda u/y \tag{2.68}$$

From eqs. (2.67) and (2.68), it results

$$\eta = \tfrac{1}{3} n m v_{av} \lambda = \tfrac{1}{3} \rho v_{av} \lambda \tag{2.69}$$

where $\rho = nm$ is the density of the gas.

This equation is approximate only. When the distribution in random velocities and the distribution in free paths are taken into account the result of the calculation (for rigid, elastic spherical molecules) gives (Dushman 1949);

$$\eta = 0.499 nmv_{av}\lambda \tag{2.70}$$

By using eqs. (2.56) and (2.38), it results that

$$\eta = \frac{0.499mv_{av}}{2^{1/2}\,\pi\xi^2} = \frac{0.998}{\pi\xi^2}\left(\frac{mkT}{\pi}\right)^{1/2} \tag{2.71}$$

From eqs. (2.67), (2.70), (2.71) it can be seen that the dimensions of the *coefficient of viscosity* η are $[M]\,[L]^{-1}\,[T]^{-1}$. In the CGS system the unit of viscosity is 1 poise = 1 g·cm^{-1}·s^{-1} = 1 dyne·s·cm^{-2} while the SI unit is 1 Pa·s = 1 kg·m^{-1}·s^{-1} = 10 poise. Table 2.9 lists values of η for various gases.

Since η is proportional to λ, the *Sutherland* equations (2.59), (2.60) also apply to the viscosity. Thus

Table 2.9.
Coefficients of viscosity η of gases at 0°C, 760 Torr and computed values of molecular diameters ξ and mean free path λ, according to eqs. (2.71), (2.70).

Gas	η micropoises	ξ cm	λ^* cm
Helium	186.9	2.20×10^{-8}	1.32×10^{-2}
Neon	312.4	2.55×10^{-8}	9.82×10^{-3}
Argon	208.8	3.69×10^{-8}	4.67×10^{-3}
Krypton	224.9	4.27×10^{-8}	3.49×10^{-3}
Xenon	216.5	4.87×10^{-8}	2.68×10^{-3}
Hydrogen	84.7	2.68×10^{-8}	8.83×10^{-3}
Nitrogen	166.6	3.78×10^{-8}	4.45×10^{-3}
Oxygen	191.0	3.65×10^{-8}	4.77×10^{-3}
Air	171.2	3.76×10^{-8}	4.49×10^{-3}
Chlorine	124.0	5.51×10^{-8}	7.61×10^{-3}
Ammonia	88.9	4.57×10^{-8}	3.05×10^{-3}
Carbon monoxide	165.8	3.79×10^{-8}	4.42×10^{-3}
Carbon dioxide	137.6	4.66×10^{-8}	2.93×10^{-3}
Methane	103.2	4.18×10^{-8}	3.64×10^{-3}
Acetylene	93.5	4.96×10^{-8}	2.59×10^{-3}

*λ for 0°C and 1 Torr.

$$\eta = \frac{0.998}{\pi\xi^2}\left(\frac{mkT}{\pi}\right)^{1/2}\bigg/\left(1 + \frac{c}{T}\right) \tag{2.72}$$

where c is Sutherland's constant.

According to this equation the *viscosity of gases increases with temperature*, whereas in the case of liquids the viscosity is known to decrease as the temperature is increased.

These *predictions are valid in a given range of pressure*. At both very high and very low pressures, the viscosity of a gas departs from this prediction. At very high pressures the average distance between the molecules is so small that the intermolecular forces become important and the momentum transfer is very different from that assumed here. At very low pressures, when the mean free path exceeds the distance between the walls, collisions between molecules almost do not occur. In this case the transfer of momentum is only between gas molecules and walls.

The mean free path determines the behavior of the gas, and whether the gas exhibits the property of *viscous* or *molecular flow*.

The theory of viscous and molecular flow is treated in detail in Chapter 3.

2.5.2. Diffusion of gases

Experience has shown that two gases placed in the same vessel diffuse into each other until the relative concentrations are the same everywhere in the vessel.

Meyer has established that the coefficient of interdiffusion of two gases is given by

$$D_{12} = \tfrac{1}{3}[\lambda_1 v_{av1} n_2 + \lambda_2 v_{av2} n_1]/(n_1 + n_2) \tag{2.73}$$

The *coefficient of diffusion* D_{12} is expressed in $[L]^2 [T]^{-1}$ units. It is defined (Dushman, 1949) by

$$D_{11} = \tfrac{1}{3} v_{av} \lambda \tag{2.74}$$

if the diffusion of molecules in the same gas (self-diffusion) is considered.

By combining eqs. (2.74) and (2.69) it results that

$$D_{11} = \eta/\rho \tag{2.75}$$

In fact by introducing various distribution factors, it was determined that

$$D_{11} = 1.34 \ \eta/\rho \tag{2.76}$$

Table 2.10.
Coefficients of interdiffusion D_{12} and the computed
values of molecular diameters ξ, according to eq. (2.77).

Gas	D_{12} (cm^2/s)	ξ_{12} (cm)
H$_2$–air	0.661	3.23 × 10^{-8}
H$_2$–O$_2$	0.679	3.18 × 10^{-8}
O$_2$–N$_2$	0.174	3.74 × 10^{-8}
CO$_2$–H$_2$	0.538	3.56 × 10^{-8}
CO$_2$–air	0.138	4.03 × 10^{-8}
CO–O$_2$	0.183	3.65 × 10^{-8}
CO$_2$–CO	0.136	4.09 × 10^{-8}

If concentration of one of the gases is very small (traces), $n_2 \ll n_1$, eq. (2.73) becomes

$$D_{12} = \tfrac{1}{3}\lambda_2 v_{av} = \frac{2}{3\,\xi^2 P}\left(\frac{k^3 T^3}{\pi^3 m}\right)^{1/2} \tag{2.77}$$

Table 2.10 gives the diffusion coefficients D_{12} (cm²/s) observed for several pairs of gases at 0°C and 1 atmosphere.

The diffusion process found its application in the *diffusion pumps*, which are the most extensively used systems for achieving high vacuum. In 1915 *Gaede* published a description of a high vacuum pump, which involves no mechanical motion but depends for its operation on diffusion of residual gases through a slit or fine opening into a high velocity stream of mercury vapour travelling in front of the opening (Dushman, 1949; Dunkel, 1975).

Gaede's apparatus (fig. 2.12) consists of a stream of mercury vapour AB passing in front of the opening C of a tube connected to the volume E to be evacuated. At R the side tube is cooled by water with the result that any mercury vapour passing into the tube is condensed at D. The residual pressure of the vapour at D is thus reduced to less than 10^{-3} Torr. The vapour stream in the vertical tube entrains any molecules of gas that get into the stream, and consequently there is a constant *diffusion* of gas from E towards C.

If v is the velocity of the mercury vapour in the direction CD, and n the concentration of gas molecules at any point x along the length l (where $x=0$ at D, and $x = -l$ at C), then the rate at which gas passes from D to C is given by

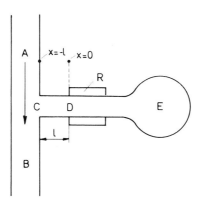

Fig. 2.12 Principle of Gaede's diffusion pumping.

$$\phi_1 = -D_{12}(dn/dx)$$

(D_{12} is the diffusion coefficient of the gas in mercury vapour), and the rate at which gas molecules are returned from C to D, is

$$\phi_2 = nv$$

In the stationary state these rates must be equal, thus

$$D_{12}(dn/dx) + nv = D_{12}(dP/dx) + Pv = 0 \qquad (2.78)$$

since n is proportional to the pressure P. From eq. (2.78) it results

$$dP/P = -(v/D_{12})dx \qquad (2.79)$$

and by integrating over l,

$$\int_0^l (dP/P) = \ln(P_c/P_D) = -vl/D_{12} \qquad (2.80)$$

where P_c and P_D denote the pressures at C and D, while D_{12} is the coefficient of diffusion. This equation shows that the gas flows from point D towards point C, since u, l and D_{12} are all positive.

2.6. Transport phenomena in molecular states

2.6.1. *The viscous and molecular states*

At low pressure, when the mean free path of the molecules of the gas becomes large compared with the dimensions of the enclosure, the energy transport from wall to wall does not include the collisions between molecules, thus it is no longer a function of the viscosity.

In a vessel of volume V the number X of intermolecular collisions per unit time, is given according to eq. (2.56) by

$$X = nVv_{av}/\lambda = \sqrt{2}\,\pi\,\xi^2\,n^2 v_{av}\,V \tag{2.81}$$

If the vessel has an internal surface A, the number of molecules striking the walls is given (eq. 2.46) by

$$N = A\phi = \tfrac{1}{4}nv_{av}A \tag{2.82}$$

The ratio between the number of intermolecular collisions X and that of the collisions molecule–wall, is

$$X/N = 4\sqrt{2}\,\pi\,n\,\xi^2\,V/A \tag{2.83}$$

This ratio will show the limit between viscous state and molecular state, and according to eq. (2.83) this ratio is a function of n, thus of the pressure, of ξ (nature of the gas), and of the dimensions of the vessel V/A.

Considering the model of a cylindrical vessel having a diameter D, and a length L large compared to D, the ratio V/A will be

$$V/A = (\pi\,D^2\,L/4)/(\pi DL) = D/4$$

thus

$$X/N = \sqrt{2}\,\pi\,n\,\xi^2 D \tag{2.84}$$

or for air

$$X/N = 6.2 \times 10^{-19}\,nD, \text{ in SI units (see table 2.7).}$$

In table 2.11 it can be seen that at atmospheric pressure the number of mole-cule–molecule collisions is 15 million times that of molecule–wall collisions. The

Table 2.11.
X/N as a function of P for $D=1$ m.

P (Torr)	λ (m)	n (mole/m³)	X/N	State
760	6.7×10^{-8}	2.46×10^{25}	1.5×10^7	Viscous
10	5.0×10^{-6}	3.24×10^{23}	2×10^5	
10^{-1}	5.0×10^{-4}	3.24×10^{21}	2×10^3	
10^{-3}	5.0×10^{-2}	3.24×10^{19}	20	
10^{-5}	5.0	3.24×10^{17}	2×10^{-1}	Molecular
10^{-7}	500	3.24×10^{15}	2×10^{-3}	
10^{-9}	5.0×10^4	3.24×10^{13}	2×10^{-5}	

pressure must drop to 5×10^{-5} Torr in order that their number be equal, $(X/N=1)$.

The viscous state does not change suddenly to molecular, between them an *intermediate state exists*. This will be analyzed in §3.6 in connection with the flow equations.

2.6.2. Molecular drag

In the viscous state, all the collisions were assumed to be perfectly elastic, thus the molecules striking a surface would be reflected as elastic balls. At low pressures, this image does not cover the experimental results. Experiments show phenomena, which can be explained by the image that the molecule "condenses" on the surface, rests on it a given time, and then it is "reevaporated" in a direction which is independent of that of incidence.

We assume a surface which is "free" of any adsorbed layer of molecules. In the presence of a gas, the molecules of the gas will "condense" on the surface, and will "rest" on the surface a given time, before being reevaporated. The number of molecules striking the unit surface being (eq. 2.46) $\phi = \frac{1}{4} n \, v_{av}$ and that necessary to form a *monolayer* being $\phi_m = (2/\sqrt{3}) \xi^{-2}$ the *time required to form this layer* will be

$$\tau_m = \phi_m/\phi = (8/\sqrt{3})(n \xi^2 v_{av})^{-1} = \sqrt{8\pi/3} \, (\xi^2 n)^{-1} \tag{2.85}$$

and in order to form the monolayer the molecules must rest on the surface at least this time. From eq. (2.85), for nitrogen, at 20°C, we find that

$$\tau_m = 1.99 \times 10^{-6} / P \tag{2.86}$$

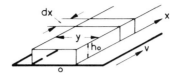

Fig. 2.13 Principle of pumping by molecular drag.

where τ is in seconds, and P is in Torr. Thus for $P = 10^{-4}$ Torr, $\tau_{N_2} = 2.0 \times 10^{-2}$ seconds, a time which is sufficient to transfer energy to the molecule. If the *surface is in motion it can transfer a velocity component to the molecule*. This is the principle on which the *molecular pumps* are designed. In such pumps the gas is pumped by a groove (fig. 2.13) having a depth h_o, and a width y. The groove is at rest, while the cover (bottom) is moving at a speed v, in the direction x (positive direction). In this case, for steady state, (zero flow), the pressure will increase in the v direction. The result of the forces applied to a volume comprised between x and $x+dx$ will be

$$df = (dP/dx) \, dx \, h_o y \tag{2.87}$$

The number of molecules striking the unit surface being $nv_{av}/4$ (eq. 2.46), the surface being $y dx$, and the momentum received by the "reevaporated" molecule being (maximum) mv, the force applied on the gas is

$$dq = \tfrac{1}{4} n \, v_{av} \, mv \, y dx \tag{2.88}$$

From $df = dq$, we obtain

$$(dP/dx) \, h_o = \tfrac{1}{4} n \, m \, v_{av} \, v \tag{2.89}$$

from which by using eqs. (2.21) and (2.38), it results

$$dP/P = [M/(2 \pi R_o T)]^{1/2} \, (v/h_o) \, dx \tag{2.90}$$

and by integrating

$$\int_o^x dP/P = \ln (P/P_o) = [M/(2 \pi R_o T)]^{1/2} (v/h_o) x \tag{2.91}$$

which shows that the pressure ratio P/P_o, which can be achieved by molecular drag (zero flow), is an exponential function of the length of the path x, and of the relative velocity v of the moving surfaces, and inverse to the distance h_o between the surfaces. Eq. (2.91) also shows that the ratio P/P_o is greater when M is greater (heavy gases). The momentum transfer has been measured by Comsa et al. (1977). The molecular/turbomolecular pumps are discussed in §5.2.8.

Molecular gauges were constructed using the principle of molecular drag. These gauges use either the method of the "decrement" or that of the "torque". In the decrement type of gauges, a surface is set in oscillation and the rate of decrease of the amplitude of oscillation is taken as a measure of pressure. Physically, the damping may be explained as due to the gradual equalization of energy between the moving surface and the molecules of gas striking it.

In the torque type, a surface is set in continuous rotation, and the amount of twist imparted to an adjacent surface is used to measure the pressure. The molecules striking the moving surface acquire a momentum in the direction of motion which they tend in turn to impart to the other surface. If that surface is suspended on a filament, the filament will have a torsion. Molecular gauges are discussed in more detail in §6.4.

2.7. Thermal diffusion and energy transport

2.7.1. Thermal transpiration

The rate at which molecules leave a chamber through an opening in a thin wall was shown, eq. (2.47), as being

$$q = \phi A = \tfrac{1}{4} n \, v_{av} \, A$$

thus the mass of gas leaving the chamber (rate of efflux) is given by

$$W = mq = \tfrac{1}{4} n \, m \, v_{av} \, A \tag{2.92}$$

and since $nm = \rho$ (specific gravity),

$$W = \tfrac{1}{4} \rho \, v_{av} \, A = \rho \, [R_o \, T/ \, (2\pi M)]^{1/2} A \tag{2.93}$$

If we have two chambers A and B, separated by a porous plug, and the gas in the chambers is at different temperatures T_A and T_B, thermal transpiration will occur until an equilibrium state is established at which

$$\rho_A \sqrt{T_A} = \rho_B \sqrt{T_B} \tag{2.94}$$

and since ρ is proportional to P, and inversely proportional to T, eq. (2.18), it results*

$$P_A/P_B = (T_A/T_B)^{1/2} \tag{2.95}$$

This is of importance in vacuum systems where low temperatures are used (traps, cryogenic pumping). Thus if a chamber A is a part of a system at liquid air temperature ($T_A = 90°K$), and the pressure is measured by means of a gauge at room temperature ($T_B = 300°K$), then the real value of P_A is given by

$$P_A = (90/300)^{1/2} P_B = 0.55 P_B \tag{2.96}$$

When the two chambers are connected with a large bore tube or the pressure is higher, so that the mean free path is much smaller than the diameter, and collisions between molecules become predominant, the *condition of equilibrium* is $P_A = P_B$ (instead of eq. 2.94), thus

$$\frac{\rho_A}{\rho_B} = \frac{T_B}{T_A} \tag{2.97}$$

The correction for thermal transpiration in capacitance gauges is discussed by Poulter et al. (1983).

Hobson (1970, 1973) found experimentally that eq. (2.95) is accurate for apertures and tubing with rough surfaces (e.g. leached Pyrex), but for tubes with vitreous (smooth) surfaces eq. (2.95) must be written $P_A/P_B = a(T_A/T_B)^{1/2}$ where $a \approx 1.1 - 1.3$. This means that in such tubes a molecule has a greater probability of traversing the tube from the hot end to the cold end than in the opposite direction.

Based on this effect Hobson (1970) proposed the *accommodation pumping*. Two volumes (fig. 2.14) A and B, both at the same temperature T_2, are joined to a third volume at temperature T_1 through tubes of leached (atomically rough) and smooth Pyrex, respectively. For volume A, $P_1/P_{2A} = (T_1/T_2)^{1/2}$, while for volume B, $P_1/P_{2B} = a(T_1/T_2)^{1/2}$, thus the equilibrium pressure ratio which develops between A and B is $P_{2A}/P_{2B} = a$. If A and B are joined by a leached–smooth combination of n tubes arranged in series, then the pressure ratio can be $P_{2A}/P_{2B} = a^n$. By using an experimental setup with 28 stages, Hobson (1970) obtained $P_{2A}/P_{2B} = 23.3$.

Hemmerich (1988) describes pumping arrays of such pumps, called by him "thermomolecular pumps". He measured in a single-stage experiment a pumping speed of 2.5 liter/second per cm^2 of pumping surface, and calculated that in a 50-stage array a compression ratio of about 100 can be achieved.

---

* The thermal transpiration exponent is always less than $\frac{1}{2}$, as it was experimentally determined by Baker et al. (1973) and calculated by Siu (1973); Dadjburjor and Sandler (1976).

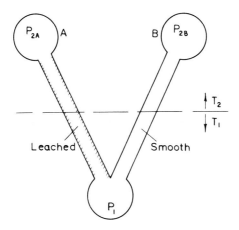

Fig. 2.14 Principle of the accommodation pumping. After Hobson (1970).

2.7.2. *Thermal diffusion*

If a temperature gradient is applied to a mixture of two gases of uniform concentration there is a tendency for the heavier and larger molecules (mass m_1, diameter ξ_1), to move to the cold side, and for the lighter and smaller molecules (m_2, ξ_2) to move to the hot side. The separating effect of thermal diffusion (coefficient D_T) is ultimately balanced by the mixing effect of ordinary diffusion (coefficient D_{12}), so that finally a steady state is reached and a concentration gradient is associated with the temperature gradient.

The *coefficient of thermal separation* is defined by

$$k_T = D_T/D_{12} \qquad (2.98)$$

It was established that if k_T is a constant then the *amount of separation* is given by

$$\Delta f = k_T \ln(T_1/T_2) \qquad (2.99)$$

The practical device utilized for the separation of mixtures of different gases and of isotopes uses a long vertical tube with a hot wire along the axis. Because of thermal diffusion, the relative concentration of the heavier molecules is greater at the cold wall. Convection causes the gas at the hot surface to rise to the top, where it is deflected to the cold wall. As a result, the heavier component concentrates at the bottom, and the lighter at the top.

2.7.3. *Heat conductivity of rarefied gases*

As in the case of viscosity, the process of heat transfer by gases is different in the case of the *viscous state* and in that of *molecular state*. In the first case the totality of molecules is responsible for the heat transfer, while in the second case the individual molecules carry the heat from wall to wall.

Heat conductivity in viscous state As in the case of viscosity (fig. 2.11) we can consider layers of thickness λ (mean free path), between two plates whose temperatures are T_1 and T_2, and distance apart y. The relative temperature drop between the layers is $2(T_1 - T_2)\lambda/y$ similarly to eq. (2.68).

If c_v is the specific heat at constant volume, the heat transferred per unit area is

$$E = \tfrac{1}{6} n v_{av} (2m) c_v (T_1 - T_2) \lambda/y = \tfrac{1}{3} \rho v_{av} c_v \lambda (T_1 - T_2)/y \qquad (2.100)$$

Therefore the heat conductivity K is given by

$$K = \tfrac{1}{3} \rho v_{av} \lambda c_v \qquad (2.101)$$

and comparing eq. (2.101) with eq. (2.69) it follows that

$$K = \eta \, c_v \qquad (2.102)$$

As in the case for the relation for η (viscosity), a more careful consideration of the mechanism of energy transfer leads to the relation

$$K = \tfrac{1}{4} (9\gamma - 5) \eta \, c_v \qquad (2.103)$$

where $\gamma = c_p/c_v$ is the ratio of the specific heat of the gas at constant pressure to that at constant volume. K is expressed in $\mathrm{cal \cdot cm^{-1} \cdot s^{-1} \cdot {}^\circ K^{-1}}$, if c_v is expressed in $\mathrm{cal \cdot g^{-1} \cdot {}^\circ K^{-1}}$.

For monoatomic gases (A, Hg, etc.) $\gamma = \tfrac{5}{3}$, for diatomic gases (O_2, H_2, N_2, etc.) $\gamma = \tfrac{7}{5}$, while for triatomic gases (e.g. CO_2) $\gamma = \tfrac{4}{3}$. Values of K are listed in table 2.12.

Since the viscosity is not a function of the pressure (eq. 2.71), it follows (eq. 2.103) that *the thermal conductivity of a gas is independent of pressure. This is valid as long as the pressure is higher than the range in which molecular state exists.*

Heat conductivity in molecular state When the gas pressure is so low that the molecular mean free path is about equal to or greater than the distance between

Table 2.12.
Heat conductivity of gases K at $0°C$ ($cal \cdot cm^{-1} \cdot s^{-1} \cdot degK^{-1}$).

Gas	K	Gas	K
Hydrogen	4.19×10^{-4}	Helium	3.43×10^{-4}
Nitrogen	5.7×10^{-5}	Neon	1.09×10^{-4}
Oxygen	5.8×10^{-5}	Argon	3.9×10^{-5}
Air	5.8×10^{-5}	Krypton	2.1×10^{-5}
Carbon monoxide	5.3×10^{-5}	Mercury	1.2×10^{-5}
Carbon dioxide	3.4×10^{-5}		

the walls of the containing vessel, the gas is no longer characterized by a viscosity. In that case eq. (2.103) is no longer valid, and the conductivity is then found to *depend upon the pressure*. The process of heat transfer under these conditions is called *free molecular conduction*. In order to express the heat conductivity at low pressures, *Knudsen* introduced the concept of the *accommodation coefficient* (Dushman, 1949).

The *accommodation coefficient* is defined as the ratio of the energy actually transferred between impinging gas molecules and a surface, and the energy which would be theoretically transferred if the impinging molecules reached complete thermal equilibrium with the surface.

When molecules originally at a temperature T_i strike a hot surface at temperature T_s ($>T_i$), complete interchange does not occur at the first collisions, and it may require many collisions for this to occur. The molecules reemitted from the hot surface consequently possess a mean energy which corresponds to a temperature lower than T_s, which we shall designate as T_r. The *accommodation coefficient* α is defined by

$$\alpha = (T_r - T_i)/(T_s - T_i) \tag{2.104}$$

If the molecules reach thermal equilibrium with the surface before escaping, $T_r = T_s$, then $\alpha = 1$. On the other hand, if the molecules are elastically reflected without undergoing any change in energy, $T_r = T_i$ and $\alpha = 0$. Table 2.13 lists some values of α. These values include both the transational and the rotational energy modes; their separate values are discussed by Ramesh and Marsden (1973, 1974).

According to eq. (2.46) the number of molecules having a velocity between v and $v + dv$, and which strike the unit surface in the unit time is

Table 2.13.
Values* of the accommodation coefficient α.

Gas \ Surface	W	Pt Ordinary	Pt black	Ni	Fe
Hydrogen	0.36 (5)	0.28 (5)			
	0.2 (1)	0.36 (1)	0.71 (1)	0.29 (5)	
Nitrogen	0.57 (1)	0.89 (1)	—	0.82 (5)	—
	0.87 (5)	0.81 (5)			
Air	—	0.90 (1)	—	—	—
Oxygen	0.9 (5)	0.85 (5)	0.95 (1)	0.86 (5)	—
Carbon dioxide	—	—	0.97 (1)	—	—
Mercury	0.95 (1)	—	—	—	—
Helium	0.016(4)	—	—	—	—
Neon	—	—	—	0.82 (5)	0.1 (3)
					0.4 (3)
Argon	0.85 (1)	0.89 (1)	—	0.93 (5)	—
	0.09 (6)				
Krypton	—	0.69 (2)	—	—	—

*References : (1) Dushman (1949), (2) Thomas and Brown (1950), (3) Eggleton and Tompkins (1952), (4) Thomas and Schofield (1955), (5) Amdur and Guildner (1957), (6) De Poorter and Searcy (1963).

$$d\phi = \tfrac{1}{4} (dn/dv) \, v \, dv = \tfrac{1}{4} \, v \, (dn) \tag{2.105}$$

Since each molecule has a kinetic energy equal to $\tfrac{1}{2} mv^2$, the energy transferred is

$$dE = \tfrac{1}{8} m \, v^3 \, (dn) \tag{2.106}$$

thus

$$E = \tfrac{1}{8} m \int_{v=0}^{v=\infty} v^3 \, dn \tag{2.107}$$

which solved by using the Maxwell–Boltzmann distribution (§2.3.2), results in

$$E = \tfrac{1}{8} m \, \tfrac{4}{3} \, n v_{av} \, \overline{v^2} = \tfrac{1}{6} \, n m \, v_{av} \, \overline{v^2} \tag{2.108}$$

This is the energy transferred by all the molecules striking the unit surface in the unit time. Since their number is (eq. 2.46) $\phi = \frac{1}{4} n v_{av}$ it follows that the average *energy transferred per molecule* is

$$E_m = E/\phi = \frac{4}{6} n m v_{av} \overline{v^2}/(n v_{av}) = \frac{2}{3} m \overline{v^2} = 2kT \tag{2.109}$$

instead of $E = \frac{3}{2} kT$ (eq. 2.35) which is the *average energy of the molecules in a volume*.

For *monoatomic gases* at low pressures, the energy transfer from the hot to the cold surface will be according to eqs. (2.46) and (2.109)

$$E_o = \phi E_m = \frac{1}{4} n v_i \, 2k(T_r - T_i) \tag{2.110}$$

and according to eqs. (2.21) and (2.104), it follows that

$$E_o = \frac{1}{2} (P v_i/T_i) (T_r - T_i) = (\alpha/2) (P v_i/T_i) (T_s - T_i) \tag{2.111}$$

where α is the accommodation coefficient (eq. 2.104), P is the pressure of the gas, v_i is the average velocity at temperature T_i and T_s is the temperature of the hot surface. Thus *the rate of energy transfer at low pressures is proportional to the pressure and the temperature difference.*

For *diatomic and polyatomic gases*, the molecules striking the hot surface acquire not only increased translational energy but also increased amounts of both rotational and vibrational energy. The amount of the vibrational energy possessed by molecules as compared with that of translational energy is measured by the value of γ (see also eq. 2.103). A detailed calculation leads in this case to

$$E_o = \frac{\alpha}{8} \left[\frac{\gamma + 1}{\gamma - 1} \right] \frac{P v_i}{T_i} (T_s - T_i) \tag{2.112}$$

which for $\gamma = \frac{5}{3}$ (case of monoatomic gases) becomes identical with eq. (2.111).

Substituting for v_i (eqs. 2.19, 2.21 and 2.38) as function of T_i and M, eq. (2.112) becomes

$$E_o = \frac{\alpha}{2} \frac{\gamma + 1}{\gamma - 1} \left(\frac{R_o}{2\pi M (273)} \right)^{1/2} \left(\frac{273}{T_i} \right)^{1/2} (T_s - T_i) P$$

$$= \Lambda_o \alpha \left(\frac{273}{T_i} \right)^{1/2} (T_s - T_i) P \text{ ergs/sec·cm}^2 \tag{2.113}$$

in which Λ_o is the *free molecular conductivity* at $0°C$, given by

$$\Lambda_o = \frac{\gamma + 1}{2(\gamma - 1)} \left(\frac{R_o}{2\pi M (273)} \right)^{1/2} = \frac{110}{M^{1/2}} \frac{\gamma + 1}{\gamma - 1} \ \text{ergs/sec·cm}^2 \cdot {}^\circ C \cdot \mu bar$$

$$= \frac{1.47 \times 10^{-2}}{M^{1/2}} \frac{\gamma + 1}{\gamma - 1} \ \text{Watt/cm}^2 \cdot {}^\circ C \cdot \text{Torr} \tag{2.114}$$

For air ($\gamma = \frac{7}{5}$, diatomic gases; $M = 28.98$)

$$\Lambda_o = [1.47 \times 10^{-2}/(28.98)^{1/2}] \ (2.4/0.4) = 1.64 \times 10^{-2} \ \text{Watt/cm}^2 \cdot {}^\circ C \cdot \text{Torr}.$$

Thus, the heat conduction per unit area from a surface at a temperature $T_s = 100^\circ$ C to the surface at 20° C by air at 10^{-2} Torr, and $\alpha = 0.8$, will be

$$E_o = 0.8 \times 1.64 \times 10^{-2} \ (273/293)^{1/2} \ (373 - 293) \times 10^{-2} \approx 10^{-2} \ \text{Watt/cm}^2$$

Table 2.14 lists values for γ and Λ_o (eq. 2.114) for various gases.

Thermal conductivity at low pressures is used for measuring the pressure of gases by using the *thermal conductivity gauges*. These gauges operate generally under conditions in which the energy input for heating a filament is maintained constant, and the pressure is determined by the variation of the temperature.

For coaxial cylinders of radii r_1 and r_2, $(r_1 > r_2)$, the rate of energy transfer from the inner cylinder or wire at temperature T_s, is

$$E_o = \alpha_r \Lambda_o P(273/T_i)^{1/2} \ (T_s - T_i) \tag{2.115}$$

Table 2.14.
Values of free molecular conductivity Λ_o (W·cm^{-2}·$^\circ$K^{-1}·Torr^{-1})
After Dushman (1949).

Gas	M	γ	Λ_o
Hydrogen	2.016	1.41	6.07×10^{-2}
Helium	4.003	1.67	2.93×10^{-2}
Water vapour	18.016	1.30	2.65×10^{-2}
Neon	20.18	1.67	1.31×10^{-2}
Nitrogen	28.02	1.40	1.66×10^{-2}
Oxygen	32.00	1.40	1.56×10^{-2}
Argon	39.94	1.67	9.29×10^{-3}
Carbon dioxide	44.01	1.30	1.69×10^{-2}
Mercury	200.6	1.67	4.15×10^{-3}

where

$$\alpha_r = \frac{\alpha}{1 - [(1 - \alpha)(r_2/r_1)]} \qquad (2.116)$$

where α is the accommodation coefficient of the surfaces. For a given gauge Λ_o, α_r, T_i and E_o being kept constant, the temperature T_s measures the pressure P. Since Λ_o is a function of the nature of the gas (eq. 2.14), the gauges are to be calibrated for each gas separately. The various kinds of thermal conductivity gauges are described in §6.6.

CHAPTER 3

Gas flow at low pressures

3.1. Flow regimes, conductance, and throughput

3.1.1. Flow regimes

The gas in a vacuum system can be in a *viscous state*, in a *molecular state* or in a state which is *intermediate* between these two. When a system is brought from the atmospheric pressure to "high vacuum", the gas in the system goes through all these states. The mean free path of the gas molecules is very small at atmospheric pressure (see table 1.1) so that the flow of the gas is limited by its *viscosity* (see §2.5.1.). At low pressures where the mean free path of the molecules is similar to the dimensions of the vacuum enclosure, the flow of the gas is governed by viscosity as well as by molecular phenomena; this is the *intermediate flow*. At very low pressures where the mean free path is much larger than the dimensions of the vacuum enclosure, the flow is *molecular*.

In the range where the state of the gas is *viscous*, the flow can be *turbulent* or *laminar*. When the velocity of the gas exceeds certain values, the flow is turbulent, the flowing gas layers are not parallel, their direction is influenced by any obstacle in the way. In the cavities formed between layers, spaces of lower pressures appear. At lower velocities the viscous flow is *laminar*, i.e. the layers are parallel, their velocity increasing from the walls toward the axis of the pipe.

Thus the flow can be *turbulent, laminar, intermediate* and *molecular* (see table 3.1). *The limit between the turbulent and laminar flow is defined by the value of Reynold's number* while *those between laminar, intermediate and molecular flow are described by the value of the Knudsen number.*

The *Reynold number* is a dimensionless quantity expressed by

$$R_e = \rho v D / \eta \tag{3.1}$$

Table 3.1.
Flow regimes.

State of the gas	Flow regime	Condition
viscous	turbulent	$R_e > 2100$
		$Q > 200\ D$ (air)
	laminar	$R_e < 1100$
		$Q < 100\ D$ (air);
		$D/\lambda > 110$
transition	intermediate	$1 < D/\lambda < 110$
rarefied	molecular	$D/\lambda < 1$

where ρ is the density of the gas, v the velocity, η the viscosity, and D the diameter of the tube. It was established that for *Reynold's numbers, larger than 2100, the flow is entirely turbulent, while for $R_e < 1100$ it is entirely laminar.* The exact value of R_e for which the flow changes from turbulent to laminar depends upon the roughness of the surface of the tube and other experimental factors, but for most cases the mentioned range is valid.

The expression of the Reynold number can be related to the throughput Q (see §3.1.3) which is defined as the quantity of gas flowing through a pipe, expressed in *PV* (pressure × volume) units per unit time. Thus

$$Q = P\,v\,(\pi D^2/4) \tag{3.2}$$

where v is the velocity. Since (according to eqs. 2.17, 2.19)

$$\rho = nm = MP/(R_o T) \tag{3.3}$$

the expression for the Reynold number (eq. 3.1) can be written

$$R_e = [MP/(R_o T)]\,[4Q/(\pi D^2 P)]\,(D/\eta) = [4M/(\pi R_o T\eta)]\,(Q/D) \tag{3.4}$$

For air at 20°C, $\eta = 1.829 \times 10^{-4}$ poise, $R_o = 62.364$ Torr·liter/°K (table 2.4), and $M = 28.98$, so that according to eq. (3.4)

$$Q_{air} = 9.06 \times 10^{-2}\,R_e\,D \tag{3.5}$$

where Q_{air} is expressed in Torr·liter/sec, while D is in centimeters. By using the limits $R_e = 2100$ and $R_e = 1100$, it results that the flow of air (at room temperature) will be *turbulent if*

$$Q > 200\ D \tag{3.6}$$

and it will be *laminar if*

$$Q < 100\ D \tag{3.7}$$

The *Knudsen number* is the ratio λ/D between the mean free path λ and the diameter of the pipe (vessel) D. In terms of the ratio D/λ the ranges can be defined as

$D/\lambda > 110$	Viscous flow	(3.8)
$1 < D/\lambda < 110$	Intermediate flow	(3.9)
$D/\lambda < 1$	Molecular flow	(3.10)

if an error of about 10% is admitted in calculating the conductances (see §3.6.4).

By using eq. (2.58), which gives $\lambda P = 5 \times 10^{-3}$, for air at room temperature, it results that the condition for *viscous flow is*

$$D\bar{P} > 5 \times 10^{-1}\ \text{cm·Torr} \tag{3.11}$$

while that for molecular flow is

$$D\bar{P} < 5 \times 10^{-3}\ \text{cm·Torr} \tag{3.12}$$

where D is the diameter of the pipe (cm), and $\bar{P}$ the average pressure (Torr). These and other limits are discussed in detail in §3.6.

3.1.2. *Conductance*

The flow of a gas can be interpreted as the number of molecules N, passing per unit time through a cross section of the pipe.

Considering two subsequent cross sections 1 and 2 of the same pipe, the number of molecules crossing them will be

$$N_1 = A_1\ v_1\ n_1 = S_1\ n_1 \tag{3.13}$$

and

$$N_2 = A_2\ v_2\ n_2 = S_2\ n_2 \tag{3.14}$$

where A is the area of the cross sections, v is the flow velocity of the gas, n is the number of molecules per unit volume (see eq. 2.18), while $S = Av$ is the rate of flow, or the *pumping speed.*

In a permanent flow, the number of molecules crossing the various cross sections is the same. Thus $N_1 = N_2 = N$, and

$$N = S_1 n_1 = S_2 n_2 \tag{3.15}$$

which expresses Boyle's law, since n the number of molecules per unit volume is proportional to the pressure (eq. 2.17).

By writing that the drop in molecular density (or the pressure drop) is proportional to the number of molecules, it results that

$$N = C(n_1 - n_2) \tag{3.16}$$

where the factor C can be a constant or a function of the molecular density (pressure). *The factor C is called the conductance* of the pipe. From eqs. (3.16) and (3.15), we have

$$1/C = (n_1 - n_2)/N = (1/S_1) - (1/S_2) \tag{3.17}$$

Since S_1 and S_2 have volume/time dimensions, *the conductance is also expressed in V/t units.* Various such units used are listed in Table 3.2. The value of the conductance depends on the kind of flow and the geometry of the pipe; §3.2-3.6 deal with their calculation.

Table 3.2.
Conversion factors (n) for conductance and pumping speed units (1 X$=n$ Y).

Y \ X	cm³/sec	l/min	m³/hr	ft³/min	l/sec
cm³/sec	1	6×10^{-2}	3.6×10^{-3}	2.1×10^{-3}	1×10^{-3}
liter/min	16.67	1	6×10^{-2}	3.53×10^{-2}	1.67×10^{-2}
m³/hr	277.8	16.67	1	5.89×10^{-1}	2.78×10^{-1}
ft³/min	471.95	28.32	1.699	1	4.7×10^{-1}
liter/sec	1000	60	3.6	2.12	1

When two *pipes are connected in parallel* (fig. 3.1), the number of molecules N reaching the cross section 1 is divided in two parts, N_a, flowing in pipe a, and N_b in pipe b. If the molecular densities at 1 and 2 are n_1 and n_2 (fig. 3.1), then according to eq. (3.16)

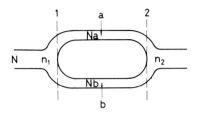

Fig. 3.1 Conductance of pipes connected in parallel.

$$N_a = C_a (n_1 - n_2)$$ (3.18)

$$N_b = C_b (n_1 - n_2)$$ (3.19)

and since

$$N_a + N_b = N = C_p(n_1 - n_2)$$ (3.20)

it results that

$$C_p = C_a + C_b + \ldots$$ (3.21)

where C_p is the conductance of the system of parallel pipes, and C_a, C_b are the conductances of the various pipes connected in parallel.

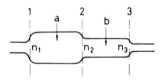

Fig. 3.2 Conductance of pipes connected in series.

When *conductances are connected in series* (fig. 3.2) and the molecular densities at 1, 2, 3 are n_1, n_2, and n_3, one can write

$$N = C_a (n_1 - n_2) = C_b (n_2 - n_3) = C_s(n_1 - n_3)$$ (3.22)

where C_a and C_b are the individual conductances of part a and b, while C_s is the conductance of the system. From eq. (3.22) it results that

$$1/C_s = (1/C_a) + (1/C_b) + \ldots \tag{3.23}$$

3.1.3. Throughput and pumping speed

The pumps used in a vacuum system remove (evacuate) gas from the system. The rate at which the gas is removed is measured by the pumping speed S_p. *The pumping speed is defined as the volume of gas per unit of time dV/dt which the pumping device removes from the system at the pressure existing at the inlet to the pump.* The pumping speed is expressed in liter/sec, m³/hr, etc. (see table 3.2).

The *throughput Q is defined as the product of the pumping speed and the inlet pressure*, i.e.

$$Q = PS_p = P(dV/dt) \tag{3.24}$$

The *throughput* is also defined (Kaminsky and Lafferty, 1980) as the quantity of gas, in pressure × volume units, at a specified temperature, flowing per unit time across a specified cross section. The throughput is expressed in Pa · m³/s (or Torr · liter/s) at a specified temperature (usually $0\,°C$ or $25\,°C$) or other units as listed in table 3.3. The throughput unit of μ (microns of Hg) × liter/sec, received the name of *lusec* ($l\mu$/sec).

The unit of mol/s includes the temperature, therefore, its use is recommended by some authors (Ehrlich, 1986; Solomon, 1986). The conversion from this unit is:

$$1 \text{ mol/s} = 2271 \text{ Pa} \cdot \text{m}^3/\text{s} = 28.964 \text{ g/s of air (at } 0\,°C)$$

Conversion factors are listed in table 3.3 and by Moss (1987).

By multiplying eq. (3.16), by kT, it results

$$NkT = C(n_1 kT - n_2 kT) \tag{3.25}$$

and with eqs. (2.21), (3.15) and (3.24) we have

$$NkT = NP/n = SP = Q = C(P_1 - P_2) \tag{3.26}$$

According to eq. (3.26), Q is the quantity of gas entering per unit of time the pipe with conductance C, at pressure P_1. If no additional gas leaks into or is removed from the pipe this same quantity of gas Q comes out of the pipe at pressure P_2. Thus *if the system is isothermic* (eq. 3.26), *Q is the same all over the system.*

By analogy with the expression (3.24), the pumping speed at any point of the vacuum system is

$$S = Q/P \tag{3.27}$$

where Q is the throughput in the system and P is the pressure at the point at which the pumping speed is defined. Substituting the values of $P_1 = Q/S_1$ and $P_2 = Q/S_2$ into eq. (3.26) we have

$$1/S_1 = 1/S_2 + 1/C$$

Table 3.3.

Conversion factors (n) for throughput or leak rate units $(1X = nY)$

X \ Y	cm³ (STP) /yr	atm·cm³/hr	μ ft³ per min.	ft³(STP)/yr	lusec
cm³(STP)/yr	1	1.12×10^{-4}	5.05×10^{-5}	3.5×10^{-5}	2.4×10^{-5}
atm·cm³/hr	8.9×10^{3}	1	4.47×10^{-1}	3.1×10^{-1}	2.11×10^{-1}
micron·ft³/min	1.98×10^{4}	2.236	1	6.9×10^{-1}	4.72×10^{-1}
ft³(STP)/yr	2.86×10^{4}	3.24	1.45	1	6.82×10^{-1}
lusec ($l \cdot \mu$/sec)	4.2×10^{4}	4.74	2.12	1.47	1
cm³(STP)/min or std.cc/min	5.36×10^{5}	60.4	27	18.6	12.7
mg of air per sec (25°C)	2.7×10^{7}	3040	1360	940	640
std.cc/sec, or cm³(STP)/sec, or cm³·atm/sec.	3.22×10^{7}	3600	1620	1120	760
Torr·l/sec	4.2×10^{7}	4738	2120	1470	1000
ft³(STP)/hr	2.5×10^{8}	2.84×10^{4}	1.27×10^{4}	8750	5970
Watt (Pa·m³/s)	3.12×10^{8}	3.55×10^{4}	1.59×10^{4}	1.10×10^{4}	7500

which is identical to eq. (3.17). This equation shows that the pumping speed at any point in the system can be obtained from the known pumping speed at some other point and the conductance of the portion of the system (pipes, holes, valves etc.) in between.

The pumping speed S obtained in a chamber, connected by a conductance C, to a pump having a pumping speed S_p, is given by

$$1/S = 1/S_p + 1/C \tag{3.28}$$

The nomogram shown in fig. 3.3, can be used to solve quickly eq. (3.28), or eq. (3.23).

If eq. (3.28) is expressed in the form

$$S/S_p = (C/S_p)/[1 + (C/S_p)] \tag{3.29}$$

the decrease of the pumping speed S/S_p results as a function of the ratio C/S_p between the conductance of the system and pumping speed (of the pump). This relationship is represented in fig. 3.4.

It can be seen (fig. 3.4) that when the value of the conductance is equal to that of the pumping speed of the pump, 50% of the pumping speed is used at the vacuum vessel. In order to use 80% of the pumping speed the ratio C/S_p must be 4, while for a ratio $C/S_p = 0.1$, only 10% of the pumping speed of the pump is felt in the vacuum enclosure. From this it results that it *is no use increasing the pump, if the conductance of the pipe is the factor which limits the pumping speed.*

Table 3.3 (contd.)

Conversion factors (n) for throughput or leak rate units ($1X = nY$)

cm³(STP) /min	mg/s (air) 25°C	cm³ atm/sec	Torr·l/sec	ft³ (STP)/hr	watt (Pa·m³/s)
1.88×10^{-6}	3.74×10^{-8}	3.1×10^{-8}	2.4×10^{-8}	4.05×10^{-9}	3.21×10^{-9}
$1 62 \times 10^{-2}$	3.2×10^{-4}	2.78×10^{-4}	2.11×10^{-4}	3.57×10^{-5}	2.81×10^{-5}
3.75×10^{-2}	7.4×10^{-4}	6.2×10^{-4}	4.72×10^{-4}	7.95×10^{-5}	9.09×10^{-5}
5.8×10^{-2}	1.07×10^{-3}	8.9×10^{-4}	6.82×10^{-4}	1.15×10^{-4}	1.33×10^{-4}
7.9×10^{-2}	1.56×10^{-3}	1.32×10^{-3}	1×10^{-3}	1.69×10^{-4}	1.69×10^{-3}
1	1.98×10^{-2}	1.68×10^{-2}	1.27×10^{-2}	2.14×10^{-3}	8.55×10^{-2}
50.4	1	8.4×10^{-1}	6.4×10^{-1}	1.08×10^{-1}	1.013×10^{-1}
58.2	1.16	1	7.6×10^{-1}	1.29×10^{-1}	1.33×10^{-1}
79	1.56	1.32	1	1.69×10^{-1}	7.97×10^{-1}
471	9.32	7.9	5.97	1	
592	11.7	9.87	7.50	1.25	1

3.2. Viscous and turbulent flow

3.2.1. Viscous flow – conductance of an aperture

A large volume where the gas is at a relatively high pressure P_1 (e.g. atmospheric) is connected to a second volume where the pressure is $P_2 < P_1$, by an aperture of area A (fig. 3.5). If the pressure P_1 is such that *the mean free path of the molecules is small compared to the dimensions of the aperture*, the gas will flow from P_1 to P_2 by a *viscous flow*. By this flow the gas reaches a velocity in the vicinity of the aperture, so that after passing it, the gas jet has a minimum cross section (fig. 3.5). After this contraction, the jet has some (about 10) successive expansions and contractions, until finally it diffuses in the mass of gas P_2. By keeping constant P_1, and decreasing P_2, the quantity of gas and its velocity are increasing, up to the state where the ratio P_2/P_1 reaches a *critical (minimum) value, corresponding to a velocity equal to that of the sound (Mach number = 1). Further decrease of the pressure P_2 does not increase the flow or the velocity.*

Based on the laws of the adiabatic expansion, it was found that the *throughput* Q of gas flowing through the aperture, is given by

$$Q = A P_1 \left(\frac{P_2}{P_1}\right)^{1/\gamma} \left\{\frac{2\gamma}{\gamma - 1} \frac{R_0 T_1}{M} \left[1 - \left(\frac{P_2}{P_1}\right)^{(\gamma - 1)/\gamma}\right]\right\}^{1/2} \qquad (3.30)$$

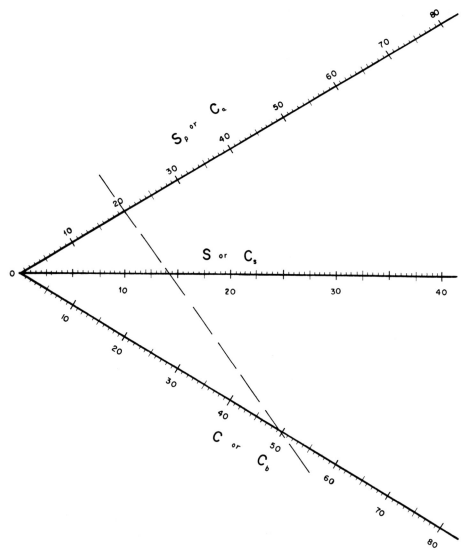

Fig. 3.3 Nomogram for calculating pumping speed or the conductances in series. After Delafosse and Mongodin (1961).

expressed in CGS units. In eq. (3.30), A is the cross section of the aperture,
$\gamma = c_p/c_v$ the ratio of the specific heat at constant pressure to that at constant
volume (see table 2.14), R_o the gas constant (see table 2.4), M molecular weight,
and T_1 temperature of the gas at the high-pressure side P_1.

Since $C = Q/(P_1 - P_2)$ (eq. 3.26), it results that the *conductance for viscous flow of an aperture is given by*

$$C = \frac{9.13\,A}{1 - (P_2/P_1)}\left(\frac{P_2}{P_1}\right)^{1/\gamma}\left\{\frac{2\gamma}{\gamma - 1}\left(\frac{T_1}{M}\right)\left[1 - \left(\frac{P_2}{P_1}\right)^{(\gamma - 1)/\gamma}\right]\right\}^{1/2} \quad (3.31)$$

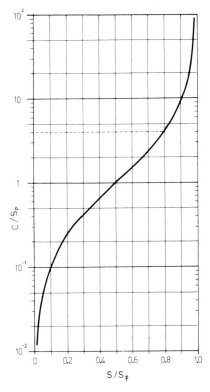

Fig. 3.4 S/S_p as a function of C/S_p.

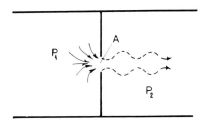

Fig. 3.5 Viscous flow through an aperture.

where A (cm²), C (liter/sec), T_1 (°K), M (g). For air at $20°C$, $\gamma = 1.4$, $T_1 = 293°K$, $M = 29$, the *conductance is*

$$C = \{ 76.6 \, A/ \, [1 - (P_2/P_1)] \} (P_2/P_1)^{0.712} [1 - (P_2/P_1)^{0.288}]^{1/2} \qquad (3.32)$$

The throughput Q (eq. 3.30) is $Q = 0$ for $P_2/P_1 = 1$, and is maximum for

$$P_2/P_1 = [2/(\gamma + 1)]^{\gamma/(\gamma - 1)} = r_c \qquad (3.33)$$

this is called the *critical value*.

For air at $20°C$, this value is $r_c = 0.525$, and

$$Q_c = 20 \, AP_1 \qquad (3.34)$$

where A (cm²), P_1 (Torr) and Q (Torr·liter/sec). Therefore the conductance for $P_2/P_1 \leqslant 0.525$, is given by

$$C = \frac{20 \, A}{1 - (P_2/P_1)} \qquad (3.35)$$

while eq. (3.32) gives the conductance for the range $1 \geqslant P_2/P_1 > 0.525$.

When $P_2/P_1 < 0.1$, eq. (3.35) can be written

$$C \approx 20 \, A \qquad (3.36)$$

and in this range (only) C can be considered independent of the pressures.

If the aperture is considered by its pumping effect on volume 1 (fig. 3.5), its pumping speed is given by

$$S = Q/P_1 = C(P_1 - P_2)/P_1 = C[1 - (P_2/P_1)] \qquad (3.37)$$

Figure 3.6 shows the values of C/A for air, as well as the value $S/A = Q_1/(AP_1)$. It can be seen that the pumping speed S/A is maximum and constant up to $P_2/P_1 = r_c$, higher values of P_2/P_1, the pumping speed S/A drops, and tends to zero for P_2/P_1 approaching one. The conductance C/A is always higher than S/A, and tends to infinity when P_2/P_1 approaches one.

3.2.2. *Viscous flow – conductance of a cylindrical pipe*

Poiseuille's law In a long tube of uniform circular cross section (fig. 3.7), gas

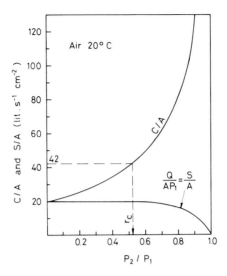

Fig. 3.6 Conductance C and pumping speed S of apertures (viscous flow). A is the cross section area of the aperture.

flow occurs from the region of higher pressure P_1 to that of lower pressure P_2. The gas contained within a thin-walled cylinder of radius r, a wall thickness dr, and within a differential length dx, experiences a force in the direction of flow given by the cross sectional area $2\pi r dr$, and the pressure difference dP, so that

$$dF_1 = -(dP/dx)\ 2\pi r dr dx \tag{3.38}$$

The minus sign appears since the pressure gradient is $-dP/dx$ (the pressure decreases in the direction of the flow).

Due to the viscosity, the velocity of the gas at the internal surface of the cylinder is greater than that on its external surface. The force due to the viscosity is eq. (2.67)

$$F = \eta A\ (dv/dr)$$

where η is the coefficient of viscosity, and A the surface of the cylinder $A = 2\pi r dx$. Therefore on the internal surface of the cylinder, this force will be

$$F_2 = -\eta\ 2\pi r dx\ \frac{dv}{dt} = -2\pi\eta\left[\ r\ \frac{dv}{dr}\ \right]dx \tag{3.39}$$

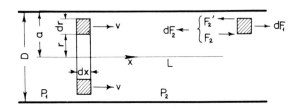

Fig. 3.7 Viscous flow in pipes (principle).

dv/dr being negative this force is directed in the direction of the flow. On the outside surface of the element, the force due to the viscosity is

$$F'_2 = -\left[F_2 + \frac{\delta F_2}{\delta r}\,dr \right] = 2\pi\eta dx\left[r\,\frac{dv}{dr} + \frac{d}{dr}\left(r\,\frac{dv}{dr} \right)dr \right] \qquad (3.40)$$

which is directed opposite to the direction of the flow. Therefore, the resulting (viscosity) force applied on the element is

$$dF_2 = 2\pi\eta\,\frac{d}{dr}\left(r\,\frac{dv}{dr} \right)drdx \qquad (3.41)$$

which is directed opposite to the flow.

Equilibrium will occur when the force due to the pressure difference given in eq. (3.38) is balanced by that due to the viscosity given in eq. (3.41). Thus

$$-\frac{dP}{dx}\,2\pi r\,drdx + 2\pi\eta\,\frac{d}{dr}\left(r\,\frac{dv}{dr} \right)drdx = 0 \qquad (3.42)$$

i.e.

$$\frac{d}{dr}\left| r\,\frac{dv}{dr} \right| = \frac{r}{\eta}\left(\frac{dP}{dx} \right) \qquad (3.43)$$

Two subsequent integrations will give dv/dr, and v :

$$r\,(dv/dr)=(r^2/2\eta)(dP/dx)+K_1 \qquad (3.44)$$

thus

$$dv/dr=(r/2\eta)\,(dP/dx)+(K_1/r) \qquad (3.45)$$

and

$$v = (r^2/4\eta)\ (dP/dx) + K_1 \ln r + K_2 \qquad (3.46)$$

The constants K_1 and K_2 can be determined by the boundary conditions. The velocity is a maximum for $r=0$, thus for this value $dv/dr=0$ (eq. 3.45), which results in $K_1=0$. The velocity is zero near the wall, thus for $r=a$; $v=0$, which gives $K_2 = -\ (a^2/4\eta)\ (dP/dx)$. Finally eq. (3.46) is written

$$v = -(1/4\eta)\ (dP/dx)\ (a^2 - r^2) \qquad (3.47)$$

which shows:
– that the velocity of the gas is directed in the direction of the pressure drop, and
– that the velocity of the gas is a parabolic function of the radius, with a maximum velocity $v = K_2 = -\ (a^2/4\eta)\ (dP/dx)$ on the axis $(r=0)$; and $v=0$ at the wall $(r=a)$, as shown in fig. 3.8.

The *volume of gas flowing* through the cross section of the tube per unit time is obtained by integrating eq. (3.47) across the cross section of the tube, i.e.

$$dV/dt = \int_0^a 2\pi r v dr = -\ (\pi a^4/8\eta)\ (dP/dx) \qquad (3.48)$$

The *throughput* is given by

$$Q = P(dV/dt) = -(\pi a^4/8\eta)\ (dP/dx)P \qquad (3.49)$$

and by integrating for a length L, corresponding to a pressure drop from P_1 to P_2, it results

$$Q = -\ (\pi a^4/8\eta) \left[\int_{P_1}^{P_2} P dP \right] \Big/ \left[\int_0^L dx \right] = [\pi a^4/(16\eta L)]\ (P_1^2 - P_2^2) \qquad (3.50)$$

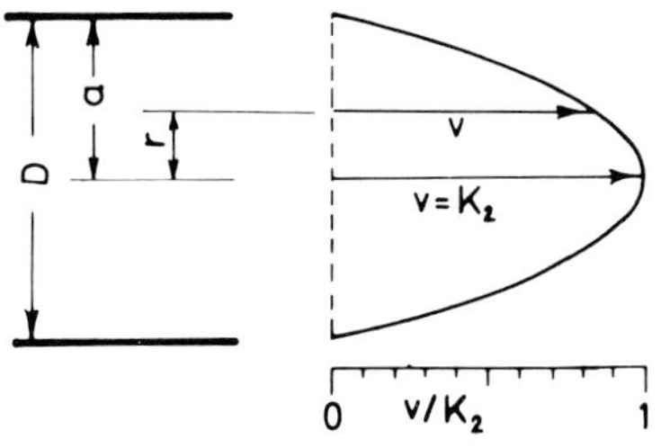

Fig. 3.8 Velocity distribution in viscous flow.

Since $P_1{}^2 - P_2{}^2 = (P_1 - P_2)(P_1 + P_2)$ and the average pressure $\bar{P}$, is $\bar{P} = (P_1 + P_2)/2$, eq. (3.50), is written

$$Q = [\pi a^4/(8\eta L)]\,\bar{P}(P_1 - P_2) = [\pi D^4/(128\eta L)]\,\bar{P}(P_1 - P_2) \tag{3.51}$$

where $D = 2a$ is the diameter of the tube. This equation is known as the *Poiseuille law*. In its form expressed in eq. (3.51), all factors are in CGS units.

The conductance is given (eq. 3.26) by

$$C = Q/(P_1 - P_2) = [\pi D^4/(128\eta L)]\,\bar{P} \tag{3.52}$$

in which D and L (cm), η in poises, $\bar{P}$ (dyne/cm²), and C (cm³/sec).

A practical form of eq. (3.52) is

$$C = 3.27 \times 10^{-2}\,[D^4/(\eta L)]\,\bar{P} \tag{3.53}$$

where $\bar{P}$ (Torr), L (cm), D (cm), η (poise), and C (liter/sec).

For *air at 20°C*, this equation is

$$C_{\text{air}} = 182(D^4/L)\,\bar{P} \tag{3.54}$$

where the units are as in eq. (3.53).

The exit loss in tubes with viscous flow is discussed by Santeler (1986b).

3.2.3. *Viscous flow – surface slip*

In the derivation of eq. (3.51), it was assumed that the velocity of the gas is zero at the tube wall. Some gas molecules in striking the wall experience specular reflection and thus retain the same component of velocity in the direction of flow as before the impact. Other molecules strike irregularities on the wall and bounce several times, the molecule being adsorbed on the wall and then reemitted later with a random distribution in angle and velocity. These molecules represent a layer of gas which is at rest next to the wall, and provide the *viscous drag*. This effect is described by the *slip-coefficient* ε, which is given by

$$\varepsilon = P\,[\,2M/\,(\pi R_o T)\,]^{1/2}\,[f/(2 - f)] \tag{3.55}$$

where f is the fraction of molecules which are adsorbed and reemitted and $1 - f$ is the fraction which are specularly reflected (Van Atta, 1965; Dawe, 1973).

Since the velocity of the gas is not zero at the wall, eq. (3.51) is written

$$Q = [\pi D^4/\,(128\eta L)]\,P\,(P_1 - P_2)\,\{\,1 + [8\eta/(\varepsilon D)\,]\,\} \tag{3.56}$$

thus from eqs. (3.56) and (3.55), it results that

$$Q = (c_1 \bar{P} D^4 + c_2 D^3) (P_1 - P_2)/L \tag{3.57}$$

where $c_1 = \pi/(128\eta)$

$$c_2 = (\pi/16) [(\pi/2) (R_0 T/M)]^{1/2} [(2 - f)/f]$$

If the pressure $\bar{P}$ is sufficiently high the term in D^4 dominates and the flow follows Poiseuille's law. When the pressure $\bar{P}$ is such that the terms in D^4 and D^3 are equal, the character of the flow departs from Poiseuille's law. The pressure for which this condition occurs is referred to as the *transition pressure* P_t, which is

$$P_t = c_2/(c_1 D) \tag{3.58}$$

This can be considered as the lowest limit of Poiseuille flow (see §3.6.3).

3.2.4. *Viscous flow – rectangular cross section*

Guthrie and Wackerling (1949) give for the conductance in viscous flow of ducts with *rectangular cross section*, the expression

$$C = 3.54 \times 10^{-2} \, Y[A^2/(\eta L)] \, \bar{P} \tag{3.59}$$

in CGS units, where Y is a correction factor with values listed in table 3.4.
If eq. (3.52) for the circular cross section is written in the form

$$C = (\pi/128) (4/\pi)^2 (\pi D^2/4)^2 \bar{P}/ (\eta L) = 4 \times 10^{-2} [(\pi D^2/4)^2/ (\eta L)] \bar{P} \tag{3.60}$$

Table 3.4.
Correction factor Y, (for eq. 3.61), as a function of the
shape a/b of the rectangle.

a/b	Y	a/b	Y
1	1	0.5	0.82
0.9	0.99	0.4	0.71
0.8	0.98	0.3	0.58
0.7	0.95	0.2	0.42
0.6	0.90	0.1	0.23

it can easily be compared to eq. (3.59).

It can be seen that the conductance of a duct with square cross section ($Y=1$, eq. 3.59) is less than that of a pipe with circular cross section of same A, in the ratio $3.54/4 = 0.88$.

Equation (3.59) is written in practical units as

$$C = 4.71 \times 10^{-2} \ Y[A^2/(\eta L)]\bar{P} \tag{3.61}$$

where $\bar{P}$ (Torr), L (cm), $A = ab$ (cm^2), η (poise), C (lit/sec) and Y (table 3.4).

For *air* at 20°C, this equation is

$$C_{air} = 260 \ Y(A^2/L)\bar{P} \tag{3.62}$$

in the same units as (3.61).

Heinze (1955) gives the expression for the conductance of ducts with rectangular cross section in laminar (viscous) flow as

$$C = (4/48) \ (a^3 b/\eta) \ (\bar{P}/L) \ \psi \tag{3.63}$$

in units as in eq. (3.61), where ψ is expressed by

$$\psi = 1 - [192a/(\pi^5 b)] \ [\text{tg hyp} \ (\pi b/2a) + (1/3^5) \ \text{tg hyp} \ (3\pi b/2a) + ...] \tag{3.64}$$

and a is the small side and b the large side of the rectangle.

Steckelmacher (1976) pointed out to the author that the values of ψ plotted by Heinze (1955) and reproduced in fig. 3.9 of the first edition of this book, are not correct. The correct graphical evaluation is given by Williams *et al.* (1968). The correct values are plotted on fig. 3.9 of the present edition.

Flow in long rectangular channels in transition (intermediate) regime is analyzed by Loyalka *et al.* (1976). The equations of viscous flow in thin rectangular channels are calculated by O'Hanlon (1987).

3.2.5. *Viscous flow – annular cross section*

For a long duct having an annular cross section between the radius of the tube r_o, and that of a concentric core with radius r_i, the conductance in viscous flow is given by

$$C = (\pi/8\eta) \ (\bar{P}/L) \ [r_0^4 - r_i^4 - (r_0^2 - r_i^2)^2/\ln (r_0/r_i)] \tag{3.65}$$

in CGS units, or *for air* this will be

$$C_{air} = 2900 \ (\bar{P}/L) \ [r_0^4 - r_i^4 - (r_0^2 - r_i^2)^2/\ln (r_0/r_i)] \tag{3.66}$$

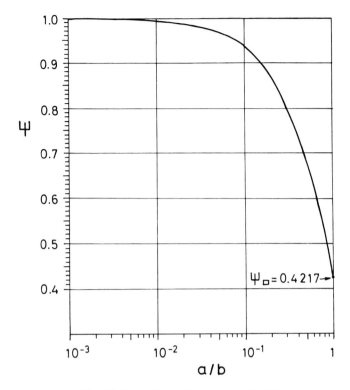

Fig. 3.9 Correction factor ψ for eq. (3.63).

where $\bar{P}$ (Torr), L (cm), r_o (cm), r_i (cm), and C (lit/sec).

3.2.6. Turbulent flow

Considering eq. (3.6), which indicates that turbulent flow occurs only for throughputs $Q > 200\ D$ (Torr·lit/sec; air), it can be shown that such situations are very rare in vacuum systems.

One of such cases occurs when *air is admitted* into a system which was previously evacuated to a low pressure. If the air is admitted through a pipe having a diameter D and length L, the condition for the existence of turbulent flow is

$$Q = 182(D^4/L)\ (P_1^2 - P_2^2)/2 \geqslant 200\ D \qquad (3.67)$$

Since $P_1 = 760$ Torr, this condition gives that turbulent flow exists if

$$P_2 \leqslant [760^2 - 2.2(L/D^3)]^{1/2} \qquad (3.68)$$

which shows that by admitting air through the usual pipes, valves, etc., the flow is turbulent practically until the pressure in the vacuum vessel reaches 760 Torr.

Considering a large diffusion pump with $S_p = 10\,000$ lit/sec at 10^{-3} Torr, thus $Q = 10\,000 \times 10^{-3} = 10$ Torr·lit/sec, the flow is turbulent if

$$10 \geqslant 200\ D$$

thus $D \leqslant 10/200 = 0.05$ cm, *which is never the case in vacuum systems.*

A *rotary pump* can give $Q = 60\,000$ Torr· lit/sec. In this case the flow is turbulent if

$$60\,000 \geqslant 200\ D$$

thus $D \leqslant 60\,000/200 = 300$ cm, *which is always the case in vacuum systems.*

3.3. Molecular flow

3.3.1. *Molecular flow – conductance of an aperture*

A volume where the pressure is P_1 is connected through an aperture (area A) to a second volume where the pressure is $P_2 < P_1$. If the pressure P_1 is low enough for molecular flow (eq. 3.12), the rate at which the gas passes through the aperture from P_1 to P_2 is (eqs. 2.48 and 3.24)

$$Q_1 = P_1(\mathrm{d}V/\mathrm{d}t) = 3.64 \times 10^3\ (T/M)^{1/2}\ AP_1\ \mu\text{bar·cm}^3/\text{sec} \qquad (3.69)$$

while the gas passing from P_2 to P_1 is

$$Q_2 = P_2(\mathrm{d}V/\mathrm{d}t) = 3.64 \times 10^3\ (T/M)^{1/2}\ AP_2\ \mu\text{bar·cm}^3/\text{sec} \qquad (3.70)$$

In molecular flow, where there is no collision between molecules, they *pass through the aperture in both directions without any influence on each other.* The throughput is the difference:

$$Q = Q_1 - Q_2 = 3.64 \times 10^3\ (T/M)^{1/2}\ A(P_1 - P_2) \qquad (3.71)$$

which is directed from P_1 toward P_2, since $P_1 - P_2 > 0$.

Thus *the conductance of an aperture of area A (in molecular flow)* is :

$$C = Q/(P_1 - P_2) = 3.64 \times 10^3 \ (T/M)^{1/2} \ A \ \text{cm}^3/\text{sec}$$
$$= 3.64 \ (T/M)^{1/2} \ A \ \text{liter/sec} \tag{3.72}$$

where A (cm²) is the area of the aperture.

For air, at 20°C, $(T/M)^{1/2} = 3.181$, and eq. (3.72) becomes

$$C_{\text{air}} = 11.6 \ A \ \text{liter/sec} \tag{3.73}$$

If the opening is of *circular cross section*, $A = \pi D^2/4$, thus

$$C = 2.86 \ (T/M)^{1/2} \ D^2 \ \text{liter/sec} \tag{3.74}$$

and

$$C_{\text{air}} = 9.16 \ D^2 \ \text{liter/sec} \tag{3.75}$$

From eq. (3.72) it can be seen that *the conductance (molecular flow) is independent of the pressure.*

The "pumping speed" of the aperture is given (eq. 3.27) by

$$S = Q/P_1 = C(P_1 - P_2)/P_1 = C[1 - (P_2/P_1)] \tag{3.76}$$

and for air at 20°C

$$S = 11.6 \ A[1 - (P_2/P_1)] \tag{3.77}$$

where A (cm²), and S (liter/sec); and for the usual case where $P_2 \leqslant 0.1 \ P_1$

$$S = C = 11.6 \ A \tag{3.78}$$

It can be seen that *the pumping speed is a function of P_2/P_1, up to a maximum value of 11.6 A*. Comparing this to eq. (3.36) and figure 3.6, it results that the maximum pumping speed of an aperture at low pressure (11.6 A) is smaller than that at high pressure (20 A).

3.3.2. *Molecular flow – conductance of diaphragm*

Consider the diaphragm of aperture A as shown in fig. 3.10. Here 1 and 3 are large volumes, connected by a pipe of cross section A_0. The pipe 2 is connected to volume 3 by a diaphragm of aperture A, which is small compared to 3, but of the same order of magnitude as A_0.

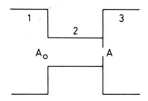

Fig. 3.10 Diaphragm effect.

The conductance of the system (eq. 3.23) in the direction 1–2–3 will be

$$1/C = (1/C_{A_0}) + (1/C_2) + (1/C_e) \tag{3.79}$$

where C_e is the conductance of A in the direction 2–3.

The conductance of the same system in the direction 3–2–1 will be

$$1/C = (1/C_A) + (1/C_2) \tag{3.80}$$

The conductances of the system in both directions must be equal, if not, a flow should exist even if the pressures in 1 and 3 are equal, which is impossible. Thus

$$1/C = (1/C_{A_0}) + (1/C_2) + (1/C_e) = (1/C_A) + (1/C_2) \tag{3.81}$$

which leads to

$$1/C_e = (1/C_A) - (1/C_{A_0}) \tag{3.82}$$

i.e.

$$C_e = C_A / [1 - (A/A_0)] \tag{3.83}$$

and using eq. (3.73), *for air at 20°C*

$$C_e = 11.6 \, A / [1 - (A/A_0)] \tag{3.84}$$

For $A \ll A_0$ eq. (3.83) gives $C_e \approx C_A$ (aperture); for $A = A_0$, $C_e \approx \infty$ (no resistance to the flow); while e.g. for $A = 0.5 \, A_0$, $C_e = 2 \, C_A$ (*diaphragm effect*).

3.3.3. *Molecular flow – long tube of constant cross section*

Knudsen derived the equations of the conductance of long tubes for low pressures (molecular flow). In this flow the molecules move in random straight lines, between collisions with the wall. The number of molecules impinging on the unit surface per unit time is (eq. 2.46) $\phi = n v_{av}/4$ and the number of molecules striking the wall each second is

$$q = \phi BL = BL\, n\, v_{av}/4 \tag{3.85}$$

where B is the periphery of the cross section, and L the length of the tube. The molecules arrive on the surface having an energy corresponding to their v_{av} velocity and the drift velocity v in the direction of the flow. They are stopped at the surface and reemitted randomly with their velocity v_{av}, thus the momentum transferred to the wall is mv. The momentum transferred by all the molecules to the wall is then

$$q' = qmv = BL\, n\, v_{av} mv/4 \tag{3.86}$$

The number N of molecules crossing the cross section A of the pipe per unit time is (eq. 3.13)

$$N = Avn \tag{3.87}$$

and the pressure difference ΔP achieved corresponds to a force

$$\Delta F = A \Delta P = AkT \Delta n \tag{3.88}$$

For equilibrium condition $q' = \Delta F$, thus

$$4AkT\, \Delta n = BL\, n\, v_{av}\, mv \tag{3.89}$$

From eqs. (3.87) and (3.89), we have

$$N/\Delta n = [4A^2/(BL)]\, [kT/(m\, v_{av})] \tag{3.90}$$

According to eq. (3.16) $N/\Delta n = C$, and using the value of $v_{av} = (2/\sqrt{\pi})\, (2kT/m)^{1/2}$ (eq. 2.38) we obtain

$$C = [2A^2/(BL)]\, [\pi\, kT/(2m)]^{1/2} = [2A^2/(BL)]\, [\pi R_o T/(2M)]^{1/2} \tag{3.91}$$

This equation contains the assumption that a uniform drift velocity v is superimposed upon the random Maxwell–Boltzmann distribution of the molecules. *Knudsen* has shown that it should be better to assume that the superimposed drift velocity of a molecule is proportional to its random velocity. On this modified assumption *Knudsen* found that the numerical factor in eq. (3.91) must be multiplied by $8/3\pi$, so that the *conductance* will be

$$C = \frac{8}{3\sqrt{\pi}}\left(\frac{2kT}{m}\right)^{1/2}\left(\frac{A^2}{BL}\right) = \frac{3.44\times 10^4}{\sqrt{\pi}}\left(\frac{T}{M}\right)^{1/2}\left(\frac{A^2}{BL}\right) \tag{3.92}$$

in CGS units.

The conductance of a tube of uniform *circular cross section* with $A = \pi D^2/4$, and $B = \pi D$, is

$$C = 3.81\ (T/M)^{1/2}\ (D^3/L) \tag{3.93}$$

where D (cm), L (cm) and C (liter/sec). For air at $20°C$, $(T/M)^{1/2} = 3.18$, thus

$$C_{\text{air}} = 12.1\ D^3/L \tag{3.94}$$

where D (cm), L (cm) and C (liter/sec). It can be seen that the conductance (molecular flow) is *independent of the pressure*.

For a tube of *rectangular cross section*, with sides a and b, $(b \leqslant a)$,

$$A = ab \quad \text{and} \quad B = 2(a+b),$$

thus eq. (3.92) becomes

$$C = K\frac{8}{3}\left(\frac{kT}{2\pi m}\right)^{1/2}\frac{a^2 b^2}{(a+b)L} = \frac{3.44\times 10^4}{2\sqrt{\pi}}\left(\frac{T}{M}\right)^{1/2}\frac{a^2 b^2}{(a+b)L}K \tag{3.95}$$

in CGS units, where K is an experimental correction factor taking into account the asymmetry of the cross section. Values of K are listed in table 3.5, as a function of b/a.

For air at $20°C$, eq. (3.95) will be written

$$C_{\text{air}} = 30.9\ a^2 b^2 K/[(a+b)L] \tag{3.96}$$

where L, a, b (cm), and C (liter/sec).

For a *slot* $a \gg b$, $A = ab$ and $B \approx 2a$, thus

$$C_{\text{air}} = 30.9\ ab^2\ K/L \tag{3.96a}$$

Table 3.5.
Correction factor K for eq. (3.95).
After Guthrie and Wakerling (1949).

b/a	K
1	1.108
0.667	1.126
0.5	1.151
0.333	1.198
0.2	1.297
0.125	1.400
0.1	1.444

For a *triangular cross section*, with side a (equilateral triangle) it was found by *Barett* (from Heinze, 1955) that $K = 1.24$, and since $A = \sqrt{3}\, a^2/4$; $B = 3a$, thus

$$C = 0.413\ [kT/(2\pi m)]^{1/2}\ (a^3/L) \tag{3.97}$$

in CGS units, and for air at $20°C$

$$C_{air} = 4.79a^3/L \tag{3.98}$$

where a, b, L (cm), and C (liter/sec).

For an *annular cross section* between two concentric tubes with diameters D_o, D_i : $A = \frac{1}{4}\pi\ (D_o{}^2 - D_i{}^2)$; $B = \pi\ (D_o + D_i)$, thus the conductance is

$$C = (\pi/3)\ [kT/(2\pi m)]^{1/2}\ [(D_o{}^2 - D_i{}^2)^2/\ (D_o + D_i)]\ (K_o/L) \tag{3.99}$$

in CGS units. The factor K_o is given in table 3.6.

Table 3.6.
Correction factor K_0, for eq. (3.99).

D_1/D_0	0	0.259	0.5	0.707	0.866	0.966
K_0	1	1.072	1.154	1.254	1.430	1.675

For *air at 20°C*, eq. (3.99) becomes

$$C_{air} = 12.1 \ (D_o - D_i)^2 \ (D_o + D_i) \ (K_o/L) \tag{3.100}$$

where L, D_0, D_i (cm), and C (liter/sec).

Calculations of conductance in tubes with annular cross sections are given by Onusic (1980), while those in tubes with elliptical cross sections were published by Steckelmacher (1978).

3.3.4. *Molecular flow – short tube of constant cross section*

If the length of the tube is decreased to zero, the conductance must decrease to that of an aperture. Thus the correct way of writing the equation of the conductance of a tube is (eq. 3.23)

$$1/C = (1/C_L) + (1/C_e)$$

or

$$C = C_L \ C_e/(C_L + C_e) = C_L/[1 + (C_L/C_e)] \tag{3.101}$$

where C_L is the conductance of the tube (eq. 3.92) and C_e is the conductance of the aperture (eq. 3.84).

From eqs. (3.84) and (3.72), we have

$$C_e = 3.64 \times 10^3 \ (T/M)^{1/2}A/[1 - (A/A_o)] \tag{3.102}$$

where A is the cross section of the tube, and A_o the cross section of the upstream vessel.

By using the value of C_L given in eq. (3.92), it results that

$$C_L/C_e = 5.3 \ [A/(BL)] \ [1 - (A/A_o)] \tag{3.103}$$

Equation (3.101) can be written

$$C = C_L K'' \tag{3.104}$$

where K'' is *Knudsen's factor*, which is expressed by

$$K'' = \left(1 + \frac{C_L}{C_e}\right)^{-1} = \left[1 + 5.3 \frac{A}{BL}\left(1 - \frac{A}{A_o}\right)\right]^{-1} \tag{3.105}$$

For *circular cross sections* : $A = \pi D^2/4$; $B = \pi D$; $A_o = \pi D_v^2/4$, and Knudsen's factor is given by

$$K'' = 1/\{1 + 1.33 \ (D/L) \ [1 - (D/D_v)^2]\} \tag{3.106}$$

For cases where $D < 0.2 \ D_v$ (tube diameter D small compared to vessel diameter D_v), eq. (3.106) can be expressed in its simplified form

$$K'' = 1/[1 + 1.33 \ (D/L)] \tag{3.107}$$

thus the *conductance of a short tube* will be

$$
\begin{aligned}
C &= 3.81 \ (T/M)^{1/2} \ (D^3/L) K'' \\
&= 3.81 \ (T/M)^{1/2} \ (D^3/L)/[1 + 1.33 \ (D/L)] \\
&= 3.81 \ (T/M)^{1/2} \ D^3/(L + 1.33 \ D) \tag{3.108}
\end{aligned}
$$

where L (cm), D (cm), and C (liter/sec).

This equation expresses the fact that the "end effect" can be taken into account by considering the pipe as being longer by 1.33 diameters. Obviously for air at 20°C, the *conductance of a short pipe is*

$$C_{air} = 12.1 \ D^3/(L + 1.33 \ D) = 12.1 \ (D^3/L) K'' \tag{3.108a}$$

Haefer (1980c) discusses the "addition theorem" for the resistance to flow of composite systems.

From a more detailed investigation, using the kinetic gas theory, Clausing (1932) found that the correction factor K'' is only approximate. A more correct value is obtained by using (instead of K'') a factor K' (Clausing's factor) which is given by

$$K' = \frac{15(L/D) + 12(L/D)^2}{20 + 38 \ (L/D) + 12(L/D)^2} \tag{3.109}$$

Figure 3.40 shows a nomogram using Clausing's factor. *

Santeler (1986a) compares the various factors published and the errors resulting in their use (up to a maximum of 12%, at $L/D = 2$). It is proposed to use the transmission probability (see also eq. 3.189)

$$P_r = 1/[1 + (3/4)(L'/D)]$$

where $L' = L\{1 + 1/[3 + (6/7)(L/D)]\}$. The use of these equations produces errors only up to 0.6%.

For a pipe of *rectangular cross section* (eqs. 3.95, 3.104, 3.105) the conductance will be given by

$$C = 9.71 \times 10^3 \ (T/M)^{1/2} \ a^2 \ b^2 \ K/[(a+b) \ L + 2.66 \ a \ b] \tag{3.110}$$

---

* An historical review was published by Venema (1973/74). Accurate determinations and calculations of this factor, e.g. De Marcus and Hopper (1955), Helmer (1967a, b), Moore (1972) found values close to those given by Clausing (1932).

in CGS units. This equation becomes for *a slot in which* $a \gg b$, and in which the length L of the slot in the direction of the flow is not large compared with b

$$C = 9.71 \ (T/M)^{1/2} \ a \ b^2 \ K/(L + 2.66 \ b)$$ (3.111)

where a (cm), b (cm), L (cm), C (liter/sec), while *for a narrow slot where* $a \gg b$, $L \gg b$, the conductance will be

$$C = 9.71 \ (T/M)^{1/2} \ (a \ b^2/L)K$$ (3.112)

where a (cm), b (cm), L (cm) and C (liter/sec), and K is the correction factor listed in table 3.5. For air at 20°C, this equation results in

$$C_{air} = 30.9(ab^2/LK)$$ (3.113)

a, b, L (cm), C (liter/sec).

The conductance of *a short tube of annular cross section* is given (eqs. 3.99, 3.104, 3.105) by

$$C = 3.81 \ (T/M)^{1/2} \ (D_o - D_i)^2 \ (D_o + D_i) \ K_o/[L + 1.33 \ (D_o - D_i)]$$ (3.114)

where D, L (cm), C (liter/sec), which for air at 20°C, becomes

$$C_{air} = 12.1 \ (D_o - D_i)^2 \ (D_o + D_i) \ K_o/[L + 1.33 \ (D_o - D_i)]$$ (3.115)

where D_o (cm) is the diameter of the outer cylinder, D_i (cm) is the diameter of inner cylinder, L (cm), and C (liter/sec), K_o the correction factor from table 3.6.

3.4. Conductance of combined shapes

3.4.1. Molecular flow – tapered tubes

Equation (3.92) can also be written (by using eq. 2.42)

$$C = \tfrac{4}{3} v_{av} K/(BL/A^2)$$ (3.116)

where K is the shape factor (tables 3.5; 3.6).

If the conductance results from a series connection of conductances of length dL, it is written

$$1/C = [\tfrac{3}{4} / (v_{av} K)] \int_0^L (B/A^2) \, dL$$ (3.117)

thus

$$C = \tfrac{4}{3}\, v_{av}\, K / \int_0^L (B/A^2)\, dL \tag{3.118}$$

Equation (3.118) is a general formula which gives the conductances of pipes with constant cross section, as well as those in which B and A are a continuously increasing or decreasing function of L.

For constant cross sections, B and A are not functions of L, $\int_0^L dL = L$ and eq. (3.118) results in eq. (3.116).

For a *tapered* (conical, pyramidal) *pipe*, having at the small end the cross section with perimeter B_1 and area A_1, and at the large end B_2, and A_2, at a distance x, these values will be

$$B_x = B_1 + (B_2 - B_1)\ (x/L) = k_B[a_1 + (a_2 - a_1)\ (x/L)]$$

$$A_x = k_A[a_1 + (a_2 - a_1)\ (x/L)]^2 \tag{3.119}$$

where a is one of the sides (or radius), and k_B is the constant ratio between the perimeter and this side (e.g. for a circle $k_B = 2\pi a/a = 2\pi$), while k_A is the constant ratio between the area and the square of the side (e.g. for a circle $k_A = \pi a^2/a^2 = \pi$).

From eq. (3.119) it results that

$$\frac{B_x}{A_x^2} = \frac{k_B/k_A^2}{[\,a_1 + (a_2 - a_1)\ (x/L)\,]^3} \tag{3.120}$$

thus (for eq. 3.118);

$$\int_0^L \frac{B_x}{A_x^2}\, dL = \frac{k_B}{k_A^2} \int_0^L \frac{dx}{[a_1 + (a_2 - a_1)\ (x/L)]^3} = \left(\frac{k_B}{k_A^2}\right)\left(\frac{L}{2}\right) \frac{a_1 + a_2}{a_1^2\, a_2^2} \tag{3.121}$$

and the conductance (eq. 3.118) will be given by

$$C = \tfrac{8}{3}\, v_{av}\, (k_A^2/k_B)\, [a_1^2\, a_2^2/(a_1 + a_2)]\, (K/L) \tag{3.122}$$

For a *circular cross section*

$$B = 2\pi r;\ A = \pi r^2;\ k_B = 2\pi;\ k_A = \pi;\ k_A^2/k_B = \pi^2/2\pi = \pi/2$$

and $K=1$. Therefore, the *conductance of a tapered pipe of circular cross section* will be (eq. 3.122)

$$C=(4\pi/3)\ [r_1^2\ r_2^2/(r_1+r_2)L]\ v_{av} \tag{3.123}$$

in CGS units; or for $D=2r$

$$C = 7.62\ (T/M)^{1/2}\ D_1^2\ D_2^2/\ [(D_1 + D_2)\ L] \tag{3.124}$$

where D, L (cm), and C (liter/sec), D_1 and D_2 being the diameters of the tapered pipe at its ends.

Comparing eq. (3.124) with (3.93) it results that the *equivalent diameter for a tapered tube* is

$$D_e = [2D_1^2\ D_2^2/\ (D_1 + D_2)]^{1/3} \tag{3.125}$$

For a *rectangular cross section* : $B=2(a+b); A=ab; k_B=2(a+b)/a=2[1+(b/a)];$ $k_A=ab/a^2=b/a$

$$k_A^2/k_B = (b/a)^2/2\ [1 + (b/a)]$$

The *conductance of a tapered pipe of rectangular cross section* will be (eq. 3.122)

$$C = \tfrac{4}{3}\ v_{av}\ \frac{(b/a)^2}{1 + (b/a)}\ \frac{a_1^2\ a_2^2}{(a_1 + a_2)\ L}\ K \tag{3.126}$$

The conductance of tapered pipes of *triangular cross section* and *annular cross section* can be calculated in the same way.

Approximate calculations of the conductance of non-uniform tubes were published by Yu-guo (1981).

3.4.2. *Molecular flow – elbows*

The molecules (in a molecular flow) through an elbow (fig. 3.11) can be divided in two categories: molecules (1) which collide with the wall in the region of the elbow, and molecules (2) which pass across the elbow. Molecules having the path (1) will see the opening of the tube as an impedance, thus the conductance of the elbow will be given for path (1) by

$$C=3.81\ (T/M)^{1/2}D^3/[L_1+L_2+1.33D] \tag{3.127}$$

Molecules having the path 2 (fig. 3.11) will pass the elbow without feeling its influence. Thus the conductance for these molecules is

$$C = 3.81 \ (T/M)^{1/2} \ D^3/(L_1 + L_2) \tag{3.128}$$

According to eqs. (3.127) and (3.128), an elbow can be represented as a tube with the diameter D, having an equivalent length L_e, which will be situated between

$$L_{ax} < L_e < (L_{ax} + 1.33 \ D) \tag{3.129}$$

where $L_{ax} = L_1 + L_2$ is the length as measured on the axis of the elbow.

For a more precise evaluation it can be considered that all the molecules will travel according to path 1 (fig. 3.11), when the shape of the elbow is that of a hairpin, thus the bend is at $\theta = 180°$. Considering that the number of molecules having path (1) is proportional to the angle θ of the *elbow*, it results that the *equivalent length* is:

$$L_e = L_{ax} + 1.33 \ \frac{\theta}{180} \ D \tag{3.130}$$

3.4.3. Molecular flow – traps

Dushman and Lafferty (1962) considered the conductances of traps such as shown in fig. 3.12, in which the diameter of the outer cylinder is $D_o = 2a_2$, and that of the inner cylinder is $D_i = 2a_1$.

The conductance of such a trap is that of the series connection of conductances C_1 of the inner cylinder (I), and C_2 of the annular space between the two cylinders (II). According to eq. (3.108) the conductance of the (short) inner cylinder is

$$C_1 = 3.81 \times 10^3 \ (T/M)^{1/2} \ D_i^3/(L + 1.33 \ D_i) \tag{3.131}$$

and that of the annular space

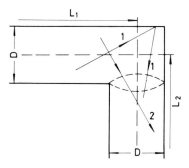

Fig. 3.11 Molecular flow through an elbow.

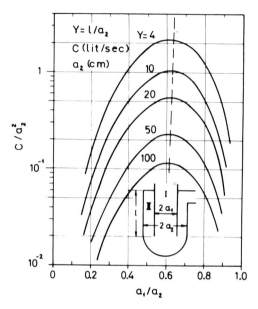

Fig. 3.12 Conductance of traps, for air, 25°C. From Dushman and Lafferty (1962), by permission of J. Wiley & Sons Inc., New York.

$$C_2 = 3.81 \times 10^3 \, (T/M)^{1/2} \, (D_o - D_i)^2 (D_o + D_i) K_o / [L + 1.33 \, (D_o - D_i)] \quad (3.132)$$

The *conductance of the trap is given* by $1/C = (1/C_1) + (1/C_2)$. By neglecting the correction K_o, and putting $X = D_i/D_o = a_1/a_2$; $Y = l/a_2 = 2l/D_o$, it results that

$$C = 3.81 \times 10^3 \, (T/M)^{1/2} \, (3/4) \, D_o^2 \, f(X, \, Y)$$
$$= 3.81 \times 10^3 \, (T/M)^{1/2} \times 3a_2^2 \, f(X, \, Y) \, (cm^3/sec)$$

i.e.

$$C/a_2^2 = 1.14 \times 10^4 \, (T/M)^{1/2} \, f(X, Y) \quad (3.133)$$

where

$$f(X, Y) = \frac{X^3(1 - X)(1 - X^2)}{X(1 - X) + \frac{3}{8} \, Y \, [X^3 + (1 - X)(1 - X^2)]} \quad (3.134)$$

Figure 3.12 plots the value

$$C/a_2{}^2 = 36.7 \; f(X, Y) \tag{3.135}$$

as it results from eq. (3.133), for air at 25°C, if a_2 (cm), C (liter/sec). The plot of C/a_2^2 is presented as a function of $X = a_1/a_2$, using values of $Y = l/a_2$ (4, 10, 20...) as a parameter. The dotted line shows the values of X for maximum conductance at different values of Y. For large values of Y, the maximum conductance corresponds to $X^2 + X = 1$; that is $X = 0.618$.

For a rigorous calculation the value of K_o (table 3.6) must be taken into account, and the conductances of the extension of the inner tube and that of the side connection must also be considered. Besides all this the real calculation must also take into account that the *trap is cooled and the temperature is not equal in all the parts*.

Figure 3.13 shows a trap where these various details are also considered (Henry, 1971). The trap is immersed in liquid nitrogen on an effective depth L_3, considered from the level of the liquid nitrogen to the outlet of the inner (inlet) tube. In such a case it may be assumed that the temperature of the inlet (inner) tube decreases

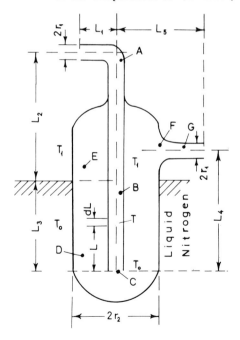

Fig. 3.13 Parts of a liquid nitrogen trap, for calculating its conductance. After Henry (1971).

linearly from T_1 at the level of the liquid nitrogen, to T_0 at the bottom of the inner tube, so that at a height L, the temperature is

$$T = T_0 + (T_1 - T_0)\, L/L_3 = T_0 + gL \tag{3.136}$$

where $g = (T_1 - T_0)/L_3$.

The outer wall of the trap may be considered at T_1 above the liquid nitrogen, and at T_0 below this level.

The trap shown in fig. 3.13 is constituted by the various parts listed in table 3.7, which are connected in series (see eq. 3.151)

Table 3.7.
Parts of trap in fig. 3.13.

Part	Description	Diameter	Length	Temperature
A	inlet–elbow	$2r_1$	L_1+L_2	T_1
B	straight pipe	$2r_1$	L_3	T (eq. 3.136)
C	diaphragm	$2r_2$	—	T_0
D	annular pipe	$2r_2/2r_1$	L_3	T_0
E	,,	$2r_2/2r_1$	L_4-L_3	T_1
F	aperture	$2r_1$	—	T_1
G	exit tube	$2r_1$	L_5-r_2	T_1

Part A has a conductance expressed (eq. 3.93) by

$$C_A = 3.81\ (T_1/M)^{1/2}\ (8r_1^3/L_e) \tag{3.137}$$

where L_e is the equivalent length for an elbow (eq. 3.130) of $90°$:

$$L_e = L_1 + L_2 + 1.33\ \frac{90}{180}\ 2r_1 = L_1 + L_2 + 1.33\ r_1 \tag{3.138}$$

thus

$$C_A = 3.81 \left[\frac{T_1}{M} \right]^{1/2} \frac{8r_1^3}{L_1 + L_2 + 1.33\ r_1} \tag{3.139}$$

Part B has a temperature distribution according to eq. (3.136), thus a length dL at a distance L from the bottom, will have a temperature T, and the conductance of this portion will be

$$dC_B = 3.81 \left(\frac{T}{M} \right)^{1/2} \left(\frac{8r_1^3}{dL} \right) = 3.81 \left(\frac{T_1}{M} \right)^{1/2} \left(\frac{8r_1^3}{\sqrt{T_1}} \right) \frac{(T_o + gL)^{1/2}}{dL} \tag{3.140}$$

The conductance of the whole length L_3 will be given by

$$\int_0^{L_3} \frac{1}{dC_B} = \frac{1}{3.81 \, (T_1/M)^{1/2}} \frac{\sqrt{T_1}}{8r_1^3} \int_0^{L_3} \frac{dL}{(T_o + gL)^{1/2}} \tag{3.141}$$

from which it results that

$$C_B = 3.81 \left(\frac{T_1}{M} \right)^{1/2} \left(\frac{8r_1^3}{2L_3} \right) \frac{\sqrt{T_1} + \sqrt{T_o}}{\sqrt{T_1}} \tag{3.142}$$

Part C is a diaphragm, which has to be considered since the molecules coming from the inner tube will collide with the round bottom and from here they have to pass by the annular diaphragm.

According to eqs. (3.72) and (3.83), the conductance of this diaphragm is given by

$$\begin{aligned} C_C &= 3.64 \, (T_o/M)^{1/2} \, \pi \, (r_2^2 - r_1^2) \, [\, 1 - (r_2^2 - r_1^2)/r_2^2 \,]^{-1} \\ &= 3.64 \, (T_o/M)^{1/2} \, \pi \, (r_2^2 - r_1^2) \, r_2^2/r_1^2 \end{aligned} \tag{3.143}$$

Part D is an annular cross section, having the outside wall at temperature T_o, and the inner wall at temperature T, which varies according to eq. (3.136). It can be assumed that the relative number of molecules having average velocities corresponding to the various temperatures is proportional to the ratio of surfaces having these temperatures. Thus the average temperature T_a can be considered

$$\begin{aligned} T_a &= \frac{r_1 \, T + r_2 \, T_o}{r_1 + r_2} = \frac{r_1 \, (T_o + gL) + r_2 \, T_o}{r_1 + r_2} \\ &= \frac{T_o \, (r_1 + r_2) + r_1 \, gL}{r_1 + r_2} \end{aligned} \tag{3.144}$$

This is the average temperature of the outer and inner wall at a distance L from the bottom. The conductance of a portion dL at distance L will be

$$dC_D = 3.81 \, (T_a/M)^{1/2} \, [\, 8 \, (r_2 - r_1)^2 \, (r_1 + r_2)/dL \,] \, K_o \tag{3.145}$$

thus the conductance of the whole length L_3 is given by

$$\int_0^{L_3} \frac{1}{\mathrm{d}C_D} = \frac{1}{3.81 \, (T_1/M)^{1/2}} \left[\frac{\sqrt{T_1}}{8 \, (r_2 - r_1)^2 \, (r_1 + r_2) \, K_o} \right] \int_0^{L_3} \frac{\mathrm{d}L}{\sqrt{T_a}} \qquad (3.146)$$

From eq. (3.146), it results that the conductance of part D is expressed by

$$C_D = 3.81 \left(\frac{T_1}{M} \right)^{1/2} \frac{8K_o \, (r_2 - r_1)^2 \, (r_1 + r_2)}{2L_3 \, \sqrt{T_1}}$$
$$\times \frac{T_1 - T_o}{(r_1 \, T_1 + r_2 \, T_o)^{1/2} - [T_o \, (r_1 + r_2)]^{1/2}} \qquad (3.147)$$

Part E of the annular space emerging above the level of the liquid nitrogen, has a conductance

$$C_E = 3.81 \left(\frac{T_1}{M} \right)^{1/2} \frac{8K_o \, (r_2 - r_1)^2 \, (r_1 + r_2)}{L_4 - L_3} \qquad (3.148)$$

Part F is the exit aperture, and has the conductance (eq. 3.74)

$$C_F = 2.86 \, (T_1/M)^{1/2} \, 4r_1^2 \qquad (3.149)$$

Part G is the outlet tube, with a conductance

$$C_G = 3.81 \, (T_1/M)^{1/2} \, 8r_1^3/(L_5 - r_2) \qquad (3.150)$$

The conductance of the trap (fig. 3.7) is given by

$$\frac{1}{C} = \frac{1}{C_A} + \frac{1}{C_B} + \frac{1}{C_C} + \frac{1}{C_D} + \frac{1}{C_E} + \frac{1}{C_F} + \frac{1}{C_G} \qquad (3.151)$$

For a trap having $r_1 = 1$ cm; $r_2 = 3$ cm; $L_1 = 3$ cm; $L_2 = 7$ cm; $L_3 = 8$ cm; $L_4 = 10$ cm; $L_5 = 6$ cm, $T_1 = 293°K$; and $T_o = 77°K$, it results that the conductance for air of the whole trap is $C = 3.1$ liter/sec, mainly determined by C_A, C_B and C_F, since the other conductances are relatively large.

3.4.4. *Molecular flow – optical baffles*

Baffles are systems of cooled walls, or plates placed near the inlet of vapour pumps to condense back-streaming vapour, and return the liquid to the pump. In order to increase their efficiency for condensing the vapour, the baffles are constructed in such a way that no molecule can traverse them without colliding with the wall. They are called *optical baffles*, since they are *opaque* for light rays (transmitted in a straight line in any direction). Baffles are constructed with straight *parallel plates* (fig. 3.14), or with *concentric plates* (fig. 3.15). The conductance of cryopumping arrays (baffles) was calculated by Hamacher (1976).

Conductance of baffles with straight plates. Consider a baffle having straight V shaped plates (chevron), inclined at an angle γ to the vertical (fig. 3.14) and spaced from each other at a distance p. The plates cross the circle of their enclosure, thus their lengths differ with their position. In order to calculate the conductance of the baffle, each space between adjacent plates can be considered as a duct with rectangular cross section. These ducts have a cross section of sides a and b while their length is L. We have

$$L = L'/\cos \gamma \qquad (3.152)$$

$$b = p \cos \gamma \qquad (3.153)$$

The side a_n will be of a length

$$a_n = \{1 - [1 - (2n/N)]^2\}^{1/2} D \qquad (3.154)$$

where n is the serial number of the plate 1, 2, 3. . . (fig. 3.14), N the total number of equidistant plates and D the diameter of the baffle.

The conductance of an opening between adjacent plates will be (eq. 3.110)

$$C_n = 9.71 \times 10^3 \left(\frac{T}{M} \right)^{1/2} \frac{a_n^2 b^2 K_n}{(a_n + b) L_e + 2.66 a_n b} \qquad (3.155)$$

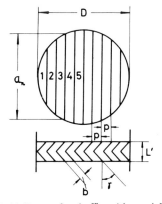

Fig. 3.14 Parts of a baffle with straight plates.

where a_n (eq. 3.154), b (eq. 3.153) and K_n, the correction factor K (table 3.5), for $b/a=b/a_n$. L_e is the equivalent length of the elbow, given as in eqs. (3.127) and (3.130) by

$$L_e = L + 1.33[(180 - 2\gamma)/180] \, [2 \, a_n \, b/(a_n + b)] \, K_n \tag{3.156}$$

where L is the axial length (eq. 3.152).

By combining eqs. (3.152–3.156), the conductance C of a baffle as shown in figure 3.14 is calculated as

$$C = 2\sum_1^{\frac{1}{2}N} C_n \tag{3.157}$$

The application of eq. (3.157) to a baffle of $D=25$ cm, $N=10$ plates, bent at an angle of $90°$, ($\gamma=45°$) and $L'=5$ cm, leads to a conductance for air at room temperature of 2400 liter/sec.

Conductance of baffles with concentric plates. Consider a baffle constituted of concentric plates as shown in fig. 3.15. The conductance of each annular opening left between the concentric plates, can be calculated by using eq. (3.118).

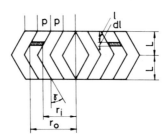

Fig. 3.15 Parts of a baffle with concentric plates.

The annulus number n (from the center), has an inner radius

$$r_i = np + l \, \mathrm{tg} \, \gamma \tag{3.158}$$

and an outer radius

$$r_o = (n+1) \, p + l \, \mathrm{tg} \, \gamma \tag{3.159}$$

thus the perimeter is

$$B_n = 2\pi \ [p(2n + 1) + 2l \ \text{tg} \ \gamma] \tag{3.160}$$

and the cross section

$$A_n = \pi \ (r_i + r_o) \ p = \pi p \ [p \ (2n + 1) + 2l \ \text{tg} \ \gamma) \tag{3.161}$$

The conductance is given by eq. (3.118)

$$c_n = \tfrac{4}{3} \ v_{av} / \int_0^L [B_n/(K_{on} A_n^2)] \ \text{d}L \tag{3.162}$$

The factor K_{on} (table 3.6) depends on the value of $D_i/D_o = r_i/r_o$. The value r_i/r_o varies from 0 to $\tfrac{1}{2}$ for the first annulus, from $\tfrac{1}{2}$ to $\tfrac{2}{3}$ for the second, from $\tfrac{2}{3}$ to $\tfrac{3}{4}$ for the third, etc. According to table 3.6, K_o varies in such a way that for each annulus it can be assumed as varying linearly, i.e.

$$K_{no} = \alpha_n + \beta_n \ (l/L) \tag{3.163}$$

Thus the quantity in the integral can be expressed as

$$\frac{B_n}{K_{on} A_n^2} = \frac{2\pi \ [p \ (2n + 1) + 2 \ l \ \text{tg} \ \gamma]}{[\alpha_n + \beta_n \ l/L] \ \pi^2 \ p^2 \ [\ p \ (2n + 1) + 2 \ l \ \text{tg} \ \gamma]^2}$$

$$= \frac{2}{\pi \ p^2 \ [\alpha_n + \beta_n \ l/L] \ [p \ (2n + 1) + 2 \ l \ \text{tg} \ \gamma]} \tag{3.164}$$

It results that

$$\int_0^L \frac{B_n \text{d}l}{K_{on} A_n^2} =$$

$$= \frac{2}{\pi \ p^2 \ [2 \ \alpha_n \ \text{tg} \ \gamma - (\beta_n/L) \ p \ (2n + 1)]} \left| \ln \frac{p \ (2n + 1) + 2 \ l \ \text{tg} \ \gamma}{\alpha_n + \beta_n \ l/L} \right|_0^L \tag{3.165}$$

Thus

$$C_n = \tfrac{4}{3} v_{av} \frac{2\,\alpha_n \mathrm{tg}\,\gamma L - \beta_n\,p\,(2n+1)}{2L} \times$$

$$\times \frac{\pi p^2}{\ln\{[\alpha_n/(\alpha_n + \beta_n)][1 + 2L\mathrm{tg}\,\gamma/p\,(2n+1)]\}} \tag{3.166}$$

In order to account for the elbow, instead of L, the effective length L_e is to be considered (eq. 3.130)

$$L_e = (L/\cos\gamma) + 1.33\,[(180 - 2\gamma)/180]\,p\,K_{no} \tag{3.167}$$

The total conductance of the baffle is the sum of the conductances C_n.

3.4.5. Molecular flow – seal interface

Roth (1967) has given a model for the mechanism of the sealing process between two surfaces which are compressed on each other. In this model the roughness of the sealing surfaces is considered to be constituted of flat equilateral pyramids which penetrate into the opposite surface or are flattened and leave between them leakage paths, as shown in fig. 3.16.

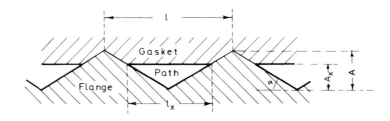

Fig. 3.16 Model of the interface sealing process. After Roth (1967).

The groove between two adjacent pyramids is called *unit groove*. The total conductance of the seal is regarded as the result of the series–parallel connection of all the unit grooves.

The cross section of the *unit groove* varies along the groove as shown in fig. 3.17. The unit groove consists of two parts connected in series. The first part (between fronts 1 and 3) has a profile changing with the distance L. The second part of the unit groove (between fronts 3 and 4, fig. 3.17) has a constant cross section. If C_1 is the conductance of the first part of the groove, and C_2 is that of the second part, the conductance of the unit groove of length l, will be given by

$$1/C_l = 2[(1/C_1) + (1/C_2)]$$ (3.168)

since the length l of the unit groove is formed by the series connection of two parts C_1 and of two parts C_2.

For *part 1* of the groove the perimeter B_1, and the cross section area A_1 are given by

$$B_1 = (4/\text{tg } \alpha) A_x + 2[(1/\cos \alpha) - 1] L$$

$$A_1 = 2 A_x^2/\text{tg } \alpha - 2 A_x L + L^2 \text{ tg } \alpha$$ (3.169)

thus

$$\int_0^{A_x/\text{tg}\alpha} B_1/A_1^2 \, dL =$$

$$= \frac{1 + 1/\cos\alpha}{A_x} \left| A_x^{-1} \text{ arc tg } [(L\text{tg}\alpha - A_x)/A_x] \right|_0^{A_x/\text{tg}\alpha} + \frac{1}{A_x^2}$$

$$= [\tfrac{1}{4} \pi (1 + 1/\cos\alpha) + 1]/A_x^2$$ (3.170)

and the conductance is

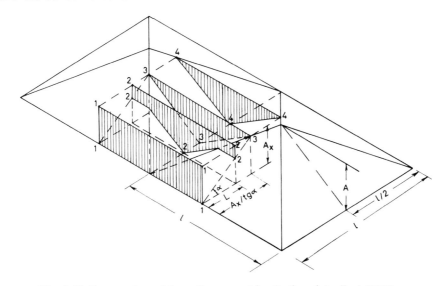

Fig. 3.17 Cross section of the unit groove. After Roth and Amilani (1965).

$$C_1 = \tfrac{4}{3} v_{av} \left(\int_{A_1^2}^{B_1} dL \right)^{-1} = \tfrac{4}{3} v_{av} A_x^2 / [\tfrac{1}{4} \pi (1 + 1/\cos \alpha) + 1] \tag{3.171}$$

For *part 2* of the groove the perimeter is

$$B_2 = \frac{2A_x}{\text{tg } \alpha} \left(1 + \frac{1}{\cos \alpha} \right) \tag{3.172}$$

and the cross section area

$$A_2 = A_x^2 / \text{tg } \alpha \tag{3.173}$$

hence the conductance C_2 of this part is given by

$$C_2 = \tfrac{4}{3} v_{av} \frac{A_2^2}{B_2 (\tfrac{1}{2}l - A_x/\text{tg } \alpha)} = \tfrac{4}{3} v_{av} \frac{A_x^3}{2 (1 + 1/\cos \alpha) A (1 + A_x/A)} \tag{3.174}$$

the conductance of the unit groove will be (eq. 3.168)

$$C_l = \tfrac{4}{3} v_{av} \frac{A^2 (A_x/A)^3}{4 (1 + 1/\cos \alpha)} \frac{1}{1 - 0.36 (A_x/A)} \tag{3.175}$$

It was found that the total conductance of a contact seal is

$$C = 2\pi C_l / \ln (r_o/r_i) \tag{3.176}$$

where r_o and r_i are the outer and inner radius of the annular sealing interface. Therefore

$$C = \tfrac{4}{3} v_{av} \frac{2\pi}{\ln (r_o/r_i)} \frac{A^2 (A_x/A)^3}{4 (1 + 1/\cos \alpha)} (1 - 0.36 A_x/A)^{-1} \tag{3.177}$$

The penetration A_x/A was found to be a function of the tightening pressure P, and the sealing factor R of the gasket material

$$A_x/A = \exp (-P/R) \tag{3.178}$$

Equations (3.178) and (3.177) permit us to derive the *basic equation of the sealing process*

$$C = 1.93 \times 10^4 \left(\frac{T}{M} \right)^{1/2} \frac{2\pi}{\ln (r_o/r_i)} \frac{A^2}{8.12} \left[\frac{e^{-3P/R}}{1 - 0.36 \, e^{-P/R}} \right] \tag{3.179}$$

where A is the peak to valley height of the initial surface roughness (fig. 3.16; 7.16; 7.17). Values of R are given in fig. 7.19.

3.5. Analytico–statistical calculation of conductances

Davis (1960) and Levenson et al. (1960; 1963) as well as Carette et al. (1983) used the Monte–Carlo calculational method, for determining the conductance of simple and complex shapes. To make such calculations, individual histories of molecules entering the model are generated from a set of random numbers. When enough histories have been generated to satisfy the accuracy required, the calculation is terminated.

The entering molecule is followed over its probable path. At each collision with the wall the molecule is assumed to be stopped and promptly reemitted. The molecule is then assigned random numbers to specify the velocity and direction after leaving the wall. The selection of direction is based upon *Lambert's law* of emission, i.e. the molecules leaving a unit area of the wall are distributed according to

$$I_\theta/I_n = \cos \theta \tag{3.180}$$

where I_θ is the number of molecules leaving per second in a direction at an angle θ with respect to the normal to the surface, and I_n is the total number of molecules leaving the surface per second. The history of the molecule is followed until it either leaves the geometry or returns to the entrance opening.

Davis, Levenson and *Milleron* use the conductance C_o of the aperture to the geometrical configuration being investigated, as their reference. The computed and measured conductance C is related to C_o by the probability factor P_r

$$C/C_o = P_r \tag{3.181}$$

The assumptions made in the calculations and the experimental conditions provided are :

(a) Steady molecular flow exists.
(b) Molecules enter the inlet aperture uniformly distributed over its surface.
(c) The geometries under study connect effectively large volumes.
(d) The probability of the molecules entering a solid angle is proportional to the cosine of the angle to the normal to the surface of the opening.

(e) The walls are microscopically rough, so that molecules are diffusely reflected according to the cosine law.

Consider a tube with two openings of area A_1 and A_2, through which particles are diffusing from infinite volumes at a net rate of N particles per second under steady-state conditions. If ϕ_1 and ϕ_2 are the numbers of molecules striking unit area per second at A_1 and A_2 then the number of molecules per second entering the tube at orifice 1 and orifice 2 are $\phi_1 A_1$ and $\phi_2 A_2$. Let P_{r1} be the probability that a particle entering orifice 1 will leave through 2, and P_{r2} the probability that a particle entering orifice 2 will leave through 1. The equation for the net flow of particles is then

$$\phi_1 A_1 P_{r1} - \phi_2 A_2 \ P_{r2} = N \tag{3.182}$$

When ϕ_1 is equal to ϕ_2 and the system is isothermal, $N=0$, thus

$$A_1 P_{r1} = A_2 P_{r2} \tag{3.183}$$

Equation (3.183) shows that the probability P_r for a molecule transmission is dependent not only on the geometry of the pipe, but also on the area of the orifice with which P_r is associated. When the orifices at each end of a pipe have equal areas (independent of their shape)

$$P_{r1} = P_{r2} = P_r \tag{3.184}$$

If eq. (2.46) is applied, eq. (3.182) can be written

$$\tfrac{1}{4} \, v_{av1} \, n_1 \, A_1 \, P_{r1} - \tfrac{1}{4} \, v_{av2} \, n_2 \, A_2 \, P_{r2} = N \tag{3.185}$$

Since $\tfrac{1}{4} \, v_{av} A$ is the volume flow rate of gas $S = \mathrm{d}V/\mathrm{d}t$ (eq. 2.48), through an orifice of area A, eq. (3.185) is written

$$n_1 \, S_1 \, P_{r1} - n_2 \, S_2 \, P_{r2} = N \tag{3.186}$$

When n_1 is equal to n_2, then $N=0$, and

$$S_1 \, P_{r1} = S_2 \, P_{r2} \tag{3.187}$$

From eqs. (3.186) and (3.187), it results that

$$S_1 \, P_{r1} = S_2 \, P_{r2} = S = N/(n_1 - n_2) = C \tag{3.188}$$

The quantity C is the conductance of the system. The Monte–Carlo method uses eq. (3.182) by assuming $\phi_2=0$. Let $\phi_1 A_1 = K_1$ the number of particles that enter the geometry per unit time through opening 1; then this method counts K_1 and N_1 to find $P_{r1} = N/K_1$.

The first geometry investigated is one which can also be determined by simple calculations. It is that of a tube of circular cross section, where the conductance can be calculated by eq. (3.108). The value of C_o is in this case given by eq. (3.74). The value of P_r is in this case

$$P_r = C/C_o = (3.81/2.86)\,[D^3/(L+1.33D)]\,(1/D^2)$$

$$= \tfrac{4}{3}\,D/[L+\tfrac{4}{3}\,D] = \tfrac{4}{3}\,(D/L)\,K'' \tag{3.189}$$

The Monte–Carlo computation, and experimental results shown in fig. 3.18, are in good agreement with eq. (3.189), especially if K' (eq. 3.109) is used instead of K'' (eq. 3.106).

The second geometry investigated by Davis, Levenson and Milleron, is a $90°$ elbow (fig. 3.19). The Monte–Carlo computation and the experimental results show that the conductance of the elbow does not differ significantly from those of a straight tube (see eq. 3.130). Xu and Wang (1982) determined that the transmission probability P_r has a minimum value for elbows of $50°–70°$.

Results for the transmission probability of a cylindrical annulus are shown in

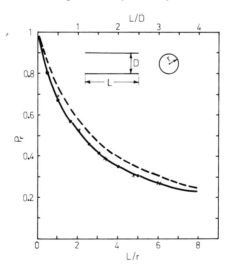

Fig. 3.18 Transmission probability P_r for tubes with circular cross section. Continuous curve: calculated by using eq. (3.109); dotted curve: calculated by using eq. (3.107), points: Monte–Carlo calculations; x–experimental values. After Levenson et al. (1960).

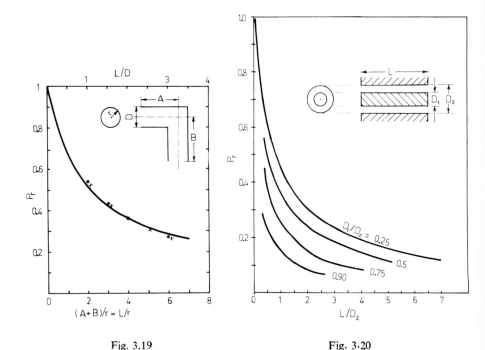

Fig. 3.19   Fig. 3·20

Fig. 3.19 Transmission probability P_r for a 90° elbow. Curves: calculated by using eq. (3.109); points: Monte–Carlo calculation; x–experimental values. After Levenson *et al.* (1960).

Fig. 3.20 Transmission probability P_r of annular pipes. After Levenson *et al.* (1963).

fig. 3.20. Eq. (3.72) becomes for an annulus

$$C_o = 3.64 \, (T/M)^{1/2} \, (\pi/4) \, (D_2{}^2 - D_1{}^2) = 2.86 \, (T/M)^{1/2} \, (D_2{}^2 - D_1{}^2) \qquad (3.190)$$

while eq. (3.99) gives

$$C = 3.81 \, (T/M)^{1/2} \, (D_2 - D_1)^2 \, (D_2 + D_1) \, K_0 / [L + 1.33 \, (D_2 - D_1)] \qquad (3.191)$$

thus

$$P_r = C/C_o = (3.81/2.86) \, (D_2 - D_1) \, K_0 / [L + 1.33 \, (D_2 - D_1)] \qquad (3.192)$$

Values of P_r computed by the Monte–Carlo method are shown in fig. 3.20.

The results for two geometries commonly used in *optical baffles* are shown in

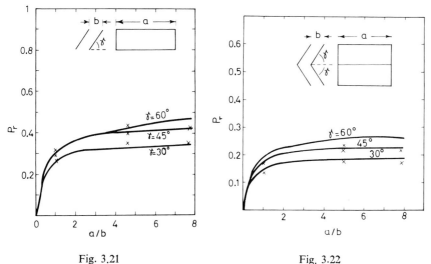

Fig. 3.21   Fig. 3.22

Fig. 3.21 Transmission probability P_r of baffles with straight plates. Curves: calculated; x-experimental values. After Levenson *et al.* (1960).

Fig. 3.22 Transmission probability P_r of baffles with chevron geometry. Curves: calculated; x-experimental values. After Levenson *et al.* (1960, 1963).

figs. 3.21 and 3.22. It can be seen that the transmission probability P_r does not increase significantly for geometries having ratios a/b greater than 5. Furthermore, the $60°$ angle in both these cases is shown to have better conductance properties than the $45°$ and $30°$ angles. The chevron geometry, fig. 3.22, has approximately half the value of P_r, for corresponding values of angle and a/b. *Transmission probability P_r has a maximum value* of 0.28 (fig. 3.22) for an aperture that is completely covered with the best chevron arrangement. In a practical case cooling tubes for liquid nitrogen or other refrigerant will be attached to the baffles with the result that the effective area will be somewhat reduced. Considering these factors, a realistic value of $P_r = 0.2$, results for a carefully designed chevron baffle.

The results for another geometry are shown in fig. 3.23. This geometry offers considerable improvement in transmission probability P_r, over the value obtainable from the chevron type geometry.

Furthermore, the blocking plate between the two openings *provides a possibility of using this configuration in a baffle–valve combination*. This allows us to avoid the additional impedance that would be brought into the system by adding a valve in series. The valve action can be obtained by moving the blocking plate to either end of the tube and sealing it over an opening.

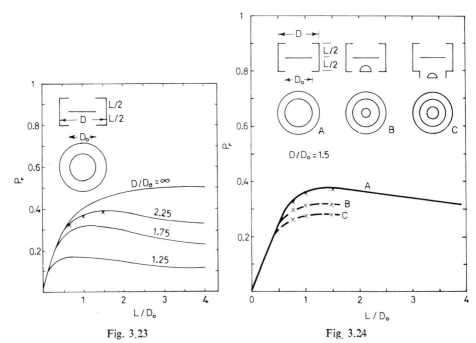

Fig. 3.23   Fig. 3.24

Fig. 3.23 Transmission probability P_r of baffles with a circular blocking plate and two restricted ends. Curves: calculated; x–experimental values. After Levenson *et al.* (1960).

Fig. 3.24 Transmission probability P_r of baffles (fig. 3.23) as used on diffusion pumps. Curves: calculated; x–experimental values. After Levenson *et al.* (1960).

Figure 3.24 shows the values of P_r for various arrangements of this geometry in a diffusion pump system. Case A (fig. 3.24) is the same geometry as in fig. 3.23 and case B shows the probability P_r for the particles to pass through the annulus between the jet–cap cover of the diffusion pump and the edge of the trap opening. Case C (fig. 3.24) simulates the use of this baffle geometry on an oil diffusion pump.

Figure 3.25 shows results for another geometry which offers possibilities as a baffle-valve. The effect of using this geometry in a diffusion pump is shown in fig. 3.26. The value of P_r reaches a maximum for a value $M/L=0.26$.

The results for another geometry that may be used as a baffle–valve combination are shown in fig. 3.27.

The circular blocking plate has the same diameter as the orifices. This plate, in principle, can be swung to cover either of the orifices. The experimental results for this geometry show that high values of P_r can be obtained for values of $W/D > 1.3$. The variation of P_r with the arrangement of this geometry in a diffusion pump is shown in fig. 3.27. A bulged elbow containing a chevron array is also shown in fig. 3.27.

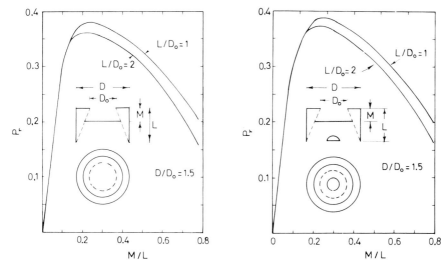

Fig. 3.25   Fig. 3.26

Fig. 3.25 Transmission probability P_r of baffles with a circular backing plate and one restricted end. After Levenson et al. (1960).

Fig. 3.26 Transmission probability P_r of baffles (fig. 3.25) as used on diffusion pumps. After Levenson et al. (1960).

From fig. 3.28 it can be seen that the bulged and cubic elbows have approximately the same values of P_r, for similar values of W/D.

Figure 3.29 shows the *average number of collisions* in some geometries that require a minimum of *one* collision (optically opaque), as it results from the Monte–Carlo calculations. It can be seen that this number is at least 4.

Calculations of the probability factor for various geometries and conditions of the flow have been published by Berman (1965, 1969), Moore (1972), Edenburn (1972), Berceanu and Ignatovich (1973), Gottwald (1973), Steckelmacher (1974), Van Essen and Heerens (1976), Füstös (1977), Henning (1978a), Steckelmacher and Henning (1979), Füstöss (1981, 1983), Xu and Wang (1982), Santeler (1986b), Santeler and Boekmann (1987b), Sagsaganskii (1988).

A computer program for the design of vacuum systems was published by Santeler (1987a).

3.6. Intermediate flow

3.6.1. *Knudsen's equation*

The conductance of a tube in viscous regime (high pressures) is directly proportional (eq. 3.52) to the average pressure $\overline{P}$, and therefore at $\overline{P}=0$ the conduc-

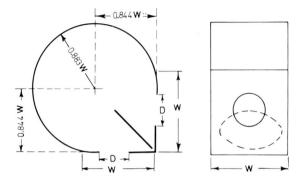

	w/D	P_r
	2.00	0.44
	1.33	0.39
	1.00	0.32
	2.00	0.33
	1.33	0.30
	2.00	0.32
	1.33	0.27
	2.00	0.38
	1.66	0.35
	1.33	0.31

Fig. 3.27 Transmission probability P_r of bulged elbow baffles. After Levenson *et al.* (1960).

tance should fall to zero. Equation (3.57) shows that as the pressure decreases, an additional term $c_2 D^3$ should be added. Thus the conductance is given by the general equation

$$C = (c_1 D^4 \bar{P} + c_2 D^3)/L \qquad (3.193)$$

Knudsen gave to this equation the form

$$C = \frac{\pi}{128\eta} \frac{D^4}{L} \bar{P} + \tfrac{1}{6} (2\pi R_o T/M)^{1/2} D^3/L \times$$
$$\times [1 + (M/R_o T)^{1/2} D\bar{P}/\eta]/[1 + 1.24 (M/R_o T)^{1/2} D\bar{P}/\eta] \qquad (3.194)$$

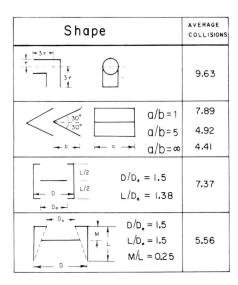

W/D	P_r
2.00	0.43
1.50	0.38
2.00	0.36
1.50	0.28
2.00	0.30
1.50	0.27

Fig. 3.28   Fig. 3.29

Fig. 3.28 Transmission probability P_r of cubic elbow baffles. After Levenson *et al.* (1960).

Fig. 3.29 Average number of molecule–wall collisions for molecules passing through baffles of various geometries. After Levenson *et al.* (1960).

thus

$$c_1 = \frac{\pi}{128\eta} \tag{3.195}$$

$$c_2 = 3.81 \times 10^3 \, (T/M)^{1/2}$$

$$\times \, [1 + 1.10 \times 10^{-4} \, (M/T)^{1/2} \, D\bar{P}/\eta]/[1 + 1.36 \times 10^{-4} \, (M/T)^{1/2} \, D\bar{P}/\eta] \tag{3.196}$$

At small values of the average pressure $\bar{P}$, the terms in the fraction in eq. (3.196) become small compared with unity, so that

$$c_2 = 3.81 \times 10^3 \, (T/M)^{1/2} \tag{3.197}$$

If this relation is compared with that resulting from eq. (3.57)

$$c_2 = \frac{\pi}{16} \left(\frac{\pi}{2} \frac{R_0 T}{M} \right)^{1/2} \frac{2-f}{f} = 2.2 \times 10^3 \left(\frac{T}{M} \right)^{1/2} \frac{2-f}{f} \tag{3.198}$$

it results that $f=0.74$. This means that at low pressures 74% of the molecules are adsorbed and reemitted randomly, and 26% are specularly reflected.

At high pressures eq. (3.196) gives

$$c_2 = 3.07 \times 10^3 \ (T/M)^{1/2} \tag{3.199}$$

Comparing eq. (3.199) with (3.57), it results that $f=0.85$.

Knudsen's results imply that the fraction of molecules adsorbed and reemitted changes slowly in the intermediate region.

Eq. (3.194) gives the conductance in any regime, *assuming that over the whole length of the tube L, the regime is the same* (molecular, intermediate or viscous).

The equations for flow in the transition range from slip to molecular flow were summarized and discussed by Thomson and Ovens (1975).

3.6.2. *The minimum conductance*

Eq. (3.194) can be written (Van Atta, 1968)

$$y = ax + b \ \frac{1+cx}{1+dx} \tag{3.200}$$

where

$$
\begin{array}{ll}
y = CL/D^3 & x = \bar{P}D/\eta \\
a = 3.27 \times 10^{-2} & b = 3.81 \ (T/M)^{1/2} \\
c = 0.147 \ (M/T)^{1/2} & d = 0.181 \ (M/T)^{1/2}
\end{array}
$$

C (liter/sec); $\bar{P}$ (Torr); D (cm); L (cm); η (poise); M (g); T ($^\circ$K).

Differentiating eq. (3.200) with respect to x and setting the resultant equation equal to zero, the value of x at which y has a minimum value is determined; this is

$$x_{min} = [\, b(d-c)/a - 1 \,]/d \tag{3.201}$$

thus

$$(\bar{P}D/\eta)_{min} = 5.47 \ (T/M)^{1/2} \tag{3.202}$$

and

$$\bar{P}_{min} \ D = 5.47 \ (T/M)^{1/2} \ \eta \ \text{Torr·cm} \tag{3.203}$$

From eqs. (2.70), (2.18), (2.19) and (2.42) it results that

$$\eta = 0.117 \, \overline{P} \, (M/T)^{1/2} \, \lambda \tag{3.204}$$

and from eqs. (3.204) and (3.203), we have

$$D/\lambda_{\min} = 0.63 \tag{3.205}$$

Since for *air* at room temperature (eq. 2.58)

$$\lambda_{\mathrm{air}} = 5 \times 10^{-3}/P$$

it results that *for air*

$$\overline{P}_{\min} D = 0.63 \times 5 \times 10^{-3} = 3.15 \times 10^{-3} \ \mathrm{Torr \cdot cm} \tag{3.206}$$

According to eq. (3.205) the minimum conductance occurs when the mean free path of the molecules is 1.59 times the diameter of the tube. For λ larger

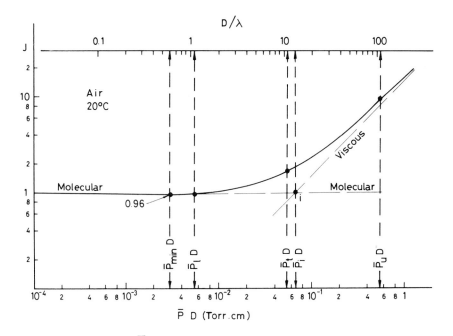

Fig. 3.30 J as a function of $\overline{P}D$ (eq. 3.225), and the various characteristic values of the intermediate flow range.

than this value, the conductance increases asymptotically toward that given for *molecular flow* (eq. 3.93), and for λ values less than λ_{min} the conductance increases with increasing pressure and tends to become proportional to the pressure in the *viscous flow* range (eq. 3.53). This evolution is shown in fig. 3.30.

3.6.3. *The transition pressure*

In eq. (3.58), the transition pressure $\bar{P}_t$ is defined as the value of the pressure for which the viscous term $c_1 \bar{P} D^4$ in eq. (3.193) is equal to the non-viscous term $c_2 D^3$. In the notation of eq. (3.200) this means

$$ax = b \, \frac{1 + cx}{1 + dx} \tag{3.207}$$

from which

$$x = \frac{1}{2ad} \left\{ (bc - a) \pm [(bc - a)^2 + 4\,abd]^{1/2} \right\} \tag{3.208}$$

By using the positive sign (the negative is meaningless) and substituting the values for x, a, b, c, d as specified for eq. (3.200), it results that the *transition pressure* is given by

$$\bar{P}_t \, D = 95.7 \, (T/M)^{1/2} \, \eta \; \text{Torr·cm} \tag{3.209}$$

By using eq. (3.204), we have

$$D/\lambda_t = 11.1 \tag{3.210}$$

and for *air* at room temperature (eq. 2.58), we have

$$\bar{P}_t \, D = 11.1 \times 5 \times 10^{-3} = 5.55 \times 10^{-2} \; \text{Torr·cm} \tag{3.211}$$

This means that for a tube of $D = 2$ cm the transition pressure is

$$\bar{P}_t = 2.77 \times 10^{-2} \; \text{Torr}$$

According to the condition expressed in eq. (3.207) the transition pressure includes a mixture of viscous and non-viscous flow, where both are significant. As it can be seen in fig. 3.30, this point is situated somehow in the middle of the range of intermediate flow.

3.6.4. *Limits of the intermediate range*

The limits of the intermediate range can be considered as being those where the contribution of one of the flow conditions predominates, e.g. where the contribution of one of them is an order of magnitude more important than that of the other.

Therefore the *upper limit of the intermediate range*, i.e. that *above which the flow can be considered viscous*, is given by

$$ax = 10b\frac{1+cx}{1+dx} \tag{3.212}$$

thus

$$x = \frac{1}{2ad}\left\{ (10\,bc - a) \pm [(10\,bc - a)^2 + 40\,abd]^{1/2} \right\} \tag{3.213}$$

Using again the positive sign, and the values of x, a, b, c, d (eq. 3.200), it results that the *upper limit of the intermediate range* is given by

$$\overline{P}_u D = 942\,(T/M)^{1/2}\,\eta \text{ Torr·cm} \tag{3.214}$$

and (eq. 3.204)

$$D/\lambda_u = 111 \tag{3.215}$$

and *for air* at room temperature (eq. 2.58)

$$\overline{P}_u D = 111 \times 5 \times 10^{-3} = 5.55 \times 10^{-1} \text{ Torr·cm} \tag{3.216}$$

This means that for pipes of $D = 2$ cm the flow can be considered viscous, for pressures $\overline{P}_u \geqslant 5.55 \times 10^{-1}/2 = 2.8 \times 10^{-1}$ Torr.

The *lower limit of the intermediate range*, i.e. that *below which the flow can be considered molecular*, is given by

$$10\,ax = b\,\frac{1+cx}{1+dx} \tag{3.217}$$

which gives (similarly to eq. 3.213)

$$\overline{P}_l\,D = 10\,(T/M)^{1/2}\,\eta \text{ Torr·cm} \tag{3.218}$$

thus

$$D/\lambda_l = 1.1 \tag{3.219}$$

and *for air* at room temperature

$$\overline{P}_l\, D = 1.1 \times 5 \times 10^{-3} = 5.5 \times 10^{-3}\ \text{Torr·cm} \tag{3.220}$$

From eqs. (3.220), (3.211), and (3.216), it results that

$$\overline{P}_\text{u} = 10\,\overline{P}_\text{t} \quad \text{and} \quad \overline{P}_l = 0.1\,\overline{P}_\text{t}$$

thus *the intermediate range extends over two orders of magnitude of the pressure.* A comparison is shown in fig. 3.30.

3.6.5. *General equation of flow*

Equation (3.194) can be written

$$C = C_\text{m}J \tag{3.221}$$

where C_m is the conductance for molecular flow (eq. 3.92

$$C_\text{m} = (\tfrac{1}{6})\,[2\pi\, R_oT/M]^{1/2}\,(D^3/L)\ \text{CGS units} \tag{3.222}$$

and

$$J = \frac{C_\text{v}}{C_\text{m}} + \frac{1 + (M/R_oT)^{1/2}\,D\overline{P}/\eta}{1 + 1.24\,(M/R_oT)^{1/2}\,D\overline{P}/\eta} \tag{3.223}$$

where C_v is the conductance for viscous flow (eq. 3.52)

$$C_\text{v} = \frac{\pi}{128\eta}\ \frac{D^4}{L}\ \overline{P} \tag{3.224}$$

in CGS units.

For *air at 20°C*, the value of J becomes

$$J = \frac{1 + 271\,D\overline{P} + 4790\,(D\overline{P})^2}{1 + 316\,D\overline{P}} \tag{3.225}$$

Figure 3.30 shows the values of J for air at 20°C in a log–log diagram. On the same diagram the various $\overline{P}D$ values, and D/λ values are also plotted. In this diagram it can be seen that $J=1$ for very low pressures, drops slowly to $J=0.96$ for $\overline{P}_{min} D$, increases to $J=1.7$ at $\overline{P}_t D$, and further increases to $J=9$ at $\overline{P}_u D$. The diagram also shows that from a constant value $J=1$ for the *molecular* range (where the conductance is independent of the pressure), the value of J tends at high pressures to be proportional to the average pressure (*viscous* flow).

3.6.6. *The molecular–viscous intersection point*

As it can be seen in fig. 3.30, the line representing the viscous flow intersects that of molecular flow at a point i. Roth (1972) has shown that the position of this point is specific for the kind of gas and its temperature.

The molecular–viscous intersection point i corresponds to $C_v = C_m$, where C_v is the conductance for viscous flow (eq. 3.224) and C_m is that for molecular flow (eq. 3.222). Therefore

$$C_v/C_m = [6\pi/(128\ \eta)]\ [M/(2\pi\ R_o T)]^{1/2}\ D\overline{P}_i = 1$$

from which

$$\overline{P}_i D = 17\ \eta\ (R_o T/M)^{1/2} \text{ in CGS units} \tag{3.226}$$

or

$$\overline{P}_i D = 116\ (T/M)^{1/2}\ \eta \text{ Torr·cm} \tag{3.227}$$

and by using eq. (3.204)

$$D/\lambda_i = 13.5 \tag{3.228}$$

For *air* (eq. 2.58) at 20°C

$$\overline{P}_i D = 13.5 \times 5 \times 10^{-3} = 6.7 \times 10^{-2} \text{ Torr·cm} \tag{3.229}$$

Figure 3.30 shows point i at this value. Roth (1972) expressed the whole range of molecular–intermediate–viscous flow, in terms of the $\overline{P}_i D$ value, with the aid of the ratio

$$\delta = \overline{P}D/(\overline{P}D)_i \tag{3.230}$$

By including eq. (3.230) in eq. (3 223) the factor J becomes

$$J = \delta + \frac{1 + 17\,\delta}{1 + 21\,\delta}$$

(3.231)

an equation which is valid for any gas at any temperature.

Figure 3.31 shows the plot of J as a function of $\bar{P}D$ for air and helium (according to eq. 3.223), as well as a scale of δ. It can be seen that the same δ scale is valid for all the gases, thus for $\delta = 1$; $J = 1.82$ both for air and helium. Since for air at 20°C, $\bar{P_i}D = 6.7 \times 10^{-2}$, $\delta = 2$ will correspond to

$$\bar{P}D = 2 \times 6.7 \times 10^{-2} = 1.34 \times 10^{-1} \; \text{Torr·cm}$$

the value at which $J = 2.81$.

For helium at 20°C, $\bar{P_i}D = 1.9 \times 10^{-1}$ Torr·cm (eq. 3.227), thus for $\delta = 2$, $\bar{P}D = 2 \times 1.9 \times 10^{-1} = 3.8 \times 10^{-1}$ Torr·cm. At this value (fig. 3.31) the J for helium is also $J = 2.81$.

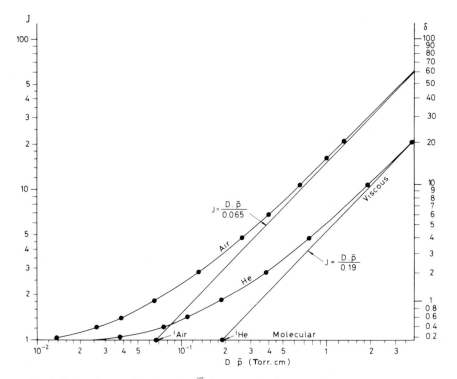

Fig. 3.31 J and δ as a function of $\bar{DP}$ for air and helium at 20°C. After Roth (1971).

3.6.7. *Integrated equation of flow*

Equations (3.194), (3.223), (3.225), and (3.231) are valid only in cases where the flow is of the same kind (molecular, intermediate, viscous) over the entire length of the tube. If the case is not this, the equations can be used for each portion of the tube where the flow regime is the same.

Considering a short portion of the tube where the variation of the pressure is dP, eqs. (3.221), (3.230), and (3.231) give the throughput as

$$Q \mathrm{d}L = C_m L J \mathrm{d}P = C_m \bar{P}_i L \left(\delta + \frac{1 + 17\,\delta}{1 + 21\,\delta} \right) \mathrm{d}\delta \tag{3.232}$$

By integrating it results

$$\frac{Q}{C_m \bar{P}_i} = \tfrac{1}{2}\delta^2 + \tfrac{17}{21}\,\delta + 9 \times 10^{-3} \ln\left(1 + 21\delta\right) \tag{3.233}$$

instead of the nonintegrated value which results from eq. (3.231), which is

$$\frac{Q}{C_m \bar{P}_i} = \left(\delta + \frac{1 + 17\,\delta}{1 + 21\,\delta} \right) \delta = \delta^2 + \frac{1 + 17\,\delta}{1 + 21\,\delta}\,\delta \tag{3.234}$$

Equations (3.233) and (3.234) are plotted on fig. 3.32. on which it can be seen that while the molecular–viscous intersection point corresponds to $\delta = 1$ using nonintegrated values, this point appears at $\delta = 2$ if integrated values are considered. Since measured values of the throughput always reflect integrated values, *experimentally obtained data should be compared only with eq. (3.233)*.

Based on eq. (3.230), we obtain

$$\bar{P}_i = P_i/2 = P_{1i}/4 \tag{3.235}$$

where $\bar{P}_i$ is the average pressure at point i (nonintegrated), P_i is the inlet pressure (nonintegrated; outlet pressure negligible), and P_{1i} is the effective inlet pressure (integrated).

From eqs. (3.226), (2.71), (2.41) it results that

$$\bar{P}_i D = 3.1 \times 10^{-19}\ T/\xi^2\ \text{Torr·cm} \tag{3.236}$$

where ξ is the molecular diameter of the gas (cm). By using eq. (3.235), we have

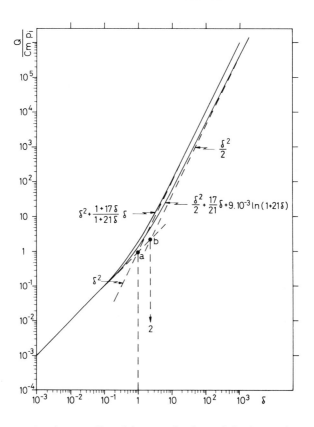

Fig. 3.32 Usual (nonintegrated) and integrated values of the flow. After Roth (1971).

$$P_{1i}D = 1.24 \times 10^{-18} \ T/\xi^2 \tag{3.237}$$

where P_{1i} is *the inlet* pressure (the outlet pressure being negligible). From eqs. (3.237) and (3.222) it results that the *position of the molecular–viscous* intersection point i is given by

$$LQ_i = I/P^2_{1i} \tag{3.238}$$

where

$$I = 7.2 \times 10^{-54} \ T^{3.5}/M^{1/2} \ \xi^6 \tag{3.239}$$

Figure 3.33 shows the values of the *intersection constant I,* for various gases as a function of their temperature.

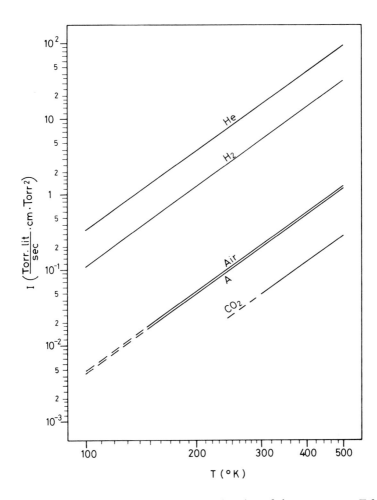

Fig. 3.33 Value of the intersection constant I as a function of the temperature T for various gases. After Roth (1971).

By plotting LQ [Torr·liter/sec·cm] as a function of the effective inlet pressure P_1 and representing the lines of molecular flow (slope 1), those of viscous flow (slope 2) and the lines on which the molecular–viscous intersection point should be (eq. 3.238), a graph as that shown in fig. 3.34 is obtained. In the example shown on fig. 3.34 for a capillary of diameter $D = 10^{-3}$ cm, the intersection point for air is at a, which corresponds to $P_{1i} = 260$ Torr. For helium the same pipe has the intersection point at b (about 760 Torr). For hydrogen the intersection point is at c.

The relative positions of the curves illustrate the ratios which exist between the

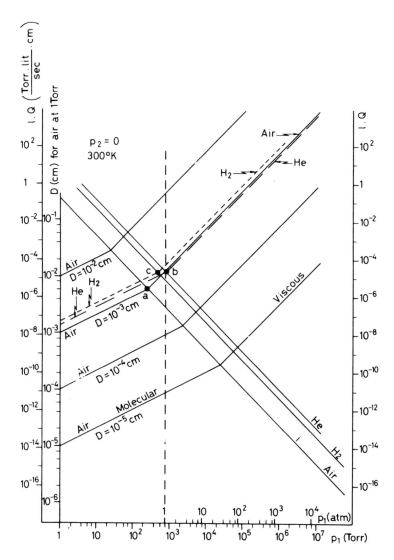

Fig. 3.34 Flow diagram for air, hydrogen and helium showing the places of their molecular-viscous points. After Roth (1971).

Table 3.8.
Conductance for various gases.

Gas	C_{gas}/C_{air}	
	Molecular	Viscous
Hydrogen	3.78	2.1
Helium	2.67	0.93
Water vapour	1.26	1.9
Argon	0.85	0.82
CO_2	0.81	1.30
Mercury vapour	0.38	—

conductances of a duct for flow of various gases (table 3.8). It can be seen that while in molecular flow the throughput (conductance) for helium is higher than that for air, in viscous flow the opposite is true. For CO_2 the conductance ratio C_{gas}/C_{air} is < 1 in molecular flow, and > 1 in viscous flow.

3.7. Calculation of vacuum systems

3.7.1. Sources of gas in vacuum systems

A vacuum system is the assembly of the components used to obtain, to measure and to maintain the vacuum in a chamber, or device. Any vacuum system is made up of a pump (or pumps), gauges and pipes connecting them together. The system contains also valves, traps, motion seals, electric lead-throughs, etc. Figure 3.35 shows a typical vacuum system.

In order to express the behavior of a vacuum system, the various sources of gas existing in it must be considered as being at any moment in equilibrium with the pumping action of the pumps on the system.

It can be considered that the sources of gas in a vacuum system are:

(a) The gas molecules of the initial atmosphere enclosed in the system (Q),

(b) The gas which penetrates into the system as a result of leakage (Q_L) (see §3.7.6; 7.3; 7.4),

(c) The gas provening from the outgassing of the materials in the system (Q_D) (see §3.7.6; 4.4),

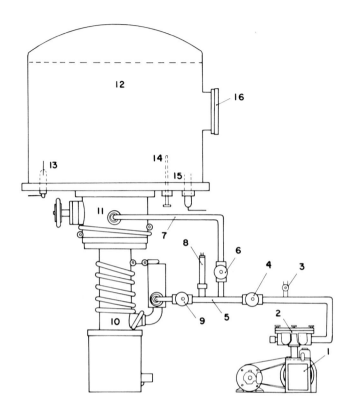

Fig. 3.35 Vacuum system. (1) rotary (backing) pump; (2) moisture trap; (3) air admittance valve; (4) throttling valve; (5) backing line;(6) roughing valve; (7) roughing line; (8) Pirani gauge; (9) backing valve; (10) diffusion pump; (11) baffle valve; (12) vacuum chamber; (13) electric lead-through; (14) shaft seal; (15) Penning gauge; (16) window.

(d) The gas (or vapours) resulting from the vapour pressure of the materials (Q_v) (see §4.1),

(e) The gas entering the system by permeation through walls, windows (Q_P) (see §3.7.6; 4.2).

The quantities of gas resulting from sources (b) to (e) are functions of the construction of the system. For the present discussion we consider the totality of the gas resulting from these sources (Q_G)

$$Q_G = Q_L + Q_D + Q_v + Q_P \tag{3.240}$$

as *being constant* (for the time interval which we consider).

Hoffman (1979) evaluates the sources of contamination (gas load) in diffusion-pumped systems. References for the evaluation of the gas load due to electron or ion bombardment are given in §4.4.3.

3.7.2. *Pumpdown in the viscous range*

We assume that the pumping speed of the pumps S_p is a constant in the range considered. According to eq. (3.28), the pumping speed obtained through the conductance C of the pipes connecting the pump to the chamber is

$$S = S_p C / (S_p + C) \quad \text{or} \quad 1/S = (1/S_p) + (1/C) \tag{3.241}$$

If the pressure is high enough for viscous flow the conductance of the pipe is given by equations of the form of eq. (3.52);

$$C = (\pi/128) \ [D^4/(\eta L)] \ \bar{P} = E\bar{P} = E \ (P + P_p)/2 \tag{3.242}$$

where $E = (\pi/128) \ [D^4/(\eta L)]$; P is the pressure in the vessel and P_p is the pressure at the inlet to the pump.
By substituting eq. (3.242) in (3.241), we have

$$S = S_p E[(P + P_p)/2]/[S_p + E(P + P_p)/2] \tag{3.243}$$

and the throughput (eq. 3.24)

$$Q = PS = P \ S_p \ E \ [(P + P_p)/2]/[S_p + E(P + P_p)/2] = -V \ (dP/dt) \tag{3.244}$$

where we used $P(dV/dt) = -V(dP/dt)$ when $PV = \text{const}$. Q_G is neglected, since it is very small compared to Q.
Since P_p is also a function of P we have to write

$$Q = P_p \ S_p = -V(dP/dt) \tag{3.245}$$

based on the fact that the throughput is the same in the chamber and at the pump. From eq. (3.245), it results that

$$P_p = -\frac{V}{S_p} \left(\frac{dP}{dt} \right) \tag{3.246}$$

By introducing this value in eq. (3.244), we obtain

$$-\left(\frac{V}{S_p}\right)^2 \left(\frac{dP}{dt}\right)^2 + \frac{2V}{E}\left(\frac{dP}{dt}\right) + P^2 = 0 \tag{3.247}$$

By putting $A = 2V/E$ and $B = (V/S_p)^2$ and solving eq. (3.247) we have

$$\frac{dP}{dt} = \frac{-A \pm (A^2 + 4BP^2)^{1/2}}{-2B} \tag{3.248}$$

Since the pressure decreases in time, only the solution $dP/dt < 0$ is real, thus

$$2B \frac{dP}{A - (A^2 + 4BP^2)^{1/2}} = dt \tag{3.249}$$

or

$$-2B \frac{A + (A^2 + 4BP^2)^{1/2}}{-4BP^2} \, dP = dt$$

By integrating eq. (3.249) we obtain

$$t = \frac{A}{2P} + \sqrt{B}\left\{\frac{(A^2/4B + P^2)^{1/2}}{P} - \ln[P + (A^2/4B + P^2)^{1/2}]\right\} + K \tag{3.250}$$

For $t = 0$, $P = P_i$ (initial pressure), and

$$K = \sqrt{B}\left\{\ln[P_i + (A^2/4B + P_i^2)^{1/2}] - \frac{(A^2/4B + P_i^2)^{1/2}}{P_i}\right\} - \frac{A}{2P_i} \tag{3.251}$$

Since

$$\frac{A^2}{4B} = \left(\frac{2V}{E}\right)^2 \Big/ \left(\frac{2V}{S_p}\right)^2 = \left(\frac{S_p}{E}\right)^2$$

it results

$$\frac{t}{V} = \frac{1}{E}\left[\frac{1}{P} - \frac{1}{P_i}\right] + \frac{1}{S_p}\left\{\frac{[(S_p/E)^2 + P^2]^{1/2}}{P} - \frac{[(S_p/E)^2 + P_i^2]^{1/2}}{P_i}\right\} +$$

$$+ \frac{1}{S_p}\left\{\ln\frac{P_i + [(S_p/E)^2 + P_i^2]^{1/2}}{P + [(S_p/E)^2 + P^2]^{1/2}}\right\} \tag{3.252}$$

This equation is plotted for air in fig. 3.36, by using as a parameter the value

$$D^4/L = (128/\pi)\,\eta\,E$$

and considering $P_i = 10^6$ dyne/cm^2 (760 Torr), and $P = 10^2$ dyne/cm^2 (7.6×10^{-2} Torr), i.e. the pressure range in which usually the flow is viscous. If a volume of $V = 100$ liter is evacuated by a pump of $S_p = 2$ lit/sec through a pipe $D = 2$ cm and $L = 200$ cm, then $D^4/L = 8 \times 10^{-2}$. On the curve 8×10^{-2}, for $S_p = 2$, it results $t/V = 6$ sec/liter. Thus the time required for 100 liter is $t = 600$ sec. If the volume is connected directly to the pump, the line $D^4/L = \infty$ gives $t/V = 4.5$ sec/liter, thus $t = 450$ sec.

It is interesting to mention that if the pump is connected directly to the vessel, $L = 0$, thus $E = \infty$, eq. (3.252) becomes

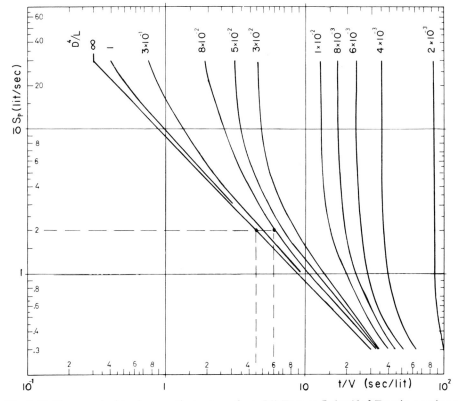

Fig. 3.36 Time required to decrease the pressure from 760 Torr to 7.6×10^{-2} Torr in a volume $V(l)$, connected by a pipe of diameter D(cm) and length L(cm) to a pump of pumping speed $S_p(l/s)$. After Delafosse and Mongodin (1961).

$$t = (V/S_p) \ln (P_i/P)$$

an equation which also describes the pumpdown time in molecular flow where the conductance is not a function of the pressure (eq. 3.269).

Empirical equations for the pumpdown in the pressure range 0.5–10 Torr were presented by Heap (1973). Using equations and computer programs, pumpdown curves were calculated by Kraus (1982), Liversey (1982) and Santeler (1987a).

3.7.3. Pumpdown in the molecular range

The pumpdown in the molecular range is limited by the equilibrium between the gas load and pumping speed in the pump itself, as well as by this equilibrium in the vacuum chamber (see §3.7.6).

The gas load Q_o of the pump itself is constituted by the leakage into the pump, and the backstreaming of the pumping fluid. If the theoretical pumping speed of the pump is S_t, the throughput will be

$$Q = S_t P_p - Q_o = S_t P_p \left(1 - Q_o/S_t P_p\right) \tag{3.253}$$

where P_p is the pressure at the inlet of the pump. The lowest pressure of the pump P_o will be obtained when $Q = 0$, thus (eq. 3.253)

$$Q_o = S_t P_o \tag{3.254}$$

The real pumping speed of the pump S_p is given by

$$S_p = \frac{Q}{P_p} = S_t \left(1 - \frac{Q_o}{S_t P_p}\right) = S_t \left(1 - \frac{P_o}{P_p}\right) \tag{3.255}$$

At the vacuum chamber, the pumping speed is $S = S_p C/(S_p + C)$, and the gas load in the chamber is Q_G.

The throughput at the chamber is

$$Q = SP = PS_p \, C/(S_p + C) = -V(dP/dt) + Q_G \tag{3.256}$$

From these equations it results that

$$Q = -V(dP/dt) + Q_G = S_t (P_p - P_o) \tag{3.257}$$

and since

$$S_p P_p = S P = [S_p C/(S_p + C)] P \tag{3.258}$$

it results that

$$-\frac{dt}{V} = \frac{1 + (S_p/C)}{S_t} \frac{dP}{P - [1 + (S_p/C)] P_o - [1 + (S_p/C)] (Q_G/S_t)} \tag{3.259}$$

from which the time required for lowering the pressure in the chamber from P_i (initial) to P is

$$t = (V/S_t) [1 + (S_p/C)] \ln \frac{P_i - [1 + (S_p/C)]P_o - [1 + (S_p/C)] (Q_G/S_t)}{P - [1 + (S_p/C)]P_o - [1 + (S_p/C)](Q_G/S_t)} \tag{3.260}$$

and the pressure reached after time t is

$$P = \{P_i - [1 + (S_p/C)]P_o - [1 + (S_p/C)] (Q_G/S_t)\}$$
$$\times \exp \{ - (S_t/V) t/ [1 + (S_p/C)]\} + [1 + (S_p/C)]P_o + [1 + (S_p/C)](Q_G/S_t) \tag{3.261}$$

If the *conductance* C *is very large*, i.e. $S_p/C \ll 1$, and the ultimate pressure at the pump due to the gas load is $P'_u = Q_G/S_t$, then eqs. (3.260) and (3.261) can be written

$$t = \frac{V}{S_t} \ln \frac{P_i - P_o - P'_u}{P - P_o - P'_u} \tag{3.262}$$

$$P = (P_i - P_o - P'_u) \exp [- (S_t/V)t] + P_o + P'_u \tag{3.263}$$

When a conductance C connects the pump to the chamber, the ultimate pressure in the chamber due to the gas load is

$$P_u = Q_G/S = Q_G (S_p + C)/(S_p C) = [1 + (S_p/C)]P'_u/[1 - (P_o/P_p)] \tag{3.264}$$

Introducing eq. (3.264) in eq. (3.260)

$$t = (V/S_t) [1 + (S_t/C)] \ln \frac{P_i - [1 + (S_p/C)]P_o - [1 - (P_o/P_p)]P_u}{P - [1 + (S_p/C)]P_o - [1 - (P_o/P_p)]P_u} \tag{3.265}$$

and for a system where the lowest pressure of the pump P_o is very small compared to P_u (and P_p)

$$t = (V/S_t) [1 + (S_t/C)] \ln \frac{P_i - P_u}{P - P_u}$$
(3.266)

By integrating eq. (3.256) where we consider S_p a constant, and independent of P, it results that

$$t=(V/S_p) [1 + (S_p/C)] \ln \frac{P_i - P_u}{P - P_u}$$
(3.267)

and

$$P=(P_i-P_u) \exp \{-(S_p/V)t/[1+(S_p/C)]\} + P_u$$
(3.268)

This equation shows that after a very long pumping time, the pressure tends toward the ultimate pressure P_u, determined (eq. 3.264) by the gas load. Thus eq. (3.268) describes the *transient* pumpdown

$$P=P_i \exp [-(S/V)t]$$
(3.269)

as well as the *steady state*

$$P=P_u=Q_G/S$$
(3.270)

This later can result in a constant ultimate pressure (fig. 3.37) if Q_G=const

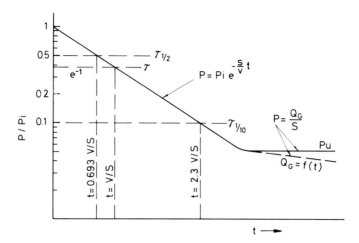

Fig. 3.37 Pumpdown and steady state.

(e.g. leak), or in a pressure decreasing with time (e.g. outgassing).

Figure 3.37 also shows the meaning of the various *time constants*, i.e. the values of the pumping time which reduces the pressure by a given ratio. The time required to reduce the pressure to $e^{-1} = 0.367$ from its original value is

$$\tau = V/S \qquad\qquad (3.271)$$

The "half-life" or the time to reach one half of the initial pressure is given by

$$\tau_{1/2} = 0.693 \ (V/S) \qquad\qquad (3.272)$$

while the time required to reduce the pressure by a decade is

$$\tau_{1/10} = 2.3 \ (V/S) \qquad\qquad (3.273)$$

In vacuum systems, where the gas load is due to outgassing, it was determined (Edwards 1977a, b, 1979a) that the upper limit of the pressure P_{max} (Torr) can be expressed by $P_{max} = 1.7 \times 10^{-5} A/(S \cdot t)$, where A (cm^2) s the geometric wall area, S (lit/sec) the pumping speed and t (sec) the pumping time.

Pumpdown computations in space chambers were published by Hamacher (1976). The Spacelab permits the evacuation by connecting the vessel directly to the outer space; the ultimate pressure which can be achieved is much higher (Hamacher, 1977) than that in the outer space, since the pumping has to be done through connecting pipes, orifices, etc. Hobson (1977) discusses the methods which can improve this ultimate pressure.

3.7.4. Steady state with distributed gas load

The steady state pressure in a vacuum chamber is given by the simple relation eq. (3.270). If the gas load is distributed along the pipe, a case which appears due to the outgassing of the surface, then the steady state is characterized by a pressure gradient along the pipe.

Consider that the pump (fig. 3.38a) evacuates a pipe of conductance C, closed at the end. Let the specific outgassing rate be q_D (e.g. Torr·liter/sec·cm^2). The gas load due to an elementary length dx (fig. 3.38) will be

$$-dQ = q_D B dx \qquad\qquad (3.274)$$

where B is the perimeter of the tube cross section, and the minus sign shows that the gas flows towards $-x$.

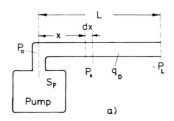

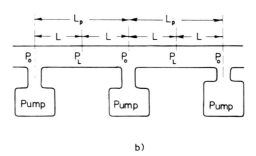

Fig. 3.38 Distributed gas load (a) in system closed at one end; (b) in long (open) system.

The throughput through the length dx is

$$Q = C(L/dx)\ dP \tag{3.275}$$

thus

$$dQ = CL\ \frac{d^2P}{dx^2}\ dx \tag{3.276}$$

By the equality between eqs. (3.276) and (3.274) we have

$$d^2P/dx^2 = -q_D B/(CL) \tag{3.277}$$

thus

$$dP/dx = -[q_D\ B/(CL)]\ x + K_1 \tag{3.278}$$

Since at the closed end of the pipe $x=L$, and $(dP/dx)_L=0$, it results that $K_1 =$

$q_D B/C$, and therefore

$$dP/dx = -[q_D \, B/(CL)] \, x + q_D \, B/C \tag{3.279}$$

and

$$P_x = -[q_D \, B/(2 \, CL)]x^2 + (q_D \, B/C) \, x + K_2 \tag{3.280}$$

At the inlet of the pump, where $x=0$, the pressure is

$$P_o = q_D \, B \, L/S_p = K_2 \tag{3.281}$$

Finally, the pressure at a distance x along the pipe is

$$P_x = q_D \, B[(L/S_p) + (x/C) - x^2/(2 \, CL)] \tag{3.282}$$

which shows that the distribution is parabolic, being maximum at the closed end, i.e.

$$P_L = q_D \, BL \, [(1/S_p) + 1/(2C)] \tag{3.283}$$

The pressure drop is

$$P_x - P_o = q_D \, B[(x/C) - x^2/(2 \, CL)] \tag{3.284}$$

$$P_L - P_o = q_D \, BL/(2C) \tag{3.285}$$

It can be seen that *the pressure drop is independent of the pumping speed*, i.e. even if the pump is extremely large, the pressure will not drop below the values given by eqs. (3.284) and (3.285).

For this reason, the *pumping of long ducts* (e.g. accelerators) must be done by *placing a number of pumps* along the duct. The number of pumps, the kind of pumps and the pressure drop are connected by eqs. (3.281–3.285).

If the kind of pump is known, the spacing L_p between adjacent pumps (fig. 3.38b) results from eq. (3.281), as $L_p = P_o \, S_p/(q_D B)$. By using then $L \doteq L_p/2$ in eq. (3.285) the pressure drop is obtained.

If the pressure drop (fig. 3.38b) $P_L - P_o$ is given, the distance between pumps $L_p = 2L$, results from eq. (3.285) by introducing in this equation the conductance C according to eq. (3.92). The value L_p determines then in eq. (3.281) the kind of pump (P_o, S_p) to be used.

If the pressure drop ratio $(P_L - P_o)/P_o$ is given, it results from eqs. (3.285) and

(3.281) that $(P_L - P_o)/P_o = S_p/(2C)$, which determines the pumps (P_o, S_p) for a given conductance (dimensions, distance) or the distances for a given kind of pump.

The calculation of the pressure drop in systems with distributed outgassing (e.g. storage rings) is discussed by Welch (1973), Calder (1974), Blechschmidt (1977, 1978), Blechschmidt and Unterlerchner (1977), and Miyahara (1986). The calculations of the pressure profiles in systems with "pumping" walls, have been discussed by Gottwald (1973) for cases of chemical reactions, by Gajewski and Wisniewski (1977) for condensation, and by Smith and Lewin (1966); Paul (1973a) for wall sorption; Paul (1973a) and Saksaganskii (1988) for adsorbing walls.

Complex vacuum systems such as those having distributed gas loads can be analysed by using the electronic circuit simulation. The subject is discussed by Ohta (1983), Kendall (1983b). Wilson (1987) summarizes the correspondences between vacuum quantities/units and electrical quantities/units as listed in table 3.9.

Table 3.9.
Vacuum vs electrical quantities/units.

Vacuum				Electrical			
Quantity	Symbol	Unit	Equation	Quantity	Symbol	Unit	Equation
Pressure	P	Torr		Potential	E	V	
Flow	Q	Torr·ℓ/s	$Q = S \cdot P$	Current	I	A	$I = (1/R) \cdot E$
Pumping speed	S	ℓ/s		Conductance (1/Resist.)	$1/R$	$1/\Omega$	
Conductance			$Q = \dfrac{V\,dP}{dt}$				$I = \dfrac{C\,dE}{dt}$
Volume	V	ℓ		Capacitance	C	F	
Speed constant	K	$\dfrac{\text{Torr·s}}{\text{Torr·}\ell/\text{s}}$	$P = \dfrac{K\,dQ}{dt}$	Inductance	L	$H =$ Vs/A	$E = \dfrac{-L\,dI}{dt}$

3.7.5. Nomographic calculation of conductances and pumpdown time

Many nomograms were built for the calculation of the conductances, pumping speed and pumpdown times. We will show here just the most typical ones, built

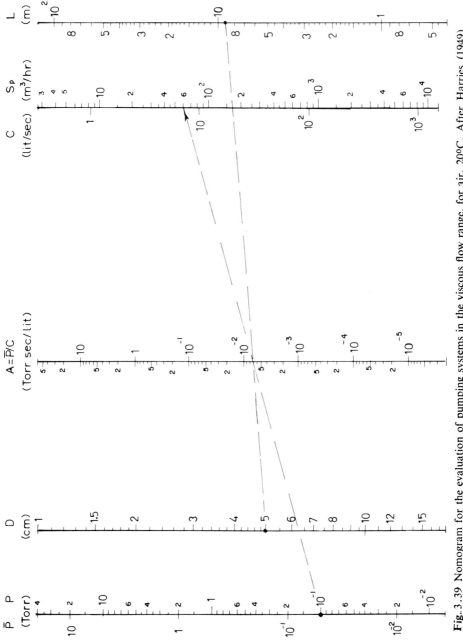

Fig. 3.39 Nomogram for the evaluation of pumping systems in the viscous flow range, for air, 20°C. After Harries (1949).

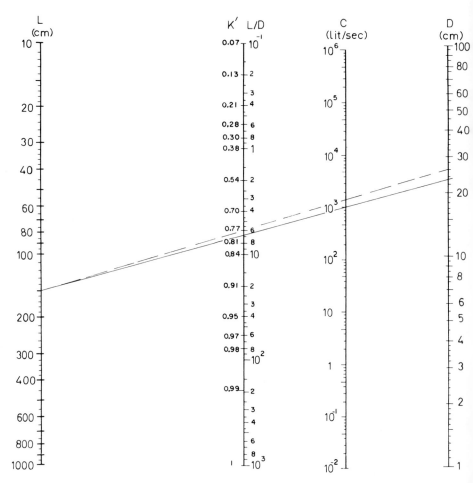

Fig. 3.40 Nomogram for determining the conductance of pipes in the molecular flow range, for air, 20°C. After Delafosse and Mongodin (1961).

for the evaluation of the system in viscous flow, molecular flow and the inter-mediate range respectively.

For the evaluation of the pumping speed in *viscous flow*, the nomogram of Harries (1949) shown in fig. 3.39 is the most known one. This graph is based on eq. (3.54). It shows how the pipe (with a diameter D, and length L) must be dimen-sioned for a pump with a pumping speed S_p at inlet pressure P, so that the pumping speed at the vacuum chamber is 0.3 S_p. If three of the factors S_p, P, L, D are known, the fourth can be found from the nomogram (fig. 3.39). The example shows that the line joining $L=9$ m and $D=5$ cm, intersects the A scale at a given point. For an inlet pressure of $P=0.1$ Torr, a second straight line extend-

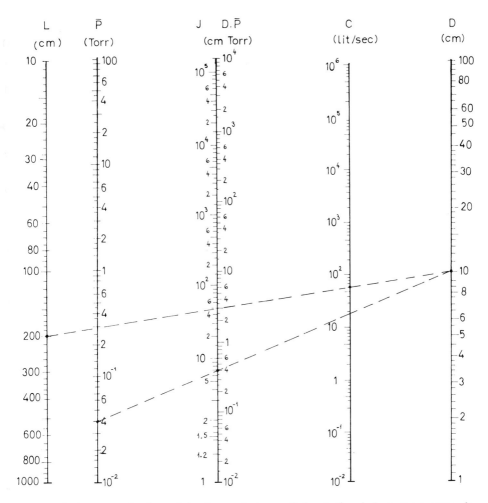

Fig. 3.41 Nomogram for determining the conductance of pipes in the whole pressure range, for air, 20°C. After Delafosse and Mongodin (1961).

ed from 0.1 Torr, through the point found by the first line on scale A, shows that the maximum admissible pumping speed S_p should be 60 m³/hr.

For the evaluation of the *conductance* of (short) pipes in the molecular range, Delafosse and Mongodin (1961) published the nomogram shown in fig. 3.40. The nomogram is based on eqs. (3.94) and (3.104), and gives the correlation beween the length L of the pipe, its diameter D, its conductance C, and the corresponding correction K' (eq. 3.109) for the short pipe. The example shown in the nomogram (fig. 3.40) evaluates the required diameter D, of a pipe $L=1.5$ m long, in order to have a conductance of about $C=1000$ liter/sec, in the molecular flow

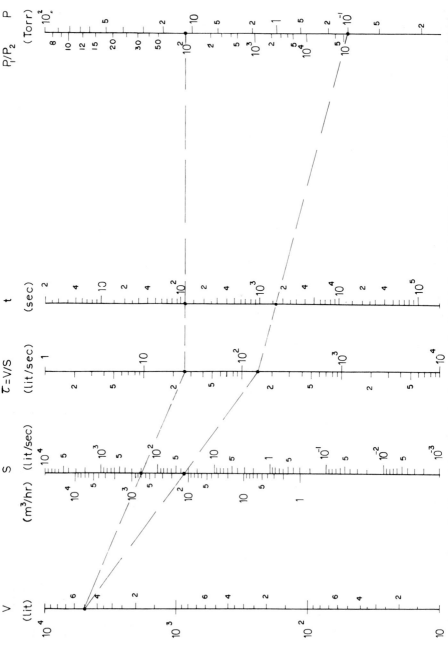

Fig. 3.42 Nomogram for determining the pumpdown time (molecular flow, air, 20°C). After Delafosse and Mongodin (1961).

range. The line through $L = 1.5$ m, and $C = 1000$ liter/sec cuts scale D, at $D = 24$ cm. The conductance at the entrance is included by the correction factor K'. It is a function of the ratio D/L, and in the example it is given at the intersection of the line L–D, with scale K', where it shows $K' = 0.79$. Thus the real conductance of the pipe considered is

$$C = 1000 \times 0.79 = 790 \text{ liter/sec} \tag{3.286}$$

When D is increased to $D = 26$ cm, the value of the conductance (dotted line) will be

$$C = 1400 \times 0.77 = 1078 \text{ liter/sec.}$$

Delafosse and Mongodin (1961) also present a nomogram for the evaluation of the conductance in the intermediate range. This nomogram is shown in fig. 3.41, and is based on eqs. (3.94), (3.221) and, (3.225).

The examples shown on fig. 3.41 refer to a pipe of length $L = 2$ m, and diameter $D = 10$ cm. The straight line joining these two points intersects scale C at $C = 60$ liter/sec. If the average pressure is $\overline{P} = 4 \times 10^{-2}$ Torr, from this point and $D = 10$ cm, the correction factor J (eq. 3.225) results in $J = 6.8$. Thus the conductance will be

$$C = 60 \times 6.8 = 408 \text{ liter/sec.}$$

The *pumpdown time* t can be evaluated from the nomogram in fig. 3.42. This nomogram is based on eq. (3.269). The first example (lower lines) shows that for a volume $V = 5000$ liter and a pumping speed of $S = 120$ m³/hr, the time constant $\tau = V/S = 140$ s. If the final pressure to be reached is $P = 10^{-1}$ Torr (initial pressure 760 Torr), then it results that the required pumpdown time is $t = 1800$ sec.

The second example (upper lines) shows that if the volume $V = 5000$ liter has to be evacuated by a pump with a pumping speed $S = 700$ m³/hr, then the time constant which results is about 25 sec. In order to decrease the pressure from 10^{-1} Torr to be 10^{-3} Torr, thus $P_1/P_2 = 100$, it results that the pumpdown time is $t = 120$ sec.

3.7.6. *Evaluation of the gas load and pumping requirements*

The *gas load* is a result of the process (e.g. vacuum drying, degassing, etc.) or a "by-product" of the vacuum chamber, and the components (materials) used. In any case the gas load is the sum (eq. 3.240) of the residual gas remaining from the initial atmosphere, the vapour pressure of the materials present in the chamber, and the leakage, outgassing and permeation.

The pumpdown from the initial (atmospheric) pressure to the working pressure of the system is discussed in §3.7.2–3.7.5. The vapour pressure of materials can be the factor limiting the ultimate pressure of the system if the precautions required for high vacuum (§7.1.4) are not respected. Usual high and ultra-high vacuum systems, or systems which are tested "empty", have their ultimate pressure determined by the leakage, the outgassing and the permeation. All these factors depend on the kind of construction (Chap. 7) and the materials used, dependence which is very often complicated or not exactly known. Nevertheless the gas load and appropriate pumping requirements can be evaluated by making the proper simplifying assumptions.

The ultimate pressure P_u in the chamber can be expressed (eq. 3.264) by

$$P_u = Q_G/S = (Q_G/V)(V/S) = (Q_G/V) \tau = (\Delta P/\Delta t) \tau \qquad (3.287)$$

where Q_G is the total gas load, V is the volume of the chamber, S is the pumping speed in the chamber (eq. 3.28), $\Delta P/\Delta t$ is the rate of change in pressure (pressure rise), and $\tau = V/S$ is the time constant (eq. 3.271) of the chamber (system).

If the total gas load consists of three main components: Q_L due to leakage, Q_D due to outgassing, and Q_P due to permeation, $Q_G = Q_L + Q_D + Q_P$, the ultimate pressure can also be expressed as the sum of partial pressures due to leakage (P_L), outgassing (P_D) and permeation (P_P)

$$P_u = P_L + P_D + P_P \qquad (3.288)$$

where (according to eq. 3.287)

$$P_L = \frac{Q_L}{V} \tau = \left(\frac{\Delta P}{\Delta t}\right)_L \tau \qquad (3.289)$$

$$P_D = \frac{Q_D}{V} \tau = \left(\frac{\Delta P}{\Delta t}\right)_D \tau \qquad (3.290)$$

$$P_P = \frac{Q_P}{V} \tau = \left(\frac{\Delta P}{\Delta t}\right)_P \tau \qquad (3.291)$$

thus the gas load and pumping requirement can be evaluated separately for leakage, outgassing and permeation. The nomograms shown in figs. 3.43–3.45 were constructed (Roth, 1970b) by using eqs. (3.289–3.291), respectively. The nomograms consider two values of the time constant $\tau = 1$ sec (slow pumping) and $\tau = 0.1$ sec (fast pumping). The gas load Q_L/V (or Q_D/V; Q_P/V) is expressed in terms of the specific gas quantity entering into or evolving in a *unit volume* of the chamber (Torr·liter/sec·m³), as well as in terms of the rate of pressure rise $\Delta P/\Delta t$ (Torr/hr) which would result from that gas load.

Numbers in italics on the nomograms indicate the relevant literature reference, as listed in the captions.

Leakage (fig. 3.43) The gas load Q_L due to leakage is expressed by

$$Q_L = q'_L L \qquad (3.292)$$

where q'_L is the leak rate per unit length of seal (e.g. Torr·liter/sec·cm) and L is the length (perimeter) of the seal (cm). From eqs. (3.289) and (3.292) it results that

$$P_L = (Q_L/V)\tau = q'_L (L/V)\tau \tag{3.293}$$

Usually *the shape of the vacuum chamber* may be considered to be either a cube of side a or a cylinder of diameter d and length l. The seal length-to-volume ratio L/V is a maximum in the case of a cube having seals (e.g. welding) on all its sides, and in this case

$$L/V = 12 \ a/a^3 = 12/a^2 \tag{3.294}$$

The ratio L/V is a minimum in the case of a cylinder having a seal at one of its ends

$$L/V = \pi d/[(\pi d^2/4)l] = 4/dl \tag{3.295}$$

The nomogram in fig. 3.43 considers these cases of maximum (eq. 3.294) and minimum (eq. 3.295) seal length-to-volume ratios. For a complicated shape, which cannot be considered as a cube or cylinder, the actual value of the ratio $(L/V)_{act}$ must be evaluated, and the resulting equivalent value

$$(dl)_{eq} = 4/(L/V)_{act} \tag{3.296}$$

can be used on the nomogram.

Combining eq. (3.293) with (3.294) or (3.295) or (3.296) the specific gas load or the ultimate pressure can be expressed as a function of the size of the chamber.

The leak rate per unit length q'_L can be considered as a characteristic value in the case of permanent seals (e.g. welds) and as a first approximation in the case of demountable (gasket) seals. In the case of gasket seals the real specific value is expressed (§7.3.4) by

$$q_L \left| \frac{\text{Torr·liter}}{\text{sec}} \ \cdot \ \frac{\text{mm (sealing width)}}{\text{cm (seal length)}} \right|$$

which is given (eqs. 7.11 and 7.12) by the expression

$$q_L = hq'_L = \rho_o \exp(-3 P/R) \ A^2 \ (P_1 - P_o) \tag{3.297}$$

where h is the sealing width (marked w on fig. 7.16), ρ_o is a factor specific to the gas (e.g. for He at room temperature $\rho_o = 340$ (lit/sec)·(mm/cm³); see eq. (7.11); P/R is the tightening index (see eq. 7.12; fig. 7.18) in which P is the compression stress exerted on the sealing surface of the gasket (kg/cm²) and R is the "sealing factor" (see eq. 7.11 and fig. 7.19) defining the sealing ability of the material. A is the peak-to-valley height of the surface roughness (see figs. 3.16; 7.16). $P_1 - P_o$ is the pressure difference across the seal (Torr).

Efficient sealing is obtained up to a tightening index $P/R = 3$. At higher values considerable deformation of the bulk of the gasket occurs, which reduces further efficiency. Thus the nomogram (fig. 3.43) is constructed based on $P/R = 3$; $P_1 - P_o = 760$ Torr; $\rho_o = 79$ (lit/sec)·(mm/cm³) (for air), and two sealing surfaces (on both sides of the gasket), i.e. (according to eq. 3.297)

$$q_L = hq'_L = 79 \times 2 \times e^{-9} \times 760 \, A^2 = 14.5 \, A^2 \tag{3.298}$$

The factors involved are grouped in the nomogram (fig. 3.43) under the sections: sealing effort, seal width, surface finish, leak rate, gas load, ultimate pres-

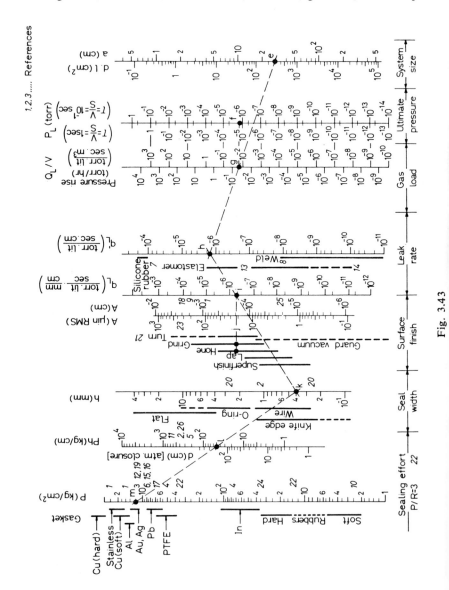

Fig. 3.43

sure and system size. The scales of the nomogram form three alignment charts:
(1) scales P, Ph, and h; (2) scales h, q_L, and q'_L; (3) scales q'_L, Q_L/V and system
size.

The *example* marked e, f, . . .m on fig. 3.43 can be expressed as follows: A
cylindrical chamber of diameter $d = 10$ cm and length $l = 30$ cm ($dl = 300$ cm²;
point e) is evacuated by fast pumping ($\tau = 0.1$ sec). The desired ultimate pressure
due to leakage is $P_L = 2 \times 10^{-6}$ Torr (point f). The horizontal through point f
shows that in order to achieve this, the gas load due to leakage must be less than
2×10^{-2} Torr· liter/sec·m³ (point g, right scale). Cut off from the pump this chamber
will have a rate of pressure rise (due to leakage) of 7×10^{-2} Torr/h (point g,
left scale).

The leak rate required per unit length of seal is given by the intersection of line
e-g with scale q'_L, i.e. $q'_L = 1.5 \times 10^{-6}$ Torr·liter/sec·cm (point h). This leak
rate can be achieved by elastomer seals, as shown by the ranges marked parallel
to scale q'_L. Since a bakeable seal is required, a wire seal is chosen, and the width
of the sealing surface is assumed to be $h = 0.4$ mm (point k). Line h-k intersects
scale q_L at $q_L = 6 \times 10^{-7}$ Torr·liter·mm/sec·cm (point i), and the horizontal j
through point i shows that a surface finish of 20 μin RMS is required (e.g. honed
surface). Gold (point m) is chosen as the wire material. The intersection of line
k-m with scale Ph determines the sealing effort $Ph = 48$ kg/cm (point l). This seal
could be closed just by the atmospheric pressure if

$$\tfrac{1}{4}\pi d^2 \geqslant \pi dPh \qquad \text{thus if } d \geqslant 4 \times 48 = 192 \text{ cm} \qquad (3.299)$$

as shown by point l on scale d (atmospheric closure). Since the seal has a diameter
of only 10 cm, it must be tightened mechanically by a force of $10 \, \pi \times 48 = 1500$ kg.

For reasons of clarity, the above example follows the scales in the order they
appear on the nomogram. Obviously the nomogram can be used by beginning with
any of the scales, and proceeding toward any other scale.

Gas load, pressure rise and ultimate pressure obtained from this nomogram
(leakage) have to be added to the respective values obtained for outgassing (fig.
3.44) and permeation (fig. 3.45) in order to obtain the total value of these factors
(eq. 3.288).

Fig. 3.43 Evaluation of the gas load due to leakage and the corresponding pumping requirements (Roth, 1970b). References: 1. Armand *et al.* (1964); 2. Armand *et al.* (1962); 3. Armand
and Lejay (1967); 4. Boulloud and Schweitzer (1959); 5. Bridge *et al.* (1960); 6. Fischoff (1962);
7. Gale and Machin (1953); 8. Garrod and Nankivell (1961); 9. Gitzendanner and Rathbun
(1965); 10. Guthrie and Wackerling (1949); 11. Hoch (1961); 12. Holden *et al.* (1959); 13.
Jordan (1962); 14. Kobayashi and Yada (1960); 15. Lange (1957); 16. Mark and Dreyer (1960);
17. Munday (1959); 18. Rathbun (1963); 19. Redman (1963); 20. Roth (1966); 21. Roth and
Inbar (1968);22. Roth (1967); 23. Roth (1970a); 24. Turner *et al.* (1962); 25. Weitzel (1960);
26. Wheeler and Carlson (1962).

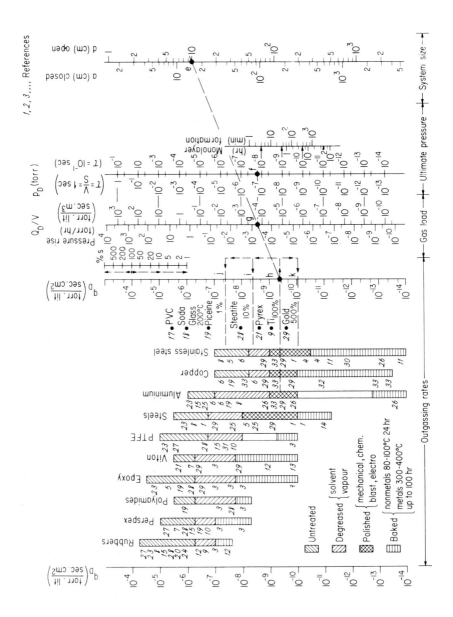

Outgassing (fig. 3.44) The gas load Q_D due to outgassing is expressed by

$$Q_D = q_D s \qquad (3.300)$$

where q_D is the specific outgassing rate (e.g. Torr·liter/sec·cm^2 – fig. 4.30), s is the area of the outgassing surface (cm^2).

From eqs. (3.290) and (3.300) it results

$$P_D = (Q_D/V)\tau = q_D \ (s/V)\tau \qquad (3.301)$$

As a basis for the evaluation, the outgassing surface is considered to be equal to the inside surface of the chamber, and the shape of the chamber is assumed to be similar either to a cube of side a or a cylinder of diameter d (and length l). The surface-to-volume ratio in the case of a closed cube is

$$s/V = 6 \ a^2/a^3 = 6/a \qquad (3.302)$$

while for an open-ended cylinder

$$s/V = \pi dl/(\pi d^2/4)l = 4/d \qquad (3.303)$$

The nomogram (fig. 3.44) for the evaluation of the gas load due to outgassing is based on eqs. (3.301–303), and on the range of q_D of various materials and treatments. The factors involved are grouped in the nomogram under the sections: outgassing rate, gas load, ultimate pressure, and system size.

The values of q_D for untreated, degreased and polished states were taken for

Fig. 3.44 Evaluation of the gas load due to outgassing and the corresponding pumping requirements (Roth, 1970b). References: 1. Amoignon and Couillard (1969); 2. Barton and Govier (1968); 3. as 2–(1965); 4. as 2–(1970); 5. Barre and Mongodin (1957); 6. Basaleva (1958); 7. Beckmann (1963); 8. Blears *et al.* (1960); 9. Boulassier (1959); 10. Brown (1967); 11. Calder and Lewin (1967); 12. Crawley and de Csernatony (1964); 13. Csernatony (1966); 14. Das (1962); 15. Dayton (1960); 16. Dayton (1963); 17. Dayton (1962); 18. Garbe and Christians (1962); 19. Geller (1958); 20. Haefer and Winkler (1956); 21. Henry (1969); 22. Jaeckel (1962); 23. Markley *et al.* (1962); 24. Munchhausen and Schittko (1963); 25. Power and Crawley (1960); 26. Power and Robson (1962); 27. Santeler (1958); 28. Schittko (1963); 29. Schram (1963); 30. Strausser (1968); 31. Thieme (1963); 32. Thibault *et al.* (1967); 33. Zhilnin *et al.* (1968). Recent data include the effects of various treatment combinations on stainless steel, Al, Ni, Ti (Patrick, 1973; Moraw and Dobrozemsky, 1974; Cost and Hickman, 1975; Elsey, 1975), the mass spectra of their residual gases (Halama and Herrera, 1976; Messer and Treitz, 1977; Nuvolone, 1977; Komiya *et al.*, 1979), or outgassing rates of special materials, such as carbon foam (Schalla, 1975), blackened Al (Rettinghaus and Huber, 1977), high voltage cables (Santhanaman and Vijendran, 1979). Other outgassing data were recently published by: Glassford and Liu (1980) on multilayer insulation materials; Beavis (1982) on estimation of bakeout characteristics; Reiter and Composilvan (1982) on Inconel; Yashimura (1985) on plastics; Chen (1985) on Aluminum; Rosenblum (1986) on epoxies; Odaka *et al.* (1987) on stainless steel (after exposure to atmosphere); Pang *et al.* (1987) on Al alloys and stainless steel; Itoh *et al.* (1980) on stainless steel (after glow discharge cleaning).

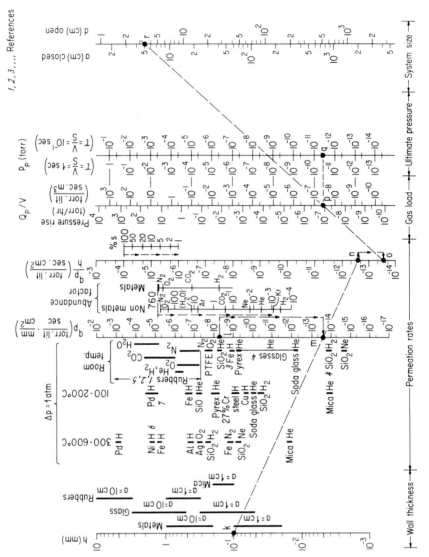

Fig 3.45

4–8 h of pumping. For baked metals the high q_D end of the range corresponds to baking at about 300°C for 24 h, the middle of the range to baking at 400°C for up to 100 h, while the lowest values for stainless steel also include an additional subsequent baking at 1000°C for 3 h.

The *example* marked e, f...k on fig. 3.44 represents the case of a cylindrical chamber $d=10$ cm (point e) which has to be pumped by slow pumping ($\tau=1$ sec), to an ultimate pressure (due to outgassing) $P_D=2\times10^{-7}$ Torr (point f). The horizontal through point f intersects scale Q_D/V at point g, thus in order to achieve the required P_D, the gas load due to outgassing must be less than $Q_D/V=2\times10^{-4}$ Torr·liter/sec·m³ (point g, right scale). Cut-off from the pump the chamber will show a rate of pressure rise (due to outgassing) of 7×10^{-4}Torr/h (point g, left scale). The specific outgassing rate required is given by the intersection of line e–g with scale q_D, i.e. $q_D=4\times10^{-10}$ Torr·liter/sec·cm² (point h). This means that the entire (100%) inside surface of the vacuum chamber may be constituted of any of the materials in any of the treatment states, which are intersected by the horizontal through point h (e.g. polished Al) or which lie below this line.

Parallel to the q_D scale, a correction scale (%s) is plotted for evaluating the distance at which the q_D value has to be shifted if the outgassing surface is smaller or larger than the inside surface of the chamber. By shifting point h, according to this correction scale, we find that 10% of the surface can also be constituted of materials (treatments) at or below the level marked i (e.g. baked Viton) and 1% of the surface area can also be of materials at or below the level j (e.g. degreased epoxy). If the outgassing surface is not only that of the chamber, but, as usual, includes parts built inside the chamber, point h has to be shifted at a level corresponding to the ratio between the total surface and that of the chamber. The example shown on the nomogram (level k) corresponds to the case when the total outgassing surface is 5 times (500%) that of the chamber. At the level k, baked Al or Cu, or polished stainless steel is required for the entire surface. In this case the distance between levels k and i, measured on scale %s, shows that only 2% of the surface can be a material (treatment) of level i (e.g. baked Viton).

Permeation (fig. 3.45) The gas load Q_P due to permeation is expressed by

$$Q_P=q_P \ (s/h) \tag{3.304}$$

Fig. 3.45 Evaluation of the gas load due to permeation and the corresponding pumping requirements (Roth, 1970b). References: 1. Crawley and Csernatony (1964); 2. Dawton (1957); 3. Eschbach et al. (1963); 4. Norton (1962); 5. Rogers (1956); 6. Waldschmidt (1954); 7. Young and Whetten (1962).

where q_P is the permeability [e.g.(Torr·liter/sec)·(mm/cm^2)], s is the permeable area (cm^2) and h the wall thickness (mm).

From eqs. (3.291) and (3.304) it results

$$P_P = (Q_P/V)\tau = (q_P/h)\,(s/V)\tau \tag{3.305}$$

The nomogram (fig. 3.45) for the evaluation of the gas load due to permeation is based on eqs. (3.302–3.305) and on the q_P range of various materials for various gases (see also figs. 4.12, 4.13). As a basis for the construction of the nomogram it is assumed that permeation takes place on the entire surface of the chamber (100%s), the appropriate fraction of permeable surface is taken by shifting the value according to scale %s. The shape of the chamber is assumed to be either a cube of side a or a cylinder of diameter d (and length l). The surface-to-volume ratio s/V is according to eqs. (3.302) and (3.303). Figure 3.45 is constituted of two alignment charts: (1) scales h, q_P and q_P/h, and (2) scales q_P/h, Q_P/V and system size. Parallel to scale h the ranges of wall thicknesses required to withstand one atmosphere pressure difference are marked for small diameter ($d=1$ cm) and medium size ($d=10$ cm) windows of various materials. Parallel to the q_P scale the ranges of permeation rates of typical materials for various gases, for a pressure difference of 1 atm are plotted within various temperature ranges. The permeation through nonmetals is proportional with the pressure difference (see §4.2.2), while for metals it is proportional with the square root of the pressure. In order to evaluate the permeation from the surrounding atmosphere, scales taking into account the appropriate corrections due to the abundance of various gases are plotted for metals and nonmetals.

The example shown on fig. 3.45 represents the case of a silica window of thickness $h=0.1$ mm (point k) included in a vacuum chamber. The q_P value of SiO$_2$ for He is $q_P = 5 \times 10^{-9}$ (Torr·lit/sec)·(mm/cm^2) (point l) at $\Delta P = 1$ atm. Since the partial pressure of He in the atmosphere is only 4×10^{-3} Torr (abundance scale), point l must be lowered on scale q_P by the distance shown by the scale "abundance" for nonmetals, to point m. Line k–m intersects scale q_P/h at point n. The area of the window is only 10% of that of the chamber, thus point n must be lowered by the distance shown by scale %s (for 10%) i.e. to point o.

The vacuum chamber is similar to a cube of side $a=5$ cm (point r); the line o–r intersects scale Q_P/V at point p, which shows that the gas load resulting from permeation is about 4×10^{-8} Torr·lit/sec·m^3. Point p shows on the scale "pressure rise" a value of 10^{-7} Torr/h, for the system cut-off from the pump. By fast pumping ($\tau=0.1$ sec), an ultimate pressure of about 4×10^{-12} Torr (point q) can be achieved. The gas load values obtained on nomogram (fig. 3.45) refer only to permeation. The *total gas load* is obtained by the sum of the values obtained on figs. 3.43–3.45 for the case which is considered.

CHAPTER 4

Physico-chemical phenomena in vacuum techniques

4.1. Evaporation-condensation

4.1.1. Vapours in vacuum systems

In addition to gases, vacuum systems also contain vapours. *The name "vapour" refers to a real gas, when it is below its critical temperature* (see §2.1).

When a substance is present, some of the molecules near its surface have sufficient kinetic energy to escape into the atmosphere and exist as a gas. Raising the temperature facilitates this process (see fig. 2.2). If a liquid is in the open, the vapour molecules rapidly diffuse away from the liquid, and in general produce what is known as an *unsaturated vapour*. If the substance is in an enclosed space, the pressure of the vapour will reach a maximum, which depends only upon the nature of the substance and the temperature. The vapour is then *saturated* and its pressure is the *saturated vapour pressure*. In this case, a dynamic equilibrium is established, between the number of molecules escaping from the surface (evaporation), and the number of molecules recaptured on the surface (condensation), in which *the net number of free molecules in the gaseous state is constant*.

Vacuum systems contain saturated as well as unsaturated vapours. All these vapours are maintaining their physical state or changing it according to the pressure-volume-temperature conditions (see fig. 2.6) existing in the system.

According to these conditions :

– Any liquid surface inside the vacuum system is a source of vapour, and as long as any liquid remains in the system, the *minimum pressure attainable is the vapour pressure of that liquid* at the existing temperature. At room temperature the presence of water limits the pressure to about 17 Torr (table 2.2), while the presence of mercury limits the pressure to about 1×10^{-3} Torr.

– If the vapour existing in the vacuum system is compressed as a result of pumping or handling operations, its pressure will increase only up to the vapour pressure. Further compression causes vapour to condense. In this way vapours which, in the system are unsaturated and of low pressure, *will condense in the pumps or gauges* where they are compressed. In order to avoid the condensation of water vapour in rotary pumps, the *gas ballast* system is used, in which a controlled amount of atmospheric air is admitted in the pump at a given stage of the compression, so that the pressure of the vapour is not increased above its saturation (see fig. 5.13).

The compression occurring in the McLeod gauge (figs. 2.7 and 6.7) condenses the vapours, and therefore this instrument does not measure accurately the contribution to the total pressure of any vapours in the system.

– A reduction in the temperature of any part of the vacuum system reduces the vapour pressure of any vapours present.

This is the principle on which the use of cold traps, refrigerated baffles, and cryogenic pumps is based.

4.1.2. *Vapour pressure and rate of evaporation*

The vapour pressure P_v of a substance is derived from the Clausius–Clapeyron equation

$$L_T = (T/J) \, (V_G - V_L) \, (dP_v/dT) \tag{4.1}$$

where L_T is the latent heat of evaporation, J is the mechanical equivalent of heat ($J_{15°C} = 4.1855 \times 10^7$ erg/cal), V_G and V_L are the specific volume (reciprocal of the density) of the gas and liquid respectively.

In the cases occurring in vacuum techniques the value of V_L is very small compared to that of V_G, while the value of V_G can be written (eq. 2.14)

$$V_G = R_o T / P_v \tag{4.2}$$

Thus eq. (4.1) becomes

$$L = \left(\frac{R_o}{J} \right) \left(\frac{T^2}{P_v} \right) \left(\frac{dP_v}{dT} \right) \tag{4.3}$$

The ratio R_o/J is equal to

$$\frac{R_o}{J} = \frac{8.314 \times 10^7 \ \text{erg}/°\text{K·mole}}{4.185 \times 10^7 \ \text{erg/cal}} = 1.987 \ \text{cal}/°\text{K·mole}$$

The latent heat of evaporation can be expressed by

$$L_T = L_o - IT \tag{4.4}$$

where L_o is the latent heat of evaporation at $T=0$, and I is a constant.
From eqs. (4.3) and (4.4) it results that

$$\int dP_v/P_v = \ln P_v = A' - (J/R_o)\ [(L_o/T)+I\ln T] \tag{4.5}$$

thus in decimal logarithms

$$\log P_v = A - \frac{B}{T} - C \log T \tag{4.6}$$

where

$$A = A'/2.302;\ B = \frac{L_o}{1.987 \times 2.302} = \frac{L_o}{4.575};\ C = I/4.575\ \text{(usually negligible)}.$$

The values of these constants were determined for a large number of materials.
Table 4.1 gives their values for the various metals.

When equilibrium exists between the solid or liquid and gaseous phases, the *rate of evaporation* W is equal to the rate of condensation, thus according to eq. (2.50)

$$W = 5.83 \times 10^{-2}\ P_v (M/T)^{1/2} f \tag{4.7}$$

where W is the rate of evaporation (g/s·cm²), P_v is the vapour pressure (Torr), M molecular weight; f is the sticking coefficient defined as the probability that an incident molecule remains on the surface (for values of W see table 4.2).

By using eqs. (4.7) and (4.6), it results that the evaporation rate can be express-ed as

$$\log W = A'' - \frac{B}{T} - C'\log T \tag{4.8}$$

where

$$A'' = A + \log 0.0583 + 0.5 \log M$$

$$C' = 0.5 + C$$

and

$$B = L_o/4.575$$

Table 4.1.
Constants for calculating the vapour pressure P_V (microns Hg, eq. 4.6) of metals
(Dushman and Lafferty, 1962).

Metal	A	B	Metal	A	B
Li	10.99	8.07×10^3	Ge	11.71	1.803×10^4
Na	10.72	5.49×10^3	Sn	10.88	1.487×10^4
K	10.28	4.48×10^3	Pb	10.77	9.71×10^3
Cs	9.91	3.80×10^3			
			Sb_2	11.15	8.63×10^3
Cu	11.96	1.698×10^4	Bi	11.18	9.53×10^3
Ag	11.85	1.427×10^4			
Au	11.89	1.758×10^4	Cr	12.94	2.0×10^4
			Mo	11.64	3.085×10^4
Be	12.01	1.647×10^4	W	12.40	4.068×10^4
Mg	11.64	7.65×10^3	U	11.59	2.331×10^4
Ca	11.22	8.94×10^3			
Sr	10.71	7.83×10^3	Mn	12.14	1.374×10^4
Ba	10.70	8.76×10^3	Fe	12.44	1.997×10^4
			Co	12.70	2.111×10^4
Zn	11.63	6.54×10^3	Ni	12.75	2.096×10^4
Cd	11.56	5.72×10^3	Ru	13.50	3.38×10^4
			Rh	12.94	2.772×10^4
B	13.07	2.962×10^4	Pd	11.78	1.971×10^4
Al	11.79	1.594×10^4	Os	13.59	3.7×10^4
La	11.60	2.085×10^4	Ir	13.07	3.123×10^4
			Pt	12.53	2.728×10^4
Ga	11.41	1.384×10^4			
In	11.23	1.248×10^4	V	13.07	2.572×10^4
C	15.73	4×10^4	Ta	13.04	4.021×10^4
Si	12.72	2.13×10^4			
Ti	12.50	2.32×10^4			
Zr	12.33	3.03×10^4			
Th	12.52	2.84×10^4			

4.1.3. *Vapour pressure data*

Vapour pressure of the various materials was measured by a large number of authors. These measurements were based either on the direct determination of the

Table 4.2.

Rate of evaporation W (g·cm^{-2}·s^{-1}) corresponding to vapour pressures P_μ (microns Hg) and temperatures $t(^\circ C)$. After Dushman and Lafferty (1962).

Metal	Data Temp. Range ($^\circ$C)	$t(^\circ$C) and W	\multicolumn{6}{c}{$P_\mu =$}					
			10^{-2}	10^{-1}	1	10	100	1000
Li	459–1080	t:	348	399	460	534	623	737
		W:	$6.17 . 10^{-8}$	$5.93 . 10^{-7}$	$5.68 . 10^{-6}$	$5.41 . 10^{-5}$	$5.13 . 10^{-4}$	$4.84 . 10^{-3}$
Na	264–928	t:	158	195	238	290	355	437
		W:	$1.35 . 10^{-7}$	$1.29 . 10^{-6}$	$1.24 . 10^{-5}$	$1.18 . 10^{-4}$	$1.12 . 10^{-3}$	$1.05 . 10^{-2}$
K	100–760	t:	91	123	162	208	266	341
		W:	$1.91 . 10^{-7}$	$1.83 . 10^{-6}$	$1.75 . 10^{-5}$	$1.66 . 10^{-4}$	$1.57 . 10^{-3}$	$1.47 . 10^{-2}$
Rb	..	t:	64	95	133	176	228	300
		W:	$2.94 . 10^{-7}$	$2.81 . 10^{-6}$	$2.68 . 10^{-5}$	$2.55 . 10^{-1}$	$2.41 . 10^{-2}$	$2.22 \quad 10^{-2}$
Cs	..	t:	46	75	110	152	206	277
		W:	$3.77 . 10^{-7}$	$3.61 . 10^{-6}$	$3.44 . 10^{-5}$	$3.26 . 10^{-4}$	$3.07 . 10^{-3}$	$2.87 . 10^{-2}$
Cu	969–1606	t:	942	1032	1142	1272	1427	1622
		W:	$1.33 . 10^{-7}$	$1.29 . 10^{-6}$	$1.24 . 10^{-5}$	$1.18 . 10^{-4}$	$1.13 . 10^{-3}$	$1.07 . 10^{-2}$
Ag	721–1000	t:	757	832	922	1032	1167	1337
		W:	$1.89 . 10^{-7}$	$1.82 . 10^{-6}$	$1.75 . 10^{-5}$	$1.68 . 10^{-4}$	$1.60 . 10^{-3}$	$1.51 . 10^{-2}$
Au	727–987	t:	987	1082	1197	1332	1507	1707
		W:	$2.31 . 10^{-7}$	$2.26 . 10^{-6}$	$2.14 . 10^{-5}$	$2.05 . 10^{-4}$	$1.94 . 10^{-3}$	$1.84 . 10^{-2}$
Be	899–1279	t:	902	987	1092	1212	1367	1567
		W:	$5.11 . 10^{-8}$	$4.93 . 10^{-7}$	$4.74 . 10^{-6}$	$4.55 . 10^{-5}$	$4.33 . 10^{-4}$	$4.08 . 10^{-3}$
Mg	736–1020	t:	287	330	382	442	517	612
		W:	$1.22 . 10^{-7}$	$1.17 . 10^{-6}$	$1.12 . 10^{-5}$	$10.8 . 10^{-4}$	$1.02 . 10^{-3}$	$0.97 . 10^{-2}$
Ca	527–647	t:	402	452	517	592	687	817
		W:	$1.42 . 10^{-7}$	$1.37 . 10^{-6}$	$1.31 . 10^{-5}$	$1.26 . 10^{-4}$	$1.19 . 10^{-3}$	$1.12 . 10^{-2}$
Sr	..	t:	342	394	456	531	623	742
		W:	$2.20 . 10^{-7}$	$2.11 . 10^{-6}$	$2.02 . 10^{-5}$	$1.93 . 10^{-4}$	$1.82 . 10^{-3}$	$1.71 . 10^{-2}$
Ba	1060–1138	t:	417	467	537	617	727	867
		W:	$2.60 . 10^{-7}$	$2.51 . 10^{-6}$	$2.40 . 10^{-5}$	$2.28 . 10^{-4}$	$2.16 . 10^{-3}$	$2.03 . 10^{-2}$
Zn	239–377	t:	208	246	290	342	405	485
		W:	$2.15 . 10^{-7}$	$2.07 . 10^{-6}$	$1.99 . 10^{-5}$	$1.90 . 10^{-4}$	$1.81 . 10^{-3}$	$1.71 . 10^{-2}$
Cd	200–260	t:	149	182	221	267	321	392
		W:	$3.01 . 10^{-7}$	$2.90 . 10^{-6}$	$2.78 . 10^{-5}$	$2.66 . 10^{-4}$	$2.54 . 10^{-3}$	$2.44 . 10^{-2}$
Hg	..	t:	-28	-8	16	45	81	125
		W:	$5.28 . 10^{-7}$	$5.08 . 10^{-6}$	$4.86 . 10^{-5}$	$4.63 . 10^{-4}$	$4.39 . 10^{-3}$	$4.14 . 10^{-2}$
B	..	t:	1687	1827	1977	2157	2377	2657
		W:	$4.33 . 10^{-8}$	$4.19 . 10^{-7}$	$4.05 . 10^{-6}$	$3.89 . 10^{-5}$	$3.73 . 10^{-4}$	$3.55 . 10^{-3}$
Al	1137–1195	t:	882	972	1082	1207	1347	1547
		W:	$8.92 . 10^{-8}$	$8.59 . 10^{-7}$	$8.23 . 10^{-6}$	$7.88 . 10^{-5}$	$7.53 . 10^{-4}$	$7.10 . 10^{-3}$

(contd.)

Table 4.2 (contd.)

Metal	Data Temp. Range (°C)	$t(°C)$ and W	$P_\mu=$					
			10^{-2}	10^{-1}	1	10	100	1000
Sc	..	t:	1058	1161	1282	1423	1595	1804
Y	..	t:	1249	1362	1494	1649	1833	2056
La	1327–1627	t:	1262	1377	1527	1697	1897	2147
		W:	$1.56 . 10^{-7}$	$1.69 . 10^{-6}$	$1.62 . 10^{-5}$	$1.55 . 10^{-4}$	$1.48 . 10^{-3}$	$1.40 . 10^{-2}$
Ce	..	t:	1004	1091	1190	1305	1439	1599
Nd	899–1279	t:	957	1062	1192	1342	1537	1777
		W:	$2.00 . 10^{-7}$	$1.92 . 10^{-6}$	$1.83 . 10^{-5}$	$1.74 . 10^{-4}$	$1.65 . 10^{-3}$	$1.55 . 10^{-2}$
Ga	957–1245	t:	757	842	937	1057	1197	1372
		W:	$1.52 . 10^{-7}$	$1.46 . 10^{-6}$	$1.40 . 10^{-5}$	$1.34 . 10^{-4}$	$1.27 . 10^{-3}$	$1.22 . 10^{-2}$
In	727–1075	t:	670	747	837	947	1077	1242
		W:	$2.04 . 10^{-7}$	$1.96 . 10^{-6}$	$1.88 . 10^{-5}$	$1.79 . 10^{-4}$	$1.70 . 10^{-3}$	$1.61 . 10^{-2}$
Tl	..	t:	412	468	535	615	713	837
		W:	$3.19 . 10^{-7}$	$3.06 . 10^{-6}$	$2.93 . 10^{-5}$	$2.80 . 10^{-4}$	$2.66 . 10^{-3}$	$2.50 . 10^{-2}$
C	2084–2597	t:	1977	2107	2247	2427	2627	2867
		W:	$4.27 . 10^{-8}$	$4.14 . 10^{-7}$	$4.03 . 10^{-6}$	$3.89 . 10^{-5}$	$3.76 . 10^{-4}$	$3.61 . 10^{-3}$
Si	..	t:	1177	1282	1357	1547	1717	1927
		W:	$8.12 . 10^{-8}$	$7.84 . 10^{-7}$	$7.54 . 10^{-6}$	$7.24 . 10^{-5}$	$6.93 . 10^{-4}$	$6.59 . 10^{-2}$
Ti	1111–1323	t:	1321	1431	1558	1703	1877	2083
		W:	$1.01 . 10^{-7}$	$0.98 . 10^{-6}$	$0.94 . 10^{-5}$	$0.90 . 10^{-4}$	$0.86 . 10^{-3}$	$0.82 . 10^{-2}$
Zr	1676–1781	t:	1837	2002	2187	2397	2647	2977
		W:	$1.21 . 10^{-7}$	$1.17 . 10^{-6}$	$1.12 . 10^{-5}$	$1.08 . 10^{-4}$	$1.03 . 10^{-3}$	$0.98 . 10^{-2}$
Th	..	t:	1686	1831	1999	2196	2431	2715
		W:	$2.01 . 10^{-7}$	$1.94 . 10^{-6}$	$1.86 . 10^{-5}$	$1.79 . 10^{-4}$	$1.71 . 10^{-3}$	$1.63 . 10^{-2}$
Ge	1237–1612	t:	1037	1142	1262	1407	1582	1797
		W:	$1.37 . 10^{-7}$	$1.32 . 10^{-6}$	$1.27 . 10^{-5}$	$1.21 . 10^{-4}$	$1.15 . 10^{-3}$	$1.09 . 10^{-1}$
Sn	1151–1415	t:	882	977	1092	1227	1397	1612
		W:	$1.87 . 10^{-7}$	$1.80 . 10^{-6}$	$1.72 . 10^{-5}$	$1.64 . 10^{-4}$	$1.56 . 10^{-3}$	$1.46 . 10^{-2}$
Pb	..	t:	487	551	627	719	832	977
		W:	$3.05 . 10^{-7}$	$2.93 . 10^{-6}$	$2.80 . 10^{-5}$	$2.67 . 10^{-4}$	$2.53 . 10^{-3}$	$2.38 . 10^{-2}$
V	1389–1609	t:	1432	1551	1687	1847	2037	2287
		W:	$1.01 . 10^{-7}$	$0.98 . 10^{-6}$	$0.94 . 10^{-5}$	$0.90 . 10^{-4}$	$0.87 . 10^{-3}$	$0.82 . 10^{-2}$
Cb	..	t:	2194	2355	2539	..	..	..
		W:	$1.16 . 10^{-7}$	$1.08 . 10^{-6}$	$1.06 . 10^{-5}$	..	..	..
Ta	1727–2997	t:	2397	2587	2807	3067	3372	3737
		W:	$1.52 . 10^{-7}$	$1.47 . 10^{-6}$	$1.41 . 10^{-5}$	$1.36 . 10^{-4}$	$1.30 . 10^{-3}$	$1.24 . 10^{-2}$
P_4	..	t:	107	130	157	187	222	262
		W:	$3.3 . 10^{-7}$	$3.24 . 10^{-6}$	$3.13 . 10^{-5}$	$3.03 . 10^{-4}$	$2.92 . 10^{-3}$	$2.81 . 10^{-2}$
Sb_2	..	t:	382	427	477	542	617	757
		W:	$2.52 . 10^{-7}$	$2.43 . 10^{-6}$	$2.35 . 10^{-5}$	$2.26 . 10^{-4}$	$2.16 . 10^{-3}$	$2.01 . 10^{-2}$

(contd.)

Table 4.2 (*contd.*)

Metal	Data Temp. Range (°C)	$t(°C)$ and W	$P_\mu=$					
			10^{-2}	10^{-1}	1	10	100	1000
Bi	409–497	t:	450	508	578	661	762	892
		W:	$3.14 \cdot 10^{-7}$	$3.02 \cdot 10^{-6}$	$2.89 \cdot 10^{-5}$	$2.76 \cdot 10^{-4}$	$2.62 \cdot 10^{-3}$	$2.47 \cdot 10^{-2}$
Cr	889–1288	t:	1062	1162	1267	1392	1557	1737
		W:	$1.15 \cdot 10^{-7}$	$1.11 \cdot 10^{-6}$	$1.07 \cdot 10^{-5}$	$1.03 \cdot 10^{-4}$	$0.98 \cdot 10^{-3}$	$0.94 \cdot 10^{-2}$
Mo	1797–2231	t:	1987	2167	2377	2627	2927	3297
		W:	$1.20 \cdot 10^{-7}$	$1.16 \cdot 10^{-6}$	$1.11 \cdot 10^{-5}$	$1.06 \cdot 10^{-4}$	$1.01 \cdot 10^{-3}$	$0.95 \cdot 10^{-2}$
W	..	t:	2547	2757	3007	3297	3647	..
		W:	$1.49 \cdot 10^{-7}$	$1.44 \cdot 10^{-6}$	$1.38 \cdot 10^{-5}$	$1.32 \cdot 10^{-4}$	$1.26 \cdot 10^{-3}$	..
U	1357–1697	t:	1442	1582	1737	1927	2157	2447
		W:	$2.17 \cdot 10^{-7}$	$2.09 \cdot 10^{-6}$	$2.01 \cdot 10^{-5}$	$1.92 \cdot 10^{-4}$	$1.83 \cdot 10^{-3}$	$1.73 \cdot 10^{-2}$
Se	..	t:	144	167	197	232	277	347
		W:	$2.54 \cdot 10^{-7}$	$2.47 \cdot 10^{-6}$	$2.39 \cdot 10^{-5}$	$2.31 \cdot 10^{-1}$	$2.21 \cdot 10^{-3}$	$2.08 \cdot 10^{-2}$
Te$_2$	..	t:	261	296	336	383	438	520
		W:	$4.03 \cdot 10^{-7}$	$3.91 \cdot 10^{-6}$	$3.78 \cdot 10^{-5}$	$3.64 \cdot 10^{-4}$	$3.50 \cdot 10^{-3}$	$3.31 \cdot 10^{-2}$
Po	438–745	t:	187	220	263	314	382	472
		W:	$3.94 \cdot 10^{-7}$	$3.81 \cdot 10^{-6}$	$3.65 \cdot 10^{-5}$	$3.49 \cdot 10^{-4}$	$3.30 \cdot 10^{-3}$	$3.10 \cdot 10^{-2}$
Mn	..	t:	697	767	852	947	1067	1227
		W:	$1.39 \cdot 10^{-7}$	$1.34 \cdot 10^{-6}$	$1.29 \cdot 10^{-5}$	$1.24 \cdot 10^{-4}$	$1.18 \cdot 10^{-3}$	$1.12 \cdot 10^{-2}$
Re	..	t:	2367	2557	2787	3057	3397	..
		W:	$1.55 \cdot 10^{-7}$	$1.50 \cdot 10^{-6}$	$1.44 \cdot 10^{-5}$	$1.38 \cdot 10^{-4}$	$1.31 \cdot 10^{-3}$	..
Fe	1092–1246	t:	1107	1207	1322	1467	1637	1847
		W:	$1.17 \cdot 10^{-7}$	$1.13 \cdot 10^{-6}$	$1.09 \cdot 10^{-5}$	$1.05 \cdot 10^{-4}$	$0.99 \cdot 10^{-3}$	$0.95 \cdot 10^{-2}$
Co	1090–1249	t:	1162	1262	1377	1517	1697	1907
		W:	$1.18 \cdot 10^{-7}$	$1.14 \cdot 10^{-6}$	$1.10 \cdot 10^{-5}$	$1.06 \cdot 10^{-4}$	$1.01 \cdot 10^{-3}$	$0.96 \cdot 10^{-2}$
Ni	1034–1310	t:	1142	1247	1357	1497	1667	1877
		W:	$1.19 \cdot 10^{-7}$	$1.15 \cdot 10^{-6}$	$1.11 \cdot 10^{-5}$	$1.06 \cdot 10^{-4}$	$1.01 \cdot 10^{-3}$	$0.96 \cdot 10^{-2}$
Ru	..	t:	1913	2058	2230	2431	2666	2946
		W:	$1.26 \cdot 10^{-7}$	$1.22 \cdot 10^{-6}$	$1.18 \cdot 10^{-5}$	$1.13 \cdot 10^{-4}$	$1.08 \cdot 10^{-3}$	$1.04 \cdot 10^{-2}$
Rh	..	t:	1587	1707	1857	2027	2247	2527
		W:	$1.37 \cdot 10^{-7}$	$1.33 \cdot 10^{-6}$	$1.28 \cdot 10^{-5}$	$1.23 \cdot 10^{-4}$	$1.18 \cdot 10^{-3}$	$1.12 \cdot 10^{-2}$
Pd	..	t:	1157	1262	1387	1547	1727	1967
		W:	$1.59 \cdot 10^{-7}$	$1.54 \cdot 10^{-6}$	$1.48 \cdot 10^{-5}$	$1.41 \cdot 10^{-4}$	$1.35 \cdot 10^{-3}$	$1.27 \cdot 10^{-2}$
Os	..	t:	2101	2264	2451	2667	2920	3221
		W:	$1.65 \cdot 10^{-7}$	$1.60 \cdot 10^{-6}$	$1.54 \cdot 10^{-5}$	$1.48 \cdot 10^{-4}$	$1.42 \cdot 10^{-3}$	$1.36 \cdot 10^{-2}$
Ir	..	t:	1797	1947	2107	2307	2527	2827
		W:	$1.78 \cdot 10^{-7}$	$1.72 \cdot 10^{-6}$	$1.66 \cdot 10^{-5}$	$1.60 \cdot 10^{-4}$	$1.53 \cdot 10^{-3}$	$1.46 \cdot 10^{-2}$
Pt	..	t:	1602	1742	1907	2077	2317	2587
		W:	$1.88 \cdot 10^{-7}$	$1.82 \cdot 10^{-6}$	$1.75 \cdot 10^{-5}$	$1.68 \cdot 10^{-4}$	$1.60 \cdot 10^{-3}$	$1.52 \cdot 10^{-2}$

mechanical effect of the *pressure* exerted by the vapour, or by measuring the evaporation rate (weight loss or increase) and calculating the vapour pressure from eq. (4.7).

Figures 4.1–4.3 give the up-to-date vapour pressure data, according to Honig and Kramer (1969).

The rate of evaporation of various elements is given in table 4.2, the vapour pressure curves of various common gases are given in fig. 4.4. Figure 4.5 shows the vapour pressures of some oils used in diffusion pumps, while fig. 4.6 gives the vapour pressures of some cleaning liquids used in vacuum techniques.

4.1.4. *Cryopumping and vacuum coating*

Evaporation and condensation phenomena are often complicating the pump-down process in vacuum systems, but they are also the basis of some vacuum technology applications. Although techniques like freeze-drying and molecular distillation are also based on evaporation phenomena, we intend to illustrate here the field of applications by techniques representing the use of extreme temperatures, i.e. the *cryopumping* and the *vacuum coating*.

Cryogenic pumping is based on the fact that if a surface within a vacuum system is cooled, vapours (gases) will tend to condense upon it, thus reducing the pressure. The *ultimate pressure* of such a pump for a given gas is determined by the vapour pressure P_v at the temperature T_v of the condenser surface. Since the quantity of gas evaporated from the surfaces of the system at temperature T is equal to that condensed on the surface at T_v, from eq. (4.7) it results that the ultimate pressure P_u for the particular gas (M) considered is

$$P_u = (f_v/f) \, P_v \, (T/T_v)^{1/2} \tag{4.9}$$

where f and f_v are the sticking coefficients at temperature T and T_v, respectively. Since both f and f_v are close to unity, it results that for $T = 300°K$, and $T_v = 4.2°K$ (liquid helium), $P_u/P_v = 8.4$. In this case, according to fig. 4.4, for most of the gases ultimate pressure below 10^{-10} Torr is readily attainable.

The difference between the number of molecules condensing and leaving the unit surface area each second is from eq. (2.46)

$$\Delta N = fP \, (2\pi mkT)^{-1/2} - f_v P_v \, (2\pi mkT_v)^{-1/2} \tag{4.10}$$

It follows from eqs. (2.15) and (2.21) that the throughput is $Q = \Delta NkT$. Thus

$$Q = A \left(\frac{kT}{2\pi m} \right)^{1/2} \left[fP - f_v P_v \left(\frac{T}{T_v} \right)^{1/2} \right] \tag{4.11}$$

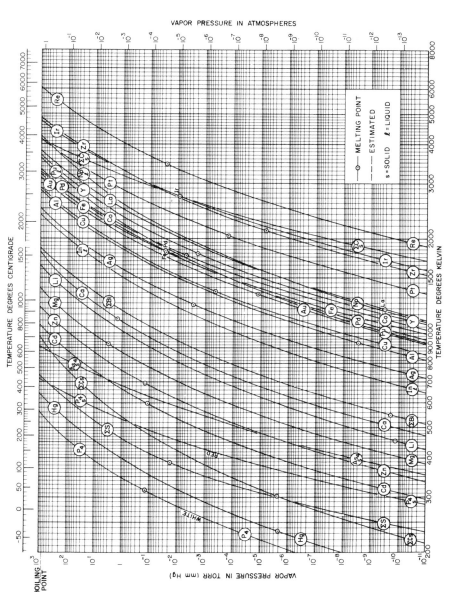

Fig. 4.1 Vapour pressure of elements. Reprinted from Honig and Kramer (1969), by permission of RCA Review, Princeton, N.J.

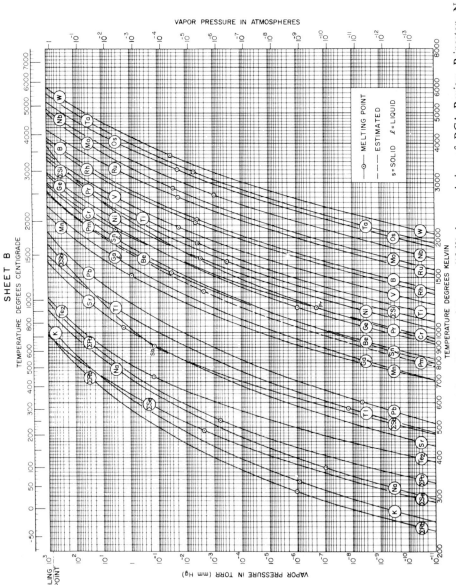

Fig. 4.2 Vapour pressure of elements. Reprinted from Honig and Kramer (1969), by permission of RCA Review, Princeton, N.J.

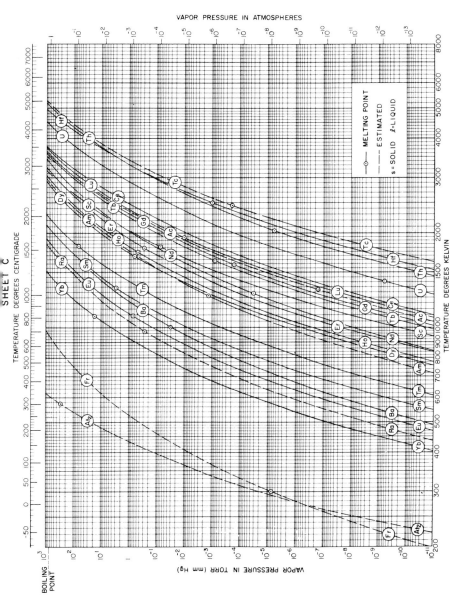

Fig. 4.3 Vapour pressure of elements. Reprinted from Honig and Kramer (1969), by permission of RCA Review, Princeton, N.J.

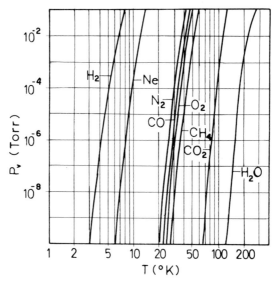

Fig. 4.4 Vapour pressure of common gases. After Honig and Hook (1960).

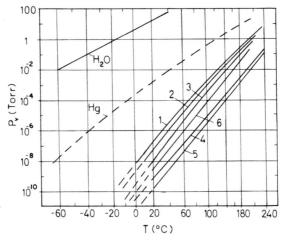

Fig. 4.5 Vapour pressure of diffusion pump oils. (1) Octoil; (2) Silicone DC–703; (3) Silicone DC–704; (4) Santovac 5–Convalex 10; (5) Silicone DC–705; (6) Apiezon C.

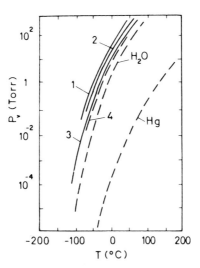

Fig. 4.6 Vapour pressure of solvents. (1) Acetone; (2) Tetrachlor carbon; (3) Trichlorethylene;
(4) Ethanol.

where A is the area of the condenser surface.

Eliminating $f_v P_v$ by eq. (4.9), the pumping speed is obtained as

$$S = 3.64 \, fA \left(\frac{T}{M}\right)^{1/2} \left(1 - \frac{P_u}{P}\right) \tag{4.12}$$

Based on this principle, the cryogenic pumps (§5.6) attain relatively high pumping speeds ($10^4 - 10^6$ liter/sec).

Vacuum coating is based on the evaporation* of the required material, and its subsequent condensation on the substrate to be coated. The process is done in a high vacuum, so that the particles do not collide with gas molecules in their way between evaporation and condensation.

Figure 4.7 shows the basic features of a vacuum coating plant. The material to be evaporated (metal or non-metal) is placed in the evaporator (a spiral or boat of tungsten, molybdenum, or tantalum), which is heated (in vacuum), up to a temperature where the vapour pressure of the material to be evaporated is high enough. To obtain admissible evaporation rates (eq. 4.7), vapour pressures of $10^{-3} - 10^{-2}$ Torr are usually required, thus the materials have to be heated up to the temperatures corresponding to these vapour pressure (figs. 4.1–4.3). For example, gold has to be heated to about $1300°C$, to obtain $P_v = 10^{-2}$ Torr.

---

* See p. 195 for vacuum coating based on sputtering.

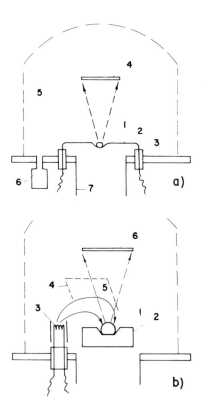

Fig. 4.7 Vacuum coating systems. (a) Resistance heating: (1) material to be evaporated; (2) Mo or Ta boat; (3) current lead-in; (4) work to be coated; (5) bell jar; (6) Penning gauge; (7) diffusion (or ion) pump. (b) Electron bombardment heating: (1) material to be evaporated; (2) copper anode; (3) electron source; (4) magnetic field; (5) electron trajectory; (6) work to be coated

The evaporated material travels in straight lines in all the directions, coating the work (substrate) as well as the bell jar. If the material to be evaporated is concentrated enough (a small filament or basket) to be considered as a *point source*, the thickness of the deposit t_o (cm) in the middle of the substrate (work) opposite the evaporator will be

$$t_o = W_o/(4\pi\rho h^2) \tag{4.13}$$

where W_o is the evaporated mass (g), ρ is the specific gravity (g/cm^3), h is the distance evaporator–work (cm).

If the material is evaporated from a boat, then the thickness in the middle of the work will be

$$t_o = W_o/(\pi \rho h^2) \tag{4.14}$$

The thickness t_δ at any point of the substrate at a distance δ from the middle is given in fig. 4.8.

Vacuum coating techniques are the subject of a very large number of publications. For a comprehensive treatment of the subject we refer to Holland (1956); Berry *et al.* (1968); Maissel and Glang (1970); Maissel and Francombe (1973).

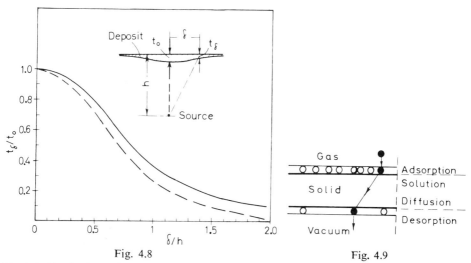

Fig. 4.8

Fig. 4.9

Fig. 4.8 Distribution of the thickness of the deposit on a plane surface for (a) evaporation from a boat (dotted line), (b) evaporation from a point source (full line).

Fig. 4.9 The process of permeation.

4.2. Solubility and permeation

4.2.1. *The permeation process*

Gases have the possibility of passing through solids, even if the openings present are not large enough to permit a regular flow. The passage of a gas into, through and out of a solid barrier having no holes large enough to permit more than a small fraction of the gas to pass through any one hole is known as *permeation*. The steady state rate of flow in these conditions is the *permeability coefficient* or simply the *permeability*. This is usually expressed in cm³ of gas at STP flowing per second through a cm² of cross section, per mm of wall thickness and 1 Torr (10 Torr, 1 atm) of pressure drop across the barrier.

The process of permeation is described by Norton (1962) as shown in fig. 4.9. It involves first the *adsorption* of the gas on the surface where the gas pressure is higher. After being *dissolved* in the outside surface layer the gas slides down the concentration gradient and *diffuses* to the vacuum side where it is *desorbed*.

Generally, gases *dissolve* in solids to a concentration c

$$c = bP^{1/j} \tag{4.15}$$

where P is the gas pressure, j is the dissociation constant of the gas, and b is the *solubility* of the gas in the solid.

The *dissociation constant* j, is $j=2$ for diatomic gases in metals, and $j=1$ for all gases in nonmetals.

The *concentration* c is the amount of gas (in Torr·cm³, or atm·cm³) at 293°K that is dissolved in 1 cm³ of the substance.

The *solubility* b is the quantity of gas (in cm³) at STP (293°K and 1 atm) that is dissolved in 1 cm³ of the substance at a pressure of 1 atm. It is dimensionless for $j=1$, but has the dimensions of atm$^{1/2}$ for $j=2$.

The *diffusion* of the gas into and through the solid obeys *Fick's laws of diffusion* which are:

Fick's first law. In the steady state, when the gas concentration is independent of time, gas diffuses across a plane of area 1 cm² in a region where the concentration gradient is dc/dx at a rate Q given by

$$Q = -D_1(dc/dx) \tag{4.16}$$

where D_1 is the diffusion coefficient, which is expressed in units of cm²/sec. The minus sign is used because the flow is opposite to the concentration gradient. D_1 varies with temperature according to

$$D_1 = D_o \exp\left(\frac{-H}{j R_o T}\right) \tag{4.17}$$

where H is the activation energy of absorption, D_o is a constant for a given gas and material, and R_o is the gas constant.

Fick's second law. In most cases, equilibrium is reached only after a long time or not at all, since D is small. In this transient period, when the concentration varies with time, Fick's second law states that

$$D_1 \frac{d^2c}{dx^2} = \frac{dc}{dt} \tag{4.18}$$

For the case of *the steady state*, the concentrations (eq. 4.15) at the two surfaces with pressures P_1 and P_2, are $c_1 = bP_1^{1/j}$ and $c_2 = bP_2^{1/j}$. From eq. (4.16) it follows that

$$Q \int_0^h dx = -D \int_{c_1}^{c_2} dc \tag{4.19}$$

thus

$$Q = D_1 b \frac{P_1^{1/j} - P_2^{1/j}}{h} \tag{4.20}$$

where h is the thickness of the material.

The *product* $D_1 b$ between the diffusion coefficient and the solubility is called the *permeation constant K*. It is commonly expressed as the amount of gas (cm^3 STP), permeating through a 1 cm^2 cross section of a slab of 1 cm thickness for a pressure difference of 1 atm. Figures 4.10 and 4.11 show a review of values of the permeation constant K.

According to Norton (1962), certain criteria must be fulfilled to distinguish, unambiguously, true permeation from gases flowing through an actual hole, and gases derived from the walls of the envelope (outgassing). After a thorough degassing by a good vacuum applied on each side of the wall, the *effect of a hole* can be distinguished from true permeation in two ways. Very rapid rise of the particular gas on the low side after application of pressure to the high side may indicate a hole. Variation of the rate with $(T/M)^{1/2}$ is shown by testing with gases of differing molecular weight. Variation following this law shows that a hole exists. The diffusion coefficient D_1 is most conveniently measured by the time lag method. In this, the effective time lag τ (sec), to attain steady state permeation through a membrane of thickness h (cm), is related to the diffusion coefficient

$$D_1 = h^2/(6\,\tau) \tag{4.21}$$

Experimental and computational techniques for determining the diffusion coefficient and the permeation constant are reviewed by Perkins (1973).

Superpermeability (activated permeability) is discussed by Livschitz (1979).

4.2.2. Permeation through vacuum envelopes

The metallic, glass or rubber walls of vacuum vessels or pipes are more or less permeable to gases. The permeation mechanism can be *atomic* or *molecular*. Hydrogen permeation through metals increases with the *square root* of the pressure ($j=2$), this fact is explained by the *dissociation of the hydrogen to atoms* and their passage as such through the metal. Recombination occurs on desorption and on the low pressure side molecular hydrogen appears. In *glasses and elastomers* the gas permeation is *proportional to the pressure* ($j=1$). Here the permeation itself

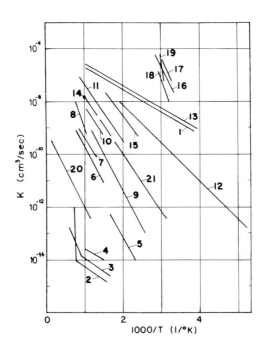

Fig. 4.10 Permeation constants for various gas–non-metal combinations as a function of $1/T$. Units are cm³/s, i.e. quantity of gas (cm³, STP), passing per second through a wall of 1 cm² area and 1 cm thickness, when a pressure difference of 1 atm exists across the wall. References: (1) He–fused silica, Altemose (1961); 2. Air–Pyroceram, Miller and Shephard (1961); (3) Air–97 % alumina ceramic, Miller and Shephard (1961); (4) Air–Pyrex, Miller and Shephard (1961); (5) He–lead borate glass G, Norton (1957); (6) He–97 % alumina ceramic, Miller and Shephard (1961); (7) Ne–Vycor, Leiby and Chen (1960); (8) N_2–SiO_2, Barrer (1951); (9) He–1720 glass, Norton (1957); (10) He–Pyroceram, Miller and Shephard (1961); (11) H_2–SiO_2, Barrer (1951); (12) He–Pyrex 7740, Rogers *et al.* (1954); (13) He–Vycor, Altemose (1961); (14) H_2–Pyrex, Norton (1957); (15) He–Pyrex 7052, Altemose (1961); (16) He–Neoprene, Dayton (1960); (17) H_2–Neoprene, Dayton (1960); (18) N_2–Neoprene, Dayton (1960); (19) A–Neopren, Dayton (1960); (20) O_2–fused silica, Norton (1962); (21) Ne–fused silica, Perkins (1971), Shelby (1972).

occurs in molecular form.

The permeation of atmospheric gases through *metal walls* does not include the rare gases (He, A, Ne, Kr, Xe) since *no rare gas diffuses through metals at any temperature* under purely thermal activation. There can be penetration of rare gas ions under a potential gradient, or rare gases can be formed *in situ* in the metal interior by nuclear disintegration processes.

The permeability of aluminium for hydrogen is very small (fig. 4.12). It is negligible in all cases except in ultra-high vacuum chambers, or with chambers heated at high temperatures and having thin walls. *Copper* is a metal with low permeability for all the gases, including hydrogen (figs. 4.11 and 4.12). Nickel has a higher

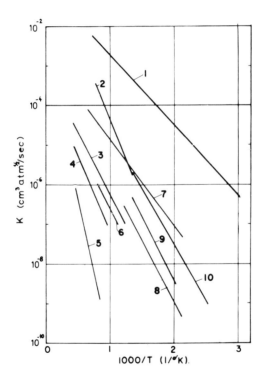

Fig. 4.11 Permeation constants for various diatomic gas–metal combinations as a function of $1/T$. Units are $cm^3 \cdot atm^{1/2}/sec$ – i.e. quantity of gas (cm^3 STP) passing per second through a wall of 1 cm^2 area and 1 cm thickness, when a pressure difference of 1 atm exists across the wall. References: (1) H_2–Pd, Barrer (1951), Holleck (1970); (2) H_2–Ni, Barrer (1951), Ebisuzaki (1967); (3) H_2–Mo, Moore and Unterwald (1964); (4) N_2–Fe, Barrer (1951); (5) N_2–Mo, Frauenfelder (1968); (6) CO–Ni, Dushman and Lafferty (1962), (7) H_2–Fe, Barrer (1951); (8) H_2–Cu, Gorman (1962), Perkins (1971); (9) H_2–stainless steel 300, Eschbach *et al.* (1963); (10) H_2–stainless steel 400, Eschbach *et al.* (1963).

permeability for hydrogen. Therefore, *for water-cooled chambers* where the danger of hydrogen permeation is greater, *copper is to be preferred to nickel. Iron* vacuum containers have high permeability (fig. 4.12) for hydrogen, especially if the hydrogen is in atomic form on the high pressure side due to chemical or electrolytic effects. Thus the cooling of iron vacuum containers should be made with liquids which do not contain hydrogen ions, or air cooling should be used. The permeation of hydrogen through *steels* increases with increasing carbon content; low carbon steels are thus preferred as vacuum containers.

Permeation of tritium is an important problem in the design of Tokamak reactors; the subject is discussed by Cecchi (1979).

The permeability of *glasses* is important only for systems in the ultra-high vacuum range ($P < 10^{-7}$ Torr). The permeation is influenced by the kind of glass and the gas involved. As a general rule the denser the structure of the glass and the larger the molecule of the gas the less the permeation. This is the reason why gases permeate easier through silica (SiO_2) than through technical glasses

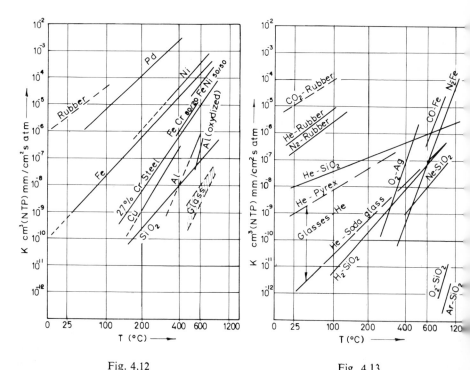

Fig. 4.12   Fig. 4.13

Fig. 4.12 Permeation of hydrogen through various materials. After Roth (1966).
Fig. 4.13 Permeation of various gases through various materials. After Roth (1966).

(fig. 4.13). In technical glasses the open meshes of silica or other glass formers are occupied by network modifiers as Na, K, Ba.

The permeation of helium through vitreous silica (or Vycor) is 10^7 times greater than through crystalline quartz. Vitreous silica has a considerable permeability also for other gases, like hydrogen, nitrogen, oxygen and argon.

Organic polymers (rubbers, plastics) are permeated by all the gases including the rare ones (Ne, He, A, Kr, Xe). There are wide variations in permeability. That of CO_2 through natural rubber is high (about 10^{-5} cm^3 STP·mm/cm^2·sec· atm.), that of air is lower (10^{-6} cm^3 STP·mm/cm^2·sec·atm.). Saran, polyethylene and Kel–F have generally low permeability (about 3×10^{-7} cm^3 STP·mm/cm^2· sec·atm. at 25°C).

The main features of the permeation processes are summarized in table 4.3.

Table 4.3.
Main features of gas permeation (Norton, 1962).

Glasses	Metals	Semiconductors	Polymers
He, H_2, D_2, Ne, A, O_2 measurable through SiO_2	No rare gas through any metal	He and H_2 through Ge and Si	All gases permeate all polymers
	H_2 permeates most, especially Pd.	Ne, A not measurable	Water rate apt to be high
Vitreous silica (fastest)	O_2 permeates Ag. H_2 through Fe by corrosion, electrolysis, etc.		Many specificities
All rates vary directly as pressure	Rates vary as $\sqrt{}$(pressure)	H_2 rate varies as $\sqrt{}$(pressure)	All rates vary as pressure

4.2.3. Consequences of permeation

The consequence of permeation is obviously the transfer of gas from the high pressure to the low pressure side. This process is *limiting the final pressure* to which a vessel can be evacuated (see fig. 3.45) but also permits us to *introduce measured quantities of specific gases* into the evacuated systems.

In order to show the importance of the *inflow of atmospheric gases through the walls of a vacuum vessel*, Norton (1962) gives some interesting examples. He considers a bulb of vitreous silica at 25°C with walls 1 mm thick, surface area 100 cm^2 and volume 330 cm^3, and assumes that the walls have been completely degassed, that the initial pressure is 10^{-16} Torr and that the steady-state flow is established at 25°C. From the abundance (partial pressure) of the various gases

Table 4.4.
Order of inflow of atmospheric gases into SiO_2 bulb at $25°C$ (for 1 mm wall thickness, 1 cm² areas).
After Norton (1962).

Gas	Atmospheric abundance C (cm of Hg, partial pressure)	Permeation P (cm³ STP/sec, for 1 cm Hg pressure difference)	Inflow $C \times P$ (cm³/sec)	Order of inflow	Atoms per sec.
N_2	59.5	2×10^{-29}	1.2×10^{-27}	5	—
O_2	15.9	1×10^{-28}	1.6×10^{-27}	4	—
A	0.705	2×10^{-29}	1.4×10^{-29}	6	—
Ne	1.8×10^{-2}	2×10^{-15}	3.6×10^{-17}	2	900
He	4×10^{-4}	5×10^{-11}	2.0×10^{-14}	1	5 00 000
H_2	3.8×10^{-5}	2.8×10^{-14}	1.0×10^{-18}	3	25

in the atmosphere, and from the permeation extrapolated to $25°C$ (table 4.4) the order of inflow of the gases, and the gas accumulation (figs. 4.14 and 4.15) are established.

It can be seen that the gases of low abundance show the highest inflow and the order of accumulation (fig. 4.14) in the silica bulb is (1) helium, (2) neon, (3) hydrogen. A large difference separates the succeeding gases oxygen, nitrogen and argon. For the vitreous silica bulb (fig. 4.14) the gases and pressures at the end of one year in air at $25°C$, would be 10^{-4} Torr helium, 10^{-7} Torr neon, and 10^{-8} Torr hydrogen. Only a few molecules of oxygen would have permeated even after a hundred years.

The increase in pressure in bulbs of various glasses is shown in fig. 4.15, for helium permeating from the atmosphere. To reach a helium pressure of 10^{-6} Torr, requires, 3 days for silica, a month for Pyrex and very long times for soda–lime glass or other glasses. From this, it is evident that if we are concerned with sealed off vacuum containers with pressures in the range of 10^{-9} Torr, it is necessary to make their envelope of a glass of low permeability, or surround it by a subsidiary evacuated chamber.

The *permeation* is used *when specific gases are to be introduced* in highly evacuated systems. Calibrated leaks of *helium* are extensively used in leak detection. These are glass bulbs filled with helium, and sealed with a graded seal and a silica tube having thin walls. Such calibrated leaks may give leak rates as low as 10^{-10} atm · cm³/sec, and are very constant for many years.

Leaks based on permeation of helium are discussed by Rubet (1983), Solomon (1986).

For the introduction of pure *hydrogen*, palladium or nickel tubes are used, while silver is the best material for the diffusion of *oxygen*. In order to increase and control the flow rate (permeability), Pd, Ni, or Ag tubes with thin walls are heated by coils, or direct electric current. The permeation of hydrogen through Pd may be used in the construction of "membrane pumps" as discussed by Murakami and Ohtsuka (1978); Livschitz and Notkin (1979).

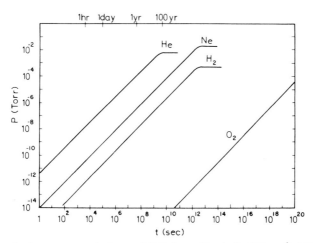

Fig. 4.14 Atmospheric gas accumulation at 25°C, in a silica bulb, 330 cm³, 100 cm² wall area, 1 mm wall thickness. After Norton (1962).

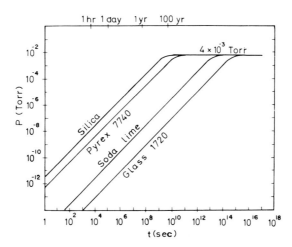

Fig. 4.15 Helium accumulation from the atmosphere in bulbs of various glasses, at 25°C. Bulbs: 330 cm³, 100 cm² wall area, 1 mm wall thickness. After Norton (1962).

4.3. Sorption

4.3.1. Sorption phenomena

In the kinetic theory (Ch. 2) and the flow (Ch. 3) of gases, it was assumed that the interactions between gas molecules and the walls of the containing vessel are mainly elastic collisions. In fact other types of interaction occur which have a profound effect upon the degree of vacuum obtained and upon the processes used to achieve the vacuum. The group of interactions in which the gas is retained by the solid (or liquid) received the name of *sorption*. This includes two mechanisms: The *adsorption* and the *absorption*.

The term *adsorption* refers to the process whereby molecules are attracted to and become attached to the surface of a solid, the resulting layer of adsorbed gas being one (or a few) molecule(s) thick. The attracting forces of the solid may be physical – *physisorption,* or chemical – *chemisorption*.

The term *absorption* refers to gas which enters into the solid in much the same manner as gas dissolving in a liquid.

The solid which takes up the gas is known as *adsorbent* or *absorbent;* the gas removed is known as *adsorbate* or *absorbate*. Terms as *adatom* or *admolecule* are also used to refer to the specific particles involved in the process. For reviews on sorption phenomena we refer to Redhead *et al.* (1968); Schrieffer (1972); Malev (1973a); Johnson and Messmer (1974).

4.3.2. Adsorption energies

Any surface of a solid or liquid exhibits forces of attraction normal to the surface, hence gas molecules impinging on the surface are adsorbed. When the pressure in the system is low enough (high vacuum) the molecules adsorbed at the wall exceed those in the volume, thus the pumping is directed toward evacuating the adsorbed gas. Conversely gas can be removed from the volume by adsorption, a process utilized in the *sorption pumps*.

Adsorption phenomena are schematically represented (Lewin, 1965) by their potential energy–distance diagram (figs. 4.16–4.19).

A molecule impinging on the surface is attracted and will assume an equilibrium position at minimum potential energy, called the *heat of adsorption,* H_A. The heat of adsorption is equal (in this simple case) to the energy of adsorption E_D.

If the adsorption is purely *physical*, it involves Van der Waals, intermolecular forces, like those occurring in liquefaction of gases. In this two-dimensional liquid, H_A is larger than the heat of liquefaction. If additional layers are adsorbed, H_A decreases until the layer becomes a three-dimensional liquid. In physical adsorption, the *attracting forces* are comparatively *weak*, and the heat of adsorption is small (max. 8 kcal/mole). Since the forces are attractive, work is done in adsorbing molecules and *heat is generated*, thus the adsorption is an *exothermic phenomena.*

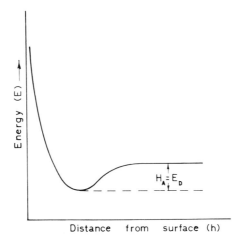

Fig. 4.16 Potential energy of a molecule in (nonactivated) adsorption.

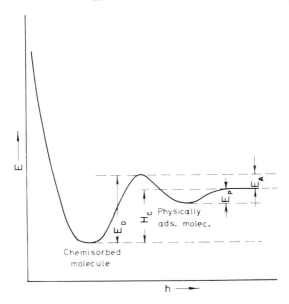

Fig. 4.17 Potential energy for activated chemisorption, with physical molecular adsorption.

In *chemisorption* the process is similar to the formation of a chemical compound with transfer of electrons. In this case the attractive forces are much larger than in the physical adsorption, *heats of chemisorption* being correspondingly higher (as large as 250 kcal/mole). The process of chemisorption does not always occur directly from the gaseous state; molecules may be initially adsorbed physically (fig. 4.17) and then, with the provision of a certain minimum energy (activation energy E_A) they may become chemisorbed. This is known as *activated chemisorption*. The energy of desorption is the sum of the heat of chemisorption H_C, and the energy of activation E_A

$$E_D = H_C + E_A$$

The process readily occurs during adsorption at a heated surface, and the total amount of gas which can be adsorbed in this manner is higher than that by non-activated processes.

The overall *chemisorption* process of molecules is *exothermic*.

The inert gases cannot be chemisorbed and they are therefore only weakly held on a surface.

Molecules may dissociate and be chemisorbed as atoms (figs. 4.18 and 4.19). This process can be endothermic or exothermic. If twice the activation energy

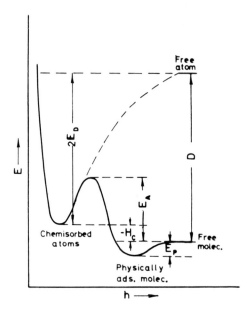

Fig. 4.18 Potential energy for activated chemisorption, endothermic adsorption.

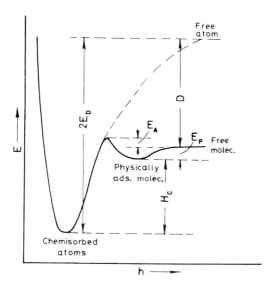

Fig. 4.19 Potential energy for activated chemisorption, exothermic adsorption.

for desorption of an atom $2E_D$ is less than the energy of dissociation D, the process is endothermic (fig. 4.18). If $2E_D > D$, the reaction is exothermic (fig. 4.19). Some values of heats of adsorption are listed in table 4.5.

The atoms have a certain mobility on the surface and they may migrate over the surface, as the necessary activation energy for this motion is only $0.2 - 0.4$ from the value of E_D. When two adsorbed atoms collide in their motion on the surface, they may recombine and desorb as a molecule. This process is known as *second-order desorption*.

For a review on chemisorption we refer to Degras (1968).

4.3.3. *Monolayer and sticking coefficient*

According to eq. (2.46) the number of molecules adsorbed on unit area per unit time is given by

$$\mathrm{d}n_a/\mathrm{d}t = 3.51 \times 10^{22} \, [P/(MT)^{1/2}] \, f \qquad (4.22)$$

where $\mathrm{d}n_a/\mathrm{d}t$ is molec./cm²·sec, f is the sticking coefficient and P is the pressure of the gas (Torr).

In addition to adsorption, molecules are desorbing from the surface at a rate given by

Table 4.5.
Heats of adsorption H (kilocalories/mole).*

Kind	Gas	Surface	H	Gas	Surface	H
Physical adsorption	Xe	W	8–9	A	C	1.8
	Kr	W	4.5	Xe	Mo	8
	A	W	1.9	Xe	Ta	5.3
Chemisorption	Rb	W	60	O_2	Ni	115
	Cs	W	64	H_2	Fe	32
	B	W	140	N_2	Fe	40
	Ni	Mo	48	H_2	Ir	26
	Ag	Mo	35	H_2	Co	24
	H_2	W	46	H_2	Pt	27
	O_2	W	194	O_2	Pt	67
	CO	W	100	H_2	Pd	27
	N_2	W	85	H_2	Ni	30
	CO_2	W	122	H_2	Rh	26
	H_2	Mo	40	CO	Ni	35
	H_2	Ta	46	H_2	Cu	8
	O_2	Fe	136	—	—	—

*References : McIrvine (1957), Hughes (1959), Kisliuk (1959), Gomer (1959), Ehrlich (1961, 1962), Young and Crowell (1962), Brennan and Hayes (1965).

$$\frac{dn_d}{dt} = \frac{N_0 \theta}{t_s} \tag{4.23}$$

where N_0 is the total number of molecules required to form a complete monolayer (see eq. 2.85), θ is the *coverage* (i.e. the fraction of possible adsorption sites which are actually occupied), and t_s is the average time spent by an adsorbed molecule at a particular site (known as *sojourn time*).

The sojourn time is shown by Frenkel (1924)

$$t_s = t' \exp [E_D/(R_o T)] \tag{4.24}$$

where t' is the period of oscillation of the molecule normal to the surface (approx. 10^{-13} sec), and E_D is the energy for desorption. A sojourn time $t_s \approx 5$ sec is typical for vacuum systems (Edwards, 1977a); a method for its measurement is described by Bailitis (1975).

Eq. (4.23) is only valid for less than a complete monolayer. Similar but more complex equations were deduced for multilayer adsorption (see Redhead *et al.*, 1968).

From eqs. (4.23) and (4.24) it results that

$$\mathrm{d}n_\mathrm{d}/\mathrm{d}t = (N_0\theta/t') \exp \left[-E_\mathrm{D}/(R_0T)\right] \qquad (4.25)$$

The exponential dependence of t_s and hence of $\mathrm{d}n_\mathrm{d}/\mathrm{d}t$, upon both E_D and T means that t_s varies over a wide range, from about 10^{-13} to 10^7 sec for small values of E_D (i.e. physical adsorption) and low temperatures ($77°\mathrm{K}$), and from 10^{-5} to 10^{30} sec for chemisorption ($E_\mathrm{D} = 10-200$ kcal/mole) at room temperature.

The equilibrium between adsorption (on the noncovered area $1-\theta$) and desorption (from covered area θ) is found from eqs. (4.25) and (4.22).

$$N_0\theta = 3.51 \times 10^{22} \left[P/(MT)^{1/2}\right] ft' \exp \left[E_\mathrm{D}/(R_0T_\mathrm{s})\right] (1 - \theta) \qquad (4.26)$$

where T is the temperature of the gas, and T_s is the temperature of the surface.

This equation can be used to express the amount adsorbed $N_0\theta$ as a function of P for constant T and T_s (the *adsorption isotherm*), as a function of T at constant P (the *adsorption isobar*), and $P=f(T)$ for constant coverage $N_0\theta$ (the *adsorption isostere*). Unfortunately, the sticking coefficient f and the desorption energy E_D are not constants. In practice, isotherms are observed experimentally and used to determine f and θ.

However eq. (4.26) predicts the following general features:

(a) The quantity of gas adsorbed increases with pressure.

(b) Very little gas can remain physically adsorbed under high vacuum conditions at room temperature.

(c) At low temperatures the quantities adsorbed (even for low E_D values) are considerable.

Typically, *sticking coefficients* at room temperature lie between $0.1-1$, and decline when monolayer coverage ($1-7 \times 10^{14}$ molec./cm²) is approached. Figures 4.20 and 4.21 illustrate this.

As it is shown in fig. 4.21, Alpert (Lee *et al.*, 1962) found that the sticking coefficient is also low at the low coverages. The explanation has been suggested that this can be attributed to the need for nucleation centers.

A review of the values of sticking coefficients is presented by Harra (1976); the (relatively high) values obtained in continuously deposited titanium films, is analyzed by Grigorov and Tzatzov (1977).

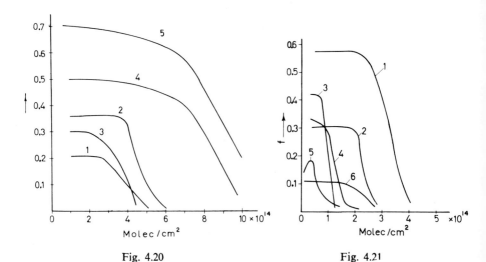

Fig. 4.20   Fig. 4.21

Fig. 4.20 Sticking coefficient f of CO on W (curves 1–4), and CO on Zr (curve 5). References (1) Eisinger (1957)–300°K; (2) Becker (1958)–300°K; (3) Ehrlich (1961)–336°K; (4) Redhead (1961)–315°K; (5) Hansen and Littman (1967)–400°K.

Fig. 4.21 Sticking coefficients f of N_2 on W. References: (1) Becker (1958)–300°K; (2) Eisinger (1959)–300°K; (3) Schlier (1961)–300°K; (4) Ehrlich (1961)–290°K; (5) Lee, Tomaschke and Alpert (1962)–300°K; (6) Madey and Yates (1967).

4.3.4. Adsorption isotherms

Langmuir (1918) used eq. (4.26) to express the adsorption isotherm as

$$\theta = \frac{bP}{1+bP} = \frac{P}{1/b+P} \tag{4.27}$$

where

$$b = 3.51 \times 10^{22} f \frac{t' \exp (E_D/R_o T_s)}{(MT)^{1/2}} \tag{4.28}$$

(for notations see eq. 4.26).

For constant f, T_s and T, b is a constant thus eq. (4.27) is an isotherm; b is a constant, expressed in Torr^{-1} units. Figure 4.22 shows such isotherms for various values of b. Since the value of b is decreasing with increasing T and T_s, the

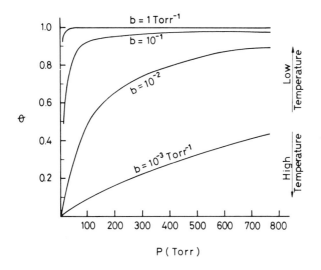

Fig. 4.22 Langmuir's isotherms.

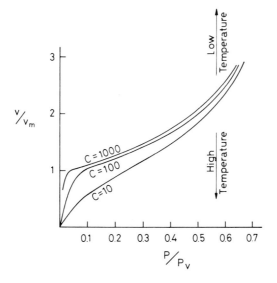

Fig. 4.23 BET isotherms.

curves for high values of b correspond to low temperatures, those for low b values
to high temperatures.

If the pressure is small compared with $1/b$, the coverage θ is proportional with P

$$\theta \approx bP$$

Langmuir's isotherm was derived for layers less than monomolecular.

The BET-isotherm (Brunauer, Emmett and Teller, 1938) was derived for *multi-molecular adsorption*. This isotherm is described by the equation

$$\frac{P}{V(P_v - P)} = \frac{1}{V_m C} + \frac{(C-1)}{V_m C} \frac{P}{P_v} \qquad (4.29)$$

where V is the volume of the gas adsorbed at a given value of θ, V_m and C are constants at any temperature. P_v is the saturation pressure of the gas at the given temperature (eq. 4.6). The BET isotherms are represented by curves of the shape shown in fig. 4.23. V_m is the volume of gas in a complete monolayer, and the constant C was found to be a function of the temperature, given by

$$C = \exp \left[(E_I - E_L)/(R_o T) \right] \qquad (4.30)$$

where E_I is the energy of adsorption of monolayer, and E_L the energy of condensation of the adsorbed gas. V_m can be determined using the method described by Genot (1975).

Various shapes of adsorption isoterms are discussed by Redhead *et al.* (1968), Elsey (1975), while measured values in specific cases were also published by Outlaw *et al.* (1974), Halama and Aggus (1974, 1975), Benvenuti, Calder and Passardi (1976).

4.3.5. *True surface*

Relating the quantity of gas which was found to be adsorbed on surfaces, it was concluded that the *true surface* is usually much larger than the apparent one. This true surface is also known as *physical surface* (A_p), while the apparent one has often the name of *geometrical surface* (A_g).

The BET equation (4.29) is a convenient method to evaluate the true surface area. A plot of the expression $PV^{-1}(P_v - P)^{-1}$ versus P/P_v will give a straight line for which the intercept is $1/(V_m C)$ and the slope $(C-1)/(V_m C)$. From these two data, the values of the constants C and V_m can be calculated. V_m indicates the volume of gas in a monolayer, thus the number of molecules forming the complete layer.

The BET method was used by Schram (1963), who determined A_p/A_g ratios as large as 900 (table 4.6). The measurement of A_p is discussed by Mikhail and Brunauer (1975).

The ratio A_p/A_g of the physical surface to the geometric one was also determined by the method of electrolytic polarization (Dushman, 1949) and these values are also listed in table 4.6.

When a metal is made the cathode in a dilute acid, and current is passed through the solution, the potential changes, due to accumulation of hydrogen on the cathode.

The phenomenon is described by the equation

$$-E = KV/A_p + \text{const.} \tag{4.31}$$

Table 4.6.
Ratio of physical (true) surface A_p, to geometric (apparent) surface A_g

Metal	Surface/shape	A_p/A_g	Reference
Pt	Bright foil	2.2	
	Bright foil, acid cleaned, flame	3.3	
	Platinized	1830	
Ni	Polished, new	75	
	Polished, old	9.7	Dushman (1949)
	Oxidized and reduced	46	
	Rolled, new	5.8	
Ag	Freshly etched dilute nitric acid	51	
	Etched, after 20 hr.	37	
	Finely sandpapered	16	
Al	Very thin foil	6	
	Anodically oxidized (20 μ)	900	
Cu	Plate (1 mm)	14	Schram (1963)
Steel	—	16	
Stainless steel	Plate (1 mm)	8	
Mo	Foil	173	
Ta	Foil	38	
			Brennan and
W	Foil	40	Graham (1965)
Ti	Foil	15	

where E is the potential, $K = - \Delta E/\Delta V$, V is the amount of hydrogen on the surface of the cathode, and A_p is the physical surface.

4.3.6. Sorption of gases by absorbents

The main absorbents used in vacuum technology are: *activated charcoal, zeolites* (molecular sieves), *silica gel*, and *alumina*. The absorption of these absorbents is explained by an adsorption, followed by the penetration of the adsorbed gas into the solid by diffusion. It can be considered that the absorption is a phenomenon similar to permeation, but having no desorption surface.

Sorption by activated charcoal. Before the development of other means of pumping to very low pressures, the technique of producing a high vacuum by absorption of the gases on activated charcoal was very frequently used. In the period 1900–1950 there are more than 4000 publications on this subject. The most frequently used charcoal is that prepared from *coconut shell*. Pieces of the shell are destructively distilled at 500–700°C, in iron containers, until vapour evolution is no longer apparent. The charcoal produced contains tarry residues, which are then removed to increase the gas absorption efficiency. This process is known as "activation", and consists in heating in steam at 800–1000°C for about one hour. The water in the activated charcoal is then driven off by heating the charcoal in a rough vacuum.

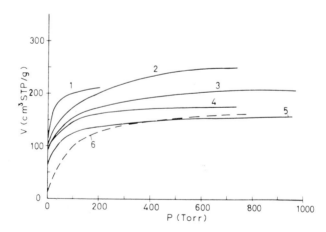

Fig. 4.24 Low temperature adsorption on charcoal. V–volume of adsorbed gas (cm³, STP) per gram of charcoal. Sample: activated charcoal, 0.2g (0.5 cm³). Curves: 1.A, −195.8°C; 2.O_2; −183°C; 3.A, −183°C; 4.N_2, −195.8°C; 5.N_2 −183°C; 6.CO −78°C. After Dushman (1949).

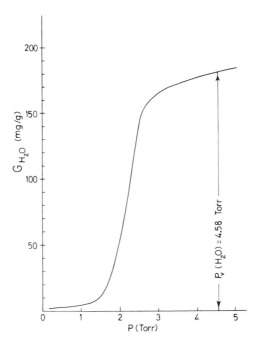

Fig. 4.25 Sorption of water vapour on charcoal at 0°C, G_{H_2O} —mg of water vapour, sorbed per gram of charcoal. After Dushman (1949).

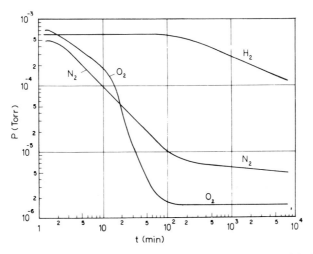

Fig. 4.26 Pressure against time curves on pumping H_2, N_2, O_2 by a liquid air cooled charcoal trap. After Espe (1955).

The ratio A_p/A_g was found to be 600–850, which corresponds to specific surfaces of the order of 1000 m²/g, (Stern *et al.*, 1965). Fig. 4.24 shows the volumes of various gases absorbed.

The sorption of *water vapour* by charcoal exhibits a behavior quite different from that observed for the less readily condensible gases (fig. 4.25). Below 1.5 Torr the sorption is small, between 1.5 and 2.5 Torr it suddenly increases. Above 2.5 the increase is slow again.

The pumping effect which can be obtained by using liquid-air cooled ($-183°C$) activated charcoal traps, is shown in fig. 4.26.

Charcoal sorbents are very effectively used in cryopumps (see also §5.6) for helium and/or hydrogen, where the charcoal is coated on surfaces cooled to 4–10 °K. The helium pumping speed obtained was about 7.7 liter/(s · cm²) of charcoal (Sedgley *et al.*, 1987; 1988). The cryopumping using charcoal was recently discussed by Coupland *et al.* (1987), Liu *et al.* (1987), Sedgley *et al.* (1987, 1988), Tobin *et al.* (1987).

Zeolites are alkali metal aluminosilicates, having tetrahedral lattices. Unlike ordinary crystals containing water of crystallization they can be dehydrated without any change in the form of their crystal lattice. As a result, molecules of different gases can occupy the spaces left vacant by the removal of water, and the zeolites are therefore very good absorbents. This is however true only for certain gases since these materials exhibit the property of *persorption*. The *persorption* may be defined as *adsorption in pores* that are only slightly wider than the diameter of the adsorbate molecules. An example of sorption curves is given in fig. 4.27. Stern *et al.* (1965) found that Linde molecular sieves present specific surfaces of 500–600 m²/g. Bonded 5A molecular sieve (1.8 mm layer) absorbs (at 4.2°K)

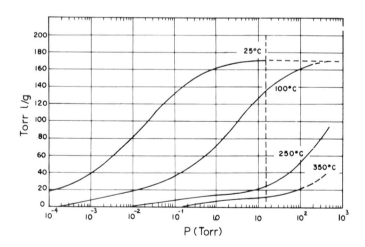

Fig. 4.27 Water vapour sorption by molecular sieve 5A.

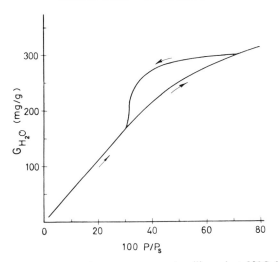

Fig. 4.28 Sorption–desorption curves for water vapour by silica gel at 20°C. In a specific range of pressures, a hysteresis loop appears. P_s-saturation pressure of H_2O. After Dushman (1949).

about 10^8 times more helium than bare copper of the same (apparent) area and temperature (Halama and Aggus, 1974).

Based on the action of zeolites (molecular sieves), the sorption pumps (§5.5) are able to pump down vacuum systems of 1–50 liters, from atmospheric pressure to the range of 10^{-2} Torr. They are in use especially in applications where the system has to be kept oil free, or where the gas which is pumped is dangerous (e.g. radioactive). The desorption of gases from molecular sieves has been analyzed by Durm and Starke (1972), Durm et al. (1972), Miller (1973a).

Silica gel is a particularly dehydrated jelly of silicic acid. It is used especially as drying agent for gases (fig. 4.28), having a specific surface of 700–800 m^2/g. *Alumina* pellets present a specific surface of about 300 m^2/g (Stern et al., 1965). Such pellets were used by Fulker et al. (1969) to prevent back streaming of oil vapours.

4.4. Desorption – outgassing

4.4.1. *Desorption phenomena*

When a material is placed in a vacuum the gas which was previously ad- or absorbed begins to *desorb*, i.e. to leave the material. The desorption is influenced by the pressure, the temperature, the shape of the material, and the kind of its surface.

The pressure has a basic influence on the desorption phenomenon since according to its tendency of increasing over or decreasing below the equilibrium, the phenomenon of sorption or that of desorption appears. Nevertheless the function between the desorption rate and the pressure is not proved at pressures much lower than the equilibrium. The difficulty consists in separating the effect of the pressure from that of the pumping time to which it is usually connected.

The *temperature* has a clear influence on desorption phenomena. Desorption is endothermic, thus it is accelerated by increasing the temperature. Electron (or photon) bombardment increases the desorption, e.g. 50–100 times when the wall is Ti or Al (Malev, 1973a).

The *shape* of the material influences the desorption either if the gas is ad- or absorbed. If the gas is adsorbed, then only the amount of the surface is the influencing factor, but if the gas has to diffuse from the interior of the material to the surface, then the third dimension (thickness) is also influencing the rate of desorption.

Since desorption phenomena are related to the *physical surface* (A_p, see table 4.6), the desorption must always be correlated to the history of treatments (polish, cleaning, etc.) of the surfaces.

4.4.2. *Outgassing*

The generation of gas resulting from the desorption is known as *outgassing*, and is expressed in terms of the outgassing constant. The outgassing constant (or specific outgassing rate) is defined as the rate at which gas appears to emanate from unit area of surface (geometric), and is usually measured in units of Torr·liter·sec^{-1}·cm^{-2}.

The experimental observations of outgassing rates can be represented by the empirical equation of the form

$$K_h = K_u + K_1 t_h^{-\gamma} \tag{4.32}$$

where K_h and K_1 are the outgassing rates at h hours and one hour respectively after the start of pumping; t_h is the time in hours after the start of pumping. K_u is the limiting value of K_h and is generally negligible unless t_h is very large.

At the beginning of the pumping, γ is large and outgassing rates fall very rapidly, but after a few minutes the fall becomes less marked, with values of γ lying between 0.5 and 2, depending upon the material. For metals, the value of γ is usually near 1. For nonmetals, γ is lying between 0.5 and 1. Values of γ greater than 1 are usually associated with an unusual surface condition, such as a porous material or rusty surface. After long pumping times (rarely less than 10 h) outgassing rates show a tendency to fall exponentially with time, until limited at K_u. A curve showing a typical time variation of outgassing constant is presented in fig. 4.29.

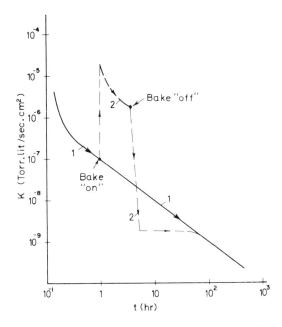

Fig. 4.29 Outgassing during pumping: (1) at room temperature, (2) using a baking cycle.

If the temperature of the material is raised (baking) the outgassing rate rises rapidly to a peak value (fig. 4.29), followed by a slower fall back to a $t_h^{-\gamma}$ variation, but at a higher level corresponding to the elevated temperature. If after a sufficiently long time, the temperature is allowed to fall to its original value the outgassing rate falls rapidly to a level which is significantly lower than that which would have existed if pumping had been at the lower temperature throughout.

Together with the acceleration of desorption, heating may also have the effect of causing activated chemisorption of physically adsorbed gas (in particular, water vapour) which can then be desorbed only by prolonged heating at much higher temperatures. Chemisorbed water vapour continues to be evolved at temperatures in excess of 300°C. It should therefore appear that a degassing programme should begin with pumping at room temperature to remove physically adsorbed water vapour, before baking is commenced.

The theory of the outgassing process was derived and summarized by Dayton (1960, 1962, 1963). The complete theory of the outgassing includes both the adsorption and the absorption simultaneously. However, in most cases the rate of diffusion is so small compared with that of desorption of adsorbed gas that the two processes may be analyzed separately and the resulting outgassing rates subsequently added. The whole complex process is considered by Carter *et al.* (1973), Elsey (1975).

The *outgassing rate* resulting from *absorbed gases* is based on the laws of diffusion, and can be obtained from solutions to eqs. (4.16) and (4.18).

In general the solution consists of the sum of an infinite series, but may be approximated to give the outgassing rate from the wall of a vessel of thickness h_o (cm) as

$$K_h = K_u + K_1/t_h{}^\gamma \tag{4.33}$$

where

$$t_h{}^\gamma = t_h{}^{1/2} - \tfrac{1}{2}\zeta^{1/2}\,[1-\exp{(t_h/2\zeta)}] \tag{4.34}$$

and

$$\zeta = \frac{\pi h_o{}^2}{5.76 \times 10^4 D_1} \tag{4.35}$$

where ζ is the diffusion time constant (hours), D_1 the diffusion coefficient (eq. 4.16).

When $t_h < \zeta/4$, then $t_h{}^\gamma = t_h{}^{1/2}$, and thus K_h varies initially as $t_h{}^{-1/2}$ but eventually falls more rapidly to approach an exponential dependence as t_h becomes large. The theoretical values of K_1 and K_u are given by

$$K_1 = [2.79 \times 10^{-3}/(3600)^{\gamma_1}]\,T\,\varepsilon_o(D_1/\pi)^{1/2} \tag{4.36}$$

where γ_1 is the value of γ when $t_h = 1$, and ε_o is the gas concentration when $t_h = 0$, measured in cm³ (STP)/cm³ of material.

$$K_u = 2.79 \times 10^{-3}\,T(D_1 b/h_o)P_o \tag{4.37}$$

The product $D_1 b$ is the permeation constant (eq. 4.20) which is measured in cm³(STP)/cm² of cross section for a thickness of 1 cm, and pressure differential of 1 Torr. The pressure P_o is the partial pressure *outside* the enclosure of the gas considered.

Considering the outgassing at 27°C of hydrogen from a steel vessel of wall thickness 1 cm, the various parameters are: $h_o = 1$ cm; $D_1 = 5 \times 10^{-9}$ cm²/sec; $\varepsilon_o = 0.1$ cm³(STP)/cm³; $b = 10^{-3}$ cm³(STP)/cm³·Torr; $j = 2$ (eq. 4.15); $P_o = 4 \times 10^{-4}$ Torr (partial pressure of hydrogen in the atmosphere).

From eq. (4.35) it results that $\xi \simeq 10^4$ hours. $t_h < \xi/4 = 2500$ hours; thus up to $t_h = 2500$ h; γ can be considered $\gamma = \frac{1}{2}$.

Substituting in eqs. (4.36) and (4.37) we have

$$K_1 = 5.6 \times 10^{-8} \text{ Torr·lit·s}^{-1}\text{·cm}^{-2} \text{ and}$$

$$K_u = 2 \times 10^{-13} \text{ Torr·lit·s}^{-1}\text{·cm}^{-2}$$

In this particular case the permeation outgassing rate will almost certainly be greater than the value of K_u calculated from eq. (4.37) because of the liberation of hydrogen at the outer surface of the vessel by action of water vapour on iron. The real value of K_u is about 5×10^{-12} Torr·lit·s^{-1}·cm^{-2}. Experimental values of K_1 are about one order of magnitude larger than that calculated above, whereas γ observed experimentally for this case is in the region of $\gamma = 1$. Thus, factors other than diffusion of gases from the interior play a considerable part in the outgassing of metals. The outgassing due to water vapour is believed to be the main additional factor.

The *outgassing rate* resulting from an *adsorbed* monolayer can be found by using eq. (4.26); but the results obtained are not always meeting the experimental values, since the coverage θ is also a function of the pressure.

For an approximation, the following equation can be used

$$K_t = [10^{-7} \ T\theta_o/t_s] \exp\left(-t/t_s\right) \tag{4.38}$$

where K_t is the outgassing rate at time t, T is the temperature, t_s the sojourn time, θ_o the coverage when $t = 0$. If $\theta_o = 1$ (monolayer), then for small values of t_s (physical adsorption), the initial outgassing rate is very high, but falls rapidly with time. On the other hand for strong chemisorption (t_s high) the initial outgassing rate is low and falls only very slowly with time.

For *water vapour*, adsorbed in *several layers* the above approach does not give consistent results. Dayton (1962) suggests that water vapour is held in the pores of the layer of oxide that is inevitably present on the surface of most metals. A semi-empirical analysis of the distribution of pore size and layer thickness leads to an expression for the outgassing rate which varies as t^{-1}.

Edwards (1977c) established an "upper bound" for the outgassing rate of "low surface-area metals" (e.g. copper, stainless steel) as $Q_{max} = 1.7 \times 10^{-5}/t$, Torr · lit · s^{-1} · cm^{-2}, where t (seconds) is the pumping time.

Schalla (1980) discusses the influence of the area/volume configuration of the material on the outgassing rates (desorption) of gas in porous materials. Flécher (1982) studied the outgassing process of metal powders. Pang et al. (1987) found that the change of grain size of Al alloys by recrystallization influences considerably the bulk outgassing of CO and CO_2. Pang et al. (1987) also found that on the surface of 304 stainless steel there are two types of water adsorption: one originating from exposure to air and the other from water rinse. These desorb at different temperatures (328 K and 366 K respectively). On the surface of 6063 Al alloy, only one type of water adsorption state exists, regardless if it is from air exposure or water rinse (it desorbs at 353 K).

4.4.3. *Outgassing rates*

Outgassing rates were determined by various authors. Figure 4.30 shows the values obtained for various metals and plastics, and their decrease with time of pumping.

These values are meant to be "true" rates, but in most of the cases are in fact "net" outgassing rates. The "net" rates (q_n) are calculated from the throughput of gas measured (which leaves the measuring vessel). The "true" rates (q_t) must include the gas desorbed but readsorbed on the surface. These are related (Hobson, 1979) by

$$q_t = q_n + 10^3 fP/(MT)^{1/2} \tag{4.39}$$

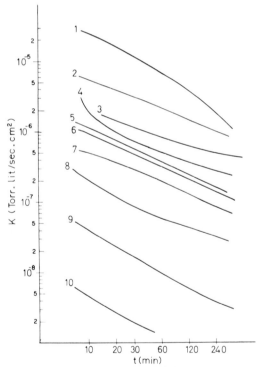

Fig. 4.30 Outgassing rates of various untreated materials at room temperature: (1) Silicone rubber; (2) Polyamide; (3) Plexiglas; (4) PVC; (5) Araldit; (6) Viton; (7) Teflon; (8) Hostaflon; (9) Copper; (10) Stainless steel. References: Geller (1958), Santeler (1958), Jackel (1962), Henry (1969), Chen (1987a), Erikson *et al.* (1988).

where M is the molecular weight and T the absolute temperature of the gas, f is the sticking coefficient, and P (Torr) is the pressure in the test chamber. It can be seen that for $q_n = 3 \times 10^{-14}$ Torr lit·s^{-1} cm^{-2}; $M = 2$ (hydrogen); $T = 300°$K; and $f = 0.1$, it results $q_t = 1.2 \times 10^{-10}$ Torr lit·s^{-1} cm^{-2}, much higher than the measured q_n value.

Methods for measuring the outgassing rates are described by Kutzner and Wietzke (1972); Elsey (1975); Messer and Treitz (1977); Komiya et al. (1979); Beavis (1982); Yoshimura (1985); Horikoshi (1987).

The gas evolution from glasses is shown in fig. 4.31. By heating the glass to 150°C in vacuum, the greatest part of the adsorbed gases is given off. The curves representing the gas evolution (fig. 4.31) have a maximum point at about 140°C for soda–lime glasses, at 175°C for lead glasses and at about 300°C for borosilicate glasses. At still higher temperatures the gas evolution is reduced, but after the temperature range between 350° and 450°C is exceeded additional gases are given off due to the decomposition of the glass.

The *outgassing rates* of various materials *depend on the state of their surface*. Figure 3.44 shows a summary of these values, for *untreated* surfaces, for *degreased*, *polished* and *baked* ones. The values for *degreased* surfaces correspond to surfaces cleaned by usual methods using liquid degreasing agents; the lowest end of the ranges correspond to vapour degreasing. *Polished* surfaces include mechanical polishing, blasting, chemical or electrochemical polishing. From the published data it was not possible to conclude if one or another of these methods systematically results in the lower outgassing rates; it rather appears that each of them can give the low values in the range, if the process is carried out carefully.

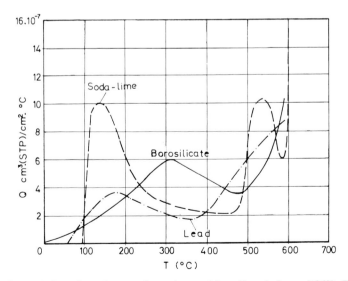

Fig. 4.31 The evolution of gas from various glasses. After Von Ardenne (1962), Espe (1966).

The values for untreated, degreased and polished states were taken for 4-8 h of pumping.

Schalla (1975) established equations expressing the influence of the surface finish, temperature and humidity on the outgassing rates.

The *baking* of nonmetals (fig. 3.44) is at temperatures of 80-100°C and baking times up to 24 h. For baked metals the upper ranges correspond to baking at about 300°C for 24 h, the middle ranges to baking at 400°C for up to 100 h, while the lowest values for stainless steel also include an additional subsequent baking at 1000°C for 3 h.

Fischer (1972) and Samuel (1970) used a pretreatment process which consists of heating the stainless steel vacuum chamber in air at atmospheric pressure for a few hours, and then while hot (about 200°C) evacuating the system.

Moraw and Dobrozemsky (1974) concluded that the actuating mechanism of the air bake-out consists in changing the activation energies required for thermal desorption. Santhanaman and Vijendran (1979) found that exposure of plastics (epoxy, high voltage cables) to nitrogen, considerably reduces their outgassing rates.

Chen (1987a) measured the outgassing rate of a chamber of Al 6063-Ex alloy and one of SUS 304 stainless steel, before and after filling the chambers with water for one day. The outgassing rates were measured initially, after bakeout, after filling with water (and drying) and after a new bakeout. At the final bakeout (24 h at 150°C) the Al alloy chamber reached 1.5×10^{-13} Torr · liter/(s · cm^2) while the stainless steel chamber (8 h at 180°C) reached only 7×10^{-13} Torr · liter/(s · cm^2). Chen et al. (1987b) explain the better behavior of the Al alloy chamber by the *diffusion* of water vapour out of the porous oxide layer, which is the main process governing the outgassing of Al alloys. In the case of the stainless steel the outgassing is a surface controlled *desorption* process.

Odaka *et al.* (1987) found that the outgassing rate of 316 L stainless steel is reduced by repeated baking/air-exposure cycles, and reaches a constant (minimum) value (e.g., 1.3×10^{-10} Pa · m^3/(s · m^2) or 1×10^{-13} Torr · liter/(s · cm^2)) after a few such cycles.

Radiation (e.g. gamma), electron or ion bombardment increase the desorption rate, usually by orders of magnitude (e.g. Calder, 1974; Fischer, 1974, 1977; Hilleret and Calder, 1977; Graham and Ruby, 1979; Edwards, 1979b).

In electron storage rings the main source of gas is a result of the synchroton radiation photons, which first produce photo-electrons and these later desorb surface held gas molecules. The mechanism, the resulting rates and the methods for decreasing the radiation induced desorption were extensively discussed, e.g. by Cummings *et al.* (1971), Falland *et al.* (1974), Benvenuti, Calder and Passardi (1976), Fischer (1977), Kouptsidis (1977), Trickett (1977, 1978), Dean *et al.* (1978), Archard *et al.* (1979), Gröbner *et al.* (1983), Kobari and Halama (1987), Kobayashi *et al.* (1987), Mathewson *et al.* (1987).

4.5. Interaction of electrons and ions with surfaces

4.5.1. *Electron scattering*

Electrons which strike a surface may be elastically reflected or may produce inelastic processes. The ratio of emitted electrons to incident current depends on the energy and direction of the incident and scattered electrons, and on the surface on which they act. Reviews on electron scattering are given by Hachenberg and Brauer (1959), Redhead *et al.* (1968), Hobson (1974), Holland *et al.* (1974), Wolsky and Czanderna (1975), Polaschegg (1978/79), Grosse *et al.* (1987a).

The general relationship between the various types of backscattered electrons is shown in fig. 4.32. If the energy E_0 of the primary electron beam is 100–500 eV, the characteristic curve (fig. 4.32, lower curve) may be divided in:

I – the peak of elastically reflected electrons,
II – the region of characteristic energy losses,

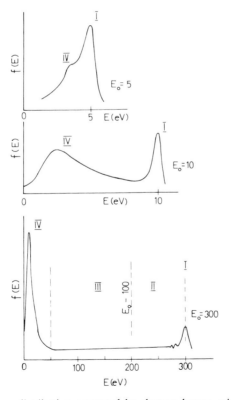

Fig. 4.32 Schematic energy distribution generated by electron beams with E_0 primary energy. After Harrower (1959); Redhead *et al.* (1968); Bauer (1972).

III – the region of inelastically scattered and emitted electrons,
IV – the region of true secondary electrons.

By decreasing the energy of the primary electron beam, the relative value of peak I compared to IV increases (upper curves, fig. 4.32) so that at very low energies only peak I appears.

At primary electron energies of 100–500 eV, peak I is used in LEED studies (see §4.5.3), while very low energy electron beams (<50 eV) are used in direct scattering measurements (Hagstrum and D'Amigo, 1960) on atomically clean surfaces in ultra-high vacuum.

The region II (fig. 4.32) of characteristic energy losses extends in the range $E_o - 100 < E < E_o$ (eV), where E_o and E are the energies of the primary and scattered electrons, respectively. This range represents reflected electrons which have lost specific amounts of their energy by: plasma losses (Klemperer and Shepherd, 1963), shell ionization (Robins and Swan, 1960), or – at low impact energies – by transfer to the vibrational state of adsorbed molecules (Schulz, 1962; Prost and Piper, 1967). Region II is used in electron loss spectroscopy (ELS) at high energies (Weber and Webb, 1969) and with some limitations (Bauer, 1972) in low energy ELS, (LEELS).

Region III (fig. 4.32) extends in the range $50 < E < E_o - 100$ (eV), and includes inelastically scattered electrons, and emitted (Auger) electrons (see §4.5.4).

Region IV extends from zero to 50 eV (arbitrary value), and includes the true secondary emission (Dekker, 1958; Hachenberg and Brauer, 1959). This group has a maximum (peak IV) at a few electron volts (scattered electrons) and forms the largest fraction of scattered electrons, except at very low primary energies (see upper curve, fig. 4.32). This range gained interest in the last time, forming the field of true secondary spectroscopy (Jenkins and Chung, 1971; Willis and Filton, 1972; Mathewson, 1974).

When electrons impinge on the surface of the solid they may desorb atoms, ions or molecules of the adsorbed gases. This process is known as EID, electron-impact desorption (Menzel and Gomer, 1964; Redhead et al., 1968; Leck and Stimpson, 1972; Moore, 1980).

4.5.2. Ion scattering

The processes discussed in §4.1–4.4 are produced by neutral particles; their thermal energy is about 0.026 eV at room temperature.

If ions having energies of $1 - 1 \times 10^5$ eV are hitting a surface, they produce elastic collision effects as: scattering, penetration, entrapment and sputtering, or inelastic processes as: neutralization and electron ejection. Reviews of these phenomena were published by Colligon (1961), Kaminsky (1965), Snoek and Kistemaker (1965), Redhead et al. (1968), Kistemaker et al. (1968), Buck and Poate (1974), Hobson (1974), Holland et al. (1974), McCracken (1974), Holm and Storp (1976), Polaschegg (1978/79), Akaishi et al. (1987).

In the low energy range (<30 eV) the predominant effect is the *scattering* of the bombarding ion, back into the vacuum space. When an ion of moderate energy ($<10^3$ eV) and sufficiently high ionization potential approaches a conducting (metal) surface, an electron from the conduction band of the solid can combine with the ion, producing its *neutralization*, but a second electron could be sufficiently excited by this process to surmount the work function barrier, and be *ejected* from the solid. These electrons are known as Auger electrons (see §4.5.4). In the energy range $30 - 3 \times 10^4$ eV, in addition to *scattering* (Goff, 1973) and electron ejection, the ejection of atoms of the solid, known as *sputtering* becomes important (Behrisch, 1964). The penetration of the ions into the solid is 1–10 lattice constants at moderate energies (up to 10^3 eV) and increases to about 100 lattice constants at higher energies (3×10^4 eV) The probability of *entrapment* of ions increases with the energy of the parucle, approaching unity at energies of about 3×10^4 eV (Grant and Carter, 1965).

The *ion scattering spectroscopy* as used for surface composition analysis is discussed e.g. by Goff (1973), Rusch and Erickson (1976). Ion entrapment is the phenomenon on which the *ion pumping* (see §5.4) is based. Ion entrapment also produces pumping effects (§6.7.2) in ion gauges (Alpert, 1953; Alpert and Buritz, 1954). This effect was analyzed by Kornelsen (1960).

Sputtering is the phenomenon on which *sputter-ion pumps* (§5.4.4) are based. The sputtering effect was studied by Wehner (1955, 1974), Moore (1960), Kay (1962), Logan *et al.* (1977), and is extensively used as one of the *vacuum coating* methods; these coating techniques are discussed by Holland (1956), Maissel and Glang (1970), Maissel and Francombe (1973), Varga and Bailey (1973), Reiber and Lantaires (1973), Laville Saint-Martin (1973), Holland and Cox (1974), Wehner (1974), Fraser and Cook (1977), Harding (1977), Holland (1978).

Ion bombardment is an effective means for *cleaning* surfaces for high and ultra-high vacuum studies, or before vacuum coating (Hagstrum and D'Amigo, 1960; Grant and Carter, 1967; Govier and McCracken, 1970).

Ion bombardment cleaning is extensively used in accelerators, storage rings and plasma machines; the cleaning processes used in various applications are discussed by Jones *et al.* (1973), Calder (1974), Mathewson (1974), Lambert and Comrie (1974), O'Kane and Mittal (1974), Winter (1975), Schiller *et al.* (1976), Blechschmidt (1977, 1978), Calder *et al.* (1977), Fischer (1977), Hartwig and Kouptsidis (1977c), Hilleret and Calder (1977), Logan *et al.* (1977), Mathewson *et al.* (1977), Sørensen and Whitton (1977), Bouwman *et al.* (1978), Dean *et al.* (1978), Gomay *et al.* (1978, 1979), Dylla *et al.* (1979), Störi (1983), Dylla *et al.* (1984), Waelbroek *et al.* (1984), Hseuh *et al.* (1985), Kobari and Halama (1987), Mathewson *et al.* (1987), Suemitsu *et al.* (1987), Dylla (1988), Itoh *et al.* (1988).

Ion bombardment (glow discharge) cleaning of a vacuum vessel requires one or more electrodes placed according to the geometry of the walls. The surfaces to be cleaned are usually grounded relative to the excitation source. The power source can be a d.c. supply (800–1000 V), an r.f. supply (e.g. 200 kHz) or a microwave power supply (e.g. 1 to 3 GHz). Hydrogen and argon are the more often used gases for the discharge, but Ar/O_2, O_2 and He/O_2 are also used (Dylla, 1988). The pressure of the gas (during the discharge) is in the range $5 \times 10^{-4} - 1 \times 10^{-2}$ Torr. A high gas throughput has to be maintained during the discharge, to remove the products of the cleaning. Turbomolecular pumps are the most appropriate for the purpose. The current densities required are 10–25 $\mu A/cm^2$ for hydrogen glow discharge cleaning (Waelbroek *et al.*, 1984), and 100 $\mu A/cm^2$ for argon cleaning (Calder *et al.*, 1977). The required ion dose for stainless steel systems is in the range $10^{18} - 10^{19}$ ions/cm^2 (Dylla, 1988), while for the discharge cleaning of Al systems a total argon ion dose of 2×10^{18} ions/cm^2 is mentioned (Mathewson *et al.*, 1987). It was found (Itoh *et al.*, 1988) that about 3×10^{-3} atoms/ion are trapped in the surfaces, which explains the relatively high reemission of argon occurring later in the cleaned systems.

4.5.3. *Low energy electron diffraction* (LEED)

Low energy electrons (100–500 eV) have a penetration depth up to 1–2 mono-layers, thus such beams are diffracted from the plane array of atoms forming the surface of the crystal. This permits us to use such electrons in order to obtain information about the surface.

The diffraction equipment for LEED studies (fig. 4.33) consists of an electron source (filament), a means of collimating and then focusing the primary beam onto the crystal, and a system for detecting the diffracted electrons. The diffracted beams pass through grids, and then are accelerated to a fluorescent screen which

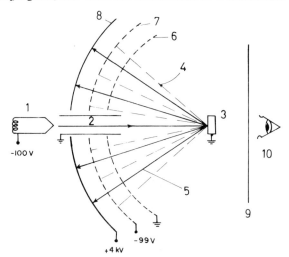

Fig. 4.33 Schematic of a LEED system : (1) Filament; (2) primary beam; (3) specimen; (4) slow secondary electrons; (5) elastically scattered electrons; (6) shield grid; (7) suppressor grid; (8) fluorescent screen; (9) quartz window; (10) observer. After Maissel and Francombe (1973).

produces a pattern that can be observed (photographed) through the front window. One function of the grids is to shield the incident and diffrated beams from the 4 kV on the fluorescent screen. The grid closest to the screen is a suppressor grid and serves as a filter to reject electrons that have lost energy after interaction with the target crystal, thus only the elastically scattered electrons, which produce the diffraction patterns can reach the screen.

In order to be able to study clean surfaces LEED systems use ultra-high vacuum $(10^{-9} - 10^{-10}$ Torr).

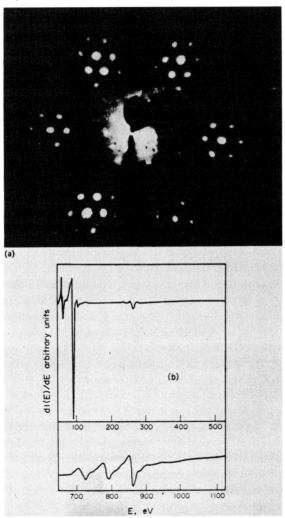

Fig. 4.34 (a) LEED pattern of Si(111) surface, $E_0 = 45$ eV; (b) Auger spectrum of the same surface ($E_0 = 2$ keV, grazing incidence). Lower curve 10 times amplified. Reprinted from Bauer (1972), by permission of Pergamon Press Ltd., Oxford.

The principles of electron diffraction were treated by Pinsker (1953), the LEED systems are discussed in many publications, e.g. Germer (1965), McRae (1966), Ducros (1968), Davison (1971), Prutton (1971), Cunningham and Weinberg (1978). A LEED pattern is shown in fig. 4.34.

Electron diffraction at normal incidence becomes less and less sensitive to scattering by surface atoms as electrons are more energetic and thus penetrate deeper into the material. This is counteracted by using glancing incidence (0.5–3°) at high energies (10–100 keV) in the method known as *reflection high energy electron diffraction* or RHEED (Sewell and Cohen, 1965). Facilities for LEED and RHEED are conveniently combined in the same instrument (Heppel, 1967), and there is a growing tendency to add to this the equipment for AES (§4.5.4) as well (Todd, 1973).

4.5.4. *Auger electron spectroscopy* (AES)

The method of surface analysis by Auger electron spectroscopy was suggested by Lander (1953), improved by Harris (1968) and treated in a large number of publications, e.g. Siegbahn (1965), Taylor (1969), Chang (1971), Davison (1971), Bauer (1972), Todd (1973), Noller et al. (1974), Sickafus (1974), Mathewson (1974), O'Kane and Mittal (1974), Fäber (1976), Gomay et al. (1978), Archard et al. (1979).

The AES method is based on the peaks which appear in the electron energy spectra of various materials bombarded by electrons (or ions), and which are due to electrons ejected by the process first explained by Auger (1925). If an atom is ionized in an inner shell, the vacancy is filled very rapidly ($10^{-17} - 10^{-12}$ sec) by an electron of one of the outer shells. The energy liberated in this transition can be radiated off in the form of characteristic X-rays or can be transferred to another electron. This electron (the Auger electron) is emitted from the surface if the transferred energy is larger than the ionization energy of this electron.

The basic requirement of an AES system is the ability to energy analyze electrons in the range from a few electron volts to about 1500 eV. The most common geometry of the analyzer used is shown in fig. 4.35. This analyzer is based upon LEED optics (see fig. 4.33). The incident beam can either pass along the axis of the optics, striking the sample at normal incidence, or enter from a direction outside the angle of optics striking the sample at a glancing angle. With the glancing angle arrangement the sensitivity is increased since the excitation is more concentrated near the surface, thereby increasing the chance of escape of the low-energy Auger electrons. Various grid arrangements and modulation techniques may be used to obtain the derivative of the energy distribution curve showing the Auger peaks (figs. 4.36; 4.34b). Employing four concentric grids and a collector (fig. 4.35) the grid nearest to the specimen is grounded as are the specimen and all neighboring components, in order to give a field-free region between the grid and specimen. This ensures that electrons emitted at the centre of curvature of the optics will travel in radial paths toward the first grid. The next two grids are retarding grids which prevent electrons with very low energy from passing them.

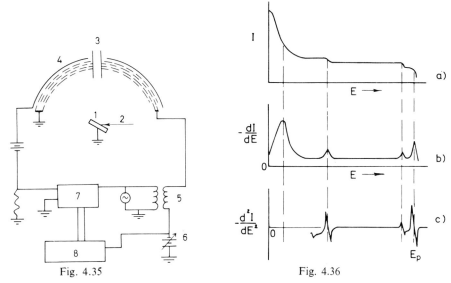

Fig. 4.35   Fig. 4.36

Fig. 4.35 Schematic diagram of a retarding field analyzer: (1) specimen; (2) electron beam; (3) LEED gun (if present); (4) collector (fluorescent screen); (5) a.c. modulation; (6) d.c. sweep supply; (7) lock-in amplifier; (8) x-y recorder. After Taylor (1969).

Fig. 4.36 Collector currents in the analyzer (fig. 4.35) in different modes of operation. (a) Retarding field plot; (b) Energy distribution (fundamental); (c) Derivative of energy distribution (second harmonic). E_p—primary beam energy. After Taylor (1969).

Two grids of this kind are used, in order to sharply define the radial retarding field, and obtain a high resolution. To these two grids is also applied the ac modulation voltage which together with the dc retarding field enables energy analysis to be carried out (see fig. 4.36). The fourth grid is held at ground potential and serves primarily as an ac shield to reduce the capacitive coupling to the collector of the ac voltages applied to the retarding grids. The collector is a fluorescent screen (in order to be used for LEED as well) biased a few hundred volts with respect to ground.

If the specimen is excited by primary electrons of energy E_p (fig. 4.36) and the current to the collector is recorded as a function of the retarding field voltage, sweeping from ground potential to the potential of the cathode, a retarding field plot is obtained (fig. 4.36). To obtain the energy distribution it is necessary to differentiate this plot with respect to retarding voltage (fig. 4.36). This is accomplished by applying a small ac modulation voltage to the retarding grids and tuning the detector to the frequency of the modulation. In order to obtain the derivative of the energy distribution curve the detector is tuned to the second harmonic of the modulation frequency. This gives the significant peaks (fig. 4.36) showing the characteristic Auger electron energies, which permit determination of the elemental composition of the surface of the specimen, including adsorbed molecules on the surface. An example of an Auger spectrum is shown in fig. 4.34, together with the LEED pattern of the surface.

CHAPTER 5

Production of low pressures

5.1. Vacuum pumps

5.1.1. Principles of pumping

Since vacuum technology extends on so many ranges of pressure (§1.1), no single pump has yet been developed, which is able to pump down a vessel from atmospheric pressure to the high vacuum or ultra-high vacuum range.

Although all the vacuum pumps are concerned with lowering the number of molecules present in the gas phase, several different principles are involved in the various pumps which are used to attain low pressures. Vacuum pumping is based on one or more of the following principles:

Compression–expansion of the gas, in piston pumps, liquid column or liquid ring pumps, rotary pumps, Root's pumps;

Drag by viscosity effects, in vapour ejector pumps;

Drag by diffusion effects, in vapour diffusion pumps;

Molecular drag, in molecular pumps;

Ionization effects, in ion pumps;

Physical or chemical sorption in sorption pumps, cryopumps and gettering processes.

5.1.2. Parameters and classifications

The selection of the pumping principle or of the pump to be used is defined by its specific parameters. The main parameters are: the *lowest pressure*, the *pressure range*, the *pumping speed*, the *exhaust pressure*. In the ultra-high vacuum range two other parameters are added: the *selectivity* of the pump and the composition of the *residual gas*.

200

The *lowest pressure* which can be achieved by a pump at its inlet, is determined either by the leakage in the pump itself, or by the vapour pressure of the fluid utilized in the pump. This pressure determines the low pressure end of the pressure range in which the various pumping types are effective (fig. 5.1).

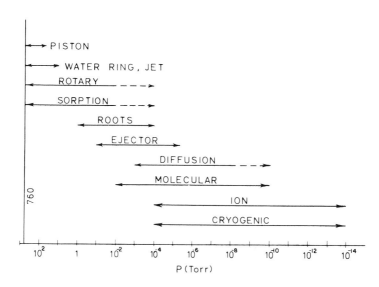

Fig. 5.1 Pressure ranges of vacuum pumps.

The *pressure range* of a single pump is that range in which the pumping speed of that pump can be considered useful (fig. 5.2). Pumps of the same type but of different sizes or constructions may have adjacent pressure ranges, so that the *pressure range of a specific pumping method* can be larger (fig. 5.1) than that of an individual pump (fig. 5.2).

The *pumping speed* of the pumps is not constant (as it was considered in §3.7), but is a function of the pressure. The pumping speed vs. pressure curve of pumps has either a shape of a curve decreasing as the pressure decreases (e.g. rotary pumps), or of a curve increasing first with decreasing pressure, reaching a maximum and then decreasing as the pressure decreases (e.g. diffusion pumps, Root's pump).

The classification of the pumps, according to the *pressure range*, is summarized in fig. 5.1, while the typical variation of the pumping speed is shown in fig. 5.2, expressed as percents of the maximum pumping speed of each type of pump.

The *exhaust pressure* is the pressure against which the pump may be operated.

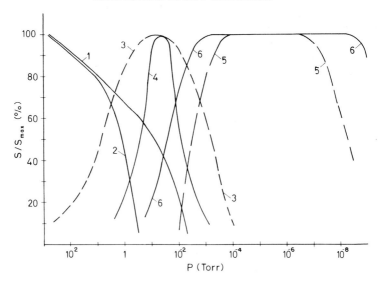

Fig. 5.2 The pumping speed range S of several pumps, in terms of their maximum pumping speed S_{max}. 1. Single stage rotating-vane pump (without gas ballast); 2. Single stage gas ballast pump; 3. Root's pump; 4. Ejector pump; 5. Diffusion pump; 6. Molecular pump.

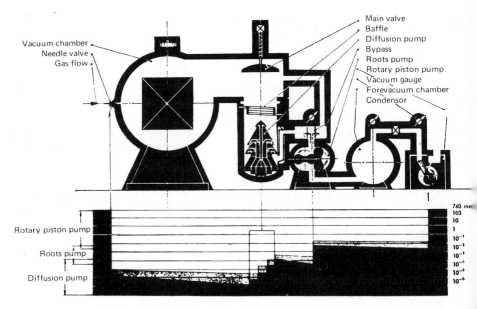

Fig. 5.3 Schematic cross section of an industrial vacuum system, and the graph of the pressure at the various points in the system. Reprinted from Lucas (1965), by permission of Pergamon Press Ltd., Oxford.

From this point of view the vacuum pumps may be broadly divided into three classes:

– Pumps which *exhaust to atmosphere*, usually known as *roughing* or *backing* pumps. The removal of the atmospheric air from the system to some acceptable operating pressure is referred to as *roughing* out the system. The maintenance of a required low pressure at the outlet of another pump, is referred to as *backing*. Mechanical rotary pumps, and ejectors are the typical roughing and backing pumps.

– Pumps which *exhaust* only to *sub-atmospheric* pressures, *require* a backing pump (in series) to exhaust to atmosphere. Diffusion, Root's pumps, and molecular drag pumps are of this type, which require a backing pump.

– Pumps which *immobilize* the gases and vapours within the system require no outlet. These are the pumps based on ionization or on sorption.

A typical laboratory vacuum system, with roughing and backing stages is shown in fig. 3.35. Figure 5.3 shows an industrial vacuum system, in which the process chamber is maintained at a low pressure by a pumping system consisting of three vacuum pumps in series: a 3-stage diffusion pump, backed by a Root's type, which is backed by a rotary plunger pump. The rotary piston (plunger) pump, being capable of unassisted discharge to atmosphere, is used initially to reduce the system pressure to about 10^{-1} Torr and then is used to back up the diffusion pump. The diffusion pump can reduce the pressure in the clean chamber to less than 10^{-5} Torr, but when the process is outgassing or there is an admission of control gas, the pressure can be maintained at only about 10^{-4} Torr.

Almost all the books mentioned in §1.4.1 have chapters dedicated to vacuum pumps. A very detailed discussion of the subject is presented by Power (1966). Pumps for ultra-high vacuum are reviewed by Weston (1978).

Pumps and pumping processes were discussed by Adam *et al.* (1980), Nelson (1980), Bhatia and Chéremisinov (1981), Currington *et al.* (1982), Liversey and Budgen (1982), Hablanian *et al.* (1987), Ota and Hirayama (1987), Wycliffe (1987), Hablanian (1988), Wong *et al.* (1988).

Hablanian (1984) comments on the history of vacuum pumps, while Hobson (1984) predicts the future of the pumps and their applications. The vocabulary to be used in connection with vacuum pumps and pumping is stated in ISO/DIS-1981.

5.2. Mechanical pumps

5.2.1. *Liquid pumps*

Most of the vacuum pumps using liquid to compress and exhaust have only historical interest. We will mention here only the Sprengel pump, the water-jet pump, and the Toepler pump, the latter two being still used in laboratory.

The *Sprengel pump* has only the historical interest of being used in the first lamp factories. It was recently discussed by Myer (1972). This pump was based on the principle shown in fig. 5.4. The mercury drops introduced in the vertical capillary T, capture between them air bubbles. In this way the system evacuates air from the side tube C and exhausts it through the mercury at the bottom, to the atmosphere.

The *water–jet* pump is a familiar practice in laboratory work, especially in filtering operations. Water supplied from a fast-running tap is fed into the nozzle at A (fig. 5.5). This water stream then emerges at high velocity from the converging jet B. The jet is surrounded by a cone to prevent splashing and also guide the water stream to waste at C. A side tube D is connected to the vessel to be evacuated. Molecules of the gas are trapped by the high speed jet and forced out into the atmosphere. By this means, pressures down to 10–17 Torr are attainable, the limit being due to the vapour pressure of the water (see table 2.2).

The principle of the *Toepler pump* is fundamentally the same as that applied by Torricelli in his famous experiment. The air from E (fig. 5.6) is "pumped" by alternately raising and lowering the mercury reservoir R, which is connected to the tube of barometric length placed below B. At each upward "stroke" the gas in B is closed from E and forced, through the tube F, into the atmosphere at M. Then, on the downward "stroke", the pressure in E is lowered by expansion of the gas into B. The glass valve G (see fig. 7.57) prevents the mercury from entering the vessel E, in the upward stroke. With the Toepler pump, pressures down to 10^{-5} Torr can be obtained, except the mercury vapour pressure which is about 10^{-3} Torr (see table 2.2). The great disadvantage of the Toepler pump is its very low pumping speed. This was somehow increased recently in the "automatically operated" Toepler pumps, e.g. Moore and Frahm (1974), Kanellopoulos (1979).

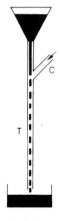

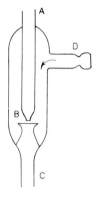

Fig. 5.4 Sprengel pump.   Fig. 5.5 Water ejector pump

5.2.2. *Piston pumps*

The piston pumps (fig. 5.7) have valves so arranged that air is pumped out of vessel A. As the piston is raised from the lowest position, the valve V_2 closes, and the motion of the piston then reduces the pressure in B. The pressure difference between A and B will open valve V_1 and gas will pass from A to B. As the piston descends, the pressure in B increases, V_1 closes, V_2 opens and the gas in B escapes through V_2.

In one stroke the volume of gas V_A is expanded to $V_A + V_B$, thus the pressure is reduced from P to P_1,

$$P_1/P = V_A/(V_A + V_B) \tag{5.1}$$

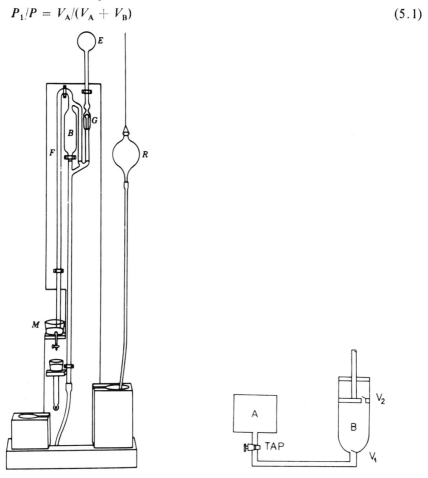

Fig. 5.6   Fig. 5.7

Fig. 5.6 Toepler pump. Reprinted from Dushman and Lafferty (1962), by permission of J. Wiley & Sons, Inc., New York.

Fig. 5.7 Piston pump (principle).

and after n strokes to P_n

$$P_n/P = [V_A/(V_A + V_B)]^n \tag{5.2}$$

The minimum attainable pressure is limited especially by the dead space below the piston i.e. the space between the valves V_1 and V_2 (fig. 5.7) when the piston is in its lowest position. If V_d represents the dead volume, then the minimum attainable pressure is

$$P_o = 760V_d/V_B \tag{5.3}$$

since at the end of the stroke the pressure in B must be atmospheric in order to open V_2. Piston pumps have $V_d/V_B = \frac{1}{8} - \frac{1}{10}$, thus their lowest pressure is about 100 Torr.

5.2.3. Water ring pumps

Water ring pumps are constituted by a multi-blade impeller, which is eccentrically mounted relative to the pump casing (fig. 5.8).

When the impeller rotates the liquid is thrown outwards to form a ring which rotates inside the pump casing. The pockets between the blades of the impeller are completely filled with liquid when at the top position, but as the pocket rotates, the liquid moves away from the axis and draws gas through the suction port. As rotation continues the liquid returns toward the axis and forces the gas out through the discharge port.

The sealing liquid, which is generally water, is heated by the action of the pump. It is either run to waste and replaced or circulated through a cooler. Water ring pumps have a nominal operating pressure of about 30 Torr, and are used in large systems where such pressures are sufficient. The range of pumping speeds of these pumps extends up to 6000 m³/h (Ebdale, 1978). Liquid ring pumps are also discussed by Powle and Kar (1983), Cole (1987).

5.2.4. Rotating-vane pumps

The rotating-vane pump, known also as "rotary pump", is constituted of a stator and an eccentric rotor which has two vanes (blades) in a diametral slot, (figs. 5.9 and 5.10). The stator is a steel cylinder the ends of which are closed by suitable plates, which hold the shaft of the rotor. The stator is pierced by the inlet and exhaust ports which are positioned respectively a few degrees on either side of the vertical. The inlet port is connected to the vacuum system by suitable tubulation usually provided with some kind of dust filter. The exhaust port is provided with a valve, which may be a metal plate moving vertically between

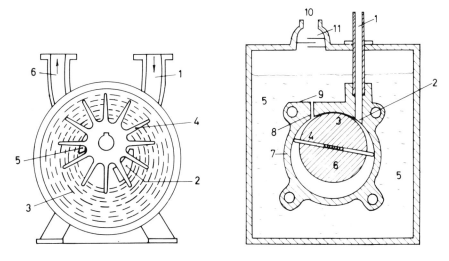

Fig. 5.8　　　　　　　　　　　　Fig. 5.9

Fig. 5.8 Cross section of a water ring pump. 1. Suction; 2. Suction port 3. Water ring; 4. Impeller; 5. Discharge port; 6. Discharge.

Fig. 5.9 Cross section of a rotating-vane pump. 1. Inlet tube; 2. Inlet port; 3. Top seal; 4. Vanes; 5. Oil; 6. Rotor; 7. Stator; 8. Exhaust port; 9. Exhaust flap valve with backing plate; 10. Exhaust outlet; 11. Oil splash baffles.

arrester plates, or a sheet of Neoprene, which is constrained to hinge between the stator and a metal backing plate.

The rotor consists of a steel cylinder mounted on a driving shaft. Its axis is parallel to the axis of the stator, but is displaced from this axis (eccentric), such that it makes contact with the top surface of the stator, the line of contact lying between the two ports. This line of contact known as the *top seal* (fig. 5.9) between rotor and stator must have a nominal clearance of 2–3 microns (see fig. 7.13). A diametrical slot is cut through the length of the rotor and carries the vanes (fig. 5.10). These are rectangular steel plates which make a sliding fit in the rotor slot and are held apart by springs which ensure that the rounded ends of the vanes always make good contact with the stator wall. The whole of the stator–rotor assembly is submerged in a suitable oil. Lubrication problems of vacuum pumps are discussed by Webb (1974), oil back-migration by Baker *et al.* (1972), Harris (1978), Hablanian *et al.* (1987), Laurenson *et al.* (1988), while selection of fluids for rotary pumps is discussed by Kuhn and Bachmann (1987). Pumps with direct drive are discussed by Nelson (1980).

The action of the pump is shown in Fig. 5.11. As vane A passes the inlet port (fig. 5.11a), the vacuum system is connected to the space limited by the stator, the top seal, the rotor and vane A. The volume of this space increases as the vane sweeps round, thus producing a pressure decrease in the system. This continues until vane B passes the inlet port (fig. 5.11b), when the volume of the gas evacuat-

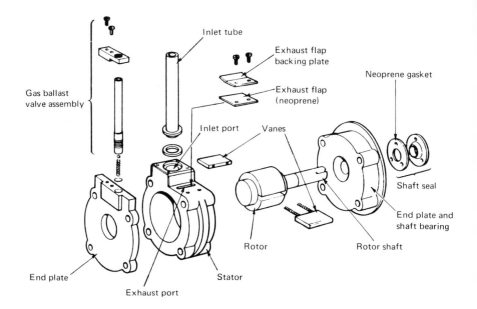

Fig. 5.10 Exploded view of a rotating-vane pump. Reprinted from Ward and Bunn (1967), by permission of Butterworths Publ. Co., London.

ed is isolated between the two vanes. Further rotation sweeps the isolated gas around the stator until vane A passes the top seal (fig. 5.11c). The gas is now held between vane B and the top seal, and by further rotation it is compressed until the pressure is sufficient (about 850 Torr) to open the exhaust valve, and the gas is evacuated from the pump.

Since both vanes operate, in one rotation of the rotor a volume of gas equal to twice that indicated in fig. 5.11b is displaced by the pump. Thus, the volume rate at which gas is swept round the pump, referred to as *pump displacement* S_t is

$$S_t = 2Vn \tag{5.4}$$

where V is the volume between vanes A and B (fig. 5.11b), and n is the number of rotations per unit time (usually 350–700 r.p.m.). Pumps with direct drive have 1500–1700 r.p.m.

The contacts of the vanes and rotor with the stator form three separate chambers each containing gas at different pressure. These contacts must therefore make vacuum-tight seals, especially for the top seals which must support more than one atmosphere pressure difference. For this reason the inner surfaces of the stator, that of the rotor and vane, are very carefully machined. Hence, great care must be taken to ensure that no abrasive material or gas which is likely to corrode the metal surfaces enters the pump chamber.

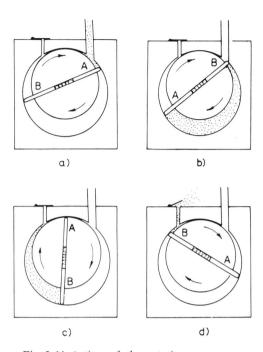

a)   b)

c)   d)

Fig. 5.11 Action of the rotating-vane pump.

In theory, the lowest pressure achieved by the pump is determined only by the fact that the gas is compressed into a small but finite "dead volume". When the system pressure becomes so low that, at maximum compression, the gas pressure is still less than that of the atmosphere it cannot be discharged from the pump. Subsequent pumping action re-expands and recompresses the same gas without further decreasing the pressure in the system. The ratio of the exhaust pressure to the inlet pressure is termed the *pump compression ratio* (see also eq. 5.3). Thus, to produce pressures of the order of 10^{-2} Torr, pumps having compression ratios of the order of 10^5 are required. In addition to lubrication and sealing, the oil also performs the function of filling the dead volume, thus increasing the compression ratio. Oil suck-back preventing systems are discussed by Harris and Budgen (1976).

The lowest (ultimate) pressure achieved by a single stage rotary pump is about 5×10^{-3} Torr, as measured by a McLeod gauge (permanent gas pressure). If the pressure is measured by a Pirani gauge (total pressure), pressures of about 10^{-2} Torr will be recorded for the same single stage pump. This higher reading is due to the vapour pressure of the sealing oil or its decomposition products in the pump. Modern two-stage pumps can achieve 1×10^{-5} Torr (about 1×10^{-3} Pa) inlet pressure while discharging to atmosphere, i.e. achieve a compression ratio near to 10^8.

Parallel connection of two identical rotor–stator systems will provide twice the displacement but the same ultimate pressure. *Series connection* provides the same displacement but greater pumping speeds at low pressures (lower ultimate pressure). A two-stage pump may reach 10^{-4} Torr (McLeod) or 2×10^{-3} (Pirani) ultimate pressure.

The *pumping speed* curves (fig. 5.12) plotted for rotary pumps do show a fairly constant speed at the higher pressures (760–10 Torr), but this speed falls off noticeably at the lower pressures and becomes zero at the ultimate pressure, as described by eq. (3.255). The performances of rotary pumps were analyzed by Kendall (1982b), Hablanian (1986, 1987, 1988).

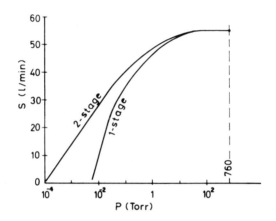

Fig. 5.12 Typical pumping-speed pressure curve of rotating-vane pumps.

Gas ballast. When a rotary pump is set to pump condensible vapour, like water vapour, the vapour is compressed and its pressure is increased, thus it condenses. The liquid (water) mixes with the pump oil, and as oil circulates in the pump, it carries some of the contaminating liquid with it to the low pressure side where it will evaporate, limiting the attainable pressure. In order to avoid this phenomenon, Gaede (1935) provided the pump with a *gas ballast* valve (fig. 5.13), which admits a controlled and timed amount of air into the compression stage of the pump. This extra air is arranged to provide a compressed gas–vapour mixture, which reaches the ejection pressure before condensation of the water vapour takes place. The principle of the gas-ballast is shown (Jaeckel, 1950; Pirani and Yarwood, 1961) in fig. 5.13, where A, B, C, D, E, F represent successive positions of the leading edge of vane V.

Consider a gas-ballast pump with:

P_b – the total pressure of the ballast air; P_h – the partial pressure of vapour in the ballast air; P_g – the partial pressure of the permanent gas (air) at the pump inlet; P_v – the partial pressure of vapour (water) at the pump inlet; P_s – the saturation vapour pressure of the vapour; P_e – the ejection pressure required to raise the exhaust valve against the spring, atmosphere and oil above it; S – the pumping speed at the inlet; S_b – the speed (rate) at which air is admitted through the gas ballast; T – the pump temperature; T_o – the ambient temperature; C_r – compression ratio.

The compression ratio C_r, i.e. the ratio of the maximum to the minimum swept volume between the rotor and the stator is

$$C_r = P_e/(P_v + P_g) \tag{5.5}$$

The maximum value that P_v can have without condensation of the vapour during compression results from

$$C_r = P_s/P_v \tag{5.6}$$

and from eqs (5.5) and (5.6), it results

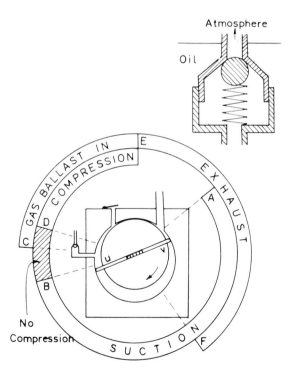

Fig. 5.13 The principle of the gas-ballast pump, and its one-way gas-ballast valve. After Jaeckel (1950), Pirani and Yarwood (1961).

$$P_v = P_s P_g / (P_e - P_s) \tag{5.7}$$

For example if the pump temperature is $60°C$, $P_s = 150$ Torr. With $P_e = 1.4$ atm $= 1060$ Torr

$$P_v = [150/(1060 - 150)]P_g = 0.155 P_g$$

It follows that in this pump condensation of water will occur if the partial pressure of water vapour at the pump inlet exceeds 16 per cent of the air pressure. If gas ballast is used, the equality between eqs. (5.5) and (5.6) gives

$$P_v = [P_s P_g + P_s P_b (S_b/S)]/(P_e - P_s) \tag{5.8}$$

and

$$S_b = (S/P_b) [(P_e P_v / P_s) - P_v - P_g] \tag{5.9}$$

Equation (5.8) is slightly changed if the vapour content of the gas ballast (P_h) and temperatures T, T_o are also considered. In this case

$$P_v = \left(\frac{P_s}{S}\right)\left(\frac{T}{T_o}\right)\left[\frac{P_g S + (P_b - P_h)S_b}{P_e - P_s} - \frac{S_b P_h}{P_s}\right] \tag{5.10}$$

The use of gas ballast increases the ultimate pressure of the pumps (fig. 5.14). However this disadvantage is unimportant in practice because the gas ballast valve is usually open only during the initial stages of pumping.

5.2.5. Sliding-vane pumps

These pumps have a single vane which slides in a slot cut in the stator between the inlet and exhaust ports. There are two types of this kind of pump (figs. 5.15 and 5.16).

In one of these types (fig. 5.15) the vane slides in its casing and on the eccentric cylindrical rotor. The reciprocating vane mounted in the casing of the stator is maintained by springs in contact with the rotor, and provides a seal between inlet and outlet ports.

Another type of sliding vane pump is shown in fig. 5.16. In this type the vane is fixed by a bearing to the outer sleeve of the rotor. The rotor rotates eccentrically, which makes the vane slide in its slot in the casing.

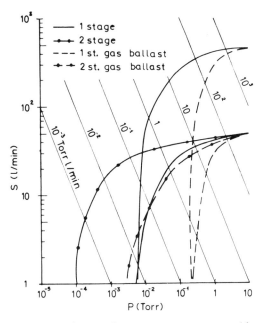

Fig. 5.14 Pumping speed curves of one and two stage rotary pumps, without and with gas ballast.

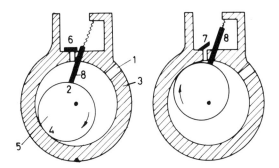

Fig. 5.15 Sliding-vane pump, with vane sliding both in casing and on rotor. 1. Inlet port; 2. Vane–rotor seal; 3. Stator; 4. Rotor–stator seal; 5. Rotor; 6. Discharge valve; 7. Discharge valve in exhaust position; 8. Sliding vane.

The whole assembly is submerged in oil which completes the vacuum seals and provides lubrication. The pumping cycle is shown in fig. 5.17. The volume of gas swept around the pump at each rotation is that between the stator and the rotor at the instant when the rotor passes the vane slot.

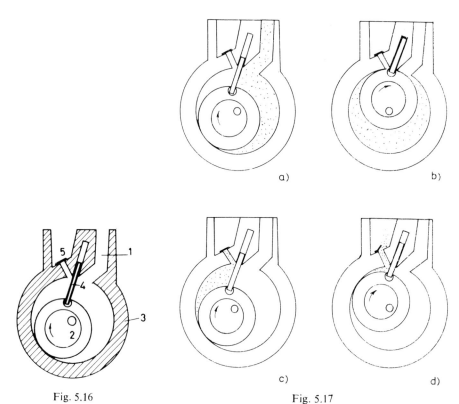

Fig. 5.16   Fig. 5.17

Fig. 5.16 Sliding-vane pump, with vane sliding in casing. 1. Inlet port; 2. Rotor; 3. Stator; 4. Vane; 5. Exhaust.

Fig. 5.17 Mode of action of a sliding-vane pump; (a) induction; (b) isolation; (c) compression; (d) exhaust.

5.2.6. *Rotating-plunger pumps*

In these pumps (fig. 5.18) the sliding vane is replaced by a hollow tube which is rigidly attached to the outer sleeve of the rotor. The tube rolls and slides in a bearing, and an appropriate hole cut in the side of the tube allows gas to be drawn into the inlet side of the pump.

These pumps are designed for large pumping speeds. The heat of compression of the gas can be considerable, so the stator is usually provided with a cooling water jacket. The shape of the pumping speed curves are similar to those of rotary vane pumps (fig. 5.14). Pumping speeds may reach 1000 m³/h (Harris, 1978).

An alternative construction of the rotating-plunger pump, the *planetary piston pump* has been recently described by Sadler (1973). Bachler and Knobloch (1972) described another kind of pump: the *trochoid vacuum* pump. The principles of the trochoid pump are discussed by Wutz (1698).

5.2.7. *Root's pumps*

The Root's pump consists of two double-lobe impellers (R_1, R_2, fig. 5.19). These are rotated in opposite directions within the pump housing. The directions of rotation being those shown by the arrows, the intake and exhaust will be as shown in fig. 5.20.

The impellers have identical cross sections and are dimensioned and arranged so that a large enough part of the surface of R_1 is a close fit to a part of the surface of R_2 through the rotation. The impellers are also a close fit inside the pump housing H (fig. 5.19). The rotating impellers *do not*, however, *touch* one another, *nor do they touch* the housing, but there is a small clearance (about 0.1 mm) at the points 1, 2 and 3 (fig. 5.19). As point 1 moves around the inside wall of the pump housing, points 2 and 3 move correspondingly (fig. 5.20).

Since the inlet port is isolated in fact from the outlet by a narrow gap (clearance between parts) there is a back flow of gas from the exhaust region to the inlet region, and therefore the efficiency of compression is much lower than in the case of oil sealed pumps. However, the absence of rubbing contacts means that higher speeds of rotation (1000–4000 rpm) are possible, leading to much higher pumping speeds.

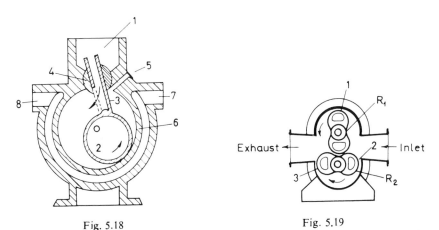

Fig. 5.18   Fig. 5.19

Fig. 5.18 Rotating-plunger pump. 1. Intake; 2. Rotating-plunger; 3. Sliding tube; 4. Bearing; 5. Exhaust ; 6. Stator; 7. Cooling water inlet; 8. Cooling water outlet.

Fig. 5.19 Root's Pump.

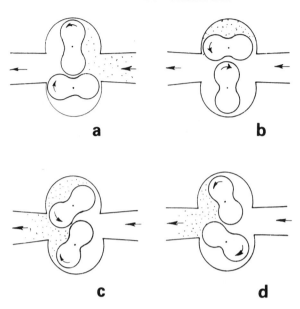

Fig. 5.20 Action of Root's pump.

The conductance of the clearance gaps decreases as the average pressure in the pump falls; the pump efficiency is expected to increase (fig. 5.21). Maximum efficiency occurs when the pump is operated at a compression ratio of about 10 at a pressure of the order of 5×10^{-2} Torr, and thus the pump must be provided with a suitable backing pump.

Root's pumps provided with cooling by gas circulation (e.g. Hamacher, 1974b; Henning and Lang, 1976; Lang, 1977) reach ultimate pressures of 70 Torr (without a backing pump). Fukotome *et al.* (1976) describe a 3-stage Root's pump with a compression ratio of 10^4 and an ultimate pressure of 1×10^{-6} Torr. Root's pumps and boosters are discussed by Budgen (1982, 1983), Henning *et al.* (1982), Bürger (1983).

In an "oil-free" system, the backing of the Root's pumps is achieved by using dry rotating pumps, e.g. of *claw-type* as described by Wycliffe (1987). In these pumps, the rotors almost touch at their cylindrical surfaces (fig. 5.20a) producing a seal in the center and the claws almost touch the stator surface (leaving a very small clearance) producing another seal. The claws enter reciprocally the depression in the mating rotor, separating a "carry over volume". In 3 stages this pump achieves an inlet pressure of 5×10^{-2} Torr (~ 7 Pa) while discharging to atmosphere (Hablanian, 1988). The pumping speed is maximal (e.g. 90 m^3/h) at about 200 Pa (~ 1.5 Torr) and is reduced, at atmospheric pressure, to about 60% of its maximum.

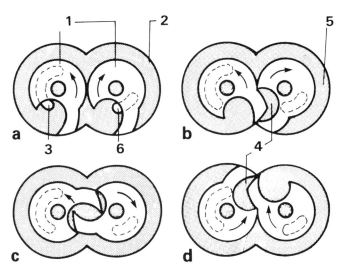

Fig. 5.20a   Action of claw-type pump. 1. Rotors; 2. Stator; 3. Inlet port; 4. Carry-over volume; 5. Swept volume; 6. Outlet port. After Wicliffe (1987); Copyright of American Institute of Physics; By permission.

5.2.8. *Molecular pumps*

The principle of the molecular drag pump, based on the directional velocity imparted to gas molecules which strike a fast moving surface is described in §2.6.2. This principle is applied in modern turbomolecular pumps (Becker, 1958, 1961, 1966; Kruger and Shapiro, 1961; Henning (1988), which contain alternate axial stages of rotating and stationary discs and plates. The discs and plates (fig. 5.22) are cut with slots (fig. 5.23) set at an angle so that gas molecules caught in the slots of the moving disc are projected preferentially in the directions of the slots in the stationary plates. The running clearances between the rotating and stationary plates are of the order of 1 mm. The rotational speed* for a pump having a rotor diameter of about 17 cm is 16000 rpm (Becker, 1958) or 42000 rpm (Osterstrom and Shapiro, 1972). The variation of the pitch angle of the slots varies the zero-flow compression ratio and pumping speed; a pitch angle of $20°$ appears to be a good compromise for many applications. Since a compression ratio per stage of about 5 can be achieved, a pump having 9 stages should maintain a zero-flow compression ratio of the order of $5^9 = 2 \times 10^6$. For this compression ratio, the pumping speed is constant below 10^{-3} Torr (fig. 5.24), but above 10^{-2} Torr the speed depends upon the size of the backing pump. The range of the pumping speeds extends up to 3500 lit/sec. Normally the presence of hydrogen, which back diffuses, limits the total ultimate pressure to about

---

* Henning (1978b) mentions 1000 Hz (60000 rpm) and Henning (1988) indicates 90000 rpm.

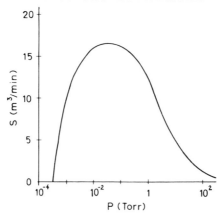

Fig. 5.21 Typical pumping speed curve of Root's pump.

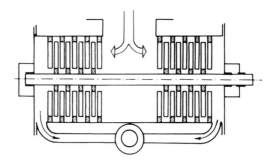

Fig. 5.22 Molecular pump. After Becker (1958).

10^{-10} Torr, although the maximum speed for hydrogen is some 20 percent more than that for air. The great advantage of molecular pumps compared to diffusion pumps is that molecular pumps are free of (hydrocarbon) vapours.

The performance and construction of several turbomolecular pumps are discussed by Power (1966), Becker (1966), Rubet (1966), Henning (1974), Bachler et al. (1974), Becker and Nesseldreher (1974), Mirgel (1972), Falland et al. (1974), Frank (1974), Henning (1974b), Maurice (1974b), Nesseldreher (1974), Sawada (1974), Frank et al. (1975), Becker and Nesseldreher (1976a, b), Frank and Usselmann (1976a, b), Flecher (1977), Gorinas (1977), Henning (1977, 1978b), Nesseldreher (1976a, b), Osterstrom (1977), Saulgeot (1977), Schittko et al. (1977), Dylla (1978), Lange and Singleton (1978), Schulz and Usselmann (1978), Weston (1978), Dennison and Gray (1979), O'Hanlon (1979), Osterstrom and Knecht (1979), Yarwood (1979), Henning (1979, 1980), Harris (1980), Abbel et al. (1982), Danziger (1982), Duval (1982), Fischer et al. (1982), Goetz (1982), Stayst (1982), Bernhardt (1983), Goetz and Henning (1983), Keller et al. (1983), Goetz et al. (1984), Hablanian (1984, 1986), Deters et al. (1987), Goetz et al. (1987), Murakami

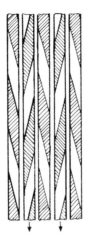

Fig. 5.23 Details of the rotor and stator plates of the molecular pump. After Becker (1958).

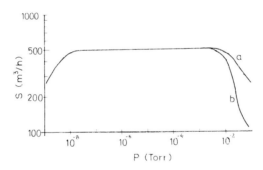

Fig. 5.24 Typical pumping speed curve of molecular pump for air. (a) with 45m³/h backing pump; (b) with 10m³/h backing pump.

et al. (1987), Rava et al. (1987), Yang et al. (1987), Casaro (1988), Chu (1988), Duval et al. (1988), Henning (1988), Mase et al. (1988), Tu et al. (1988).

Turbomolecular pumps have low compression ratios for light gases (see eq. 2.91) sometimes as low as 300 (for hydrogen). This limitation can be corrected by adding another stage, changing the blade angles and/or increasing the speed (Hablanian, 1986). Schittko and Schmidt (1975) improved the density ratio of light gases by using gas-ballast in their molecular pump.

The very high rotational speeds require advanced bearing and balancing techniques. Besides the mechanical bearings, "gas bearings" (Maurice, 1974b) and "magnetic bearings" (Frank and Usselmann, 1977) were commercially developed.

Turbomolecular pumps used in plasma confinement devices with strong magnetic fields, experience induction of eddy currents in their metallic rotors.

Excessive heating results if the magnetic flux density exceeds 10–30 mT (Becker and Henning, 1978a, b; Bieger *et al.*, 1979; Henning, 1988). Murakami *et al.* (1987), constructed a turbomolecular pump with a *ceramic rotor*, which has been tested at a magnetic flux density of 460 mT.

Modern specialized pumping technologies (semiconductor industries, plasma fusion devices, etc.) require pumping at ultra-high vacuum level and handling (at low pressures) of corrosive or radioactive gases. Turbomolecular pumps for such applications are discussed by Abbel *et al.* (1982), Henning (1988).

5.3. Vapour pumps

5.3.1. *Classification*

The general term of *vapour pump* applies to *ejector pumps* as well as to *diffusion pumps*. The distinction between these two kinds of pumps is described fundamentally, by considering the *mean free path* of the gas molecules at the intake port (mouth) of the pump, in relation to the throat width (nozzle clearance $t/2$ fig. 5.25).

Both the ejector and the diffusion pumps have at their base a boiler (heater) which supplies the vapour (e.g. oil). In the *diffusion pump* (fig. 5.25) the vapour (oil or mercury) travels up the chimney and is deflected by the umbrella placed at the top of the chimney. The molecules of the vapour stream collide with the gas molecules entering through the intake port. Since the mean free path λ of the gas molecules is greater than the throat width (nozzle clearance, $t/2$), the interaction between gas and vapour is based on *diffusion* (see §2.5.2), which is responsible for the drag of the gas molecules toward the fore-pression region. This effect establishes a pressure gradient between the high and fore vacuum sides.

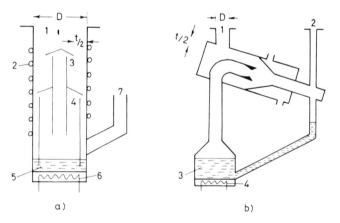

Fig. 5.25 Vapour pumps. (a) diffusion pump: 1. High vacuum; 2. Water cooling; 3. First stage; 4. Second stage; 5. Pump oil; 6. Heater; 7. Fore-vacuum; (b) vapour ejector pump: 1. Intake port; 2. Fore-vacuum; 3. Oil; 4. Heater.

In the *ejector pump* (fig. 5.25b), the mean free path λ of the gas molecules at the intake is less than the clearance $t/2$. Thus the gas is entrained by the *viscous drag* and *turbulent mixing* which carries the gas (at high speeds) down the pump chamber of diminishing cross section and through an orifice near the fore-vacuum side.

Combinations of the diffusion and ejector principles are encountered in *diffusion-ejector* pumps, sometimes called *vapour booster pumps*.

5.3.2. Vapour ejector pumps

Ejector pumps work with oil vapour or steam. Figure 5.26 shows the diagram of an *oil ejector pump*. The pump fluid is contained in the boiler (10) and the vapour flows through the jet chimneys (8, 9) into the nozzles (2.4). Due to their special shape each of these nozzles produces a supersonic jet, which enters the nozzles (diffusers 3, 5) and condenses on their cooled walls. The air to be pumped

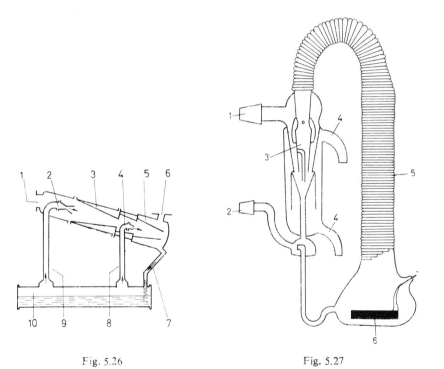

Fig. 5.26   Fig. 5.27

Fig. 5.26 Double-stage oil-vapour ejector pump.

Fig. 5.27 Glass oil-vapour ejector pump. 1. Intake port; 2. Backing pump; 3. Nozzle system; 4. Water cooling; 5. Lagging; 6. Immersion heater.

enters the pump through the high vacuum connection (1), and is carried with the jet and compressed. The process is repeated in the second stage. A compression ratio of about 10 is achieved in each stage. The air compressed to a pressure equivalent to that of the fore-vacuum line (6) is removed by the backing pump. The condensed fluid flows back through the return pipe (7) to the boiler. The fluid column in the return pipe counterbalances the vapour pressure in the boiler.

Oil vapour ejector pumps are constructed of glass (small sizes), for pumping speeds of a few liter/sec at 10^{-1} Torr, and heaters of hundreds of watts (fig. 5.27). Metal constructions reach pumping speeds of thousand(s) of liter/sec at $10^{-2} - 10^{-1}$ Torr, and have heaters of many kilowatts. Figure 5.28 shows the pumping speed curve of oil vapour ejectors.

The performance characteristics of vapour ejector pumps are discussed by Wutz (1982).

Steam ejector pumps are able to produce rough vacuum at high pumping speeds. A four stage steam ejector is able to produce 0.5 Torr with the discharge to atmosphere, a five stage arrangement with interstage water-cooled vapour condensers can be used to produce a pressure of about 3×10^{-2} Torr.

In the steam ejector (fig. 5.29) a high velocity jet of steam is discharged through the nozzle into a convergent–divergent diffuser and entrains gases and vapours

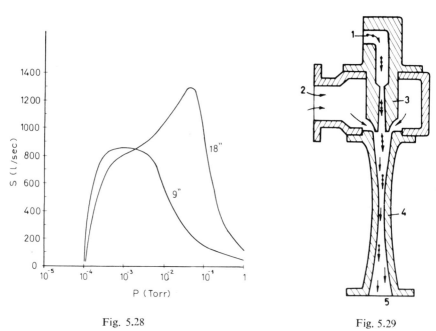

Fig. 5.28   Fig. 5.29

Fig. 5.28 Pumping speed curves of oil vapour ejector pumps.

Fig. 5.29 Steam ejector. 1. Steam inlet; 2. Air intake port; 3. Steam nozzle; 4. Diffuser; 5. Discharge.

entering the pump intake port. The discharge to intake pressure ratio is about 7.5 per stage. Steam ejector pumps are discussed by Ebdale (1978).

5.3.3. *Diffusion pumps*

The theory of diffusion pumps is discussed by Dushman and Lafferty (1962), Power (1966), Beck (1966), Roth (1968), Duval (1969), Hablanian (1974), Nöller (1977), Toth (1977), Wutz (1979), while reviews and discussions on these pumps were published by Hablanian (1974, 1980), Weston (1978), Hablanian and Landfors (1980), Harris (1980), Dennis *et al.* (1982a, b), Duval (1982), Hablanian (1984, 1986). The *pumping speed* of a diffusion pump is determined by the size of the intake clearance and the Ho-factor (Ho, 1932). The area A (cm^2) of the intake annulus is (fig. 5.25a)

$$A = \pi D^2/4 - \pi(D-t)^2/4 \tag{5.11}$$

where D is the diameter of the intake port, and $t/2$ is the throat width.

In accordance with eq. (2.48) it is impossible for a gas of molecular weight M and temperature T to pass through this area at a flow rate exceeding

$$S_{max} = 3.64(T/M)^{1/2}A$$

or for air at 20°C exceeding

$$S_{max} = 11.6 \, A \text{ liter/sec}$$

The ratio between the admittance (i.e. the true pumping speed S across the throat of the pump) and the maximum flow rate S_{max} is known as the *Ho-factor* or speed factor (H). This is usually $H = S/S_{max} = 0.3 - 0.45$. Best modern pumps have $H = 0.5$ (Hablanian and Maliakal, 1973).

The pumping speed of the pump is thus given by

$$S = HS_{max} = 3.64 \, (T/M)^{1/2} \, H(\pi/4)t(2D-t) \tag{5.12}$$

With $H=0.4$ and $t=D/3$ (large diffusion pumps) the pumping speed for air at room temperature is

$$S = \frac{11.6\pi}{4} \, 0.4 \tfrac{1}{3} \, (2-\tfrac{1}{3}) \, D^2 \approx 2D^2 \text{ liter/sec} \tag{5.13}$$

A pump with a pumping speed of 1000 liter/sec would need to have an inlet port diameter of

$$D = \left(\frac{1000}{2}\right)^{1/2} = 22.4 \text{ cm} \approx 9 \text{ in}$$

and this is about the diameter of diffusion pumps having such pumping speeds.

Assuming that the Ho-factor is independent of the molecular weight of the gas being pumped, eq. (5.12) implies that the pumping speed of a diffusion pump should be inversely proportional to the square root of the molecular weight of the gas. This proves to be approximately true for some pump designs (fig. 5.30b).

Equation (5.12) also implies that the pumping speed of a diffusion pump is independent of the pressure. This is indeed the case for a range of pressures of many decades, the pumping speed curve being as shown in fig. 5.30. At the maximum pressure at which the vapour pumping action begins (A, fig. 5.30a) this action reduces the system pressure, thus decreasing the density of gas molecules entering the vapour stream. This in turn reduces back-diffusion and hence the pumping speed rises with decreasing pressure. The pumping speed continues to increase until the pressure is such (B, fig. 5.30a) that the rate of back-diffusion at the top jet is not determined by the inlet gas density but by the rate at which gas is removed from the jet. At this, and lower system pressures, the speed remains

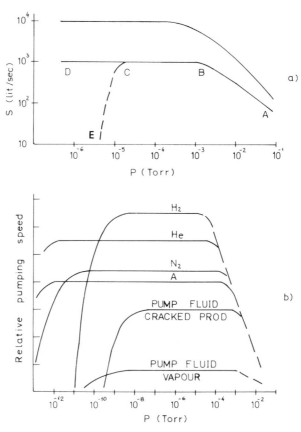

Fig. 5.30 Typical diffusion pump characteristics. (a) Shape of pumping speed-pressure curve for air; (b) Relative pumping speed for various gases. After Hablanian and Maliakal (1972).

constant at a maximum value of S. Modern pumps produce throughputs of more than 1 Torr lit/sec per kilowatt of heating power (Hablanian and Maliakal, 1973).

The *ultimate pressure* obtainable is theoretically determined by the vapour pressure of the pump fluid. With additional refrigerated baffles much lower pressures can be obtained. In practice however the ultimate pressure is governed by the characteristics of the system, in that the pumping speed in the system must necessarily become zero when the gas load (leaks, outgassing) is equal to the maximum rate at which the pump can handle gas. This is illustrated by fig. 5.30a. Curve ABCD shows the theoretical characteristic, while curve ABCE shows a typical practical characteristic in which the pumping speed becomes very small at 5×10^{-6} Torr, when the throughput is 5×10^{-4} Torr·lit/sec. The characteristic of a larger pump ($S = 10\,000$ lit/sec) is also shown in fig. 5.30a. The section AB of the characteristic is reasonable linear on a log plot and hence may be represented by the empirical equation

$$S = S_m (P/P_m)^Z \tag{5.14a}$$

where S_m is the maximum pumping speed (B,C), while P_m is the pressure corresponding to point B. The slope Z takes values between 0.8 and 1. The section B C E can be represented by an equation similar to that governing the variation of a rotary pump (eq. 3.255)

$$S = S_m [1-(P_u/P)] \tag{5.14b}$$

where P_u is the ultimate pressure (point E).

Pump sizes. A wide range of sizes of diffusion pumps are available with inlet port diameters from 1 in to 48 in, with corresponding pumping speeds from about 10 lit/sec up to about 10^5 lit/sec. Small sizes of diffusion pumps are also constructed of glass.

Roughing and backing. The single jet pump does not function very efficiently in practice. For efficient operation two conditions must be fulfilled:

– The system pressure must be initially reduced below a certain value (roughing) which in most practical cases is of the order of 10^{-1} Torr.
– The pressure below the jet must be kept reasonably low (backing pump) to reduce the probability of back-diffusion.

To make these conditions easier to achieve by the external roughing and backing pumps, vapour pumps are constructed with several jet stages in series, one acting as a backing pump to another. The main function of the top jet (fig. 5.31) is to give a large pumping speed and thus this jet has a large admittance area. On the other hand the lower jets have smaller admittance areas, and hence smaller escape areas for the vapour stream. Consequently the pumping speeds of the lower jets become successively smaller whilst the pressure differences which they can support become larger. It should be noticed that the throughput is necessarily constant throughout the pump.

Table 5.1.
Diffusion pump fluids.

Fluid	Chemical name	Molecular weight	Vapour pressure 25°C (Torr)	Boiling point at 10^{-2} Torr (°C)	Viscosity at 25°C (centistokes)	Ref.
Amoil S	i–diamyl sebacate	343.3	10^{-6}	111	—	
Octoil	d–2–ethyl hexyl phtalate	390.3	3×10^{-7}	122	75	(1)
Octoil S	d–2–ethyl hexyl sebacate	426.3	3×10^{-8}	1435	—	
Apiezon A	hydrocarbon	414	2×10^{-5}	—	—	
„ B	„	468	4×10^{-7}	110	—	(2)
„ C	parafinic hydrocarbon	574	4×10^{-9}	125(265*)	295	
Silicone DC 702	Methyl poly-siloxanes	530	10^{-6}	160	28	
DC 703	„	570	10^{-7}	200	58	(3)
DC 704	Phenyl poly-siloxanes	484	10^{-8}	216*	47	
DC 705	„	546	5×10^{-10}	243*	170	
Santovac 5/ Conwalex 10	Mixed 5-ring polyphenil ether	447	1×10^{-9}	288*	2500	(4)
Fomblin 425/9	Perfluoro-polyether	2600	5×10^{-8}	252**	180	(5)

(1) Bendix Corp., (2) Shell Chem. Co., (3) Dow Corning Corp., (4) Hickman (1962)., (5) Edwards Co.
and Laurenson et al. (1979).
* Flashpoint.
** Boiling Pt. at 1 Torr.

Pump fluids. Diffusion pumps use either mercury or organic fluids. Table 5.1 gives data on several of the organic fluids used (see also fig. 4.5). Table 5.2 lists the properties required from these fluids and the extent to which they are satisfied.

The relative advantages and disadvantages of the various oils result from table 5.2. A comparison between oils and mercury is shown in table 5.3.

The properties of some oils and their utilization in modern diffusion pumps are discussed by Hirsch and Richards (1974; polyphenylether), Luches and Perrone (1976), Laurenson and Caporiccio (1977), Caporiccio and Steenrod (1978; perfluoropolyethers); Sakuma and Nagayama (1974, eicosyl-naphthalene); Laurenson (1980, 1982), O'Hanlon (1984), D'Anna et al. (1987), Whitman (1987).

Table 5.2.
Requirements of fluids for diffusion pumps.*

Requirement		Remarks
Ability to reach low ultimate pressures	Low vapour pressure (table 5.1)	Santovac 5 and DC 705 attain 10^{-9} Torr with water cooled baffles. DC 705 reaches 10^{-10} Torr with baffle at $-20°$C.
	Freedom of volatile constituents	If present, fractionating pumps shall be used.
	Low ability to dissolve gases	All oils dissolve gases to some extent.
Ability for effective pumping	High molecular weight	Oils (table 5.1) have fairly high molecular weights.
	Suitable viscosity	All oils (table 5.1) have corresponding viscosity–temperature characteristics.
	High surface tension	Oils (table 5.1) have surface tensions of 30–50 dyne/cm, which minimizes creep.
	Low heat of vaporization	—
	Low cost	—
Good thermal stability	High flash and fire points	Silicone DC 705 has minimum combustion hazard.
	Resistance to oxidation in air	Silicones have good resistance to oxidation.
	Resistance to decomposition in vacuum	Silicones least prone to decomposition.
	Resistance to catalytic decomposition	Copper has greatest catalytic action. Cr, Ni plating useful.
	Resist. to decomposition by electr. discharge or hot filaments	All oils prone to decomposition. Octoil and Santovac 5 give less insulating deposits.
Chemical inertness	Chemical inertness	Most oils inert, except to halogens, acid vapours and alkali metals.
	Non toxicity	Requirements fulfilled by all oils (table 5.1).

*References : Pirani and Yarwood (1961), Hickman (1962), Crawley *et al.* (1963), Solbrig and
    Jamison (1965), Hablanian and Maliakal (1973).

Fractionating pumps. If the pump oil is a mixture of chemical compounds the interest is to separate them so that the top jet includes those of the lowest vapour pressure. This is achieved by incorporating into the design the principle of *fractionation* (fig. 5.32). Fluid returning from the jets to the boiler flows radially inward toward the center of the boiler. If this flow is impeded by barriers with small openings (shaded parts on base plate, fig. 5.32), the fluid is heated substantially

Table 5.3.
Comparison between oils and mercury as pump fluids.

Requirement	Advantage (A) or disadvantage (D) for		Remarks
	Oils	Mercury	
Low vapour pressure	A	D	Liquid nitrogen trap necessary for mercury.
Freedom of volatiles and easy purification	D	A	Mercury easily distilled, oil requires self-fractionating pumps.
Low ability to dissolve gases	D	A	Mercury as pump fluid sensitive to traces of oils and greases.
High molecular weight	A	D	
High density	D	A	
Resistance to oxidation in air	D	A	See table 5.2.
Resistance to decomposition in vacuum	D	A	,,
Use with high fore-pressure	D	A	Mercury can be used up to 40 Torr.
Resistance to decomposition on hot filaments or due to discharges	D	A	
Non-toxicity	A	D	
Inertness	D	D	Mercury used only with steel and glass. Oils require chromium or nickel plating

while it is still near the outer portion of the boiler, so that high vapour pressure constituents are boiled off near the outside.

As the fluid flows toward the center it is further heated and lower vapour pressure components are vaporized. The jet system is constructed of concentric tubes arranged such that each jet receives vapour from only a specific annular region of the boiler. The backing (lowest jet) receives vapour from the outer portion of the boiler where the vapour pressure is highest, and the top jet receives

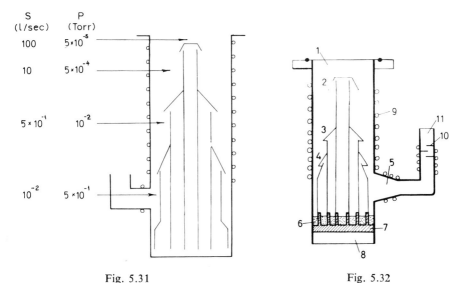

Fig. 5.31   Fig. 5.32

Fig. 5.31 Pumping speed and pressure in a multistage diffusion pump.

Fig. 5.32 Fractionating oil diffusion pump. 1. Pump inlet, high vacuum; 2. First stage; 3. Second stage; 4. Third stage; 5. Ejector; 6. Pump fluid; 7. Fractionating boiler; 8. Heater; 9. Water cooling coil; 10. Forline baffle; 11. Backing line, fore-vacuum.

vapour from the central section where the vapour pressure of the fluid is the lowest.

Cooling. The cooling of the diffusion pump is a part of its working principle, as is the heating. Diffusion pumps are usually water-cooled, but air-cooled pumps are also available (small size pumps). The cooling should be arranged so that the coolest region is where the vapour stream strikes the pump wall. The rate of cooling is also fairly critical since if it is too low the vapour is not entirely condensed and thus the back-streaming effect is increased. On the other hand, if the cooling rate is too high the vapour is not only condensed but also considerably cooled. This results in a slow flow-back to the boiler, and necessitates a high boiler power to re-evaporate it.

The instabilities of small diffusion pumps due to the variations of the pressure in the boiler and their correction are discussed by Hablanian and Landfors (1976). Specific operation problems were solved by compact pump systems (Power, 1974a, b), by quick start and stop pumps with a porous carbon boiler (Power *et al.*, 1977), or by systems having the vacuum chamber inside the diffusion pump (Kuypers, 1977).

Back-streaming and back-migration. Diffusion pumps suffer from two defects whereby the pump fluid enters the vacuum enclosure: the back-streaming, and the back-migration. These phenomena were studied and discussed by Auwarter

(1957), Power and Crawley (1954), Schulze (1967), Holland (1970), Hablanian and Maliakal (1973), Rettinghaus and Huber (1974), Meyer (1974), Power *et al.* (1974b), Harris (1977), Hoffman (1979), Maurice *et al.* (1979).

Back-streaming is due to a small fraction of the molecules of the working fluid (oil or mercury) travelling from the top jet in the wrong direction (toward the vacuum chamber). This undesired direction is either imparted to the molecules as they issue from the jet or it is acquired after impacts with gas molecules or with other vapour molecules in the stream. A well-designed nozzle should be dry during pumping.

Back-migration is due to re-evaporation of the working fluid from the walls of the pump and the connecting tube to the vacuum chamber.

Back-streaming and migration can be decreased by using *baffles* and *cold-*

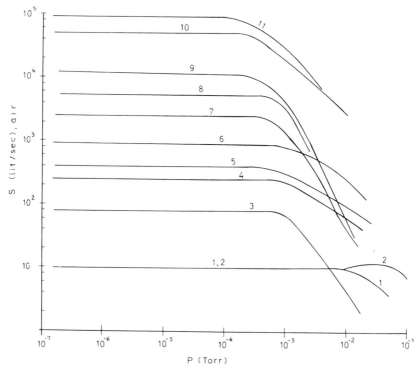

Fig. 5.33 Examples of characteristics of diffusion pumps. Ultimate pressure depends usually on the pump oil used. (1) One inch oil diffusion pump, EO–1 (Edwards, England); (2) One inch mercury diffusion pump, EM–1 (Edwards, England); (3) Two inch oil diffusion pump, E203 (Edwards, England); (4) 80 mm oil diffusion pump, Type 6080 (Alcatel, France); (5) 100 mm oil diffusion pump, Type 400/1 (Leybold–Heraeus, Germany); (6) 138 mm oil diffusion pump, DIFF 900 (Balzers, Liechtenstein); (7) 9 inch oil diffusion pump, EO–9 (Edwards, England); (8) 12 inch oil diffusion pump, HVS–10, (Varian–NRC, U.S.A.); (9) 24 inch mercury diffusion pump 24M4 (Edwards, England); (10) 1000 mm oil diffusion pump, Type 50 000 (Leybold–Heraeus, Germany); (11) 48 inch oil diffusion pump, PMC–48C (Bendix, U.S.A.)

traps (see §3.4.4. and 5.6.5), which obviously lower the net pumping speed. Hablanian and Maliakal (1973) appreciate that an optimum design should have about 40% net pumping speed and only 1×10^{-10} g/cm² · minute back-streaming.

Diffusion pumps with high-quality fluids and well-designed traps, can be used in ultra-high vacuum systems which do not require pressures lower than 10^{-11} Torr. The behaviour of diffusion pumps in ultra-high vacuum systems is discussed by Harris (1980), Dubois (1988).

Characteristic curves. The characteristic curves of diffusion pumps are generally plotted in the form of pumping speed (lit/sec) against the fine side pressure; fig. 5.33 shows some such curves.

The diffusion pump must always be backed by a forepump, and the *throughput* of the forepump must be equal to or larger than that of the diffusion pump.

For the reason of choosing the appropriate backing pump, it is useful to plot

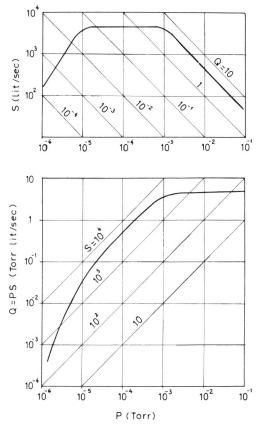

Fig. 5.34 Pumping speed and the equivalent throughput curve of a diffusion pump.

the characteristic curve of the diffusion pump in terms of throughput-pressure. Figure 5.34 shows such a pair of curves.

5.4. Ion-pumps

5.4.1. *Classification*

If a gas is ionized and the resulting positive ions are accelerated to a negatively charged plate, atoms of the gas are effectively removed from the system, and thus a pumping action is produced. The general term *ion-pump* includes those vacuum pumps in which gas molecules are pumped by being *ionized and transported* in the desired direction by an *electric field*. The ionization may be produced by collisions of the gas molecules with electrons emitted either from a *hot filament* or from a *cold cathode discharge*. The first type of pump is referred to as *hot cathode ion-pump*, while those belonging to the second type are known as *cold cathode ion-pumps*. Both the hot and cold cathode *ion gauges* (§6.7) act as pumps in this manner, and can be used to pump small closed systems, which have been previously evacuated to about 10^{-3} Torr. Pumping speeds for hot cathode gauges are typically of the order of 10^{-3} liter/sec, while cold cathode gauges can reach about 1 lit/sec. Ames *et al.* (1978, 1979) achieved pumping speeds up to 2 lit/sec by using a 60 Hz, 80 kV field.

In order to increase the pumping efficiency, sorption and gettering phenomena are combined with ionization. Pumps which combine the action of an *ion-pump* and the *sorption* of the ions in a sorbent, are known as *ion-sorption pumps*. Ion-sorption pumps in which a getter* is continuously or intermittently vaporized and condensed on the trapping surface to give a fresh deposit of sorbent, are termed *getter-ion-pumps*. The vaporization is due to thermal evaporation in *evapor-ion-pumps* and to cathode sputtering* in *sputter-ion-pumps*. Although ion-pumps without sorption or gettering were also constructed, commercial ion-pumps are either of the *evapor-ion* or of *sputter-ion* type.

Pumps based only on *sorption* are discussed in §5.5, and pumping based only on gettering in §5.7. *Sublimation* and *cryo-sublimation* pumps are discussed in §5.7.5.

The feasibility of a "solid state diffusion pump", in which ionized atoms are injected into a permeable membrane then pumped by the backing pump on the other side of the membrane, is shown by Carter *et al.* (1975).

5.4.2. *Ion-pumping*

An electrical gas discharge, in which ions are formed, is basically capable of pumping gases, and one assumes that the formed ions are either bombarded into

---

Getter = a material which is included in a vucuum device (tube) for removing gas by sorption (see §5.7).

Sputtering = process of ejecting atoms of a cathode by bombarding it with heavy positive ions (see §4.5.2).

a metallic collector provided for the purpose or that these ions are trapped within the surface atoms of such a collector, due to a chemisorption effect. The pumping efficiency is expressed by the ratio i^+/P between the ion current and the pressure of the gas in the device. The ion current i^+ is proportional to the number of molecules entering the device per unit time, thus to the throughput Q. Therefore

$$i^+/P = K(Q/P) = KS \qquad (5.15)$$

and the pumping speed S is given by

$$S = \beta(i^+/P) \text{ liter/sec} \qquad (5.16)$$

where $\beta = 1/K$ is the constant of the pump expressed in Torr·liter/sec·Amp, i^+ is the ion current (Ampere), P is the pressure (Torr).

The value of the pump constant β was determined by plotting the pumping speed S (for a determined gas), as a function of the sensitivity i^+/P (fig. 5.35).

The pump constant β of practical pumps was found (Bächler, 1965) for nitrogen $\beta = 0.07$, while the maximum value of this constant is

$$\beta_{max} = \frac{1}{en} = 0.191 \qquad (5.17)$$

where

e – charge of an electron $= 1.6 \times 10^{-19}$ A· sec,

n – number of molecules in 1 liter gas, at $P = 1$ Torr, $20°C$, $n = 3.27 \times 10^{19}$ Torr^{-1}· lit.$^{-1}$ (see eq. 2.18).

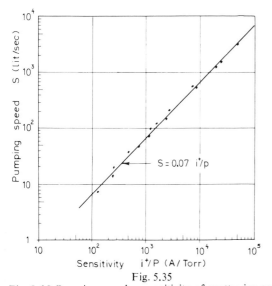

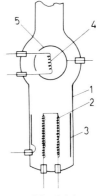

Fig. 5.35   Fig. 5.36

Fig. 5.35 Pumping speed vs sensitivity of sputter-ion-pumps, for nitrogen. After Bächler (1965).

Fig. 5.36 Small evapor-ion-pump. 1. Glass envelope; 2. Titanium coil on tungsten filament; 3. Electrode (formed by evaporated titanium); 4. Filament; 5. Anode ring.

Unlike the diffusion pump, the ion-pumps (evapor-ion and sputter-ion) do not require a forepump to pump the collected gas up to atmospheric pressure, since the pumped gas is in effect trapped. However, an auxiliary pump is needed to reduce the system pressure to the range of 10^{-3}–10^{-4} Torr, where the ion-pumping action will commence. *Mechanical* or *sorption pumps* are usually used for this initial pumpdown.

5.4.3. *Evapor-ion-pumps*

Evapor-ion-pumps combine the ion-pumping effect with the gettering process of evaporated active metal. The gettering effect is used both during evaporation (dispersal gettering), and in the form of a fresh film on a surface (contact gettering). The gas is ionized to ensure transport by electrical pumping of the inert gases (which are not gettered) to the getter-coated wall at which they are made to arrive with energies of a few hundred electron volts. At these energies, about 20% of the ions are retained, and embedded in the film as fresh getter is vaporized. The most widely employed getter in these pumps is *titanium*. Pumping and re-emission of gas in the ion-pumping process has been studied by Hobson (1963).

From the various types of evapor-ion pumps we will describe here:

(a) Small pumps, designed to perform a single pumping operation,
(b) Large high speed pumps with continuous feed of titanium, and
(c) The Orbitron pump.

Small evapor-ion pumps. These pumps (fig. 5.36) consist typically of a hot cathode ionization gauge (see §6.7) containing an extra filament around which is wrapped a wire of getter material (usually titanium or zirconium).

Usually, these pumps are first evacuated to diffusion pump pressures (below 10^{-3} Torr). On firing the titanium getter, sorption of the common gases takes place; on operating the ionization gauge, positive ions of the gas, and in particular the inert (rare) gases, are driven into the titanium-coated walls, where 10–20% of those incidents are retained and buried in any subsequently deposited titanium. A pumping speed of about 5 liter/sec for air is possible and ultimate pressures of 10^{-9} Torr are attainable. A typical application of such pumps is the improvement of the residual vacuum in special vacuum tubes. An auxiliary small getter-ion-pump in a glass bulb is attached to the main tube. After sealing-off the main tube from the conventional pumping system at a pressure of about 10^{-6} Torr, the getter-ion-pump (evapor-ion-pump) is operated to reduce the pressure to 10^{-9} Torr. The evapor-ion-pump is sealed-off, and discarded. Small evapor-ion (getter-ion)-pumps have been described by Klopfer and Ermrich (1959), Huber and Warnecke (1958), and Kornelsen (1960).

Large evapor-ion-pumps. The construction of such a pump is schematically shown in fig. 5.37 (Davis and Divatia, 1954; Swartz, 1956). A spool, carrying

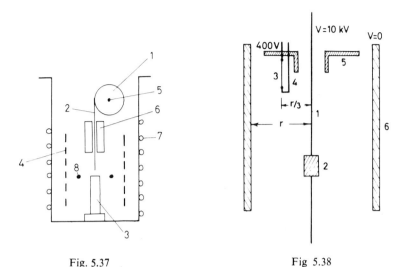

Fig. 5.37   Fig 5.38

Fig. 5.37 Evapor-ion-pump (principle). 1. Wire feed spool; 2. Titanium wire; 3. Heated post; 4. Grid; 5. Spool shaft driven from outside; 6. Wire guide; 7. Cooling water coil. 8. Filament. After Davis and Divatia (1954); Swartz (1956).

Fig. 5.38 The Orbitron pump (principle). 1. Tungsten rod (anode); 2. Titanium cylinder; 3. Filament–electron source; 4. Lead-in wire; 5. Plate; 6. Wall (cathode).

titanium wire, is externally controlled so that the wire is fed downward onto a post of refractory conducting material maintained at 1000 V positive with respect to the pump wall. Electrons produced at the circular filament (100 V positive with respect to the wall) bombard the post and heat it to about 2000°C. This causes rapid evaporation of the titanium wire which then condenses on the cooled pump walls. This continuous evaporation of the wire ensures a continuous pumping action by both dispersal and contact gettering. A wire mesh grid, at a potential of 1000 V positive (with respect to the walls) also attracts electrons from the filament and these cause ionization of the gas, resulting in positive ions travelling to and being retained by the pump walls. Using a titanium evaporation rate of about 5mg/min, pumping speeds are of the order of 3000 l/s for H_2 at 10^{-6} Torr; 2000 l/s for N_2 at 3×10^{-6} Torr; 1000 l/s for O_2 at 10^{-5} Torr; and 5 l/s for argon at 5×10^{-5} Torr.

Satisfactory maintenance-free life times of at least 1000 hours for these pumps are recorded. The component most likely to fail early is the titanium feed device due to build-up of titanium on the tip of the wire guide. For this reason in some designs (e.g. Pauly et al., 1960) multiple titanium cartridges (moved magnetically) were used, instead of wire. Difficulty is also experienced due to peeling of the titanium film on the pump walls after repeated evaporations, especially if air at atmospheric pressures is frequently admitted.

The large titanium evapor-ion-pumps have been applied in those cases where a high pumping speed, coupled with oil vapour free residual gases at low pressure (10^{-6}–10^{-7} Torr), are required in chambers of considerable sizes, such as accelerators, large X-ray devices, etc.

Bills (1973) established that evapor-ion-pumps have high pumping speeds for active gases, but their pumping speed for inert gases is lower than that of sputter-ion-pumps. In order to produce pumping of the inert gases, the ionization probability has to be increased. This is done by causing a large number of energetic electrons to travel a relatively short distance (e.g. Kuznetzov et al., 1969) or by increasing the electron path by an appropriate field (Grigorov, 1973).

Schwarz (1972) describes an ion-pump in which the ionization is based on the quadrupole principle. Electrons are emitted from a hot filament at one end and accelerated by a dc potential. At the other end of the tube a flat spiral carries the same potential as the cathode, causing the electrons to return. Hereby a back and forth movement of the electrons is achieved. A rotational high frequency potential adjusted according to the quadrupole principle (Schwarz and Tourtellotte, 1969) keeps the electrons within the center part of the tube, until they hit a gas molecule. Thus a high ionization yield is obtained. These pumps have pumping speeds of 4–100 l/sec at 10^{-5}–10^{-9} Torr. The up-to-date theory of the quadrupole ion pumps is discussed by Hora and Schwarz (1974), Schwarz (1977).

The Orbitron pump. The principles of the Orbitron have been discussed by Herb et al. (1963), Maliakal et al. (1964, 1965), Feidt and Paulmeir (1980).

The Orbitron is a device in which electrons are injected into an electrostatic field between two concentric cylinders, the central cylinder being the anode and the outside cylinder being grounded. Electrons injected into this field with sufficient angular momentum have paths of several tens of meters, thus a high efficiency to ionize inert gases.

The central cylinder (fig. 5.38) is a tungsten rod of small diameter supporting titanium cylinders of relatively large diameters. Most of the electrons are collected by the titanium cylinders, because they intercept electrons of relatively high angular momentum, and the titanium is heated to sublimation temperatures. Active gases are pumped by the fresh titanium on the walls of the pump. Inert gases are ionized in the electrostatic field, driven to the walls, and buried by fresh titanium.

The electrons must be injected with an angular momentum such that they move on orbits between anode and cathode. For this purpose tungsten filaments are placed parallel to the cylinder axis at a distance of about $\frac{1}{3}$ of the radius of the cathode. One of the lead-wires of the filaments is placed to be parallel with the filament and is situated between the filament and the anode. This arrangement avoids direct path of electrons to the anode. In order to give a slight axial component to the electrons in their orbiting motion, a plate is provided at the upper end, this plate is at the same potential as the filaments.

The pumping speed of an Orbitron pump is shown in fig. 5.39. The ultimate pressure is of the order of 10^{-10} Torr. It is difficult to obtain lower pressures, since the outgassing rate of the hot titanium is relatively high.

Naik and Herb (1968) constructed a glass Orbitron pump with a pumping speed of 11.6 l/sec for N_2, 5.6 l/sec for air and 0.1 l/sec for argon, in the range 10^{-5}–10^{-10} Torr. To avoid difficulties due to peeling of titanium, a stainless steel mesh was used on the inside wall of the glass body.

The characteristics of Orbitron pumps are discussed by Douglas et al. (1965) and Kato et al. (1974), Naik and Verma (1974, 1977), Feidt and Paulmeir (1980, 1982), Petit and Feidt (1983).

Besides the Orbitron pumps in which the titanium is evaporated by electron bombardment (e.g. fig. 5.38), pumps with resistively heated titanium sources were also constructed (e.g. Denison, 1967; Bills, 1967).

5.4.4. Sputter-ion-pumps

The pumping action of a magnetically confined discharge observed by Penning (Penning and Nienhuis, 1949) was increased by using titanium cathodes (Guren-

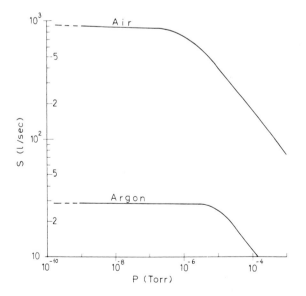

Fig. 5.39 Pumping speed of Orbitron pump.

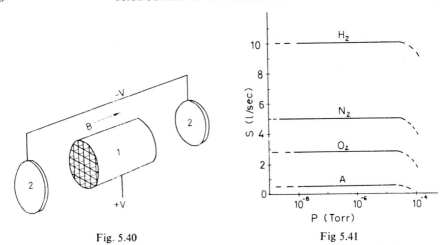

Fig. 5.40   Fig 5.41

Fig. 5.40 Sputter-ion-pump (principle). 1. Stainless steel anode; 2. Titanium cathodes.

Fig. 5.41 Pumping speed of a sputter-ion-pump (5 l/sec).

witch and Westendrop, 1954). This action was used in multicell pumps, which give higher pumping speeds and attain lower pressures (Hall, 1958), and further improved by using combined titanium–zirconium cathodes (Hall, 1969).

The sputter-ion-pumps are designed such that an electrical discharge occurs between the anode and the cathode at a potential of several thousands of volts in a magnetic field of a few thousand gauss. Since the magnetic field causes the electrons to follow a flat helical path, the length of their path is greatly increased. A high efficiency of ion formation down to pressures of 10^{-12} Torr and less is assured by this long path length. The gaseous ions so formed are accelerated to the titanium cathode, where they are either captured or chemisorbed. Due to the high energies they are propelled into the cathode plate and sputter cathode material (titanium), some of which settles on the surfaces of the anode where it also traps gas atoms.

The discharge mechanism of sputter-ion-pumps is discussed by Jepsen (1961, 1968) and Hirsch (1964). The pumping mechanism was discussed in many publications; these were reviewed by Andrew (1968), Jepsen (1968), Power (1966), Baker and Laurenson (1972), Harra (1974).

Sputter-ion pumps consist essentially of a stainless steel vessel, containing an anode of honeycomb construction (fig. 5.40), and a titanium cathode* mounted opposite each end of the anode. A potential of about 3000 V is maintained between the electrodes and a magnetic field of about 1500 G is applied by external permanent magnets along the axis of the electrode system. Positive ions of system gas which are formed in the region of the electrodes are accelerated to

---

* Cathodes of Zr–Al (Okano *et al.* 1984), Al–Ti–Zr (Ishimaru, 1984), and Al plus rare earths (Lu and Fang, 1987) are also used.

the cathode and acquire sufficient energy to sputter titanium. The sputtered titanium condenses mainly on the open structure anode and in so doing pumps active gases by both dispersal and contact gettering. Gas molecules which reach the anode by either of these processes are rapidly buried beneath succeeding layers of titanium and are thus permanently removed from the system. On the other hand, gas which reaches the cathode as positive ions has a high probability of being desorbed by succeeding ion bombardment. This is particularly so in the case of the inert gases since they can only be ion-pumped and then held at the cathode by the relatively weak forces of physical adsorption. Surprisingly, helium, for which normal sorption by any material is insignificant, is pumped quite well, apparently by being rather buried in the cathode material. Argon is most troublesome in this respect (fig. 5.41) and is the main factor governing the ultimate pressure attainable. The low pumping speed for argon is frequently of concern, since the atmosphere contains 1 percent argon. In addition to their poor argon pumping speed, simple diode sputter-ion-pumps exhibit regular pressure bursts (fig. 5.42), at time intervals of several minutes. The argon instability was studied by Jepsen et al. (1960) and Vaumoron (1970).

In order to obtain stable pumping for argon, either a third element, the sputter cathode, is interposed, or the pump cathode is slotted (Jepsen et al., 1960). In the first solution an electrode in the form of grid (cathode, fig. 5.43) is incorporated between the anode and the outer plate electrode, such that the new grid becomes the true cathode and the side plates become auxiliary electrodes (fig. 5.43). This arrangement is referred to as the *triode-pump*. Its principle is that titanium is preferentially sputtered from the cathode and ion burial and noble gas pumping occurs on the auxiliary electrode.

The triode design results in a stable pumping of argon, but brings to a loss in the cathode life time, or to a reduction of the pumping speed for other gases. The *slotted cathodes* in the diode pump (fig. 5.44) appear to bring a better solution to the argon instability.

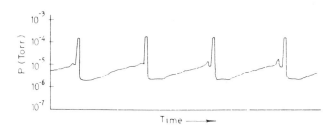

Fig. 5.42 Pressure vs time curve of a sputter-ion-pump exhibiting the argon instability. After Jepsen et al. (1960).

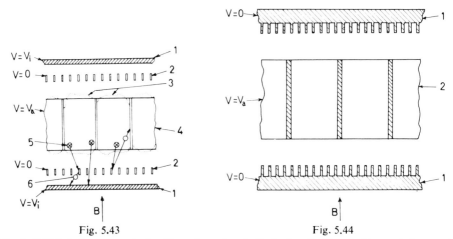

Fig. 5.43   Fig. 5.44

Fig. 5.43 Triode ion-pump. 1. Auxiliary electrodes; 2. Cathodes; 3. Cloud of trapped electrons; 4. Anode; 5. Positive ions; 6. Sputtered atoms of cathode material; B-axial magnetic field. After Jepsen et al. (1960).

Fig. 5.44 Slotted-cathode diode ion-pump. 1. Slotted cathodes; 2. Anode; B-axial magnetic field. After Jepsen et al. (1960).

As another solution to the argon pumping speed Tom and James (1966) achieved differential sputtering by use of a titanium and a tantalum cathode.

Singleton (1971) studied the difficulties produced by hydrogen. Heavy hydrogen loads produce deformations of the cathode due to titanium hydride formation.

The problems of sputter ion pumps are discussed by Tom (1972), Henning (1975), Welch (1976), Denison (1977a), Weston (1978), Pierini and Dolcino (1983), Okano et al. (1984), Audi and Pierini (1986), Lu and Fang (1987), Audi and De Simon (1988), Liu et al. (1988).

A further difficulty with ion-pumps is the care which must be taken when *starting* the pump at high pressures (max. 10^{-2} Torr). The ion current at high pressures is large and causes heating of the pump. If the pump has previously handled much gas the temperature rise leads to outgassing which in turn causes a larger ion current. Such a process rapidly becomes "run-away" leading to glow discharge between the electrodes and a rapid rise in system pressure. Even if the gas evolution is not rapid the pumping speed is reduced giving what is termed "slow starting". These troubles can be largely overcome, by initially pumping to pressures of the order of 5×10^{-4} Torr before operating the sputter-ion-pump. The problems of starting of these pumps were discussed by Bell et al. (1967), Horikoshi et al. (1974), Konishi and Mizumachi (1974).

Sputter-ion-pumps are available (fig. 5.45) from 1 liter/sec to 5000 liter/sec.

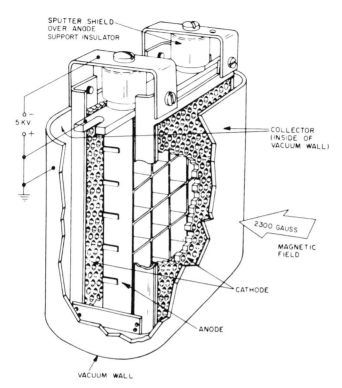

SPUTTER SHIELD
OVER ANODE
SUPPORT INSULATOR

5 KV

COLLECTOR
(INSIDE OF
VACUUM WALL)

2300 GAUSS

MAGNETIC
FIELD

CATHODE

ANODE

VACUUM WALL

Fig. 5.45 Sputter-ion-pump. Reprinted, by permission, from Lafferty (1964).

The life of a sputter-ion-pump is limited by saturation of the titanium or build-up of sputtered materials (flaking). Commercial pumps are quoted as having a life up to 50 000 h at a pressure of 10^{-6} Torr, but at 10^{-5} Torr the life is only 3000–5000 h. If pressure bursts occur the life time of the pump may be considerably less.

The pumping speeds of sputter-ion-pumps have been analyzed by Bächler (1965), Dallos and Steinrisser (1967), Paul (1973b), Halama (1977), Ho et al. (1982), Pierini (1984), Audi and Pierini (1985), while the power consumption of such pumps is discussed by Snouse (1971).

In order to achieve continuous pumping action along the (long) tubular vessels of the storage rings, thus decreasing the pressure drop discussed in §3.7.4, *distributed sputter-ion pumps* were constructed. The pumping speed of such pumps (for nitrogen) is 100 to 400 lit/sec per meter length of the ring. The distributed pumps are discussed by Cummings et al. (1971), Malev and Trachtenberg (1973, 1975), Hartwig and Kouptsidis (1974, 1977a, c) Bostic et al. (1975), Fischer (1977), Rees (1977), Reid and Trickett (1977), Laurent and Gröbner (1979), Chou (1987).

5.5. Sorption pumps

5.5.1. *Nature of sorption pumping*

The sorption pump consists of a refrigerated enclosure containing an activated sorbent (§4.3.6). On opening the pump to the system, gas is sorbed until the sorbent is saturated; therefore sorption pumping is a *batch process*.

The principles and performances of the sorption pumping process have been analyzed and reviewed by Turner and Feinleib (1962), Manes and Grant (1963), Barrington (1963), Boers (1968), Stern and Dipaolo (1969), Vijendran and Nair (1971), Dobrozemsky and Moraw (1971), Bergandt and Henning (1976), Weston (1978).

The materials used as sorbents in commercial sorption pumps are the zeolites (§4.3.6) also known as "molecular sieves". The ability of zeolites to adsorb large quantities of gases rests on their unique porous crystal structure. The structure of molecular sieves is made of very fine roughly spherical cavities, connected by minute channels of about the same diameter as that of the gas molecules (table 2.5). The ratios between the diameters of the channels and the molecular diameter of various gases may make highly selective the sorption pumping process of a gas mixture. The properties of sorbent materials and especially of molecular sieves have been discussed by Kindall and Wang (1962), Bauer and Jeffers (1965), Stern et al. (1965), Espe and Hybl (1965), Boers (1968), Halama and Aggus (1974), Durm and Starke (1972), Durm et al. (1972), Miller (1973a), Watson and Fischer (1977), Fischer and Watson (1978, 1979).

The adsorption characteristics of a typical molecular sieve (with a mean pore size of 5 Å) is shown in fig. 5.46. The curves show the quantity of gas which can be sorbed, in Torr·liter of gas per gram of sorbent, as a function of the residual gas pressure. It is apparent that as the sieve temperature is lowered, more gas molecules can be sorbed. Neon and helium, however, are sorbed to a lesser degree than nitrogen. In the atmosphere, the partial pressures of nitrogen, neon and helium are 595, 1.4×10^{-2} and 4×10^{-3} Torr, respectively. If the sieve is exposed to a volume of atmospheric air, the nitrogen partial pressure will be reduced to a much greater extent than that of neon and helium.

The isotherms (fig. 5.46) indicate that a partial pressure of 10^{-2} Torr is obtained after pumping 100 Torr/liters of nitrogen, 10^{-3} Torr·liters of neon or 8×10^{-5} Torr·liters of helium per gram of zeolite.

Figure 5.47 shows the equilibrium pressure after atmospheric air has been pumped. Commercial sorption pumps normally operate in the range 10^{-3}–10^{-1} liter (chamber volume) per gram (of sorbent). For small values of this range the limiting pressure is governed by the quantity of nitrogen sorbed at room temperature (branch A). It can be lowered by preheating the zeolite to 300°C and cooling from this temperature (branch B).

Oxygen, argon and carbon dioxide are pumped quite effectively and do not present the problems encountered with neon and helium.
Although molecular sieves are the most extensively used sorbent materials in

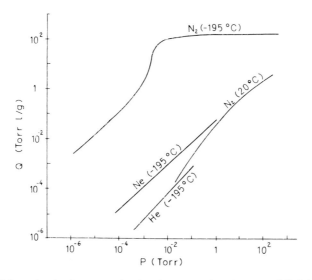

Fig. 5.46 Adsorption isotherms, molecular sieve 5A. After Turner and Feinleib (1962).

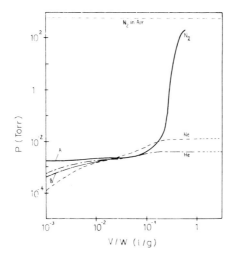

Fig. 5.47 Predicted equilibrium pressures of some constituents of air as a function of the pumping load (expressed as the chamber volume V, per absorbent weight W). The sorbent is 5A molecular sieve at -195°C. After Turner and Feinleib (1962).

sorption pumping, other materials were also used and tested. Activated charcoal was used by many workers in the early times of vacuum technology (Dushman, 1949) and restudied more recently by Jepsen et al. (1959), Stern et al. (1965). Milleron (1965) analyzes the possibilities of using porous metals, while Mason and Williams (1972) describe the use of porous silver.

The use of zeolites and alumina (at room temperature) for trapping oil vapours backstreaming from rotary pumps, has been discussed by Luches and Zecca (1972), Fulker et al. (1969), Buhl (1981).

5.5.2. The sorption pump

The sorption pump consists of a stainless steel body with internal copper fins to facilitate heat transfer to the zeolite charge. A liquid nitrogen container can be attached to three support brackets (fig. 5.48a).

The sorption pump is valved into the system, and immersed in liquid nitrogen, As the temperature of the zeolite (molecular sieve) falls, it sorbs (fig. 5.47) more gas from the system to cause a reduction in the pressure. After pumpdown to the equilibrium pressure has been accomplished, the valve to the system is closed. At this stage the molecular sieve is saturated. The re-activation can be carried out by allowing the pump to warm at room temperature, care being taken to vent the pump. A removable rubber stopper (fig. 5.48) allows easy release of the gas which is evolved from the molecular sieve as it warms. The stopper also acts as a safety pressure-release valve, since a high pressure may develop in the pump on its heating.

Normally, warming the molecular sieve to room temperature is all that is necessary to prepare it for the next pumpdown. However, since adsorbed water is not readily evolved from the molecular sieve at room temperature, it occasionally becomes necessary to bake the molecular sieve for several hours to drive off water vapour. For this purpose a bakeout unit is used, which fits tightly around the sorption pump, and heats it up to 300°C.

Sorption pumps are generally used to pump from atmospheric pressure, and ultimate pressures of the order of 10^{-2} Torr are achieved provided the sorptive capacity is correctly matched to the volume of the system. Typical performance characteristics of the pump from fig. 5.48a are shown in fig. 5.49. In order to pump larger volumes several pumps may be used simultaneously. In order to obtain a sort of continuous pumping it is possible to use two pumps alternately, one pumping whilst the other is valved off from the system and is being re-activated (fig. 5.48b).

Sorption pumps are described by Barrington (1963), Turner and Feinleib (1962), Danielson (1970), Wheeler (1974), Weston (1978). In order to increase the thermal contact to the sorbent, several authors recommend to bond the sieves to the cooling

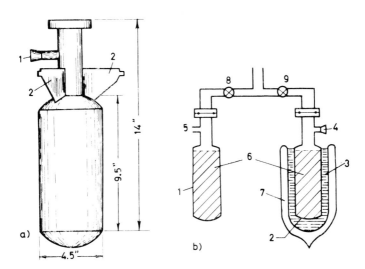

Fig. 5.48 (a) Sorption pump (Varian). 1. Rubber stopper; 2. Liquid nitrogen container support. (b) Two sorption pumps connected for multistage pumping. 1. Pump being re-activated; 2. Pump in operation; 3. Liquid nitrogen; 4. Rubber stopper; 5. Vent; 6. Molecular sieve; 7. Dewar; 8. Valve closed; 9. Valve open.

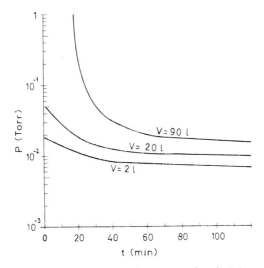

Fig. 5.49 Pressure vs pumpdown time, for sorption pump (fig. 5.48a) operating on various volumes; after 10 min pre-chill

surfaces (e.g. Sands and Dick, 1966; Stern *et al.*, 1966; Powers and Chambers, 1971; Halama and Aggus, 1974, 1975; Dillow and Palacios, 1979).

5.5.3 *Multistage sorption pumping*

The final pressure achieved by sorption pumping can be improved by prepumping the sorption pump with another sorption pump or a mechanical pump (Turner and Feinleib, 1962; Barrington, 1963; Nair and Vijendran, 1974, 1977; Dobrozemsky and Moraw, 1974; Weston, 1978).

The effects which can be achieved may be shown by considering the example of a sorption pump connected to a vacuum system, where the ratio of the chamber volume V to the sorbent weight w is $V/w=0.02$ liter/g. If the partial pressure of a particular gas in the atmosphere of the system is P_1, then the quantity of this gas present in the system is $P_1 \cdot V/w$ Torr·liter/g. As the pump is cooled to $-195°C$, a lower equilibrium pressure P_2 is achieved, since a quantity of gas $(P_1-P_2) \times V/w$ was sorbed. This quantity of gas is equal with that shown by the $-195°C$ adsorption isotherm (fig. 5.46). If we denote this latter quantity by Q_T/w, the equilibrium is described by the equation

$$(P_1 - P_2)\, V/w = Q_T/w \tag{5.18}$$

which can be arranged as

$$Q_T/w + (V/w)P_2 = (V/w)P_1 \tag{5.19}$$

Figure 5.50 shows the plot of $P_1(V/w)$ for $V/w = 0.02$ (line 1) and $V/w = 0.01$ (line 5). Curves 2 and 3 are plots of $Q_T/w + 0.02\,P_2$ for nitrogen and neon using the values of Q_T/w (at 195°C) shown in fig. 5.46. Line 4 is the value of Q_T/w nitrogen at room temperature.

Point A (fig. 5.50) results from the intersection of $P_1 = 595$ Torr (partial pressure of N_2), with line 1, thus defines the value $P_1(V/w) = 595 \times 0.02$. The horizontal through A, intersects curve 2 at B, and the vertical through B defines the pressure $P_2 = 3 \times 10^{-3}$ Torr, and intersects line 1 at C. According to eq. (5.18), the vertical distance BC is the quantity of nitrogen sorbed.

By adding an additional sorption pump V/w is practically halved to 0.01 (line 5, fig. 5.50), and for *both pumps cooled simultaneously* point A moves to E. Since $0.02\,P$ or $0.01\,P$ are small values compared to Q_T/w for N_2 at $-195°C$, the curve $0.01\,P + Q_T/w$ is practically the same with curve 2.

The horizontal through E intersects curve 2 at F, thus the new equilibrium pressure is about $P_2 = 1.3 \times 10^{-3}$ Torr, which cannot be considered as a significant improvement.

By *cooling the first pump* and leaving the second pump connected but uncooled, the pressure drops from A to B as before, and the quantity of N_2 remaining at C

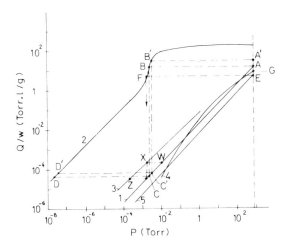

Fig. 5.50 Multistage sorption pumping with 5A molecular sieve. After Turner and Feinleib (1962).

is about five orders of magnitude smaller than the initial quantity at A. If the first pump is now valved off and the *second pump is cooled*, the pressure drops from C to D, thus the final pressure of N_2 is of the order of 10^{-7}–10^{-8} Torr, which is a serious improvement. Since the two pumps have already adsorbed a certain quantity of nitrogen at room temperature which is given by point G (line 4) this must be added to that represented by point A (fig. 5.50). Thus the quantity of N_2 present is increased to A', which raises the intermediate pressure from B to B', moves C to C' and D to D'.

If *neon* is considered, point W (fig. 5.50) represents its partial pressure (1.4×10^{-2} Torr) in the atmosphere, and the quantity of neon present. The first pump lowers the neon pressure to 2×10^{-3} Torr (point X), and the second pump to 2.6×10^{-4} Torr (point Z), which is a pressure much higher than that of nitrogen (point D, or D'). In practice, adsorption of N_2 can interfere with neon adsorption and the actual equilibrium pressure of neon may be higher than point Z.

If *helium* is considered, the small sorptive capacity of zeolite for this gas (fig. 5.46) explains the fact that its partial pressure of 4×10^{-3} Torr is practically not changed by sorption pumping.

Thus sequential pumping is able to reduce the partial pressure of N_2 (and active gases), but cannot achieve total ultimate pressures lower than a few microns due to neon and helium.

If a mechanical pump is used to prepump the system, although the result is not entirely oil-free, it reduces both the partial pressures of neon and helium. By cooling the sorption pump after this prepumping, the ultimate pressure in the system should be much lower (fig. 5.51).

5.6. Cryopumping

5.6.1. Cryopumping mechanism

Cryopumping is the process by which gases (vapours) are condensed at low temperatures in order to reduce the pressure. The ultimate pressure which can be achieved by such a process was described by eq. (4.9), while the pumping speed which can be obtained was expressed by eq. (4.12).

Cryopumping may involve three mechanisms: cryocondensation, cryosorption and cryotrapping (Schäfer and Häfner, 1987). *Cryocondensation* occurs when gas molecules impinge on molecules of the same species adhering to the cold surface; by this process condensation layers of millimeters (thickness) may grow. *Cryosorption* is the physical adsorption of gases on cold surfaces; the sorption capacity is increased by enlarging the adsorbing surface, e.g. using a layer of activated charcoal (inner surface of up to 1000 m^2 per gram of charcoal). *Cryotrapping* is the process by which gases are trapped (buried) in the growing solid condensation layer of a second gas, e.g. trapping of H_2 or He in N_2, Ar or CO_2.

Review discussions of cryopumping have been published by Gareis and Hagenbach (1965), Power (1966), Redhead *et al.* (1968), Kidnay and Hiza (1970), Hobson (1973), Benvenuti (1977), Schäfer (1978), Weston (1978), and Haefer (1981). The measurement of the sticking coefficient is discussed by Habets *et al.* (1975).

In order to understand the cryopumping phenomenon Moore (1962) analyzed it in some detail. He considered the molecular flow (see §3.1.1) of a gas between two infinite parallel planes; one a gas source and the other a cryopump condenser (gas sink), as shown in fig. 5.52. For this analysis Moore made the following assumptions: (1) The distance L between the surfaces is small compared to the mean free path (molecular flow). (2) The condensing surface is covered with a deposit of condensed solid (formed from gas from the source surface), and the deposit has an exposed surface temperature T_2. (3) Of the stream of molecules that strike the solid on the condensing surface (the mass flow rate w_1) the fraction f stick and the rest are diffusely reflected. (4) The reflected molecules leaving the condenser constitute a mass flow $(1-f)w_1$, and have a velocity distribution corresponding to the temperature T_2, i.e. the accommodation coefficient (eq. 2.104) is unity. (5) In addition to the reflected molecules, the solid deposit also emits molecules by evaporation at the temperature T_2, and at the same rate We_2 as if it were in equilibrium with a gas at temperature T_2; (6) The mass flow, w_1, from the source consists of the flow W_1 emitted by the source and that of diffusely reflected molecules w_2 which strike the source surface. The mass flow w_1 is constituted of molecules having a velocity distribution corresponding to the temperature T_1. (7) The velocity distributions of all molecular streams are Maxwellian (eq. 2.38).

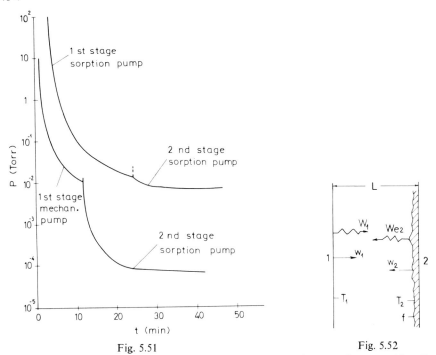

Fig. 5.51   Fig. 5.52

Fig. 5.51 Two-stage sorption pumping, with two different combinations of pumps. After Turner and Feinleib (1962).

Fig. 5.52 Model for analysis of cryopumping between infinite parallel planes. 1. Source; 2. Condenser with thin deposit. After Moore (1962).

According to these assumptions, the gas between source and condenser (fig. 5.52) can be considered composed of two streams[*] moving in opposite directions: w_1 flowing from source to condenser, and w_2 flowing from condenser to source: w_1 and w_2 have velocity distributions corresponding to temperatures T_1 and T_2 respectively.

Since the net mass flow input to the system is given by

$$W_1 = w_1 - w_2 = w_1 f - We_2 \qquad (5.20)$$

the mass flow for the two opposite streams results as

$$w_1 = (W_1 + We_2)/f \qquad (5.21)$$

$$w_2 = W_1 (1-f)/f + We_2/f \qquad (5.22)$$

[*] Directional detectors are discussed by Tuzi and Kobayashi (1983).

If w_1 and w_2 are considered as being gas streams with molecular densities n_1 and n_2, according to eqs. (2.47) and (2.38) it results

$$\frac{w_1}{A} = \frac{mn_1 v_{1av}}{4} = n_1 \left(\frac{mkT_1}{2\pi} \right)^{1/2} \tag{5.23}$$

$$\frac{w_2}{A} = \frac{mn_2 v_{2av}}{4} = n_2 \left(\frac{mkT_2}{2\pi} \right)^{1/2} \tag{5.24}$$

where w/A is the mass flow per unit area, m is the molecular mass, and v_{av} is the average molecular velocity (eq. 2.38).

Equations (5.21–5.24) can be used to express the densities of the hypothetical gas streams

$$n_1 = \frac{4}{mAv_{1av}} \frac{W_1 + We_2}{f} = \left(\frac{2\pi}{mkT_1} \right)^{1/2} \frac{W_1 + We_2}{Af} \tag{5.25}$$

$$n_2 = \frac{4}{mAv_{2av}} \frac{W_1(1-f) + We_2}{f} = \left(\frac{2\pi}{mkT_2} \right)^{1/2} \frac{W_1(1-f) + We_2}{Af} \tag{5.26}$$

The total molecular density is the average of these two expressions, but the gas between source and condenser will be far different from isotropic, *thus the usual meaning of pressure cannot be used here*. The situation can be illustrated by considering the pressure which would be sensed by an open ended probe in various orientations (fig. 5.53). It is assumed that the pressure sensed is the pressure P_p inside the probe, the gas being at the temperature of the probe T_p. The pressure P_p reaches a level so that the efflux of molecules through the opening equals the influx from the environment, which is the sum of the molecular fluxes incident on the probe opening from surfaces 1 and 2 (fig. 5.53).

Based on the expressions marked on fig. 5.53 and eqs. (5.25, 5.26) the pressure sensed in various orientations is

Case A:

$$P_p = \left(\frac{2\pi k T_p}{m} \right)^{1/2} \frac{W_1 + We_2}{Af} \tag{5.27}$$

Case B:

$$P_p = \left(\frac{2\pi k T_p}{m} \right)^{1/2} \left[\frac{W_1(2-f)}{2Af} + \frac{We_2}{Af} \right] \tag{5.28}$$

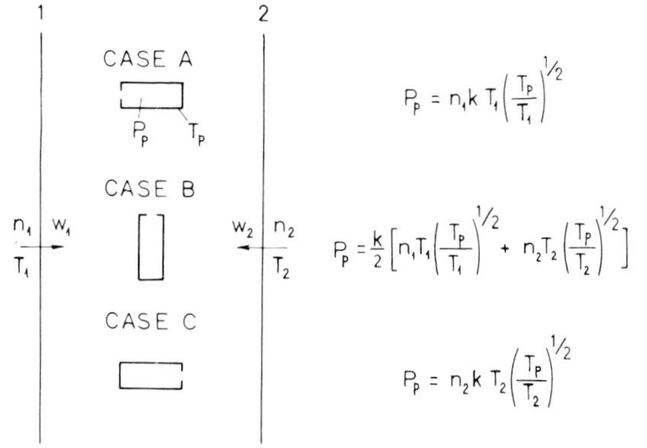

Fig. 5.53 Pressures inside an open-ended probe. After Moore (1962).

Case C:

$$P_p = \left(\frac{2\pi k T_p}{m}\right)^{1/2} \frac{W_1(1-f) + We_2}{Af} \tag{5.29}$$

Equations (5.27–5.29) show that regardless of probe orientation the contribution of the molecules evaporating from the deposit is the same and equal to

$$P_{e2} = \left(\frac{2\pi k T_p}{m}\right)^{1/2} \frac{We_2}{Af} \tag{5.30}$$

According to assumption 5

$$\frac{We_2}{A} = fP_{v2}\left(\frac{m}{2\pi k T_2}\right)^{1/2} \tag{5.31}$$

where P_{v2} is the vapour pressure of the condensed gas at T_2, and from eqs. (5.30) and (5.31)

$$P_{e2} = P_{v2}(T_p/T_2)^{1/2} \tag{5.32}$$

thus the contribution due to reevaporation from the condensing surface is equal to the vapour pressure of the deposit corrected for the probe temperature.

The contribution of the net mass flow input W_1 to the system is dependent on both f and the probe orientation (eqs. 5.27–5.29). This dependency is shown in table 5.4 for extreme values of f.

Table 5.4.
Probe pressure P_p, for $We_2 = 0$

Case	$f \ll 1$	$f = 1$
A		$\left[\dfrac{2\pi k T_p}{m} \right]^{1/2} \dfrac{W_1}{A}$
B	$\left[\dfrac{2\pi k T_p}{m} \right]^{1/2} \dfrac{W_1}{Af}$	$\left[\dfrac{2\pi k T_p}{m} \right]^{1/2} \dfrac{W_1}{2A}$
C		0

Obviously the values $f \ll 1$ represent the cases without appreciable cryopumping, thus the pressure is independent of the orientation of the probe, i.e. the gas is isotropic.

When $f = 1$ the portion of the probe pressure contributed by W_1 is highly sensitive to the orientation of the probe, varying from a maximum value for case A to zero for case C. Thus when W_1 is large compared to We_2 the gas is far from being isotropic.

The pumping speed of an isotropic system is defined by $S = Q/P$. Since the expression of the pressure in cryopumping (fig. 5.53) is different according to the orientation, the pumping speed is also different.

In the case of space simulation a *pumping speed per unit area S_{s1}/A based on the molecular density n_{s1} sensed by the source of gas* is most significant. This is equivalent to case C (fig. 5.35) with $T_p = T_1$. Thus from eqs. (5.29, 5.31 and 5.32)

$$P_{s1} = \left(\frac{2\pi k T_1}{m} \right)^{1/2} \frac{W_1 (1-f)}{A f} + P_{v2} \left(\frac{T_1}{T_2} \right)^{1/2} \tag{5.33}$$

and since

$$n_{s1} = P_{s1}/(k T_1) \tag{5.34}$$

the pumping speed per unit area is

$$\frac{S_{sl}}{A} = \frac{W_1}{A \, m \, n_{sl}} = \frac{f}{1-f} \left(\frac{k \, T}{2\pi \, m} \right)^{1/2} \left[1 - \frac{P_{v2}}{P_{sl}} \left(\frac{T_1}{T_2} \right)^{1/2} \right]$$

$$= \frac{f}{1-f} \frac{v_{avl}}{4} \left[1 - \frac{P_{v2}}{P_{sl}} \left(\frac{T_1}{T_2} \right)^{1/2} \right] \tag{5.35}$$

As P_{s1} approaches the value $P_{v2} \, (T_1/T_2)^{1/2}$ the pumping speed diminishes to zero. It results that as long as $P_{v2} \ll P_{s1}$ the pumping speed apparent to the gas source is independent of the pressure and temperature of the condenser and has the value

$$S_{s1}/A \approx [f/(1-f)] \, (v_{av1}/4) \tag{5.36}$$

When $f \ll 1$ (reflecting)

$$S_{s1}/A = f \, v_{av1}/4 \tag{5.37}$$

while when $f = 1$ (good sticking)

$$S_{s1}/A = \infty \tag{5.38}$$

This latter *is the case* which is desired to be achieved *in space environment simulation*.

A second definition of the pumping speed is that based on case B (fig. 5.53), which corresponds more directly to that *used for diffusion pumps*. In this case the pumping speed S_p is given by

$$\frac{S_p}{A} = \frac{W_1}{A \, m \, n_p} = \frac{W_1 \, k \, T_p}{A \, m \, P_p} =$$

$$= \frac{2 \, f}{2-f} \left[1 - \frac{P_{v2}}{P_p} \left(\frac{T_p}{T_2} \right)^{1/2} \right] \left(\frac{k \, T_p}{2\pi \, m} \right)^{1/2}$$

$$= \frac{2 \, f}{2-f} \frac{v_{avp}}{4} \left[1 - \frac{P_{v2}}{P_p} \left(\frac{T_p}{T_2} \right)^{1/2} \right] \tag{5.39}$$

To get maximum pumping speed it is required that $P_{v2} \ll P_p$, thus

$$S_p/A = [2f/(2-f)] \, (v_{avp}/4) \tag{5.40}$$

Here the pumping speed is independent of pressure, and the significant temperature is T_p, the probe temperature. This temperature is practically equal to the ambient temperature if conventional vacuum systems are concerned, but is usually very different in systems using cryogenic surfaces.

From eq. (5.40) and for $f \ll 1$, the pumping speed per unit area is

$$S_p/A \approx f v_{avp}/4 \qquad (5.41)$$

and for $f = 1$

$$S_p/A = v_{avp}/2 \qquad (5.42)$$

The result in eq. (5.41) is identical with that shown by eq. (5.37), because probe orientation is not important when $f \ll 1$. The result in eq. (5.42) differs markedly from that given by eq. (5.38), because the probe inlet in position B will still receive molecular flux, even when $f = 1$. Calculations of the cryopumping speed by Monte–Carlo method have been published by Chubb (1970), Akiyama et al. (1971). Performances achieved by cryopumping are discussed by Eder (1972), Flesch et al. (1975), Kleber (1975), Haefer and Kleber (1976), Bentley and Hands (1977), Thompson and Hanrahan (1977), Edwards and Limon (1978), Graham and Ruby (1979).

5.6.2. Cryopumping arrays

Cryopumping surfaces cannot generally be exposed directly to a source of gas at room temperature because the heat load due to radiation would exceed that due to the condensation of gas molecules. Therefore, the cryogenic surface is protected on the side facing the gas source by an optically opaque baffle (e.g. figs. 3.14, 3.15) at an intermediate temperature to act as a radiation shield. Figure 5.54 shows some common arrangements used in space simulation.

The radiation shields also impede the gas flow to the condenser and limit the maximum achievable pumping speed. Fig. 5.55 shows a cryopumping array which offers a reasonable compromise between pumping speed and radiation losses. Here both the chevron baffle and the back shield are cooled with liquid nitrogen to 77–100°K, while the condenser is operated at 20°K, so that nitrogen and all less volatile gases are cryopumped. Additional means, such as diffusion or ion-pumps must be provided for removing helium, hydrogen and neon. The emissivity of each surface is chosen primarily to minimize the heat load on the condenser, and secondarily to minimize that on the radiation shields. The relatively high value of the condenser emissivity takes into account the presence of frost.

Cryopumping arrays (cryopanels) and shielding methods are discussed by Forth (1965), Kienel and Wutz (1966), Turner and Hogan (1956), Haefer (1970), Omelka (1970), Holkeboer et al. (1967), Elo et al. (1979), Haefer (1980a), Schwarz (1987).

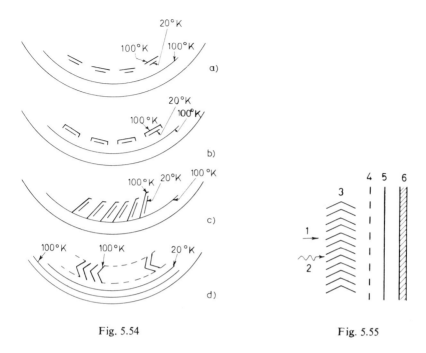

Fig. 5.54   Fig. 5.55

Fig. 5.54 Arrangements of cryopumping arrays.
Fig. 5.55 Detail of a cryopumping array. 1. Radiation; 2. Gas: 3. Chevron baffles (100°K);
4. Condenser (20°K); 5. Back shield (100°K); 6. Chamber wall (300°K). After Moore (1962).

The pumping effectiveness of the array will be described by its over-all capture probability G, rather than by the sticking coefficient of a simple condensing surface. The *capture probability* G is the fraction of the total number of molecules incident on the inlet side of the cryopumping array which is finally captured in the array. Moore (1962) analyzes the model array shown in fig. 5.55, by using: g_s – the probability that a molecule impinging on chevron shields will pass through (see §3.5); f – the condenser sticking coefficient, and g_c – the probability that a molecule impinging on the condenser will pass through (equal to the ratio of open area).

According to the notations on fig. 5.56, the conditions of equilibrium are

$$
\begin{aligned}
W_1 &= w_1 - w_2 \\
w_2 &= (1-g_s)\, w_1 + g_s w_4 \\
w_3 &= g_s w_1 + (1-g_s)\, w_4 \\
w_5 &= g_c w_3 + (1-g_c)(1-f)\, w_5 \\
w_4 &= g_c w_5 + (1-g_c)(1-f)\, w_3
\end{aligned}
\tag{5.43}
$$

and the simultaneous solution of these equations give the over-all capture probability as

$$
G = \frac{W_1}{w_1} = g_s \left[\frac{1-2(1-g_c)(1-f) + (1-g_c)^2(1-f)^2 - g_c^2}{1-(1-g_c)(1-f)(2-g_s) + (1-g_s)\left[(1-g_c)^2(1-f)^2 - g_c^2\right]} \right]
\tag{5.44}
$$

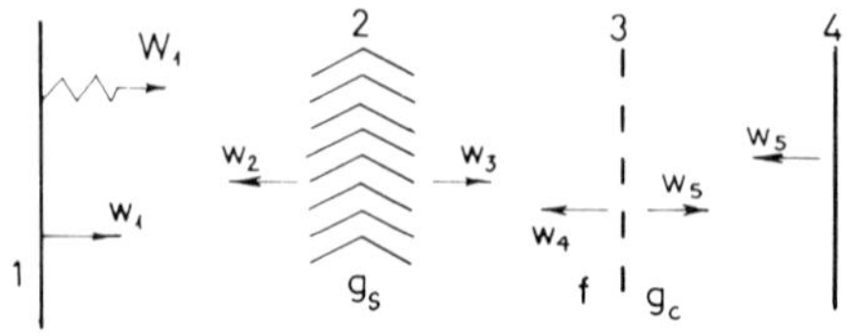

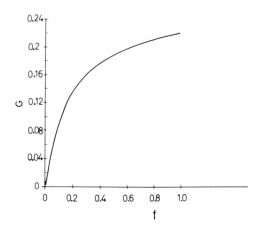

Fig. 5.56 Model for analysis of a cryopumping array. 1. Gas source; 2. Chevron baffle; 3. Condenser; 4. Back shield.

Fig. 5.57 Effect of condenser sticking coefficient on array capture probability. After Moore (1962).

Assuming $g_s = 0.23$ (fig. 3.22), and $g_c = 0.25$, the capture probability given by eq. (5.44) is a function of f, as shown by fig. 5.57. The pumping speed of such an array, as defined by probe oriented as case B (fig. 5.53), is given by eq. (5.40) with f replaced by G, thus

$$S_p/A = [2G/(2-G)] \, (v_{avp}/4) \tag{5.45}$$

5.6.3. Cryotrapping

Gases may be trapped on cooled surfaces on which a condensable vapour (eq. water) has been condensed, with the result that the partial pressure attainable may be significantly lower than the equilibrium vapour pressure at the temperature of the cooled surface. This pumping action is known as *cryotrapping*. It offers the possibility of cryopumping of gases such as N_2, H_2, Ar much more effectively in the presence of a contaminating agent (such as water vapour) than in a system from which all such agents have been carefully removed and excluded. Cryotrapping is believed to be due to the non-condensable gas being carried down by a condensable vapour and trapped within the pore structure of the condensate.

The *cryotrapping of nitrogen and argon by water vapour* (Schmidlin, 1962) condensed at $77°K$, has shown that the water vapour forms a porous deposit with an effective area of about 600 m^2/g of water. The quantity of N_2 and Ar required to saturate the surface deposit of water is proportional to the quantity of water deposited (fig. 5.58). The number of molecules of nitrogen trapped per molecule of water condensed on the surface has a constant value of about 10^{-2} for partial pressures of N_2 above about 0.1 Torr, and then decreases with decreasing partial pressure to a value of about 5×10^{-5} at 10^{-5} Torr.

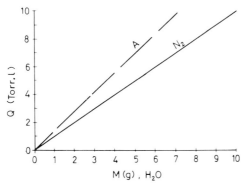

Fig. 5.58 Quantity Q of nitrogen and argon to saturate an ice coating of M grams. There results 1.0 Torr l/g for N_2 and 1.4 Torr l/g for A. After Schmidlin (1962).

The *cryotrapping of hydrogen and helium by condensed argon* at 42°K was also achieved. The results of Hengevoss and Trendelenburg (1963) are shown in fig. 5.59. For the curves (a) and (b) the *hydrogen* flow rates are so low that the equilibrium pressure is below that for saturation at 4.2°K, so that no condensation occurs and the pumping action is only that of the diffusion pumps. In each of these cases, when the argon partial pressure reaches about $\frac{1}{10}$ of that of the hydrogen, the hydrogen partial pressure drops suddenly by a factor of 10 or more and then remains at this lower value as the argon-flow rate and partial pressure are increased. The decrease in hydrogen partial pressure is due to additional pumping resulting from the trapping of hydrogen by the condensed argon. From the trapping rates and the area of the cryosurface, the sticking coefficient for hydrogen on the argon deposit was determined to be 0.4. The 45° line on fig. 5.59 corresponds to the case in which one hydrogen molecule is trapped by one condensed argon molecule. Curve (c) is taken at a hydrogen flow rate which is great enough that the hydrogen partial pressure exceeds the saturated value at 4.2°K so that the condensation on the cryostatic surface occurs even in the absence of argon. Thus in curve (c) both condensation and trapping occur at the same time. Therefore the break in the curve (cryotrapping by argon) occurs at an appreciably lower argon flow rate than that corresponding to the intercept with the 45° line. This indicates that about 10 times as many hydrogen molecules are deposited by the combination of condensation and trapping as are argon molecules.

Experiments for *cryotrapping of helium* showed that the sticking probability of helium on argon deposit is about 0.03 and that about 30 argon molecules are required to trap one molecule of helium.

Experiments on cryotrapping were reported by Schmidlin (1963), Hengevoss and Trendelenburg (1963, 1967), Muller (1966), Templemeyer *et al.* (1970), Boissin *et al.* (1972), Benvenuti and Calder (1972), Mongodin *et al.* (1974), Chou and Halama (1977, 1979), Sedgley *et al.* (1988).

The characteristics of cryotrapping by a layer of a second gas have been summarized by Haefer (1981). The pumping of H_2 at pressures of 10^{-6} Pa ($\sim 10^{-8}$ Torr) on a cryosorption pump using a condensed layer of CO_2 is discussed by Arakawa and Tuzi (1986).

5.6.4. *Cryopumps*

A practical cryogenic pump is illustrated in fig. 5.60. It consists of a helix made of stainless steel tube, which acts as the condenser surface, mounted directly in the chamber to be evacuated. The coolant (liquid nitrogen, hydrogen or helium) is supplied from a Dewar to the helix through a vacuum insulated feed tube, and is made to flow through the coil by means of a gas pump at the outlet end of the coil. The coolant boils as it passes through the coil, hence cooling the tube. The rate at which coolant passes through the system, and hence the temperature to which the condenser surface is cooled, is controlled by a throttle valve mounted in the gas exhaust line. A temperature sensing element mounted on the

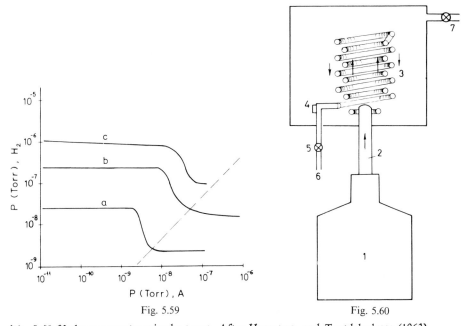

Fig. 5.59   Fig. 5.60

Fig. 5.59 Hydrogen cryotrapping by argon. After Hengevoss and Trendelenburg (1963).

Fig. 5.60 Cryopump (schematic), 1. Liquid nitrogen Dewar; 2. Insulated feed tube; 3. Condenser coil; 4. Temperature sensing element; 5. Throttle valve; 6. Connection to gas pump; 7. Connection to roughing pump.

condenser coils automatically controls the throttle valve setting.

The cryogenic pump is not used at pressures above 10^{-2} Torr, partly because of the large quantities of coolant that would be required, and partly because the thickness of solid built up during high pressure pumping would seriously reduce the pump efficiency at low pressures. The rate of built-up of solids is typically of the order of 10 cm/h at 10^{-1} Torr and 10^{-2} cm/h at 10^{-4} Torr.

Pumps having speeds for nitrogen up to 5000 l/s (liquid helium consumption = 2 l/h) are commercially available. Pumps with speeds of the order 10^6 l/s are feasible, but for these high speeds the coolant would be fed directly from a gas liquifier rather than from a storage vessel. The introduction of small, cryogenic, closed-cycle refrigerators provided possibilities to construct systems in which helium is compressed and expanded, reaching two steps of refrigeration: 80°K and 12–20°K, respectively (Schäfer, 1978; Visser and Sheer, 1979; Coupland et al., 1987; Foster, 1987).

The sorbent surface ratio A_p/A_g (see §4.3.5) can be increased by using activated charcoal, coated on the cryogenic surfaces as mentioned in §4.3.6.

Cryopumps are discussed and described by Hobson (1964, 1973), Dawson and Haygood (1965), Roussel et al. (1965), Power (1966), Davis (1968), Hengevoss et al. (1970), Fox (1970), Benvenuti and Calder (1972), Benvenuti (1973), Forth *et al.* (1972), Wössner (1972), Dobrozemsky (1973), Schäfer and Schinkmann (1973), Benvenuti (1974, 1977), Benvenuti and Blechschmidt (1974), Blinov and Minaichev (1974), Klipping (1974), Woods and Delvin (1974), Flesch *et al.* (1975), Wolgast (1975), Benvenuti *et al.* (1976), Davey (1976), Hands (1976), Becker (1977), Bottglioni *et al.* (1977), Denison (1977b), Forth and Frank (1977), Hands and Davey (1977), Siebert and Omori (1977), Visser *et al.* (1977), Rijke (1978), Schäfer (1978), Benvenuti (1979), Benvenuti and Hilleret (1979), Visser and Sheer (1979), Yarwood (1979), Charland and Frei (1980), Harra *et al.* (1980), Robens (1980), Schwenterly (1980), Wilson and Watts (1980), Benvenuti and Firth (1979), Bentley (1980), Coupland *et al.* (1982), Duval (1982), Hands (1982), Liu *et al.* (1982), Longsworth and Bonney (1982), Muhlenhaupt (1982), Rommel (1982), Sänger *et al.* (1982), Scheer and Visser (1982), Nöller (1983), Hood (1984), Klein *et al.* (1984), Bridwell and Rodes (1985), Arakawa and Tuzi (1986), Foster (1987), Liu *et al.* (1987), Sedgley *et al.* (1987), Schäfer and Häfner (1987), Tobin *et al.* (1987), Porter (1988), Sedgley *et al.* (1988).

The "cold bore" system of (future) proton storage rings represents a very large (perhaps the largest) cryopump, where the whole vacuum chamber wall is at liquid helium temperatures. These systems and their problems were analyzed by Halama and Herrera (1975), Wolgast (1975), Aggus *et al.* (1977), Benvenuti, Calder and Hilleret (1977), Blechschmidt (1977), Cho *et al.* (1977), Fischer (1977), Benvenuti and Hilleret (1979).

5.6.5. *Liquid nitrogen traps*

Liquid nitrogen traps are cryogenic devices which function primarily to prohibit the transfer of pump oil vapour into the vacuum system, and to pump by condensation, water vapours and other vapours which originate in the system.

Well-designed traps incorporate the following features:

(a) The trap offers minimum impedance to the diffusion pump (see §3.4.4 and 3.5) and to the vapour condensation or cryopumping surface (§5.6.2). However the trap should be effective in keeping pump oil from the chamber.

(b) A constant low temperature of the cryopumping surfaces is maintained to prevent pressure bursts resulting from the reevaporation of condensate.

(c) There is no warm surface path shunting the cold areas which would permit the diffusion pump oil to migrate.

(d) Additional considerations include minimum liquid nitrogen consumption, and a trap interior accessible for cleaning. A water-cooled baffle is required to prevent holdup of the pump oil on the trap during long-term continuous operation. The condensed oil should return to the pump along the trap walls rather than dripping onto the hot jet assembly.

Liquid nitrogen traps have been discussed by Milleron (1959), Holkeboer

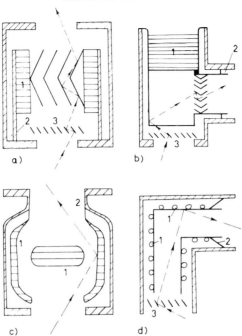

Fig. 5.61 Liquid nitrogen traps. 1. Liquid nitrogen; 2. Place of thermal gradient; 3. Water baffle.

et al. (1967), Boissin and Thibault (1971), Santeler (1971), Harris (1977), Lunelli (1978), Carpenter and Watts (1982), Singleton (1984).

Figure 5.61 illustrates several trap designs which meet some of the above criteria. Types a, b, c are generally used with pumps of small and medium sizes. An elbow trap (fig. 5.61d) may be used on large vacuum systems where a liquid nitrogen circulation system is available. A vacuum system pumped down to its operating pressure exhibits molecular flow through the trap, thus for an optically tight trap, an oil molecule must undergo at least one cold wall collision (§3.5) before entering the vacuum chamber. Since the sticking coefficient is less than unity, and some oil-to-oil and oil-to-gas molecular collisions also occur, traps can be more effective if they are designed such that each oil molecule impinges on the cold surfaces a greater number of times (Types a, b) than the minimum of one contact required by simple optical tightness (Types c, d) (see also fig. 3.29).

Jones and Tsonis (1964) calculated the oil–backstreaming rate through an (36″ diam) elbow trap, and presented the results in the diagram of fig. 5.62. The calculations were made for DC 704 silicone oil held at $300°K$ at the trap bottom, neglecting any cracking effects. Solid horizontal lines show the backstreaming due to a sticking coefficient α less than unity, each line corresponding to a value of α, resulting from the conclusion of a calculation which indicates that $1.09\,(1-\alpha)$

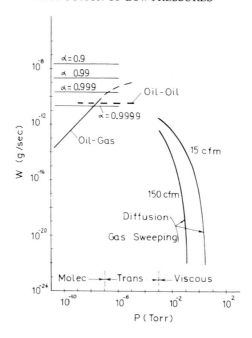

Fig. 5.62 Oil backstreaming W (calculated values) through a 36 inch diam. optically tight elbow trap. After Jones and Tsonis (1964).

percent of the entering molecules pass through the trap. Unfortunately the sticking coefficient is not accurately known, and small differences in its value can cause differences of decades in the backstreaming rate.

The interrupted line (fig. 5.62) is the oil transfer due to oil–oil collisions. From the calculation, Jones and Tsonis established that the oil transfer due to oil-oil collisions is

$$N = 2 \times 10^{14}\, Pa^3/M^{1/2} \text{ molec/second} \tag{5.46}$$

where a (cm) is the radius of the pipe of the elbow, M is the molecular weight of the oil, and P is the partial pressure of oil vapour at trap inlet (Torr). It results that this mode of transfer is significant only if $\alpha > 0.9999$.

The probability P_r of an oil molecule escaping the trap due to an oil–gas collision was found to be

$$P_r = 4.5 \times 10^{-5} \,[1-\exp\,(-l/\lambda)] \tag{5.47}$$

where l is the mean length of escape path from trap base to knee (in the case considered $l=2.7a$), and λ is the mean free path of the oil molecule. Since λ is an inverse function of the gas pressure, the transfer due to oil–gas collisions is rising with gas pressure. It results that this mode of transfer is significant only if $\alpha > 0.99$.

The right-hand side of fig. 5.62 shows the oil transfer resulting from diffusion and sweeping action in the transition and viscous flow regions.

Traps are usually not meant to operate in the transition region (see table 3.1), although conditions existing in the transition region can be expected during pumpdown. The trap configuration shown in fig. 5.61b suppresses this oil transfer by having two regions of trapping which differ significantly in dimensions, the skirt and the chevron. Thus, transition flow in the two sections will occur at two different pressure ranges, and at any pressure at least one of the sections will be efficient.

Regions of the trap where the cold surfaces are in contact with parts at ambient temperature experience temperature gradient shifts resulting from changes in ambient temperature, or in the cooling conditions. These gradient shifts result in the reevaporation of a part of the condensate. If the reevaporated gas is not retrapped prior to chamber reentry, a pressure rise will occur in the chamber. The type of trap shown in fig. 5.61c has no allowance for retrapping capability. The other designs shown in fig 5.61 have a good possibility of retrapping such molecules. The amount of gas condensing on the thermal gradient region can also be minimized by shielding these regions from the chamber with cold surfaces (fig. 5.61a, b, d). Gas condensed on the intermediate temperature regions during the higher pressure phase of a pumpdown will reevaporate at lower pressures and thus cause an increase in pumpdown time. Figures 5.61a and c indicate that troublesome temperature gradients may exist on the top of a partially empty liquid nitrogen reservoir when it is in sight of the chamber. If the reservoir walls have a sufficiently high thermal conductivity, this temperature gradient can be avoided.

5.7. Gettering

5.7.1. *Gettering principles*

Any sealed-off vacuum device (lamp, electronic tube) in which the pressure remains essentially constant, contains a chemically active *getter*.

While effort is made to effectively degas the tube parts during construction and evacuation before sealing-off, there will always be some evolution of gases during operation. To avoid a build-up of pressure *a getter*, i.e. a material able to chemisorb gases, is included in the tube. Getters and gettering are discussed by Jackson and Haas (1967), Della Porta (1972), Malev (1973), Giorgi (1974), Lange (1975), Giorgi *et al.* (1985).

Getters may be classified into three groups according to the form in which the getter material is active: *flash*, *bulk* and *coating* getters.

Flash getters are chemically active metals which can be easily volatilized. Flash getters may work by the process of *dispersal gettering* in which gas is sorbed whilst the getter is being evaporated, and *contact gettering* in which the film of getter material deposited by evaporation on the walls continues to sorb gases. Dispersal gettering is limited in time, but the getter presents its maximum possible surface to the gas; contact gettering is a continuous process where sorption on the outer layer is followed often by diffusion into the bulk.

Bulk getters are heated metals in sheet or wire form.

Coating getters consist usually of powders sintered on the surfaces of electrodes. The general conditions for getters can be summarized as:

(a) The getter must be resistant to storage prior to use.

(b) It must not be affected by the process of manufacture of the tube (sealing-in of the electrode assembly, bake-out of the tube, sealing-off of the tube).

(c) The getter should sorb during its evaporation (if flash getter) and after the tube has been processed.

(d) The vapour pressure of the getter material and its reaction products with the residual gases in the tube should be negligible.

Operating temperatures and uses of various getters are summarized in table 5.5.

Table 5.5.
Operation characteristics of getters.

Characteristic \ Material		Flash getters			
		Ba (and alloys)	Batalum	Ba–berylliate	Bato
Shape		Metal clad, wire or pellets	Paint on Ta	Paint on Ta	Metal clad pellets
Temperature (°C)	Degassing	400–700	800–1100	900–1000	—
	Flashing	900–1200	1200–1300	1300	750–900
	Gettering	20–200	20–200	20–200	20–200
Uses		Most kinds of tubes	Small and oxide cathode tubes	Small and oxide cathode tubes	Medium and high power tubes

(contd.)

5.7.2. *Flash getters*

The active ingredients in flash getters are chemically active metals that can be easily volatilized, such as Ba and Ba–Al alloys. An alloy of rare earth metals (Cerium, Lanthanum) known as *misch metal* is also used. *Phosphorus* is the getter used in most incandescent lamps.

Barium getters are made in a form that can be easily evaporated and deposited as a thin film on the vacuum envelope or on other cool inactive parts of the structure. In the processing of an electron tube, the procedure is: evacuation and bakeout at 400–500°C, cathode heating, induction heating of electrodes, and flashing the getter just before the tube is sealed-off or immediately afterwards. As barium is very active it must be protected against atmospheric action during storage and within the vacuum tube prior to flashing. The protection is done by incorporating the Ba in metal pellets or tubes, or alloying it.

Nickel, copper or iron clad tubes or pellets are obtained by filling a tube of these metals with liquid barium under vacuum. The tube is then drawn down to

Table 5.5 (*contd.*)
Operation characteristics of getters.

Flash getters			Bulk and coating getters				
Mg	Al–Mg	Red phosphorus	Ti	Zr	Ta	Th	Ceto
Ribbon, wire	Powder paint	Powder suspension	Sheet, wire	Sheet, wire, powder	Sheet, powder	Powder	Powder
400	400	—	800–900	700–1300 (−1700)	1600–2000	800–1500	800–1200
500	—	> 200	—	—	—	—	—
Only during flashing		100–200	1200–400	1400–400	1500–800	400–500	200–300
Gas filled and mercury vapour tubes	Small and oxide cathode tubes	Incandescent lamps	Sputter and sublimation pumps	Medium size and gas, (Hg) tubes	Medium and high power tubes	Medium size tubes	Special tubes

size obtaining a wire, or cut into small cushion-shaped pellets. Ni, Cu and Fe tubes with Ba filling are commercially available in a wide range of diameters and in "infinite" length under such names as *Niba*, *Cuba*, and *Feba* wire. The right amount of Ba for a particular use is nipped from the clad wire with blunt-edged cutters, which causes self-sealing of the ends and protects the Ba from atmospheric oxygen to some extent. The tube (cladding) wall is often purposely thinned along an axial strip to facilitate exit of Ba vapour and to direct it in a predetermined direction. The surface is ground flat (fig. 5.63a) on one side in the getter known as KIC (Kemet iron-clad), or given a longitudinal notch, known as Kerb getter wire (fig. 5.63b). These getters are to be degassed at about 750°C, and subsequent heating to 850°C or more causes the tube or pellet to burst.

To avoid getter spoilage during storage or processing, *alloys of Ba with Al* are used. These are known as the *Stabil* type getters of SAES–Milano, and consist of Ba–Al alloy either in a straight, grooved Fe or Ni tube which is welded to a nickel or iron alloy wire bridge (fig. 5.64a, b) or within a groove in a ring of stainless steel (fig. 5.64c). These getters are chemically stable in air up to temperatures of 600°C, but require a higher flashing temperature (above 1000°C) than the pure Ba getters.

A gettering method used in all-metal electron tubes is the *Batalum* process. In this process a thin tantalum strip which can be heated by special connecting leads, is coated with a mixture of $BaCO_3$ and $SrCO_3$ which is fully stable in air. This strip is heated first to 800–1100°C at a suitable stage of the evacuation process, whereupon the carbonates decompose to their respective oxides. CO_2 is formed which is pumped off, the strip temperature raised to 1300°C, the Ta substrate reduces the oxides to Ba which evaporates onto the tube wall. A disadvantage is that the tantalum oxide formed slowly dissociates and increases the

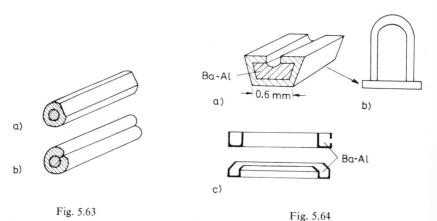

Fig. 5.63   Fig. 5.64

Fig. 5.63 Barium getter wires; (a) KIC; (b) Kerb.
Fig. 5.64 Ba–Al alloy getters.

pressure in the sealed-off tube. The replacement of the carbonate mixture by *barium berylliate* ($BaBeO_2$) is a subsequent development which avoids the evolution of CO_2. Two other types of these (reaction getters) are the *Alba* in which barium oxide is reduced in the presence of aluminum, and *Bato* in which barium is evaporated from a mixture of Ba–Al alloy, iron oxide and thorium powder. Other getters are obtained by using Sr–Al or Ca–Al alloys (Turnbull, 1977).

Red phosphorus is used as a getter in both vacuum and gas-filled lamps. The tungsten filament is coated with red phosphorous from an alcoholic suspension. At the first incandescence of the filament the phosphor is flashed and it gives rise to a deposit on the glass envelope which is transparent to light.

Magnesium is used as a getter in mercury vapour hot cathode discharge tubes, and *Al–Mg alloy* (the so-called Formier getter) is used in oxide–cathode tubes.

Della Porta (1972) describes a "getter mercury dispenser" to be used in mercury (fluorescent) lamps. The dispenser is a cathode shield coated on alternative sides with Zr 16 % Al getter and a Ti_3Hg compound. By heating (after seal-off of the lamp) at 850–950°C, the getter is activated, whereas the mercury alloy decomposes to yield the precise quantity of mercury necessary in the lamp (Della Porta and Rabusin, 1974).

5.7.3. *Bulk and coating getters*

Bulk getters are usually operated at elevated temperatures to promote diffusion of the adsorbed gas into the solid. Since evaporation is not essential to the operation of bulk getters, they are usually selected from chemically active elements with low vapour pressures and high melting points. These include Ti, Zr, Ta, Th, U, W, and Mo, used in a variety of forms: sheet, ribbon, wire, rod (bulk getters) or powder (coating getters).

Titanium is frequently used as a bulk getter in electron tubes. After Ti has been outgassed in vacuum at 800°C for a few minutes no additional gases are liberated when it is heated to higher temperatures. At temperatures above 700°C Ti continuously sorbs N_2, O_2 and CO_2, since at these temperatures C, N_2 and O_2 readily diffuse into the Ti, leaving an active surface. Sorption of these gases from 10 to 90 atomic per cent is possible, and the compounds formed do not dissociate at higher temperatures or upon subsequent reheatings. Hydrogen is sorbed by Ti from room temperature up to 400°C. Above 500°C, H_2 is released by the Ti and at 800°C it becomes practically free of H_2. Water vapour and CH_4 may be gettered by a two-step process: the Ti is heated first at 1000–1200°C where H_2O and methane are dissociated, the O_2 and C diffuse into the Ti, leaving the H_2 in the gaseous phase. The Ti temperature is then lowered below 400°C, and the H_2 is gettered.

Zirconium absorbs H_2 when heated to 400°C and re-evolves it again at 880°C or more. Water vapour is cleaned up between 200 and 250°C; O_2, N_2, CO and

CO_2 are sorbed at $1400°C$. An efficient getter can therefore be arranged by means of two Zr wires or strips; one at $400°C$ and the other at $1400°C$. Another usual procedure is to spray or paint the zirconium powder in a binder (nitrocellulose in amyl acetate) onto the electrodes. Zirconium hydride (ZrH_4) may also be applied as a paste, and at $800°C$ in vacuum the H_2 is released leaving the Zr coating.

The properties of Zr16Al getters and their applications are discussed by Barosi and Rabusin (1974), Benvenuti and Decroux (1977), Ferrario and Rosai (1977), Lange (1977), Borghi and Ferrario (1977), Rosai et al. (1978), Borghi and Rosai (1979), Cecchi et al. (1985), Emerson et al. (1986, 1987), Shen (1987), Benvenuti and Francia (1988), Ichimura et al. (1988). The use of Zr as a getter on a porous graphite (Barosi and Giorgi, 1973), or of preparing the Zr in situ by electrolysis of ZrO_2 (Fouletier and Kleitz, 1975) were also proposed.

Zirconium is also used as Zr–Ni alloy getter (Kuus and Martens, 1980) and as Zr–V–Fe getter alloy (Audi et al., 1987; Boffito et al., 1987; Ichimura et al., 1988).

The NEG (nonevaporable getter) is used as the linear pump of the LEP (Large Electron–Positron) collider (storage ring at CERN-Geneva), described by Reinhard (1983), Benvenuti and Francia (1988). Here the Zr–Al (16%) alloy is bonded as a powder (by cold pressing) on both sides of a Constantan ribbon $30 × 0.2$ mm, producing an active geometrical surface (see §4.3.5) of 540 cm^2 per meter length of ribbon. The getter is used/kept at ambient temperature and heated only intermittently for activation. It has a pumping speed of about 1000 liter/second per meter length of ribbon for CO and H_2 (at 10^{-2} Torr · liter/m) and drops at 0.1 Torr · liter/m to a level where heating for conditioning (activation) is required (Benvenuti and Francia, 1988).

Tantalum behaves like Ti at elevated temperatures. After flashing at $2000°C$, it getters H_2 at $800°C$ and O_2 and N_2 at $1500°C$.

Thorium sorbs O_2 and H_2 in the temperature range of $400-500°C$. It is also used in a getter called *Ceto* which is an alloy of 80% Th with 20% misch metal (an alloy of $50-60\%$ cerium, $25-30\%$ lanthanum and other rare earth metals). Thorium powder and Ceto powder are also used as coating getters.

Tungsten filaments getter O_2 at temperatures above $1500°C$. A fraction of the O_2 molecules striking the hot filament are converted to WO_3 and then immediately evaporated off leaving a clean surface for further gettering action. Nitrogen and CO cannot be gettered in this way, but they can be gettered by evaporating W onto a cool surface. When W is heated above $2300°C$, a fraction of the H_2 molecules striking the filament will dissociate into atoms. The atomic hydrogen produced is very active chemically and may reduce glass or ceramic in the vicinity of the filament to form metal oxides and water vapour. On striking a cool metal wall the atomic hydrogen may recombine to H_2. Water vapour is dissociated on coming in contact with a W filament above $2300°C$. The O_2 combines with the W and is distilled to the surrounding walls as WO_3 (white deposit). Atomic hydrogen evolved from the filament reduces the WO_3 on the wall to metallic W (black deposit) with the formation of water vapour. The "water cycle" can then be repeated indefinitely until the filament is etched to the point where it burns out.

An interesting inverse cycle occurs when a W filament is heated above $2800°C$ in the presence of iodine vapour. Tungsten which evaporates from the filament and condenses on a hot surrounding wall at about $500°C$ will combine with the iodine to form WI_4 vapour. This vapour is then dissociated on coming in contact with the hot filament and iodine vapour is released to react with the W deposit on the wall again. The phenomenon is used to increase the life time of incandescent lamps.

5.7.4. Gettering capacity

The sorption capacity of getters can be evaluated in terms of pumping speed per unit area of getter deposit, and in terms of total sorption per unit weight of getter. For the same getter both figures differ for various gases, and depend in some limits on the history of the getter (degassing, flashing, etc).

It is obvious that the pumping speed S_g of a getter film of area A, can be calculated by a formula of the type of eq. (3.72) or eq. (4.12) (for $P_u = 0$), thus

$$S_G/A = 3.64 \ f(T/M)^{1/2} \ \text{liter/sec·cm}^2 \tag{5.48}$$

where f is the sticking coefficient, T and M the absolute temperature and molecular weight of the gas. By comparing the maximum theoretical pumping speed ($f = 1$) with the gettering rates measured it was found that for freshly deposited barium films at room temperature, the value of the sticking coefficient is 0.4 for CO, 0.6 for CO_2, 0.02 for O_2, 0.001 for H_2 and 0.003 for N_2.

Although these values can be changed if a mixture of gases is present, it results

Table 5.6.
Sorption capacity of several getters at room temperature (Torr·liter/g).

Getter \ Gas	O_2	H_2	N_2	CO_2
Barium	15–50	45–88	3–36	5–60
Magnesium	20–200	—	—	—
Misch metal	20–50	46–64	3–16	2–45
Thorium	7–33	19–54	—	—
Aluminum	8–36	—	—	—
Uranium	9–10	8–22	—	—

that the maximum initial gettering speed for CO is 4.75 liter/sec·cm², and for O_2 is 0.2 lit/sec·cm². All measurements show that this speed will fall off with time. The gettering capacity of Ba for various gases was measured by Malev (1973a).

The total sorption capacity of several getters is shown in table 5.6.

The pumping speed and sorption capacity of some getters is discussed by Kuus and Martens (1980), Rohring et al. (1980), Parkash and Vijendran (1983), Mehrhoff and Barnes (1984), Ichimura et al. (1988).

5.7.5. Sublimation pumps

Sublimation pumping is based on the gettering action of (continuously) evaporated titanium layers. The simplest form of a sublimation pump consists of a titanium evaporator wire from which titanium is sublimed onto a glass or metal envelope (Nazarov et al., 1965).

It was found (Sweetman, 1961; Clausing, 1962) that evaporating the titanium onto liquid nitrogen-cooled substrates, the pumping speed is considerably increased. It appears that the effect is due to the microscopic crystalline form of the titanium film formed at low temperatures, which presents a greater sorption area and capacity. The pumps constructed based on this effect are known as cryogetter- or cryosublimation pumps; they are described and discussed by Rivire and Thompson (1965), Biguenet et al. (1968), Beaufils and Geller (1968), Prevot and Sledziewski (1964, 1968, 1971, 1972).

A simple titanium evaporator can be made (Redhead et al., 1968) by simultaneously winding a 0.25 mm diameter titanium wire and a 0.1 mm diameter tungsten wire onto a tungsten rod. An alternative construction consists of 0.8 mm titanium wire and a 0.6 mm molybdenum wire wound simultaneously on a 2 mm tantalum rod (Prevot and Sledziewski, 1972). The tungsten or molybdenum wire (or niobium wire, Clausing, 1962) keeps by surface tension the drops of titanium on the evaporator. McCracken and Pashley (1966), and Lawson and Woodward (1967) found that the performance of the evaporator is more reliable when an alloy of 85% titanium–15% molybdenum* is used instead of pure titanium wire. For pumping in the range of 10^{-11}–10^{-12} Torr Biguenet et al. (1968) recommend the use of tungsten wire coated with very pure titanium by vapour plating (e.g. from an iodine reaction).

Titanium sublimators (evaporators) are usually cylindrical. Harra and Snouse (1972) propose an improved sublimator which employs a spherical shell of titanium enclosing a tungsten filament heater. This sublimator can dispense 37 g of titanium at rates of 0.5 g/h.

For large sublimation pumps electron beam heating is used, evaporating from titanium bars. These sublimators can dispense about 8 g/h (Robertson, 1968).

* The outgassing of the 85 Ti–15 Mo wire was studied by Hu (1987).

Sublimation pumping and pumps have been discussed by Gupta and Leck (1975), Harra (1974b, 1975), Cyransky and Leck (1976), Haque (1976), Aggus *et al.* (1977), Halama (1977, 1979), Okano *et al.* (1977), Malinowski (1978), Edwards (1980), Strubin (1980), Grigorov and Tzatzov (1983).

Distributed getter and sublimation pumping as required in storage rings is discussed by Autin *et al.* (1977), Benvenuti and Decroux (1977), Blechschmidt and Unterlerchner (1977), Fischer (1977), Blechschmidt (1978).

Cryosublimation pumps attain specific pumping speeds for O_2, N_2, CO of about 10 l/sec·cm² of active surface, and about 40 l/sec·cm² for H_2. However, noble gases are practically not pumped by these pumps. Pumps having pumping speeds up to 10^6 l/sec were constructed (Prevot and Zledziewski, 1972) which attain ultimate pressures in the range of 10^{-10} Torr (except noble gases), and which are able to work about 2000 hours in continuous operation at pressures not higher than 10^{-6} Torr.

Murakami (1973) proposed a "catalytic pump" in which hydrogen is pumped by dissociation into atomic state (using an incandescent filament), adsorption and reaction on Cu_2O layers, and trapping the resulting water vapour on a liquid nitrogen trap. A 5000 lit/sec pump is described by Murakami *et al.* (1974).

5.8. Pumping by dilution

The partial pressure of residual active gases can be lowered very effectively by alternatively diluting the active gas with an inert one and repeating the pumping process. This procedure is known in the lamp industry as "flushing".

Suppose that the atmosphere (760 Torr) existing in a vacuum device is constituted of 20% (152 Torr) oxygen (active gas) and 80% nitrogen (inert gas). By pumping the device to 0.76 Torr, the residual gas is still 20% (0.152 Torr) oxygen and 80% nitrogen. If we fill up the device with nitrogen to a total pressure of 76 Torr, and pump it again to 0.76 Torr, the residual gas contains 1.52×10^{-3} Torr oxygen, thus only $1.52 \times 10^{-3} \times 100/0.76 = 0.2\%$ oxygen. Repeating the filling up and pumping once more, the partial pressure of oxygen is lowered to $.52 \times 10^{-5}$ Torr, and so on.

The partial pressure P_u which can be obtained by flushing is

$$P_u = P_o(P_p/P_f)^n \tag{5.49}$$

where P_o is the partial pressure at the first evacuation to P_p, P_f is the filling pressure with inert gas, P_p is the pressure to which the device is repeatedly evacuated, while n is the number of flushing cycles. For the previous example, after 3 cycles the partial pressure of O_2 will be

$$P_u = 0.152(0.76/76)^3 = 1.52 \times 10^{-7} \text{ Torr}$$

5.9. Measurement of pumping speed

5.9.1. *Methods of measurement*

The pumping speed can be defined either by the steady state condition (eq. 3.27)

$$S = Q/P \qquad\qquad (5.50)$$

or by the transient pumpdown condition (eq. 3.269)

$$S = 2.3(V/t) \ \log \ (P_1/P_2) \qquad\qquad (5.51)$$

Equation (5.50) is the basis of the *constant pressure measuring methods*, in which the pumping speed S is determined by *measuring the throughput Q* at *a constant pressure P.*

Equation (5.51) is used in the *constant volume method* in which the pumping speed S is determined by measuring *the change in pressure* (from P_1 to P_2) during the pumpdown of a *constant volume V.*

5.9.2. *Constant pressure methods*

The throughput is defined (eq. 3.24) as the product of the displaced volume of gas by its pressure, thus by measuring these two values, eq. (5.50) permits to establish the pumping speed.

Two constant pressure methods of measuring the pumping speed are based on this concept: the method of the *moving mercury pellet*, and *the inverted buret.*

In the *moving mercury pellet* test, a large vessel A (fig. 5.65) is being evacuated by the pump under test and air is admitted through the calibrated capillary tube via the needle valve. An appropriate gauge is connected to the vessel at T, and the needle valve is adjusted until the pressure indicated by the gauge is constant. The

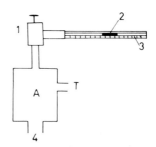

Fig. 5.65 The moving mercury pellet method for measuring pumping speed: 1. Needle valve; 2. Mercury pellet; 3. Calibrated glass capillary; 4. Connection to pump.

volume of air V pumped out of the capillary tube in time t is given by the movement of a mercury pellet in the capillary tube. In this measurement the pumping speed can be determined at various pressures P (in the vessel A) and is given by

$$S = \frac{Q}{P} = \frac{P_a V}{Pt} = \frac{P_a}{P} \frac{\pi a^2 l}{t} \qquad (5.52)$$

where P_a is the atmospheric pressure, a is the radius of the capillary and l the distance by which the mercury pellet was displaced during the time t.

It must be mentioned that S is the pumping speed in the vessel A, and includes the conductance (eq. 3.28) of the parts between the pump and the vessel.

The inverted buret measures both the displaced volume of gas and its pressure. The outlet at the top of the buret is connected to a T leading to the needle valve (fig. 5.66) on the vessel which is pumped. The other side of the T connects to atmosphere through a valve so that when this valve is closed the gas evacuated by the pump, sucks oil up into the tube from the beaker at the bottom.

If V_0 is the volume of gas at atmospheric pressure initially existing above the oil level at the instant the valve to atmosphere is closed, then when the oil level has risen h (cm), the volume is

$$V = V_o - hA \qquad (5.53)$$

where A is the cross sectional area of the buret.

The pressure P at this moment is given by

$$P = P_a - h \frac{\rho_{oil}}{\rho_{Hg}} \qquad (5.54)$$

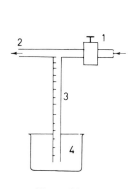
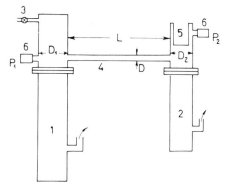

Fig. 5.66   Fig. 5.67

Fig. 5.66 The inverted buret method for measuring pumping speed: 1. Valve for gas inlet; 2. Connection to needle valve on vacuum system; 3. Buret; 4. Beaker with light oil.

Fig. 5.67 Arrangement for determining throughput by measuring the pressure drop across a known conductance; (1) 4-inch diffusion pump; (2) 2-inch diffusion pump; (3) Controlled leak; (4) Metering tube; (5) Liquid nitrogen trap; (6) Gauge head; $D = 1.27$ cm; $L = 30.5$ cm; $D_1 = 10.5$ cm; $D_2 = 5.3$ cm. After Landfors and Hablanian (1959).

where P_a is the atmospheric pressure, ρ_{oil} and ρ_{Hg} the densities of the oil and mercury respectively. The quantity of gas present is

$$PV = P_a V_0 - P_a hA - \frac{\rho_{oil}}{\rho_{Hg}}(V_0 h - h^2 A) \qquad (5.55)$$

If the oil is raised to the height h, during the time t, then the throughput is expressed by

$$Q = \frac{P_a V_0 - PV}{t} = \frac{h}{t}\left[P_a A + \frac{\rho_{oil}}{\rho_{Hg}}(V_0 - hA)\right] \qquad (5.56)$$

If h (cm), A(cm^2), t(sec), and P_a (Torr), then Q results from eq. (5.56) in Torr·cm^3/sec.

When the diameter of the buret is large (A large), the oil level is raised essentially to the top, and the volume of the connecting tubing over to the needle valve is small, eq. (5.56) can be simplified to

$$Q \approx (h/t)P_a A \qquad (5.57)$$

In usual cases $hA \ll V_0$, thus eq. (5.56) can be written

$$Q \approx \frac{h}{t}(P_a A + \frac{\rho_{oil}}{\rho_{Hg}} V_0) \qquad (5.58)$$

in which the *correction term may be quite large*, so that the use of a simple expression (eq. 5.57) will lead to a considerable error.

In order to shorten the waiting time low viscosity liquids should be used. Water is for some purpose a better choice than oil. The error due to humidity is in this case about 2.4 % of the atmospheric presssure. With a 0.2 cm^3 buret tube calibrated in units of 10^{-3} cm^3 a throughput as small as 10^{-4} Torr lit/sec can be measured with acceptable accuracy. Room temperature variations can produce errors; a change of 3°C results in 1 % change in volume.

An *alternative constant pressure measurement method* consists of measuring the *pressure difference across a known conductance** and using eq. (3.26). The conductance can be an aperture (eq. 3.72) or a tube (eq. 3.108). Fig. 5.67 shows the arrangement for the use of this method, which was investigated by Landfors and Hablanian (1959), Noller *et al.* (1960), Fischer and Mommsen (1967), Denison and McKee (1974), Denison (1974, 1975), Landfors and Hablanian (1983), McCulloh et al. (1986).

---

* Using a variable conductance and plotting P against $1/C$ a straight line is obtained, with intercept equal to $1/S$ (Oatley, 1954).

The gas entering the test dome of a vacuum pump (fig. 5.67) flows through a metering tube or aperture of known dimensions from an auxiliary chamber (D_1). A controlled leak (needle valve) for admitting the test gas and a separate diffusion pump are connected to the auxiliary chamber, so that the pressure P_1 can be adjusted to any desired value. The conductance C of the metering tube can be calculated from its dimensions (eq. 3.108), the pressures P_1 and P_2 are measured, and the throughput is given by

$$Q = C(P_1 - P_2) \qquad (5.59)$$

In order to avoid errors the appropriate gauges for P_1 and P_2 are to be used, they must be calibrated compared to each other, and the flow must be kept in a range where the same flow regime is in the whole metering tube (see §3.6.3).

5.9.3. Constant volume method

This method is based on eq. (5.51), and consists of recording the pressures P_1 and P_2 at the beginning and end of given time intervals. The method is usually less accurate than the constant pressure ones, because all pump speeds vary in fact with pressure.

The pumping speed determined by eq. (5.51) for the pressure drop from P_1 to P_2 is in fact related to the average pressure $\bar{P} = (P_1 + P_2)/2$. Since eq. (5.51) is based on the assumption that S is constant (see §3.7.3), the time intervals used in this measurement have to be as short as possible.

The constant volume method has the disadvantage of requiring quite large vessels (see eq. 5.51) and of excluding the use of McLeod gauges because of their time-lag.

5.9.4. Measurement of the pumping speed of mechanical and diffusion pumps

Typical arrangements for measuring the pumping speed of *mechanical pumps* are shown in fig. 5.68.

Some of the conditions of the test dome are: (a) it should be of same diameter D as the pump inlet; (b) the height of the test dome, and the place of the inlets and their direction should be as shown in fig. 5.68; (c) the gas inlet should not be oriented directly toward the gauge inlet; (d) McLeod gauges should be preferred; (e) the ranges of the gauges should overlap.

It was found that consistent results are more readily obtained by first pumping down the test system and then increasing the throughput from zero upward and taking pumping speed readings at successively higher values of the pressure. This method is able to check if the leaks in the system are not excessive and to drop the outgassing rate to a low enough value.

The pressure range of interest for *diffusion pumps* is generally less than 10^{-4}

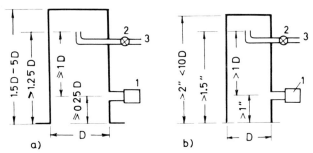

Fig. 5.68 Test domes for measuring the pumping speed of oil sealed mechanical pumps; (a) for inlet larger than 2 inch inside diameter; (b) for inlet smaller than 2 inch inside diameter; D – diameter of pump inlet; 1. Vacuum gauge; 2. Leak valve; 3. Connection to flow meter. After Van Atta (1965).

Torr (molecular flow). In this range the geometry and dimensions of the connections to the pump inlet affect critically the measured values of the pumping speed. For example, adding a tubular extension (dome) of the same diameter as the pump barrel and of length equal to 3 times its diameter to the inlet of a diffusion pump will reduce the net pumping speed to about half that measured directly at the pump inlet. The configuration of the test dome, the location and orientation of the gas inlet and gauge connections, all influence critically the measured value of the pumping speed.

The influences of the methods used for pumping speed measurements and the recommended practices regarding the measuring domes are discussed by The American Vacuum Society (1956), Dayton (1968), Steckelmacher (1968, 1974), Yarwood (1977), ISO (1978b, c), Sharma and Sharma (1982a, b, 1988b), Boeckmann (1986), Hablanian (1987). The Recommended Practice of AVS for measuring pumping speed of high vacuum pumps producing ultimate pressures less than 1×10^{-2} Pa gives the standard conditions for the measurements (Hablanian, 1987) as follows:

(a) The pumping speed is measured in m^3/s at $22°C$ or liter/s at $22°C$, and the throughput is expressed in $Pa \cdot m^3/s$ or $Torr \cdot liter/s$ at $22°C$.

(b) The pumping speed is determined either by the *flowmeter method* or by the *orifice method*, by using the test domes as shown in fig. 5.69a and 5.69b, respectively.

(c) In the *flowmeter* method, the throughput Q of the test gas (admitted through the variable leak valve) is measured by the flowmeter, and plotted as a function of the pressure P measured by the vacuum gauge (fig. 5.69a). The pumping speed S_p is derived from

$$S_p = Q/(P - P_0) \qquad (5.60)$$

where P_0 is the pressure indicated by the gauge when the leak valve is in "closed" position.

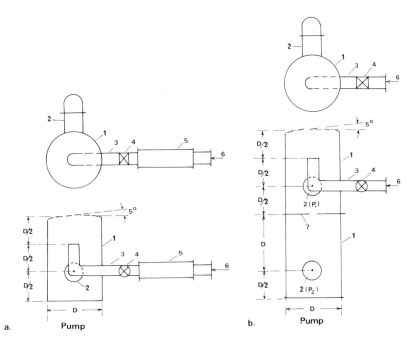

Fig. 5.69 Test domes for measuring the pumping speed of high vacuum pumps. a. for the flowmeter method; b. for the orifice method; 1. Test dome; 2. Vacuum gauges; 3. Gas inlet; 4. Variable leak valve; 5. Flowmeter; 6. Gas supply line; 7. Orifice plate.

The *test dome* (fig. 5.69a) is of circular cross section of diameter D equal to that of the inlet of the pump to within $\pm 2\%$, but not less than 50 ± 1 mm. The top of the dome is flat, or slightly dished outward or inclined to an angle of $5°$. The total mean height of the test dome is $1.5\ D$. The test dome has two cylindrical outlets projecting perpendicularly outwards from the dome wall, at a height of $0.5\ D$ above the bottom flange connecting to the pump (fig. 5.69a). One outlet is fitted with the vacuum gauge; a metal seal is used to reduce outgassing in the vicinity of the gauge. The other outlet is fitted with a gas inlet tube having a right-angle bend upwards, and extending (inside the test dome) up to a distance of $0.5\ D$ to the top of the dome. The gas flows from an outside supply through the flowmeter and variable leak valve, into the tube bent upwards.

(d) In the *orifice* method the continuous flow of the test gas via the variable leak valve (fig. 5.69b) produces an equilibrium pressure P_1 above the orifice and P_2 below the orifice. The conductance C (molecular flow) is calculated (eq. 3.74, 3.83 or 3.108 and 3.109) and the pumping speed is derived from

$$S_p = C \left[\frac{P_1 - P_{01}}{P_2 - P_{02}} - 1 \right] \tag{5.61}$$

where P_{01} and P_{02} are the pressures indicated by the gauges when the leak valve is in "closed" position. The pressure P_1 must always be low enough so that C represents true molecular conductance, i.e. the mean free path (eq. 2.57) be greater than the diameter of the orifice (7, fig. 5.69b).

The *test dome* is of circular cross section of diameter D equal to that of the inlet of the pump to within $\pm 2\%$, but not less than 50 mm. The top of the dome (fig. 5.69.b) is flat, slightly dished outward or inclined to $5°$. The total height of the dome (on the axis) is $3 D$. A flat plate with a central circular *orifice* of diameter d and a thickness $t \leqslant 0.02 \, d$ is mounted at a height of 1.5 D above the bottom flange. The diameter of the orifice should be such that during the measurement of the pumping speed P_1/P_2 be between 50 and 100.

The section of the dome above the orifice plate has two cylindrical outlets (fig. 5.69b) projecting perpendicularly outward from the dome wall at a height of 0.5 D above the orifice plate. One outlet is fitted with the vacuum gauge (2; P_1, fig. 5.69b), the other outlet is fitted with the gas inlet tube (3; fig. 5.69b) having a right-angle bend, and extending up to 0.5 D from the top of the dome. The gas flows from the inlet (6) through the leak valve (4).

The section of the dome below the orifice plate (7) has only one cylindrical outlet, projecting outward at a height of 0.5 D from the bottom flange. This outlet holds the second vacuum gauge (2; P_2, fig. 5.69b) of the same type and model as the upper vacuum gauge (2; P_1).

(e) It is recommended to measure the pumping speed by *increasing* the inlet pressure in steps (e.g. 3 steps per decade). The pressure should then be decreased in steps to check hysteresis. Equilibrium conditions are established when the speed or pressure does not vary more than $\pm 5\%$ during 5 min at 10^{-5} Pa (about 10^{-7} Torr).

5.9.5. Measurement of pumping speed of sputter-ion pumps

The pumping speed of sputter-ion pumps is specific to each kind of gas pumped (e.g. fig. 5.41). Steckelmacher (1968) recommends to use dry air which also shows the pumping capabilities for argon in air.

For measurement purposes two pumping speeds are defined: the *regenerated pumping speed* and the *saturated pumping speed*. In order to reach the *regenerated pumping speed*, the pump is conditioned by:

(a) Pre-evacuation to below the starting pressure, using a suitable clean roughing pump.

(b) Baking the sputter-ion-pump without applying any voltages (and with roughing pump in operation) up to 300°C (maximum), for 4 hours.

(c) Applying the normal operating voltages, closing the roughing pump, and continuing to bake for a further 10 hours.

(d) Measurements are made after 48 hours of continuous pumping.

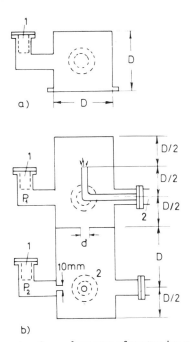

Fig. 5.70 Test domes for measuring the performance of sputter-ion-pumps; (a) for low pressure ultimate; (b) for pumping speed; 1. Ion gauges; 2. Connections to bakeable valves. The orifice $d=0.05\ D$ to $0.1\ D$. After Steckelmacher (1968).

After operation of the pump for a sufficiently long time at a high pressure, the pumping speed deteriorates to what is called the *saturated* value.

The recommended test dome is shown in fig. 5.70. The measurement is based on the pressure difference across a known conductance. Assuming linear calibration over the range of the measurement, the pumping speed is given by

$$S = C\left[\beta(P_1/P_2) - 1\right] \tag{5.62}$$

where P_1 and P_2 are the gauge indications (fig. 5.70b), C is the conductance of the orifice, corrected for its finite thickness and for the injected gas (eq. 3.108), and β is the ratio of gauge sensitivities which is determined in a separate experiment.

The regenerated pumping speed is first determined. Then the gauge sensitivity ratio is checked. The saturated pumping speed is determined after pumping a total quantity of air in excess of $S/50$ Torr·liter for a pump of S liter/sec nominal pumping speed.

The recommended methods of measurement of sputter-ion pumps are indicated by ISO (1974a); Hablanian (1987).

CHAPTER 6

Measurement of low pressures

6.1. Classification and selection of vacuum gauges

The range of vacuum technology extends nowadays about 19 orders of magnitude of pressure below atmospheric. Consequently vacuum measuring techniques have had to be developed to measure low pressures of widely differing magnitudes, from a few Torr to about 10^{-16} Torr (Lafferty, 1972). There is no single gauge which is able to cope with such a range, although the ideal of vacuum scientists and engineers is to develop such a gauge.

The measuring techniques are made all the more difficult because pressures can only be measured in the range from 760 Torr – 1 Torr by using the force resulting from the pressure to set some form of mechanism in motion. In the range below about 1 Torr it is necessary to use some other physical properties of the gas (compression, viscosity, thermal conductivity, ionization). Table 6.1 shows a classification of the gauges according to the property of the gas used to measure the pressure.

Each type and kind of gauge is sensitive to variations of pressure in a specific range (fig. 6.1). Most of the gauges measure total pressure (table 6.1), but some of them (McLeod gauges) indicate only the partial pressure of noncondensible gases. The reading of mechanical and liquid column gauges is independent of the kind of the gas, while in most of the other gauges the reading is a function of the kind of gas.

According to their over-all shape, gauges are usually of the *enclosed* (tubulated) type, which are characterized by a sensing zone located in an enclosing envelope which in turn is connected to the vacuum space to be measured. These gauges can be regarded as a means of measuring the incident flux density of molecules entering the gauge mouth. The gas entering the gauge may be modified by the envelope (adsorb, desorb, heat, cool). For the lower ranges of pressures, *nude*

Table 6.1.
Classification of vacuum gauges.

Physical property involved	Kind of gauge		Kind of pressure recorded
Pressure exerted by the gas	Mechanical	Bourdon Diaphragm	
	Liquid column	U-tube Inclined Differential	Total, independent of kind of gas
	Gas compression – McLeod		Partial; only non-condensibles
Viscosity of the gas	Decrement gauge Rotating molecular Resonance gauge		Total; depends on kind of gas
Rate of transfer of momentum	Radiometer – Knudsen gauge		Total; roughly independent of kind of gas
Thermal conductivity	Pirani gauge Thermistor gauge Thermocouple gauge		Total; depends on kind of gas
Ionization	Discharge tube		Total; depends on kind of gas
	Normal hot cathode gauge Bayard – Alpert gauge Extractor, Suppressor, Deflector, Magnetron (fig. 6.31)		
	Penning gauge (cold cathode) Inverted magnetron gauge Redhead gauge Alphatron		
	Partial pressure analyzers		Partial

gauges can be used; these gauges consist of the sensing element which is mounted inside the vacuum space to be measured without using any gauge envelope.

When *selecting* a suitable gauge for a definite purpose, consideration must be given to the following points:

(a) The pressure range for which the gauge is desired.
(b) If the total or the partial pressure is to be measured.
(c) If the gauge reading can be dependent on the kind of gas.
(d) The accuracy of measurement required.
(e) Kind of mounting (panel, table).

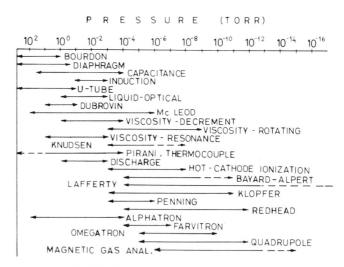

Fig. 6.1 Pressure ranges of vacuum gauges.

Most of the books mentioned in §1.4.1 have chapters dedicated to vacuum gauges. A detailed discussion of the subject is presented by Leck (1964), Berman (1985). General reviews on gauges are given by Schwarz (1960/61), Steckelmacher (1965), Sellenger (1968), Huber (1977), and general reviews on the measurement of pressures in the ultra-high vacuum range by Kornelsen (1967), Redhead et al. (1968), Leck (1970), Lafferty (1972), Weston (1979). Bibliographies on low pressure measurement have been published by Brombacher (1961, 1967).

Gauging requirements in storage rings and accelerators in the 10^{-8}–10^{-10} Torr range are discussed by Benvenuti (1977), Halama and Foester (1987), Halama and Hseuh (1987). The history of low-pressure measurements is presented by Redhead (1984), while the future developments are predicted by Hobson (1984). A possible gauging method using electrostatic levitation is discussed by Kendall et al. (1987a, b).

6.2. Mechanical gauges

6.2.1. Bourdon gauge

The Bourdon gauge consists of a helical coil of hollow tubing of elliptical cross section sealed at one end and connected at the other to the vacuum system to be measured. A pointer is attached by a mechanical linkage to the free sealed end of the helix and moves over a calibrated scale. If the pressure inside the tubing decreases below atmospheric, the tube cross section tends to become more flat, which causes the radius of the helix to decrease, and moves the pointer.

The readings of the Bourdon gauge are dependent on the pressure difference between the inside and outside of the tube thus on the external atmospheric pressure. Variations in atmospheric pressure can be up to 40 Torr. This limits the lower end of Bourdon gauges to about 20 Torr.

6.2.2. Diaphragm gauges

Diaphragm gauges measure pressure differences by the deflection of metal (or glass) diaphragms (aneroid capsules) or bellows. The reading is amplified mechanically, optically, or electrically (capacitance, strain gauge, inductance).

For the measurement of pressure in vacuum technology, the reference pressure of interest is not the atmospheric pressure, but "zero", compared to the sensitivity range of the gauge. In the gauge shown in fig. 6.2, an evacuated beryllium–copper capsule (pressure sensitive element) is mounted in the gauge chamber,

Fig. 6.2 Diaphragm gauge with mechanical indication. 1. Pressure sensitive element; 2. Push rod; 3. Geared sector; 4. Pinion; 5. Pointer; 6. Zero setting adjustment. Reprinted by permission of Wallace & Tiernan Ltd., Tonbridge, Kent, England.

which is connected to the system within which the pressure is to be measured. Distortion of the capsule due to the pressure is transmitted through a *mechanical linkage* (fig. 6.2) to a rotating pointer which indicates the pressure on a circular dial viewed through a sealed window in the front of the gauge chamber. These gauges operate over a pressure range 0–50 Torr and can be read to 0.2 Torr.

In order to measure smaller displacements of the diaphragm than is possible by using mechanical linkage, optical or electrical methods are used. With gauges using *bellows* (e.g. East and Kuhn, 1946) as sensing elements, and amplifying the deflection by a *light beam* reflected on a small mirror which is tilted by motion of the bellows, pressure changes of 5×10^{-4} Torr were detected. These gauges were not adopted for general vacuum use, probably because of their delicate nature.

By using *electrical methods* of detecting changes in the position of the diaphragm, gauges both sensitive and robust have been developed. One such method depends upon the *capacitance* between the diaphragm and a fixed flat electrode. Movements of the diaphragm, in response to the pressure, change the spacing between diaphragm and electrode, and therefore the capacitance, which can be measured with a capacitance bridge. Such gauges reach sensitivities of about 10^{-4} Torr.

Capacitance manometers have been described by Alpert *et al.* (1951), Pressey (1953), Drawin (1958), Loriot and Moran (1975), Van Zyl (1976), Norström *et al.* (1977), Dylla and Provost (1982), Buckman *et al.* (1984). The gauge constructed by Alpert (fig. 6.3) uses a *null deflection technique*, and is bakeable to 400–500 °C. The instrument uses a liquid manometer for absolute calibration, but the pressure sensitive element is the metallic diaphragm. The Kovar cup is divided by the thin corrugated metallic diaphragm into region A (fig. 6.3) connected to the vacuum system, and the region B in which the gas pressure is recorded by the liquid manometer. If the pressure in A equals that in B the diaphragm is undeflected.

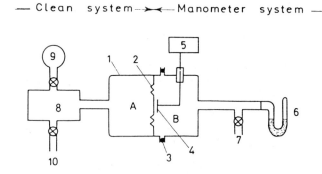

— Clean system ——><—— Manometer system ——

Fig. 6.3 Absolute manometer. 1. Kovar cup; 2. Metallic diaphragm: 3. O-ring seal; 4. Probe; 5. Capacitance bridge; 6. Liquid manometer; 7. Connection to pump; 8. Manifold; 9. Pure gas; 10. Connection to clean pump. After Alpert *et al.* (195)1.

This null position is recorded by measurement with an a.c. bridge of the capacitance of the diaphragm relative to an inserted metal probe. The pressure in A is compensated by that in B, the exact null point of the diaphragm being determined by the capacitance. The method facilitates the production of ultra-high vacuum, and subsequent introduction of pure gas samples up to 50 Torr, with an accuracy of about 0.1 Torr (by using an oil manometer).

The maximum error introduced by the flexible diaphragm is 10^{-2} Torr. The diaphragm undergoes a movement of 3×10^{-4} in for a pressure difference of 1 Torr, and the capacitance method is able to detect deflection of about 5×10^{-6} inches.

The readings of the capacitance manometers may have to be corrected (Poulter et al., 1983) for thermal transpiration (§2.7.1).

Other electrical methods are capable of measuring diaphragm deflections down to a few microinches. By using strain gauges pressure changes of 10^{-3} Torr can be detected. The mutual inductance is used in some gauges, in which two coils are arranged near the diaphragm, one coil is energized by high frequency a.c. and induces e.m.f. in the other. The magnitude of the induction depends on the movement of the diaphragm, and the electrically indicated readings are directly proportional to the pressure. Induction manometers can measure in the range 10^{-1}–10^{-3} Torr with a sensitivity of about 10^{-4} Torr. Warshawsky (1972) describes a system for calibrating vacuum gauges, in which the total pressure is determined by measuring the very small force exerted by the gas on a circular disk that is freely suspended inside an aperture. The force is measured by a microbalance. In this way pressures of the order of 5×10^{-4} Torr were accurately measured.

6.3. Gauges using liquids

6.3.1. U-tube manometers

The currently used pressure unit of vacuum technology (see §2.4.2), the Torr (millimeter of mercury) results from the concept that pressure is expressed by the height of a liquid column. Manometers using liquids consist of a U-tube partly filled with liquid, having one end connected to the system in which the pressure is to be measured. The other end is either open to some reference pressure (usually atmospheric) or is closed off. A closed-end manometer is first thoroughly evacuated and then filled to the proper level while still under vacuum so that the gas pressure over the liquid in the closed arm is negligible as compared with any pressure to be measured. The open end is connected to the system, so that the difference in level between the surfaces of the liquid in the two arms will be just proportional to the total pressure in the system. The difference in level h is related to the pressure according to

$$P = g \rho h \qquad\qquad (6.1)$$

where P is the pressure (dyne/cm^2), ρ the density of the liquid (g/cm^3), h (cm), and $g=980.7$ cm/sec^2. When the liquid is mercury, h, expressed in millimeters, is by definition equal to the pressure in Torr (§2.4.2).

Differences of level of 0.1 mm can just be detected by eye, and a pressure of 1 Torr can be read with unaided eye with a probable error of about 10%. For lower values of the pressure, various causes produce errors: liquid sticking to glass (variable capillarity), irregular light refraction in the glass, dissolution of gases in the liquid (especially in oil), and temperature differences. By using optical magnification (microscope, cathetometer) the position of the liquid can be established with an accuracy of about 10^{-2} mm. Delbart (1967) succeeded to read the position of the mercury level at an accuracy of 4×10^{-4} Torr by using the diffraction fringes produced by the total reflection on the meniscus of the liquid.

An oil manometer measuring 10^{-2} Torr is described by Sharma and Mohan; Heydemann et al. (1977) achieve a resolution of 1×10^{-5} Torr by measuring the position of the liquid with ultrasonic interferometry.

6.3.2. Inclined manometers

The scale of manometers can be extended by constructing it on an inclined side of the U-tube (fig. 6.4). In this arrangement the pressure P (Torr) is given by

$$P = n[1 + (A_i/A_p)] \sin \alpha \qquad (6.2)$$

where n is number of millimetric divisions on the inclined scale, A_i and A_p are the cross sections of the inclined branch and of vertical (pressurized) branch respectively, and α the angle of the inclined branch to the horizontal.

The ratio A_i/A_p has usually values of the order of 1/200, thus it can be neglected, and the magnification is given in fact by $\sin \alpha$.

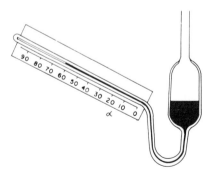

Fig. 6.4 Inclined manometer.

6.3.3. *Differential manometers*

If the tube diameter is sufficiently large (about 1 cm) and the tube and mercury are kept clean, a manometer can give accurate readings down to 10^{-3} Torr, by the use of *optical means of magnifying* small differences in level. Two arrangements of this sort are shown in fig. 6.5. The column height difference due to pressure difference between C and B (fig. 6.5a) is measured by tilting the framework on which the bulbs are fastened and observing the deflection on a mirror 0 (Rayleigh gauge). Another system (fig. 6.5b) includes the mirror M in the vacuum system. Changes in the level of the mercury act on the floater b and rotate the mirror on a (Schrader and Ryder, 1919). The range of these gauges is $1-10^{-3}$ Torr.

A differential interferometric manometer (accuracy less than 10^{-7}) is described by Truffier and Choumoff (1974).

6.3.4. *The Dubrovin gauge*

The gauge constructed by Dubrovin (1933) consists of a glass cylinder partly filled with mercury and a stainless steel tube closed at the upper end and open at the bottom, floating vertically in the mercury (fig. 6.6). The gauge is prepared for use by laying it on its side so that the open end of the steel tube is exposed and evacuating the gauge so that the residual pressure in the gauge and steel tube be very low. While still evacuated the gauge is returned to the vertical position. When low pressure gas is admitted through the connection at the top of the gauge, the steel tube is pushed down more deeply in the mercury. For a pressure P in the gauge the balance is reached when the weight of the tube plus the force exerted on the closed end of the tube by the gas pressure is equal to the change in weight of the displaced mercury. If D_1 and D_2 are the inner and outer diameters of the steel tube, and ρ_s is the steel density, then

$$\tfrac{1}{4}\pi D_1^2 P + \tfrac{1}{4}\pi (D_2^2 - D_1^2) \rho_s g L = \tfrac{1}{4}\pi (D_2^2 - D_1^2) \rho_m g (1-h) \tag{6.3}$$

where L is the length of the steel tube, and h its length protruding above the mercury, while ρ_m is the density of mercury.

From (6.3) it results that

$$\begin{aligned}
P &= [(D_2^2 - D_1^2)/D_1^2] g [\rho_m (L-h) - \rho_s L] = \\
&= [(D_2^2 - D_1^2)/D_1^2] g \rho_m \{ [(\rho_m - \rho_s)/\rho_m] L - h \}
\end{aligned} \tag{6.4}$$

which gives for the zero position, i.e. $P=0$

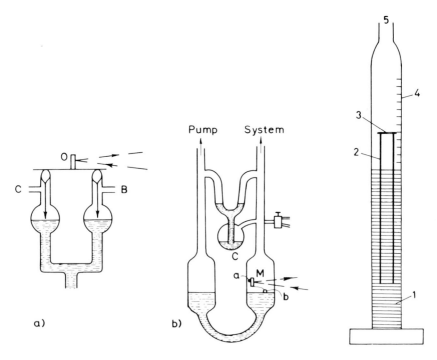

Fig. 6.5   Fig. 6.6

Fig. 6.5 Differential manometers with optical magnification; (a) tilting system; (b) lever system with mercury cut-off (c) for zero reading.

Fig. 6.6 The Dubrovin gauge. 1. Mercury; 2. Thin-wall steel tube; 3. Thin steel disc welded on top of tube; 4. Calibrated glass tube; 5. Connection to system.

$$h_o = [(\rho_m - \rho_s)/\rho_m] L \tag{6.5}$$

thus eq. (6.4) may be written

$$P = [(D_2{}^2 - D_1{}^2)/ D_1{}^2] g \rho_m (h_o - h) \tag{6.6}$$

The sensitivity of the Dubrovin gauge from eq. (6.6) is

$$dh/dP = [D_1{}^2/(D_2{}^2 - D_1{}^2)]/(g\rho_m) \tag{6.7}$$

Since $1/(g\rho_m)$ is the sensitivity of an usual U-tube manometer (eq. 6.1), the sensitivity of the Dubrovin gauge is greater by a factor of $D_1{}^2/(D_2{}^2 - D_1{}^2)$. For $D_1 = 1$ cm, $D_2 = 1.05$ (wall thickness of 0.025 cm), this gauge is ten times more sensitive than an U-tube. For such a gauge a change in pressure of 1 Torr shows a position change of h of 1 cm, so that pressure changes of 0.1 Torr can be detected easily. The Dubrovin gauge is a convenient instrument for measuring in the pressure range below that of an U-manometer and above that in which normally a McLeod gauge is used.

6.3.5. The McLeod gauge

The principle of the gauge constructed by McLeod (1874) was explained in §2.2.1 as an example of the use of Boyle's law. The *McLeod gauge*, also known as *compression gauge*, is the instrument most frequently used for absolute pressure measurements in the range 1 Torr–10^{-6} Torr. Its merits were recently discussed by Thomas and Leyniers (1974).

The sensitivity of the McLeod gauge may be defined as dh/dP, the change in height of mercury for unit change in pressure. As it was shown in §2.2.1 the McLeod gauge can be used either with a *quadratic* or with a *linear scale*. In the first case the pressure is given by eq. (2.5), thus the sensitivity will be

$$dh_1/dP = V/[A\,(\Delta h)_1] \qquad (6.8)$$

therefore it is increased as the volume V of the bulb is increased and as the cross sectional area of the capillaries (A) is decreased. The sensitivity is also a function of $(\Delta h)_1$, thus it is different at different points of the pressure scale, the maximum sensitivity occurring at small values of $(\Delta h)_1$, i.e. at low pressures.

For the case using a *linear scale*, the pressure is given by eq. (2.6), and the sensitivity is

$$\frac{dh_2}{dP} = \frac{V}{A(h_o - h_s)} \qquad (6.9)$$

thus it is independent of $(\Delta h)_2$, i.e. constant at all points over the range. The sensitivity may be increased by making $h_o - h_s$ small.

The ultimate limitation on sensitivity is determined by the practical limits to the values of V, A, $(\Delta h)_1$, and $h_o - h_s$. If V is very large the quantity of mercury required to fill the gauge is excessive, and the weight of mercury tends to distort the glass and thus falsify the readings. The reasonable value of V is about 500 cm^3. In capillaries of less than 1 mm diameter it is found that surface tension forces hold the mercury in the capillary tube even when the mercury in the bulb is lowered. The limit of $(\Delta h)_1$ and $h_o - h_s$ is set by the difficulty in measuring these lengths accurately when they are less than 1 mm. Thus maximum sensitivities of McLeod gauges are in the range of 3–6 $\times$ 10^5 mm/Torr.

The *range of pressures* which can be measured by a McLeod gauge is also determined by A, V, $(\Delta h)_1$ and the total length of the capillary tube. Since the minimum length is about $(\Delta h)_1 = 1$ mm, and the maximum is about $(\Delta h)_1 = 100$ mm, the pressure range which can be measured with a quadratic scale (eq. 2.5) is of four decades, and that of a linear scale (eq. 2.6) of only two decades.

The McLeod gauge does not measure accurately the contribution of the vapours in the system. It is only accurate for gases obeying Boyle's law. If the vapour does not become saturated during the compression in the capillary, it can be considered that its behavior is well approximated by Boyle's law, thus that the gauge will read the total pressure due to both gas and vapour. On the other hand, if condensation does take place, the gauge will read the pressure due to the gas plus the saturation vapour pressure due to the condensed vapour. A method (Pirani and Yarwood, 1961) to determine if condensation has occurred is to compress the gas–vapour mixture into a length L of the closed capillary, by a head of mercury of height h (fig. 6.7). If V_L is the volume per unit length of the closed capillary, then LV_L is the volume of the compressed gas.

With a mixture of gas (P_g) and vapour (P_v)

$$h = P_g + P_v \tag{6.10}$$

thus

$$Lh = L\,P_g + L\,P_v \tag{6.11}$$

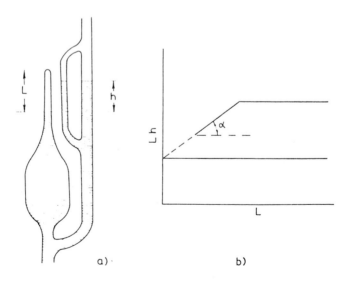

Fig. 6.7 Estimating condensation by using a McLeod gauge.

and since

$$LP_g = k/V_L = \text{const.} \tag{6.12}$$

eq. (6.11) is written

$$Lh = k/V_L + L\,P_v = k/V_L + \alpha L$$

With no vapour present ($P_v = 0$) the plot Lh vs L is a horizontal line. In the case of a gas–vapour mixture, the graph (fig. 6.7b) is a line with slope α.

The most common vapour present in a vacuum system is water vapour. Assuming that water vapour obeys Boyle's law down to its condensation point, then condensation occurs if

$$P_v^1 = P_v\, V/[A(\Delta h)_1] > \text{sat. vap. pres. at gauge temp.}$$

Similarly, the minimum value of the head (fig. 2.7) of mercury $(\Delta h)_1$ which can cause condensation is also numerically equal to the value of the saturation vapour pressure in Torr. This would be the case with only water vapour in the system, which is unlikely to arise in practice. Thus if P_v^1 and $(\Delta h)_1$ are set equal to the saturation vapour pressure (s.v.p.), the minimum partial pressure of water vapour which can condense on compression is equal to (s.v.p.)$^2 A/V$. For a gauge at $20°C$ where $A = 3.14$ mm^2 and $V = 10^5$ mm^3, this pressure is equal to about 10^{-2} Torr, since s.v.p. $= 17$ Torr. Thus any water vapour in the system exerting a partial pressure of less than 10^{-2} Torr will be recorded more or less correctly. However, the initial concentration of water vapour is less than 100% and thus the critical value of P_v for condensation to occur is greater than 10^{-2} Torr. The calculation must be performed for each individual gauge, but it can be seen that the greater the value of V or the smaller the value of A, the lower is the water vapour pressure which may result in condensation. The maximum practical value of $V = 500$ cm^3 and the minimum capillary diameter of 1 mm lead to a minimum critical value for P_v of 4×10^{-4} Torr.

Further limitations on the use of the McLeod gauge is set by the connecting tube and the outgassing of the gauge bulb.

The conductance of usual connecting tubes is not more than 0.1 liter/sec, so that a small leak at the gauge end of the tubing can give rise to an unexpectedly large discrepancy between the pressure in the system and that measured by the gauge. Such an error can be estimated by closing the gauge connection next to the system, and measuring the pressure rise in the gauge due to leakage (for a time interval of about 5 min.). As an example assume that:

P_0 (the pressure reading before closing) $= 10^{-3}$ Torr;

P_t (the reading at time t after closing the gauge) = 10^{-2} Torr;
$t = 300$ sec;
V_g (volume of the gauge) = 300 cm^3;
D (diameter of connecting line) = 0.5 cm;
L (length of connecting line) = 100 cm;

The volume of the connecting line is $V_c = (\pi\, D^2/4)\, L$ cm^3 and the leak rate is given by

$$Q = [(P_t - P_o)/t]\ (V_g + V_c)\ 10^{-3}\ \text{Torr·liter/sec} \tag{6.13}$$

and the conductance (eq. 3.94) by

$$C = 12.1\ (D^3/L)\ \text{liter/sec} \tag{6.14}$$

If a leakage exists near the gauge the pressure in the system is P_s different from P_o, and

$$Q = C\,(P_o - P_s) = C\Delta P \tag{6.15}$$

where ΔP is in fact the error of the gauge due to the leakage. Thus

$$\Delta P = Q/C \tag{6.16}$$

For the numerical values of the above example

$$V_c = \frac{\pi\ \cdot\ 0.25}{4}\ 100 = 19.6\ \text{cm}^3$$

$$Q = \frac{10^{-2} - 10^{-3}}{300}\ 319.6 \times 10^{-3} = 9.6 \times 10^{-6}\ \text{Torr·liter/sec}$$

$$C = 12.1\ \frac{0.125}{100} = 1.5 \times 10^{-2}\ \text{liter/sec}$$

and the error is

$$\Delta P = \frac{9.6 \times 10^{-6}}{1.5 \times 10^{-2}} = 6.4 \times 10^{-4}\ \text{Torr.}$$

If with this arrangement a pressure of 1×10^{-3} Torr is read, 64% of the reading is the gauge error. Errors of this magnitude or much greater frequently appear when the pressure rise test is carried out. Therefore *leakage-free and large conductance connections* are very *important for McLeod gauges.*

Outgassing of the gauge bulb and pipe connection may give errors of the same order of magnitude as leakage. Since a spherical bulb of $V_g = 300$ cm³ has a surface $S = 200$ cm², in order to obtain $Q = 9.6 \times 10^{-6}$ Torr·liter/sec from outgassing an outgassing rate of

$$q = \frac{Q}{S} = \frac{9.6 \times 10^{-6}}{200} = 4.8 \times 10^{-8} \text{ Torr·lit/sec·cm}^2$$

is necessary. According to fig. 3.44 this is quite possible for undegassed surfaces.

Another source of error is due to the cold trap which is normally interposed between the gauge head and the system to prevent mercury vapour entering the system. The trap acts as a mercury condenser causing a steady stream of mercury vapour to flow from the gauge to the trap. This streaming creates a pumping action similar to that in a diffusion pump, producing *a lower pressure within the gauge than that in the system.*

These errors and the possibilities of their correction are discussed by Ishii and Nakayama (1962), Nakayama *et al.* (1968), Colgate and Genre (1968), Berman (1974), Thomas *et al.* (1977), Sharma *et al.* (1980).

Various systems may be used to raise the mercury. These systems are based either on *changing the level of the mercury* in its container or on *changing the volume of the container* (Roth, 1966). The level of the mercury can be changed either by *changing the position* of the container relative to the McLeod gauge, or by *exerting pressure onto the surface of the mercury* in the container. The first mode may be done by raising or lowering a concentric container (fig. 6.8a), or a container

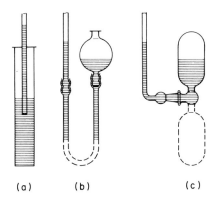

(a)   (b)   (c)

Fig. 6.8 Systems for raising the mercury, based on the change in position of the container.

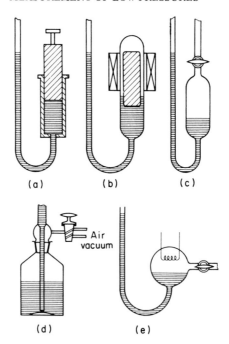

Fig. 6.9 Systems for raising the mercury, by the change of its level in the container.

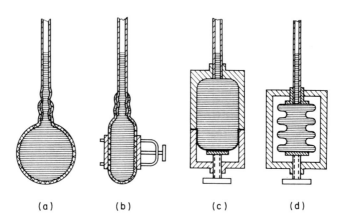

Fig. 6.10 Systems for raising the mercury, by changing the volume of the container.

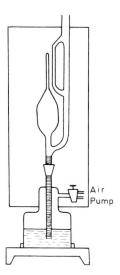

Fig. 6.11 Bench type McLeod gauge.

connected through an elastic pipe (fig. 6.8b), or by reversing the container connected by a ground joint (fig. 6.8c). Figure 6.9 shows means of exerting pressure onto the free surface of the mercury. This may be done by pressing a piston on the surface (fig. 6.9a), immersing (magnetically) a plunger into the mercury, or by changing the air pressure above the mercury (fig. 6.9c–e).

The volume of the mercury container may be changed by using a rubber ball (fig. 6.10a); a rubber pipe (fig. 6.10b); diaphragms (fig. 6.10c) or bellows (fig. 6.10d).

Forms of McLeod gauges. McLeod gauges are commercially available in bench types or shortened models.

In the *bench type* (fig. 6.11) the overall height of the gauge is reduced by using a subsidiary pump to raise and lower the mercury. To raise the mercury, air is admitted to the reservoir by suitable adjustment of the two-way stopcock. To lower the mercury the stopcock is turned to the pump position.

Two *shortened models* of the McLeod gauge are extensively used: the *Measuvac* and *Vacustat*.

The *Measuvac type* (fig. 6.12) uses only a small quantity of mercury which is contained in a flexible reservoir.

The level of the mercury is normally below the lower ends of the capillary and reference tubes. In order to raise the mercury, the lever is turned. The reference tube also acts as the tube connecting the gauge to the system and is of relatively large diameter to ensure reasonable high pumping speed. Allowance for the different capillary effects in the reference and capillary tubes is made by setting the reference mark at the appropriate distance above the closed ends of the capilla-

System

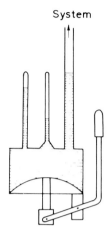

Fig. 6.12 Measuvac gauge.

ries. The use of two capillaries, respectively associated with different volumes, provides this gauge with a range $150\text{--}10^{-3}$ Torr.

The *Vacustat* is a compact McLeod gauge in which the gauge head is mounted on a panel and can be rotated about its centre point (fig. 6.13).

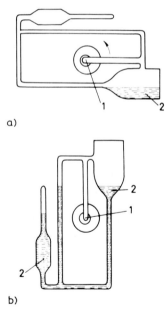

a)

b)

Fig. 6.13 Vacustat; (a) before reading pressure, gauge in horizontal position; (b) pressure recording (vertical) position; 1.Centre of rotation, connected to vacuum system by elastic tubing or ground joint; 2. Mercury.

System

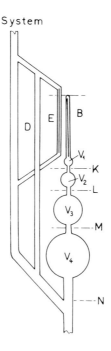

Fig. 6.14 Multirange McLeod gauge.

Because the volume of gas (which is compressed) is small and the capillary tube is of fairly wide bore, the gauge is only suitable to measure pressures in the range $1-10^{-3}$ Torr. When placed in a horizontal position, the mercury flows into the reservoir and the rest of the gauge is evacuated. On rotating to the vertical position (fig. 6.13b) the mercury rises into the capillary tubes. The quantity of mercury used in the gauge is just sufficient to rise to the fixed mark on the reference tube at the lowest measurable pressure; slight tilting may be necessary at higher pressures. The gauge uses the square law scale (eq. 2.5).

Multirange McLeod gauges (e.g. Romann, 1948), which extend the range of measurement, are constructed (fig. 6.14). The single bulb of the usual McLeod gauge is replaced by a series of bulbs of volumes V_4, V_3, V_2 ending in a bulb and a capillary tube B of volume V_1. For very low pressures the gauge is operated on the square law principle (eq. 2.5), the mercury being brought to a fixed mark on the capillary tube E and the tube B being calibrated to read the pressure.

For higher pressures the gauge is operated on the linear scale principle (eq. 2.6). Marks are provided separating each of the volumes V_1, V_2, V_3 and V_4 at points K, L, M, and N. Thus for the next pressure range, the mercury is raised to fill the gauge to point K, and the height difference Δh between K and the mercury level in D is measured. The tube D has the same diameter as the connecting tubes at K, L, M, N. The pressure is given by

$$P = V_1 \Delta h/(V_2 + V_3 + V_4) \tag{6.17}$$

A further extension of the range is made by raising the mercury to level L, where the pressure is given by

$$P = (V_1 + V_2) \Delta h/(V_3 + V_4) \tag{6.18}$$

If the mercury is raised only to level M, the pressure is

$$P = (V_1 + V_2 + V_3) \Delta h/V_4 \tag{6.18a}$$

In a gauge for which $V_1 = 300$ mm^3; $V_2 = 20\,000$ mm^3; $V_3 = 80\,000$ mm^3, and $V_4 = 120\,000$ mm^3, would cover the range of pressures 2×10^{-5} to 100 Torr.

Moser and Poltz (1957) have described a modified McLeod gauge, which extends its lower limit of measurement to about 10^{-8} Torr. In this gauge a very small (5×10^{-8} liter) cylindrical volume with accurately measured dimensions serves to define the compressed volume. The position of the mercury is accurately established by using optical reflection. The pressure of the compressed gas is read on a 20 mm diameter tube where the variation of capillary depression with meniscus height can be accurately determined. Some improvements to the Moser–Poltz gauge are discussed by Miller (1972), Cespiro (1973).

Cleaning a McLeod gauge. Before filling the McLeod gauge with double distilled mercury, the gauge must be thoroughly cleaned. A recommended procedure is to clean with nitric acid, followed by ammonium hydroxide, distilled water and then alcohol. The alcohol can then be removed by passing clean, dry air through the gauge. After filling the gauge with mercury and sealing to the vacuum system, the lowest possible pressure is obtained and the glass is heated to release water vapour.

6.4. Viscosity (molecular) gauges

6.4.1. Principles

Viscosity gauges (also called molecular gauges) were suggested by Sutherland (1897). The first such gauges were described by Hogg (1906) and Langmuir (1913). The development of viscosity gauges and their theory has been reviewed by Dushman (1949), Drawin (1965), Christian (1966), Steckelmacher (1973a) and Weston (1979).

Viscosity gauges may be based on the decrement of a motion (§6.4.2), on motion induced by molecular drag (§6.4.3) or on vibration at resonant frequency (§6.4.4).

6.4.2. The decrement gauge

The principle of this method is to observe the rate of decay in the amplitude

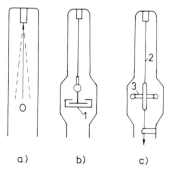

Fig. 6.15 Decrement type gauges; (a) vibrating quartz-fibre; (b) rotating suspended disc; (c) torsion system. 1. Fixed plates between which the disc rotates; 2. Fibre fixed at two ends; 3. Oscillating (torsion) suspension.

of a small light pendulum swinging freely in the vacuum. Provided the friction at the pivot is low, the damping forces, and hence the rate of decay, are functions of the gas pressure.

Decrement type gauges may use a vibrating quartz fibre (fig. 6.15a), a rotating suspended disc (fig. 6.15b) or an oscillating torsion system (fig. 6.15c).

The vertical quartz–fibre (fig. 6.15a) is illuminated from one side and viewed through a telescope with an eye-piece scale. The fibre is set vibrating by drawing the small mass of iron at the lower end (0, fig. 6.15a) to the side wall of the gauge by means of an external magnet. To calibrate the gauge, the time taken for the amplitude of vibration to decrease to half of its initial value is plotted vs. the pressure on a log–log diagram (fig. 6.16).

The theoretical study of the decrement gauge leads to the equation

$$t = A/[PM^{1/2} + B] \qquad (6.19)$$

where t is the time taken for the amplitude of vibration to decrease by half; P is the pressure of the gas; M the molecular weight of the gas; A and B are constants determined by the geometry of the fibre used.

The useful range of the gauge is 1 to 10^{-4} Torr. The low pressure limit is set by the large values of t involved. As M increases t is less for a given pressure (eq. 6.19), thus the operation of the gauge at low pressures improves with the heavier gases.

Morimura et al. (1974) measure low pressures by using the Brownian motion and the damping ratio of a suspended mirror, while Butler et al. (1977) propose to use this principle on magnetically levitated graphite particles.

6.4.3. The rotating molecular gauge

In this gauge a horizontal disc (1. fig. 6.17) (3 cm diameter) rotates about its vertical axis at controlled speeds up to 1000 rpm, with a second disc suspended

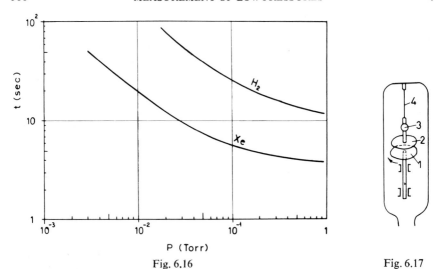

Fig. 6.16   Fig. 6.17

Fig. 6.16 Calibration curves for a viscosity gauge. Time t for amplitude to decrease to half, as a function of pressure P.

Fig. 6.17 Rotating type molecular gauge.

coaxially at a small distance above it (2. fig. 6.17). The molecular drag (§2.6.2) due to the rotation of the lower disc exerts a torque on the upper one. This torque is balanced by the restoring force of the suspension (4. fig. 6.17), the resulting torsion being measured by the displacement of the mirror 3.

The equation describing the effect is

$$\beta = KWP \, [M/(R_o \, T)]^{1/2} \tag{6.20}$$

where β is the angle of rotation shown by the displacement of the mirror 3; K is a constant of the construction for a specific gas, W is the angular speed of the rotating disk (1), P is the pressure, M, R_o, T the data of the gas (eq. 2.91). These gauges can measure pressures from 10^{-4} to 10^{-9} Torr.

Beams *et al.* (1962) and Fremerey (1972) describe viscosity gauges which are able to measure pressures as low as 10^{-10} Torr, by using the deceleration of a freely spinning magnetically suspended high speed rotor.

The *spinning rotor* is usually a steel ball of 4 mm (or 4.5 mm) diameter floating and spinning magnetically in a (horizontal) tube whose internal diameter is about twice that of the ball (Reich 1982a, b; Messer *et al.* 1987). The ball rotates with its set frequency (e.g. 400 Hz), and the gas molecules produce a drag-deceleration which is proportional to the pressure. The pressure P results from:

$$P = \frac{\pi v_{av} D_b \rho_b}{20 \sigma_a} \frac{t_n - t_{n-1}}{t_n \cdot t_{n-1}} \tag{6.20.a}$$

where v_{av} is the average molecular velocity (eq. 2.38), D_b and ρ_b are the diameter and the density of the ball, σ_a is a coefficient of impulse and energy accomodation (ideal smooth surface $\sigma_a = 1$; rough ball surface σ_a up to 1.17), t_n and t_{n-1} are the times required for completing n respectively $n - 1$ revolutions of the ball.

Spinning rotor gauges can be used in the pressure range from 1 down to 10^{-7} Pa (about 10^{-2} to 10^{-9} Torr). The construction, operation and calibration of these gauges were discussed by Comsa et al. (1980), Fremerey (1982), Reich (1982a, b), Berman and Fremerey (1987), Breakwell and Nash (1987), Messer et al. (1987), Setina et al. (1987), Winkler (1987), Miyake et al. (1988).

6.4.4. The resonance type viscosity gauge

In this gauge (Becker W., 1962) a light suspension is mounted vertically and allowed to vibrate at its resonant frequency (normally between 30 and 300 c/s), the damping forces obviously being a function of the gas pressure. By means of a photoelectric sensing device an electromagnetic driving signal is created which maintains the oscillation at a constant amplitude, the whole device forming a closed-loop servo-system. The driving signal is equal to the damping losses and hence a function of pressure. A commercial version (manufactured by Pfeiffer) operates over the pressure range 10^{-3}–100 Torr.

Austin (1969) and Austin and Leck (1972) describe gauges for the pressure range 10^{-6}–10^{-3} Torr, where an oscillating aluminum foil is maintained electrically in vibration, and the power fed into the system measures the damping losses, thus the pressure.

Hirata et al. (1987), and Kokubun et al. (1987) describe and discuss a "quartz friction" vacuum gauge, based on the variation of the resonance impedance of a quartz oscillator as a function of the frictional force of the ambient gas. The sensor head is a tuning fork shaped quartz oscillator (mass produced for wrist watches) with a resonance frequency of 33 kHz and a resonance intrinsic impedance of 22 kΩ. The pressure range which can be measured is 10^{-1} to 10^5 Pa (10^{-3} to 100 Torr).

6.5. Radiometer (Knudsen) gauge

The basic element of the radiometer gauge proposed by Knudsen (1910) consists of two parallel plates (fig. 6.18a) one of which is heated, separated by a distance. The unheated plate is supported on a sensitive suspension so that a small force acting upon it can be measured by its deflection.

The plates A_1 and A_2 at temperatures T_1 and T_2 respectively are separated by a distance d (cm) which is small compared with their linear dimensions and with the molecular mean free path (fig. 6.18a). It is assumed that the molecules coming from A_1 (travelling toward A_2) have an average velocity v_{1av} dependent upon

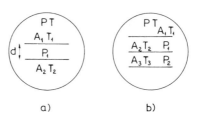

Fig. 6.18 Arrangements of the radiometer vanes.

the temperature T_1 (eq. 2.38), and those leaving A_2, have v_{2av} dependent upon T_2; v_{1r}, and v_{2r} representing the corresponding root-mean-square velocities (eq. 2.40) at any instant there are n_1 and n_2 molecules per cm³ with v_{1av} and v_{2av} respectively.

In the surrounding gas the average velocity is v_{av}, the root-mean-square velocity is v_r, and the molecular density is n.

The pressure P_1 (dyne/cm²) in the space between A_1 and A_2 is given by (eq. 2.34)

$$P_1 = \tfrac{1}{3} mn_1 v_{1r}^2 + \tfrac{1}{3} mn_2 v_{2r}^2 \tag{6.21}$$

and n_1 and n_2 can be related to n, by (eq. 2.46)

$$\tfrac{1}{4} nv_{av} = \tfrac{1}{4} n_1 v_{1av} + \tfrac{1}{4} n_2 v_{2av} \tag{6.22}$$

As

$$n_1 v_{1av} = n_2 v_{2av} \tag{6.23}$$

it follows from eq. (6.22) that

$$n_1 v_{1av} = n_2 v_{2av} = \tfrac{1}{2} nv_{av} \tag{6.24}$$

Substituting for n_1 and n_2 in eq. (6.21)

$$P_1 = \tfrac{1}{3} m \left[\tfrac{1}{2}nv_{1r}^2 v_{av}/v_{1av} + \tfrac{1}{2} nv_{2r}^2 v_{av}/v_{2av}\right] \tag{6.25}$$

For a Maxwellian velocity distribution

$v_r/v_{av} = \sqrt{(3\pi/8)}$ (eq. 2.38; 2.39), thus making

$$\frac{v_r}{v_{av}} = \frac{v_{1r}}{v_{1av}} = \frac{v_{2r}}{v_{2av}}$$

$$P_1 = \tfrac{1}{6} mn \, (v_r)^2 \left(\frac{v_{1r} + v_{2r}}{v_r}\right) =$$

$$= \tfrac{1}{2} P \left(\frac{v_{1r} + v_{2r}}{v_r}\right) = \tfrac{1}{2} P \left[(T_1/T)^{1/2} + (T_2/T)^{1/2}\right] \tag{6.26}$$

by using eqs. (2.34) and (2.40).

Since the pressure on the outside of the plates is P, the resultant pressure ΔP on either plate tending to force them apart is

$$\Delta P = P_1 - P = (P/2)\,[(T_1/T)^{1/2} + (T_2/T)^{1/2} - 2]\ \mathrm{dyne/cm^2} \tag{6.27}$$

Thus *it is independent of the kind of gas.*

By introducing a third plate A_3 (fig. 6.18b), the pressure between A_2 and A_3 is P_3

$$P_3 = (P/2)[(T_3/T)^{1/2} + (T_2/T)^{1/2}] \tag{6.28}$$

thus the resulting force on A_2 per unit area is $P_1 - P_3 = \Delta P$

$$\Delta P = (P/2)[(T_1/T)^{1/2} - (T_3/T)^{1/2}] \tag{6.29}$$

that is *the force on the central vane is independent of its temperature.*

These equations assume that the accommodation coefficients (eq. 2.104) on all the plates are unity.

A more exact treatment of this theory, taking into account the accommodation coefficients for the vane surfaces and the inside surface of the gauge tube leads to *results which differ for various gases*, and are also a function of T_2 (Wu and Dutt, 1972; Eschbach and Werz, 1976; Kreisel, 1976; Weston, 1979).

The *Knudsen gauge* consists of a light vane C (fig. 6.19) supported vertically at its centre point by a torsion wire, and of two plates A and B heated to temperature T_1. Surfaces E and F receive molecules of velocities corresponding to T_1 while surfaces G, H molecules from the walls of the vessel, at T. If $T_1 > T$ there is a net couple on the vane, and the resultant torsional twist in the suspension wire is measured by the conventional mirror, lamp and scale. This case corresponds to $T_2 = T$ in eq. (6.27).

The useful range of the Knudsen gauge is 10^{-3}–10^{-5} Torr, but can be extended

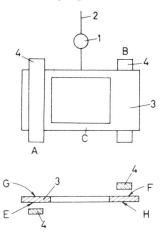

Fig. 6.19 Knudsen gauge; 1. Mirror; 2. Suspension fibre; 3. Light vane; 4. Plates.

down to 10^{-8} Torr. A gauge using magnetic suspension (e.g. Evrard and Boutry, 1969) may measure pressures down to 10^{-10} Torr.

6.6. Thermal conductivity gauges

6.6.1. Thermal conductivity and heat losses

Thermal conductivity gauges are based on a filament mounted in a glass or metal envelope attached to the vacuum system, the filament being heated by the passage of an electric current. The temperature the filaments attain depends on the rate of supply of electrical energy, the heat loss by *conductivity through the surrounding gas*, the heat loss due to *radiation* (and convection), and the heat loss *through the support leads* to the filament.

If the rate of supply of electrical energy is maintained constant, and radiation plus support lead losses are minimized, the temperature of the wire depends primarily on the loss of energy due to thermal conductivity of the gas, which (in a specific range of pressures) is a direct function of the pressure (see eqs. 2.111, 2.115). The temperature variations of the filament with pressure are measured in terms of the change of the *resistance* (Pirani gauge, Thermistor gauge), or the temperature of the filament is recorded by an attached *thermocouple* (thermocouple gauge).

In §2.7.3 it is shown that *in the viscous range* of pressures, the thermal conductivity of the gas *is independent of its pressure*, while in the *molecular range* the thermal conductivity *is proportional to the pressure* of the gas.

In connection with eq. (2.113) it was calculated that the heat conduction per unit area from a surface at temperature $T_s = 100°C$ to a surface at 20°C by air at 10^{-2} Torr (accommodation coefficient $\alpha = 0.8$) is $E_o = 10^{-2}$ Watt/cm². If the filament of the gauge is 1 mil (0.025 mm) diameter and 4 in (10 cm) long, the surface area is about 8×10^{-2} cm², and the gas heat conduction of the order of $10^{-2} \times 8 \times 10^{-2} = 8 \times 10^{-4}$ Watt.

For perfectly absorbing surfaces (black body), the energy loss by radiation* is

$$E_r = 5.6 \times 10^{-12} \ (T_s{}^4 - T_i{}^4) = 5.6 \times 10^{-12} \ (373^4 - 293^4)$$
$$= 6.8 \times 10^{-3} \ \text{Watt/cm}^2;$$

thus the total loss by radiation is

$$(6.8 \times 10^{-3}) \times (8 \times 10^{-2}) = 5.4 \times 10^{-4} \text{ Watt,}$$

which is comparable with the loss due to thermal conductivity of the gas at $P = 10^{-2}$ Torr. However, since the emissivities of surfaces of clean metals at tempe-

---

*For tungsten filaments the radiated energy is $W_r = 7.5 \times 10^{-15} \ T^{4.7}$ Watt/cm².

ratures in the range 0–100°C are generally of the order of 0.1, the true loss due to radiation would be of the order of 5×10^{-5} Watt, so that radiation loss and gas–conduction loss would become about equal at a pressure of

$$\frac{5 \times 10^{-5}}{8 \times 10^{-4}} \times 10^{-2} \approx 6 \times 10^{-4} \text{ Torr}$$

Since radiation increases faster with temperature than does gas conduction, an equality between them occurs at higher pressure if the filament temperature is higher. Therefore, for measurements down to lowest pressures, the filament should be *operated* at the *lowest temperature* for which heat loss due to gas conduction can be measured.

The third loss, that by thermal conduction to the support leads to the filament, can be kept sufficiently small by using a filament of small cross section and low heat conductivity.

The losses by end conduction are

$$W_c = 2 \times 0.239 \, kA \, \frac{\mathrm{d}T}{\mathrm{d}L} =$$

$$= 2 \, (0.239) \, (0.14) \, (4.9 \times 10^{-6}) \, \frac{3 \, (100 - 20)}{10} =$$

$$= 7.9 \times 10^{-6} \text{ Watt}$$

where k is heat conductivity (for nickel $k = 0.14$ cal·cm/°C in the temperature range 0–200°C), A is the cross section area of the wire (diameter = 0.0025 mm), and the factor 0.239 converts from calories to Watts. For the temperature gradient $\mathrm{d}T$, it is assumed that the central third of the filament is at the maximum temperature (100°C) and the third at each end have a uniform temperature gradient $3(T_s - T_i)/L$. For the example given here the loss by end conduction is much lower than that by radiation.

6.6.2. *Pirani gauge*

The gauge constructed by Pirani (1906) is one of the most widely used vacuum instruments. It consists of a glass or metal envelope containing a heated (see §6.6.1) filament of a metal with a high temperature coefficient of resistance, such as platinum or tungsten. As the pressure in the gauge tube increases, the temperature of the filament and therefore its electrical resistance tend to decrease. The usual control circuit for a Pirani gauge is the Wheatstone bridge, in which (fig. 6.20) one leg of the bridge is the filament of the gauge tube R_p and the other three legs have resistances nearly equal to that of the gauge tube. It is advantageous to use two identical gauge tubes in the circuit, one of which R_2 is evacuated to a low pressure and sealed off. If the sealed-off (dummy) tube is mounted adjacent to the gauge tube, fluctuations due to changes in ambient temperature and bridge voltage are to some degree compensated. The details of the temperature compensation

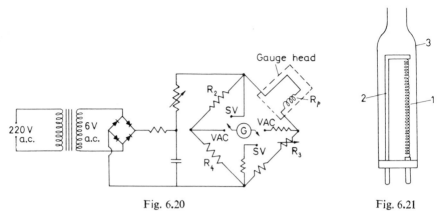

Fig. 6.20   Fig. 6.21

Fig. 6.20 Circuit for a Pirani gauge.

Fig. 6.21 Pirani gauge head; 1. Filament; 2. Filament support; 3. Glass envelope.

of these gauges are discussed by Gorski *et al.* (1977); Oguri (1977) analyzes the possibilities of reducing their "noise".

In fig. 6.20, resistances R_2 and R_4 are fixed, while R_3 and R_p are variable. With the milliamperemeter G connected in the VAC position, the balance condition of the bridge is

$$R_p = R_2 R_3 / R_4 \tag{6.30}$$

One method of measuring the pressure in the gauge head R_p is to balance the bridge by varying R_3 and calculate the resistance R_p, a previous calibration permitting to convert the values of the resistance into pressure.

Another method is to keep R_2 and R_4 constant and present R_3 and to measure the out-of-balance current through G. In this case it is essential to keep the *voltage* across the bridge *constant*. The bridge may be balanced initially at *atmospheric pressure*, then an increase in the resistance R_p causes an increase in out-of-balance current so that the lowest pressures correspond to full-scale readings of G. Alternatively the bridge can be *balanced* at a fixed *low pressure* ($<10^{-3}$ Torr) and then as the pressure falls from atmospheric the out-of-balance current decreases. In both these cases the high pressure end of the scale is very compressed and the scale becomes more open toward the low pressure end. The usual useful scale extends from 5×10^{-3} Torr to 5×10^{-1} Torr. Steckelmacher (1973b) analyzes the methods used recently to extend the range of these gauges to 10–100 Torr.

In commercial forms of the Pirani gauge the control unit (fig. 6.20) includes the appropriate power supply, which supplies the rectified a.c. By switching to the position SV (set voltage) the milliamperemeter G can be used as a voltmeter and the voltage can be set to the standard value marked on the scale. In some instruments the bridge voltage varies with the pressure in the gauge head and set voltage must be controlled before each pressure reading.

The *Pirani gauge head* includes a tungsten, nickel or platinum filament wire (0.005–0.1 mm diameter) wound in a helix of 0.5–2 mm outside diameter (fig. 6.21), with a pitch of at least 10 wire diameters to prevent any one turn from shielding its neighbours. This filament is stretched between supports usually 6–8 cm apart to which it is spot welded.

It is not possible to calibrate the Pirani gauge from first principles, and the calibration is made against another gauge (e.g. McLeod). Typical calibration curves at constant temperature (A) and constant voltage (B), are shown in fig. 6.22 (Hamilton, 1957; Povh and Lah, 1967).

Pirani gauges and their calibration curves are also discussed by Poulter *et al.* (1980b), Morel (1983).

6.6.3. *The thermocouple gauge*

The thermocouple gauge was first described by Voege (1906), and refined by Haase *et al.* (1936), Dunlap and Trump (1937), Weber and Lane (1946).

In this gauge a filament is heated electrically and its temperature is measured directly by means of a thermocouple. The heating current which is passed through the hot filament is kept constant at a standard value independent of the temperature of the filament. As the pressure increases, the heat conduction through the

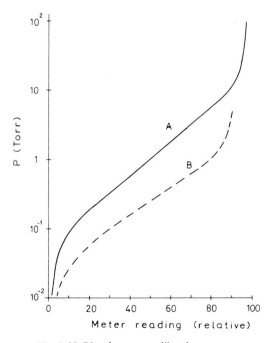

Fig. 6.22 Pirani gauge calibration curves.

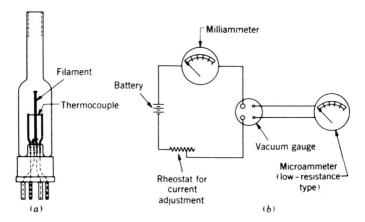

Fig. 6.23 Thermocouple gauge (General Electric type); (a) gauge head; (b) electrical circuit. Reproduced from Dushman and Lafferty (1962), by permission of J. Wiley & Sons, Inc., New York.

gas increases (eq. 2.113) and the temperature of the filament decreases. The thermocouple (usually spot welded to the midpoint of the filament) responds to the temperature of the filament and provides a direct reading of the pressure.

From the many thermocouple gauges built, it is useful to show, the original gauge of General Electric (fig. 6.23), and the more refined type manufactured by Hastings–Raydist, described by Benson (1957), (fig. 6.24). In the GE type gauge the filament consisted of a platinum–iridium ribbon 0.0234 by 0.0078 cm in cross section and 3.66 cm in length with a Nichrome–Advance thermocouple welded to its midpoint. The heating current used was 30–50 mA.

In the Hastings–Raydist gauge (fig. 6.24) the sensitive element consists of two thermocouples acting in parallel and a third thermocouple in series to compensate for variations in ambient temperature. The two thermocouples A and B are heated in series by alternating current from a transformer. Thermocouple C (fig. 6.24) connected from the midpoint between A and B to the center tap of the transformer provides temperature compensation. Since thermocouples A and B are connected "back to back" in the a.c. circuit, they act as parallel sources of electromotive force for the d.c. circuit for which the lead from C, through the d.c. meter to the center tap, is the common return path (see equivalent circuit fig. 6.24). These gauges are available for ranges of 0.1–20 Torr, 5×10^{-3}–1.0 Torr and 1×10^{-3}–1×10^{-1} Torr, and have a speed of response shorter than other heat conductivity gauges. Figure 6.25 shows some calibration curves for different gases. Calibration for various gases was carried out by Chapman and Hobson (1979), Bills (1979).

Zettler and Sud (1988) extend the sensitivity of a thermocouple gauge up to atmospheric pressure by allowing the reference temperature to vary in steps: below 10^{-2} Torr the reference temperature t_r is kept at $t_r = 500\,°C$; for 10^{-2}

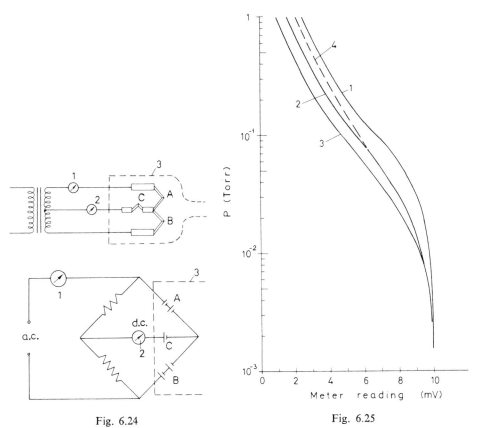

Fig. 6.24   Fig. 6.25

Fig. 6.24 Hastings–Raydist thermocouple gauge and its equivalent circuit. 1. Current meter; 2. Pressure reading; 3. Gauge tube. After Benson (1957).

Fig. 6.25 Calibration curves of Hastings–Raydist gauge, for 1. Argon; 2. Air; 3. Hydrogen; 4. Carbon dioxide.

Torr to 4 Torr, t_r is dropped to $220\,°\mathrm{C}$; from 4 Torr to 200 Torr, t_r is held nearly constant at $220\,°\mathrm{C}$; above 400 Torr, t_r is decreased and reaches $100\,°\mathrm{C}$ at atmospheric pressure.

Aldao and Löffler (1984) describe a chromel–alumel thermocouple gauge, while Kuo *et al.* (1988) discuss a miniature gauge using Cu–Cr thin films.

6.6.4. *The thermistor gauge*

The thermistor gauge was first described by Becker *et al.* (1947). This is a Pirani-type gauge which employs a semiconductor element having a high negative temperature coefficient of resistance. The principal advantage of this type of gauge is that the response curve of the bridge current as a function of the pressure may be essentially linear over a wide pressure range (e.g. 10^{-3}–1 Torr) if plotted on a log–log scale.

Bretschi (1978) describes a gauge for the range 760 to 10^{-5} Torr, using a silicon single crystal; Shioyama et al. (1978) utilized TaN thin film on glass.

Thermal vacuum sensors using integrated silicon thermocouples were studied and discussed by Van Herwaarden (1987), Van Herwaarden and Sarro (1987).

6.6.5. Combined McLeod–Pirani gauge

It was suggested to seal a Pirani filament (or a thermistor) in the top of the McLeod capillary. Thus the gas compressed in a certain ratio by the McLeod gauge is measured on the thermal conductivity gauge, recording pressures as low as 10^{-7} Torr.

6.7. Ionization gauges

6.7.1. The discharge tube

The discharge tube (fig. 6.26) is an elementary form of ionization gauge in which a potential difference of several thousand volts is applied between two electrodes

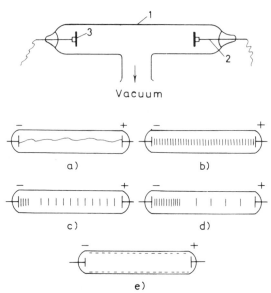

Fig. 6.26 The discharge tube and the appearance of the discharge at various pressures. 1. Glass tube; 2. Current lead-in; 3. Electrode (Ni). (a) Spark = 5 Torr; (b) Discharge of colour specific to gas = 1 Torr; (c) Striations = 5×10^{-1} Torr; (d) Diminishing striations = 1×10^{-1} Torr; (e) Fluorescence of glass wall = 10^{-2} Torr.

in a narrow glass tube connected to the vacuum system. The ionization in the tube produces a glow discharge, whose form is characteristic for the pressure existing in the tube. The colour of the discharge is characteristic for the kind of gas existing in the tube. At pressures from 1 to 20 Torr, a spark (steamer) of discharge passes from one electrode to the other (fig. 6.26a). At about 1 Torr the spark widens to a glow discharge. As the pressure is still further decreased definite regions (striations) in the glow discharge can be observed. When the pressure reaches about 10^{-2} Torr the number of collisions is not sufficient to maintain an easily visible discharge. The electrons, however, bombard the walls of the tube fluorescence of the glass may be observed. The fluorescence disappears at about 10^{-3} Torr, a condition which is known as "black-out".

To improve the correlation between the observation of the nature of the glow discharge and the pressure, two main techniques have been adopted. The first technique is to measure the applied potential difference across the discharge tube in terms of the length of the spark between polished metals spheres of a given size. The second technique is to include a fluorescent screen in the discharge tube, and use the intensity of the luminescence of this screen as the indicator of the pressure.

6.7.2. Hot-cathode ionization gauges

In any ionization gauge, the residual gas existing in the gauge head is subjected to ionizing radiation and some of the gas molecules become ionized. Hot cathode ionization gauges use the thermionic emission of a cathode, the emitted electrons being accelerated by the electrostatic field through the grid of radius r_g (fig. 6.27) set at a positive potential V_g ($\approx$200 V) relative to the cathode. The anode of radius r_a is set at a negative potential V_a ($\simeq -20$ V) relative to the cathode. The grid is made of fine wire, thus most of the electrons coming from the cathode miss the grid wires and continue toward the anode until they reach a point at which the electrical potential is the same as that of the cathode. From this point (shaded area, fig. 6.27) the electrons are turned back to oscillate radially, through the grid until they finally strike a grid wire and are captured. The oscillating electrons will eventually collide with gas molecules, and ionization of the gas molecules may occur. The positive ions which are created in the annulus between grid and anode are driven to the anode, and produce an *ion current*. In order to produce ionization by impact with a molecule an electron must have a kinetic energy at least equal to the *ionization potential* of the gas (12.6 V for water vapour and oxygen; 15–15.6 V for N_2, H_2, Ar; and 24.6 V for He). The *probability of ionization* P_r is defined as the fraction of electrons at a given energy producing an ionizing collision (the number of ions produced per electron) per centimeter of path and per Torr of gas pressure (fig. 6.28), at 0°C (Tate and Smith, 1932).

Since at a pressure of 1 Torr and temperature of 273°K the molecular density is $n_1 = 3.54 \times 10^{16}$ cm^{-3} (eq. 2.18) the probability of ionization P_r is

$$P_r = n_1 \, \sigma_i = 3.54 \times 10^{16} \, \sigma_i \qquad (6.31)$$

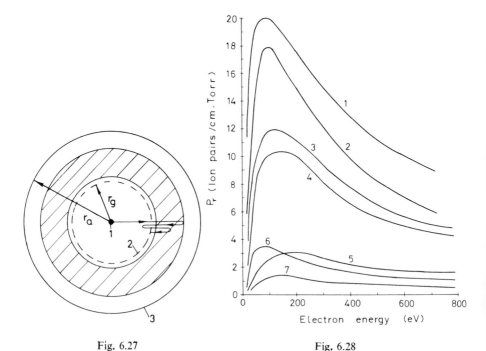

Fig. 6.27   Fig. 6.28

Fig. 6.27 Typical electron trajectory in a hot-cathode ionization gauge. 1. Cathode; 2. Grid; 3. Anode. After Leck (1964).

Fig. 6.28 Number of positive ions produced per electron, per cm of path at 1 Torr, 0°C, as a function of the electron energy. 1. Mercury; 2. Acetylene; 3. Argon; 4. Oxygen, Nitrogen; 5 Neon; 6. Hydrogen; 7. Helium. After Tate and Smith (1932); Lafferty (1972).

where σ_i is the cross section for an ionizing collision by an electron. The *number of ions* produced by an electron per cm of path is according to eqs. (6.31) and (2.21)

$$n^+ = n\,\sigma_i = (273/T)\,P\,n_1\,\sigma_i = (273/T)P_r P \tag{6.32}$$

where n is the molecular density at P and T. For an electron stream of current i_- Amperes*, the positive ion current i_+ (assuming all the ions are collected) is

$$i_+ = (273/T)P_r\,Pi_- \tag{6.33}$$

According to this equation the ion current is a function of the pressure, the temperature, the electron current, and the probability of ionization (fig. 6.28) thus of the electron energy, the path length and the kind of gas. For all other factors specified, the pressure is measured in terms of the ion current (Close and Yarwood, 1970).

* $1A = 6.24 \times 10^{18}$ electrons/sec.

The sensitivity s of an ionization gauge is defined by

$$s = (i_+/i_-)/P \tag{6.34}$$

thus the value of s could be calculated from eq. (6.33) as a function of P_r, which is expressed per cm of path (fig. 6.28). Since the average length of path of the electrons is not easily estimated for practical tube geometries, and the energy of the electrons varies from the maximum value of the grid voltage V_g to zero, the practical way of sensitivity calibration of ionization gauges is against a McLeod gauge. Measurement of the sensitivity s as a function of the pressure in the range $10^{-4}-1$ Torr shows that s increases with increasing pressure until a maximum is reached and then decreases. The maximum sensitivity for nitrogen is ($10-20$ Torr^{-1}) at about 2×10^{-3} Torr, while that for helium at about 2×10^{-2} Torr^{-1}, due to the difference in the ionization probabilities (fig. 6.28). The rise in sensitivity in the vicinity of 10^{-3} Torr for nitrogen is caused by multiple ion production by each electron when the mean free path becomes small compared with the average electron path. The decrease in sensitivity at higher pressures beyond the maximum is attributed by Nottingham and Torney (1961) to ion–electron or positive–negative ion recombinations. Operation of the ionization gauge at pressures above 10^{-3} Torr greatly shortens the life of the cathode, thus the *upper limit of operation* is set at about this pressure. *The lower limit* of pressure which can be measured with a conventional ionization gauge is determined by the production of soft X-rays by the electrons (Nottingham, 1947). These X-rays possess sufficient energy to cause the photoemission of electrons from the anode. Electrically the emission of an electron by the anode is equivalent to the capture of a positive ion, leading to a total current in excess of that due to the positive ions. The photocurrent appears to be independent of pressure and is of the same order of magnitude as the ion current at about 10^{-8} Torr, thus at lower pressures this background will be always shown by the gauge (Lafferty, 1972).

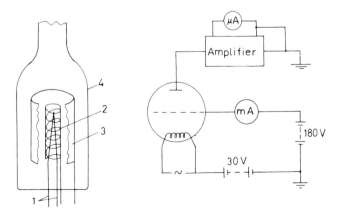

Fig. 6.29 Ionization gauge head and control circuit. 1. Filament; 2 Grid; 3. Collector; 4. Envelope.

The common ionization gauge. Descriptions of ionization gauges have been first published by Buckley (1916), Hauser–Ganswindt and Rukop (1920), Simon (1924). Reviews were published by Huber (1977), Reid (1978), Weston (1979).

An *ionization gauge* for the range $10^{-3} - 10^{-8}$ Torr (fig. 6.29) is similar to a triode valve. In the centre is a tungsten filament (cathode) and surrounding it is a helix of nickel wire (grid). The anode (collector) is a cylinder of nickel concentric with the grid and filament.

The voltages used in the gauge must be highly stabilized to prevent spurious variations in current, which would give apparent changes in pressure as is shown by the microammeter in the anode circuit (fig. 6.29). In this connection it is particularly important to regulate the emission of electrons from the cathode, which is done by some form of feedback circuit.

The gauge records the presence of all gases and vapours (total pressure), but the sensitivity is different for the various gases (fig. 6.30). Figure 6.30 shows a graph of calibration curves for a specific gauge. In fact the values measured by various authors (e.g. Wagener and Johnson, 1951; Riddiford, 1951; Moesta and Renn, 1957; Schulz, 1957; Close *et al.* 1979) differ from each other. Hollanda (1973) analyzed the data obtained by various authors, and found that the relative sensitivity of the ionization gauges for various gases has to be correlated to the value of the ionization cross section existing in the conditions in which the gauge operates. Further measurements of sensitivities were published by Bartness and Georgiadis (1983), Bills *et al.* (1984), Poulter (1984), Tilford (1985). Haefer (1980b) discusses the behaviour of ionization gauges in chambers with cryopumping surfaces; Stansfield *et al.* (1986) describe an ionization gauge for use in a magnetic field.

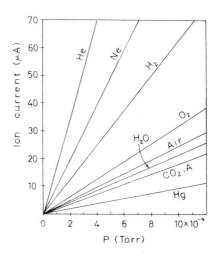

Fig. 6.30 Calibration curves of an ionization gauge for various gases (Grid current 5 mA).

Initial operation of an ionization gauge results in the heating of the electrodes and the emission of large quantities of adsorbed gases from the surfaces. Unless the gauge elements are heated vigorously to *outgas* them, the *reading will remain high* as compared with the system pressure. The grids can usually be heated electrically, the anode can be heated by electron bombardment by connecting the anode and grid together at the same positive potential. Finally, it is necessary to heat the glass or metal envelope of the gauge tube.

After the gauge head *has been outgassed*, gas entering the tube is readily *adsorbed* especially on the tube walls. Chemical reactions induced by the hot filament produce further sorption. These processes are responsible for the *pumping action* of gauges, which causes the *pressure at the gauge to be lower* than that in the system (Langmuir, 1915; Riddiford, 1951; Close and Vaughan-Watkins, 1976; Gear, 1976; Poulter and Sutton, 1981; Berman, 1982; Chapman and Hobson, 1985).

If diffusion pump vapour is present in the system, these vapours react with the hot tungsten and change the electron emission, which produces an apparent change in pressure. It is therefore essential to employ a liquid nitrogen trap between the diffusion pump and the gauge. If diffusion pump vapour is present in a system operating at very low pressures, a normal tubulated ionization gauge *indicates lower pressures* than the nude gauge. This phenomenon is known as the *Blears effect*. Blears (1947) found that this phenomenon lies in the vastly different conductance of the small connecting tube for oil vapour and permanent gases in conjunction with cracking of the oil molecules by the gauge. Haefer and Hengevoss (1961) confirmed the original results of Blears.

High pressure ionization gauges. Although the usual range of hot cathode ionization gauges extends up to 10^{-3} Torr, their upper limit can be extended to about 1 Torr, by using very small electrode spacings and nonoxidizing cathodes (e.g. thoriated iridium). Such gauges are described by Schulz and Phelps (1957), Cleaver (1967), Blauth *et al.* (1970). The Schulz–Phelps gauge utilizes a straight iridium filament mounted parallel to and midway between two parallel molybdenum plane electrodes set only 2–3 mm apart. One of these two electrodes acts as the grid (+60 V) whilst the other acts as the ion collector (−60 V). The sensitivity of this gauge is only 0.6 $Torr^{-1}$ for nitrogen. Beeck and Reich (1974) discuss the methods for linearizing the characteristics of such gauges up to about 1 Torr. A miniature ionization gauge for the range $10^{-1} - 10^{-5}$ Torr, using reverse biased SiC p–n junction "hot electron" emitter, is described by Dobrott and Oman (1970).

The behaviour of ionization gauges at their high-pressure range (around 1 Torr) is discussed by Kudzia and Stowko (1981), Kuo (1981), Tilford and McCulloh (1982), Wood and Tilford (1985).

Low pressure (ultra-high vacuum) ionization gauges. The low pressure limit of about 10^{-8} Torr of conventional hot cathode ionization gauges is given by the equality of the ion current at the collector to the photoelectric current. In order to extend the low pressure limit, the ratio of the ion current to the X-ray photo-current has to be increased. This ratio is increased by various methods resulting in the various gauge constructions (listed below). A study of the principles of the X-ray limit was carried out by Schutze and Ehlberg (1962), and reviews of the methods used to achieve lower pressures were published by Redhead et al. (1968), Groskowski (1968), Leck (1970), Lafferty (1972), Redhead (1987). The methods used may be classified in the following groups:

(a) *Reducing the solid angle* subtended by the ion collector at the X-ray source has been achieved by changing the position of the ion collector and/or by making its area smaller. The fundamental step in this direction was made by Bayard and Alpert (1950) by interchanging the positions of the filament and ion collector (fig. 6.31a, b) and using a wire as the ion collector. This reduced the X-ray limit to about 10^{-10} Torr. Further reductions of the diameter of the ion collector (Venema and Bandriga, 1958; Scuetze and Stork, 1962; Van Oostrom, 1962; Redhead et al. 1968) brought the X-ray limit of Bayard–Alpert gauges to the range 5×10^{-11}–1×10^{-12} Torr.

The solid angle can be further reduced by *withdrawing* the ion collector from the interior of the grid. This method resulted in the group of *external collector gauges*. Gauges based on this principle are: the *extractor gauge* (fig. 6.31c) designed by Redhead (1966), the *screened collector* gauge of Groskowski (1969), and the *buried collector* gauge of Melfi (1969). External collector gauges were analyzed by Clay and Melfi (1966), Cleaver and Zakrzewski (1968), Barz and Kocian (1970), Bernardet and Choumoff (1970), Pittaway (1970), Fletcher (1970), Beeck and Reich (1972), Steckelmacher and Fletcher (1972), Blechschmidt (1973, 1974a), Pittaway (1974), Fujii et al. (1983), Tang et al. (1987).

(b) *Suppressing* the photoelectrons emitted by the ion collector, by using appro-priate *retarding electric fields*. This is done by placing a negative suppressor elec-trode (fig. 6.31d) near the ion collector to force the return of photoelectrons eject-ed by the ion collector. Schuemann (1962) achieved with his suppressor gauge an X-ray limit of about 10^{-13} Torr. Suppressor gauges are described and discussed by Hua et al. (1982), Chen et al. (1987a), Hua (1987).

(c) *Modulating the ion current* improves the signal to noise ratio of the Bayard–Alpert gauge. Redhead (1960) introduced this principle by adding an extra elec-trode (fig. 6.31e), a small wire parallel to the ion collector, inside the grid space, and applying to it a potential (relative to the cathode) which is switched (modulat-ed) from the grid voltage to the ion collector potential. The X-ray limit of the modulated Redhead gauge was about 10^{-11} Torr.

By adding a *modulator* to the *suppressor* gauge Redhead and Hobson (1965) brought the lowest measurable pressure in the 10^{-14} Torr range. Modulation methods and gauge geometries were discussed by Blazek and Hulek (1977), Gros-zowsky et al. (1977), Horikoshi and Mizuno (1977), Hulek (1977), Kanaji et al.

(1977), Mizuno and Horikoshi (1977), Weston (1979), Zhongyi *et al.* (1980), Watanabe *et al.* (1983), Kanaji *et al.* (1987).

(d) *Increasing the sensitivity* of the gauge by enhancing the probability of ionization can be achieved *by increasing the path of the electrons*. This was done by Meyer and Herb (1967) using the *orbitron* principle (fig. 6.31f), by Lafferty (1961, 1962a) using the *magnetron* principle (figs. 6.31g, 6.33), and by Schwarz and Tourtellotte (1969) using the *quadrupole* principle (see §6.9.6). The X-ray limits achieved were 9×10^{-12} Torr for the orbitron gauge, 4×10^{-14} Torr for the magnetron gauge and 1×10^{-10} Torr for the quadrupole gauge.

(e) *Collimating* the electron beam or *deflecting* the ion beam also results in a lower X-ray limit. Klopfer (1962) used a magnetically *collimated* electron beam (fig. 6.34), Helmer and Hayward (1966) *deflected** the ion beam to a remote collector (fig. 6.31h) by using an electrostatic field, while Culton and Peacock (1970) used the *omegatron* principle (§6.9.4) for total pressure measurements. The X-ray limits achieved were about 10^{-11} Torr for the Klopfer gauge, 2×10^{-14} for the deflector gauge, and about 10^{-11} Torr for the omegatron.

(f) *Using electron multipliers* in combination with external collectors or magnetrons. The *channeltron* described by Blechschmidt (1973) achieves 10^{-16} Torr, while the Lafferty gauge with electron multiplier should achieve 10^{-18} Torr.

The Bayard–Alpert gauge. This gauge was described by Bayard and Alpert (1950), and Alpert (1953). The gauge (fig. 6.32) consists of a cylindrical grid structure with a fine wire ion collector along its axis and a cathode (filament) located just outside the grid structure to one side. With this arrangement the X-ray falling on the ion collector is greatly reduced because of the reduced area of the ion collector. The original Bayard–Alpert gauge used a 0.2 mm diam. ion collector and achieved about 10^{-10} Torr. Venema and Bandriga (1958) decreased the ion collector diameter to 25 micron, Schuetze and Stork (1962) to 10 microns and Van Oostrom (1962) to 4 microns, decreasing the lowest pressure to the range of 10^{-12} Torr. The sensitivity for nitrogen of these gauges is in the range 12–25 $Torr^{-1}$.

Watanabe (1987a) describes a gauge having a "point collector" formed by a needle tip placed into a spherical grid. This gauge measures pressures down to 10^{-11} Pa ($\sim 10^{-13}$ Torr) and has a sensitivity for nitrogen of 53 $Torr^{-1}$.

Bayard–Alpert gauges were thoroughly studied, and there are many publications on the subject. Groskowski (1967) analyzed the influence of the electrode geometry and dimensions on the sensitivity of these gauges, Segovia and Martin (1967) evaluated the effects of surface ionization, Lawson (1967) and Redhead (1970) that of ion desorption, while Gopalaraman *et al.* (1970) discussed the effects of Auger emission of adsorbed layers (§4.5.4) on the sensitivity of Bayard–Alpert

---

* A (new) deflector gauge is described by Blechschmidt (1975). A deflector gauge using an electron multiplier is presented by Li and Zhang (1987).

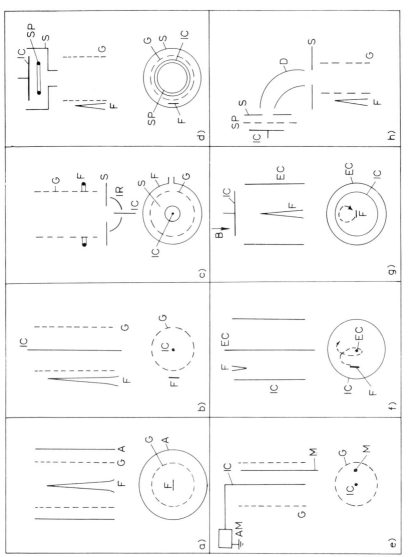

Fig. 6.31 Hot cathode ionization gauges (schematic); (a) Conventional gauge; (b) Bayard–Alpert gauge; (c) Extractor gauge; (d) Suppressor gauge; (e) Modulated Bayard–Alpert gauge; (f) Orbitron gauge; (g) Magnetron gauge; (h) Deflector gauge. F – filament; G – grid; A – anode; IC – ion collector; S – shield; IR – ion reflector; SP – suppressor; AM – amplifier; M – modulator; EC – electron collector; B – magnetic field; D – deflector.

gauges. Bayard–Alpert gauges are analyzed and discussed by Angerth and Hulek (1974), Moraw (1974), Nakayama and Hojo (1974), Suchannek and Sheridan (1974), Nakao (1975), Benvenuti (1977), Huber (1977), Laurent et al. (1977), Szwemin and Pitkowski (1977), Weston (1979), Chen and Suen (1982), Edelmann and Engelmann (1982), Hseuh (1982), Ohsako (1982), Arnold and Bills (1984), Chou and Tang (1986), Filippelli (1987a, b), Hseuh and Lanni (1987), Hua (1987), Watanabe (1987b).

The pumping speed of Bayard–Alpert gauges is quite high. Redhead (1961) mentions 2 liter/sec (for nitrogen) when the gauge was first put into operation. From this value, about 0.25 liter/sec is due to ion pumping, the balance being a result of gettering (chemisorption) on electrodes and glass envelope.

Variations in sensitivity, pumping action and other conditions leading to non-consistent response of Bayard–Alpert gauges can be alleviated by reducing the filament temperature and the electron emission current (Winters et al., 1962) or by using "cold" emitters (Windsor, 1970).

Orbitron gauges. The orbitron gauge injects the electrons from a hot filament (cathode) into the electrostatic field between the ion collector and the electron collector (fig. 6.31f) in such a way that they travel in complex orbits a path of about 90 m. around the small electron collector at the center. The nature of these orbits has been analyzed by Hooverman (1963), Deichelbohrer (1973). Orbitron gauges were described by Herb et al. (1963), Mourad et al. (1964), while Meyer and Herb (1967) analyzed and improved the performance of these gauges. Fitch et al. (1969) attached an external electron gun, while Gosselin et al. (1970) added a suppressor coil as well.

The sensitivity of these gauges is high (about 10^5 Torr^{-1}), but their low pressure limit is only 10^{-11} Torr (Blechschmidt, 1974b).

The Lafferty gauge. The low pressure limit of the hot cathode ionization gauge has been extended to 10^{-14} Torr with the arrangement designed by Lafferty (1961, 1962a, b) in his *hot cathode magnetron gauge* (fig. 6.33). This gauge consists of a cylindrical magnetron operated with a high magnetic field. The ion collector and shield are maintained at a negative potential relative to the cathode to prevent the escape of electrons. Electrons emitted by the lanthanum boride coated filament, spiral around the axial magnetic field in the region between the ion collector and shield. If the magnetic field is high enough, most of the electrons fail to reach the anode, and some of the electrons make many orbits around the cathode before being collected. Because of this increased path length, the probability of the electrons to collide with and ionize gas molecules is greatly enhanced, and the sensitivity of the gauge is improved with no increase in X-ray photoemission. At a magnetic field of 250 Oersted, the ion current is enhanced 25 000 times over what it would be without the field, and the electron emission current to the anode drops to 0.02 of its zero magnetic field value. The ratio in ion current to X-ray

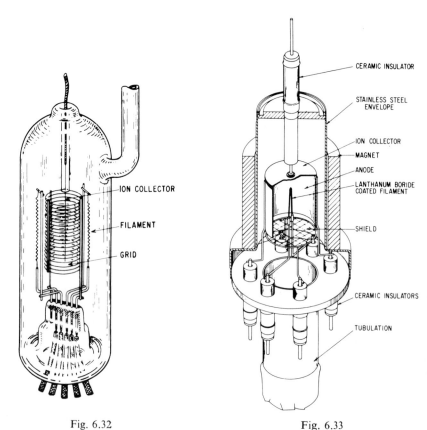

Fig. 6.32   Fig. 6.33

Fig. 6.32 The Bayard–Alpert gauge. Reprinted from Dushman and Lafferty (1962), by permission of J. Wiley & Sons, Inc., New York.

Fig. 6.33 The Lafferty magnetron gauge. Reprinted by permission from Lafferty (1964).

photocurrent is thus increased 1.25×10^6 times by the application of the magnetic field, making it possible to detect ion current at pressures as low as 10^{-14} Torr. For an emission current of 10^{-6} A, the ion current output of the gauge is about 0.1 A/Torr and is linear down to 10^{-13} Torr. This low emission current makes it possible to operate the cathode below 700°C, which avoids the difficulty in which the outgassing limits the low pressure obtained. The sensitivity of the Lafferty gauge is about 10^8 Torr^{-1}. The sensitivity of this gauge may be increased and the X-ray photocurrent reduced even further by the use of a shielded *electron multiplier* ion detector (Lafferty, 1963). In this way a pressure of 10^{-18} Torr should be detectable by counting individual ions entering the electron multiplier.

The Klopfer gauge. This is an ionization gauge constructed by Klopfer (1962)

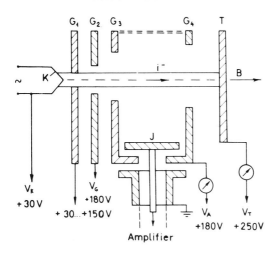

Fig. 6.34 The Klopfer gauge (schematic). After Klopfer (1962).

in which the ions are produced by a *magnetically collimated electron beam*. Electrons are emitted from a thermionic cathode K (fig. 6.34), collimated through a series of apertures at various electrical potentials, traverse an isolated chamber in which the ions are produced, and finally collected by an electron-trapping electrode T. Ions produced within the chamber between G_3 and G_4 are attracted to the ion-collector J at ground potential, which is negative relative to the electron beam and the chamber walls. The magnetic field of about 1000 Oersteds is carefully aligned relative to the series of apertures through the electrodes G_1 to G_4 so that all the electrons pass through the aperture and are caught on the electrode T. The geometry of the gauge is such that X-ray photons emitted by T cannot irradiate J directly, so that electron emission from J is minimized. The presence of the magnetic field further reduces the X-ray effect by causing any electrons emitted to move in circular orbits.

The response of this gauge is linear from about 10^{-2} Torr to 10^{-11} Torr, and has a sensitivity (for nitrogen) of 15 Torr.$^{-1}$.

6.7.3. *Cold-cathode ionization gauges*

The Penning gauge. The useful life of a hot-cathode ionization gauge is determined by that of the incandescent cathode, which is very sensitive to chemical attack and bombardment of positive ions. The cold cathode gauge (known as Penning gauge, Philips gauge or PIG) eliminates this sensitivity. This type of gauge was designed by Penning (1937) and described by Penning and Nienhuis (1949). Calibrations were carried out by Leck and Riddock (1956).

Two parallel connected cathodes (fig. 6.35) are used and midway between them is placed the anode. The cathodes are metal plates while the anode is a loop of

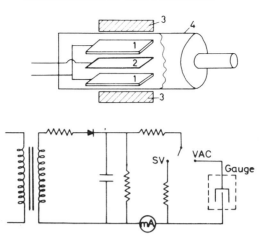

Fig. 6.35 Penning gauge and its simplified control circuit. 1. Cathodes; 2. Anode; 3. Magnet; 4. Envelope.

metal wire whose plane is parallel to that of the cathodes. A potential difference of about 2 kV is maintained between the anode and the cathodes. In addition a magnetic field of the order of 500 Oersteds is applied at right angles to the plane of the electrodes by a permanent magnet.

An electron emitted by the cathode is accelerated towards the anode by the electric field, but the action of the magnetic field causes its path to be in the form of a helix (fig. 6.35). The electron generally passes through the plane of the anode loop until its path is reversed by the electric field due to the second cathode. The electron continues to oscillate in this manner about the plane of the anode loop. Due to the very long path of the electron the ionization probability is high even at low pressures. The positive ions created are captured by the cathodes, producing an ion current in the external circuit. The gauge is operated from a control unit consisting of a rectified a.c. power supply. The voltage across the gauge head may be standardized by using the milliammeter as a voltmeter (switch at SV).

The range of the Penning gauge is about $10^{-2} - 10^{-6}$ Torr. The upper limit is set by the glow discharge which appears, and the lower limit by the smallness of the ion current. At low pressures the initiation of the ionization may be difficult, and in order to start the ionization process it is necessary to produce a few electrons near one of the cathodes (gamma or beta source, auxiliary filament). Young and Hession (1963) described a *trigger-discharge* gauge, having an additional small filament which can be flashed to start the discharge at low pressures. This gauge has been calibrated down to the 10^{-11} Torr range (Lange *et al.*, 1966; Young J.R., 1966; Bryant and Gosselin, 1966). Hayashi (1966) used a beta-emitting radioisotope (Ni^{63}) to trigger the Penning gauge.

A cold cathode gauge specially constructed for measurements on the moon is described by Johnson *et al.* (1972).

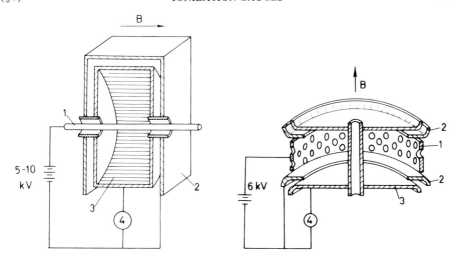

Fig. 6.36   Fig. 6.37

Fig. 6.36 The inverted magnetron gauge. 1. Anode; 2. Auxiliary cathode; 3. Ion collector; 4. Ion current amplifier. After Hobson and Redhead (1958).

Fig. 6.37 The Redhead magnetron gauge. 1. Anode; 2. Auxiliary cathode; 3. Cathode; 4. Ion current amplifier. After Redhead (1959).

The inverted magnetron gauge. This type of gauge was studied by Haefer (1953/54), and the design suitable for ultra-high vacuum was reported by Redhead (1958), Hobson and Redhead (1958), Peacock and Peacock (1988).

In this gauge – known also as the Hobson–Redhead gauge – the cathode is surrounded by an auxiliary cathode (fig. 6.36) outer shell. The auxiliary cathode acts as an electrostatic shield and protects the edge of the openings through the cathode from field concentrations, thus preventing field emission. The cathode and auxiliary cathode are both grounded, but the current to the cathode alone is taken as the measure of the true positive ion current. The anode rod is typically maintained at about 6 kV, and the magnetic field intensity at 2000 Oersteds.

The inverted magnetron gauge is effective in the range $10^{-4} - 10^{-13}$ Torr (Redhead *et al.*, 1968).

The Redhead magnetron gauge. In this cold cathode gauge, described by Redhead (1959), the anode consists of a cylinder, perforated to improve gas flow (fig. 6.37). The cathode is shaped like a spool consisting of an axial cylinder welded on to circular end disks. The gauge is normally operated with a magnetic field of 1000 Oe and an anode–cathode potential difference of about 6 kV.

The magnetron gauge is linear in the range $10^{-4} - 10^{-10}$ Torr and extends according to $i_+ = cP^{1.7}$ down to 10^{-12} Torr. This gauge has a pumping speed* of about

* Kageyama *et al.* (1974) suggested to reduce the pumping effect by using the gauge with a step voltage (pulses at low duty cycle).

0.15 liter/sec. Measurements and calibrations on this type of gauge were carried out by Rhodin and Rovner (1961), Feakes and Torney (1963), Bryant *et al.* (1966). A calculation of the characteristics is proposed by Wutz (1969).

Woods (1973) describes a miniature cold cathode magnetron gauge (18.5 mm height; 25 g weight) for the range $10^{-4}-10^{-7}$ Torr. Chapman and Hobson (1983) describe a solar powered magnetron gauge.

6.7.4. *Gauges with radioactive sources*

Any process which causes ionization can, in principle, be used as a basis for an ionization gauge. X-rays, alpha particles, beta particles, and gamma rays are all ionizing agents.

The *Alphatron gauge* utilizes a small source of alpha particles (e.g. a gold–radium alloy). This type of gauge was developed by Downing and Mellen (1946), and an improved version is described by Vacca (1957).

The Alphatron (fig. 6.38) consists of a source holder and two grid structures. The ionization current is found to be substantially a linear function of the pressure from 10^{-3} to 40 Torr. The lowest limit of these gauges is about 10^{-5} Torr; but their disadvantage is the necessity of shielding.

The alpha source can be replaced by a beta source. Gauges using tritium chemi-

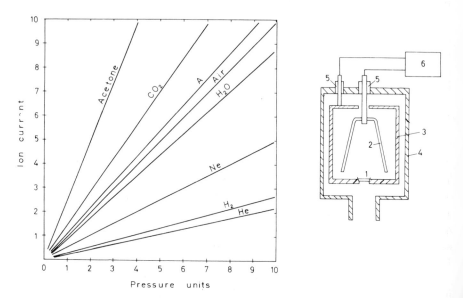

Fig. 6.38 The Alphatron gauge (schematic) and its calibration curves for various gases (arbitrary units). 1. Source and holder; 2. Grid; 3. Grid collector plate; 4. Housing (shielding); 5. Insulators; 6. Power supply and control circuit. After Downing and Mellen (1946).

cally bound to titanium (yttrium or scandium) are also reported. These are operating in the range $1.0–10^{-5}$ Torr. Such gauges are described by Roehrig and Vanderschmidt (1959), Blanc and Dagnac (1964), Rigby and Wright (1968), Berman (1975), Singleton and Yannopoulos (1975).

6.8. Calibration of vacuum gauges

6.8.1. General

In the majority of cases vacuum gauges are used merely to determine the order of magnitude of the pressure within the system. There are occasions, however, when it is necessary to check the calibration provided by the manufacturer, or to have a more precise knowledge of the pressure. The most known methods of calibration are the McLeod gauge method, the expansion method, the flow method, and the pumpdown method. Calibration rules are defined in the International Standards, ISO (1974b, c, 1975, 1976).

Calibration methods are reviewed by Leck (1964), Sellenger (1968), Steckelmacher (1974), Peggs (1976) and Poulter (1977), Poulter et al. (1980a), Grosse and Messer (1981), Hirata (1982), Sutton and Poulter (1982), Warshawsky (1982), McCulloh (1983), Nash and Thompson (1983), Reich (1983), Tilford (1983), Berman (1985). Calibration in the ultra-high vacuum range is treated by Lange and Eriksen (1966), Davis (1968), Redhead et al. (1968), Fowler and Brock (1970), Weston (1979), Hua (1987). Uncertainties and accuracy of the calibration methods are discussed by Simons (1963), Simons and King (1967), Ruthberg (1972). Procedures for calibration of the controls of ionization gauges are indicated by the American Vacuum Soc. (1972a).

6.8.2. McLeod gauge method

Calibration may be effected against a McLeod gauge in the pressure range $10–10^{-5}$ Torr. The use of this method is limited to gases which obey Boyle's law up to the maximum pressure to which it is compressed in the operation of the McLeod gauge. The usual practice is to use a glass or metal chamber (fig. 6.39) evacuated by a liquid nitrogen-trapped diffusion pump to which the McLeod gauge and the gauges to be calibrated are connected, each through a liquid–nitrogen-cooled trap. A needle valve is provided so that any chosen gas can be admitted to the system at a controlled rate to vary the pressure.

For calibrating thermal conductivity gauges in the range $1–10^{-3}$ Torr, a rather insensitive McLeod gauge is sufficient. For calibrating ionization gauges, the greatest McLeod gauge is required, and even so the calibration is possible only in the range $10^{-3}–10^{-6}$ Torr. The Moser and Poltz (1957) version of McLeod gauge (§6.3.5) may be used down to 10^{-8} Torr.

When McLeod gauges are used for calibration the errors listed in §6.3.5 have to be evaluated and corrected.

6.8.3. Expansion method

A small known volume of gas at atmospheric (or similar) pressure is allowed to expand into an evacuated vessel of large known volume, and by applying Boyle's law the new pressure can be calculated accurately. The expansion process may be repeated to produce accurately known pressures between 10^{-4} to 10^{-9} Torr, against which gauges can be calibrated. The chief source of error in this method is the outgassing of the walls of the vessel, which has to be minimized by thoroughly baking the system before use.

The expansion method is discussed by Schuhmann (1962), Edmonds and Hobson (1965), Meinke and Reich (1966), Smetana and Carley (1966), Elliot et al. (1967), Sar-el and Pellach (1976), Messer (1977), Berman (1979), Kendall (1983a), Berman and Fremerey (1987), Huang et al. (1987), Winkler (1987).

6.8.4. Flow method

This method is based on calculating the pressure difference $P_1 - P_2$ across a known conductance C at a known (measured) throughput Q, by using eq. (3.26)

$$Q = C(P_1 - P_2)$$

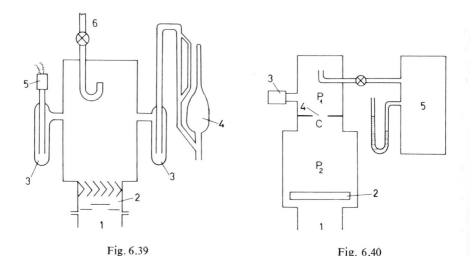

Fig. 6.39   Fig. 6.40

Fig. 6.39 System for calibrating vacuum gauges against a McLeod gauge. 1. Diffusion pump; 2. Liquid nitrogen and water-cooled baffles; 3. Liquid nitrogen trap; 4. McLeod gauge; 5. Gauge to be calibrated; 6. Dry gas inlet.

Fig. 6.40 Gauge calibration system. 1. Diffusion pump; 2. Baffle; 3. Gauge to be calibrated; 4. Calibrated orifice; 5. Gas reservoir. After Normand (1961).

The system used for this measurement (Normand, 1961) is shown schematically in fig. 6.40. The conductance is in the form of an orifice or a short tube, and the value of C is calculated from the dimensions (§3.3). The gas inlet is via a narrow bore tube from a large reservoir where the pressure is in the range measurable by a simple manometer. The throughput Q is calculated by measuring the rate of decrease in pressure of the reservoir. The gauge to be calibrated is used to indicate P_1. Providing the rate of gas flow is small the pressure P_1 changes only slowly with time, so that during the time required for a pressure observation, P_1 may be considered constant. If $P_1 \gg P_2$, then $P_1 = Q/C$. The gauge may be calibrated over a range of pressures by varying Q or C. Choumoff and Bernardet (1970) propose a method of calibration using an adjustable (variable) conductance. Warshawsky (1972) measures the pressure difference across the conductance by using a microbalance method.

The flow method is discussed by Hayward and Jepsen (1962), Florescu (1962), Owens (1965), Christian and Leck (1966), Meinke and Reich (1967), Calcatelli et al. (1974a, b), Choumoff and Japteff (1974), Buckingham (1976). Hoio et al. (1977), Messer (1977), Laurent et al. (1977), Poulter (1977, 1978), Weston (1979), Berman (1985), Sharma and Mohan (1988a). At pressures less than 10^{-7} Torr the throughput Q cannot be measured by simple flowmeters. A method used for this range consists in measuring the pressure difference $P_0 - P_1$ across a second conductance (capillary, orifice) C', then using the equation

$$Q = C'(P_0 - P_1) = C(P_1 - P_2) = SP_2$$

If C' is chosen to be small (e.g. porous plug) P_0 can be easily measured. Roehring and Simons (1962) describe this method using a cascade of differentially pumped stages connected by calibrated orifices. Close et al. (1977) use a system consisting of 3 chambers, the first two connected by a porous plug, the others by orifices.

6.8.5. Pumpdown method

This method enables calibration to be made against a McLeod gauge but at pressures beyond its normal range. The equipment is similar with that shown in fig. 6.40, but with the addition of a fast closing valve in the pipe between the gas reservoir and the chamber. The pressure P_1 in the chamber is indicated by the gauge to be calibrated and also by a McLeod gauge (fig. 6.39). A steady pressure P_1 as measured by the McLeod gauge is established and then the valve is closed. The pressure in the chamber falls exponentially (fig. 3.37) as the gas is pumped away through the conductance C, and the pressure P_t after a time t is given by

$$P_t = P_1 \exp\ [(-S/V)t] = P_1 \exp\ [(-C/V)t] \tag{6.35}$$

where V is the volume of the chamber, and C is the conductance.

The pumping speed S in the chamber is practically equal to the conductance C (eq. 3.28) if the conductance is very small compared to the pumping speed (S_p) of the pump.

Thus the pressure in the chamber after a time t can be calculated, and compared with the reading of the gauge under calibration.

The main precaution to be taken in this method is to ensure that the amount of gas from outgassing of the walls is small compared with that pumped from the closed vessel (see §3.7.3).

A combination of the expansion and the pumpdown methods was also used for calibrating Bayard–Alpert gauges as presented by Huang et al. (1987).

6.9. Partial pressure measurement

6.9.1. General

Besides the measurements of total low pressures, the importance of analyzing the residual gases in vacuum systems is increasing as the attainable lowest pressure decreases. The instruments measuring the partial pressure of the residual gases are basically ionization gauges in which the ions formed are resolved into a mass spectrum, the intensity of each component being measured separately.

The mass spectrometer has proved a most useful instrument in the measurement of partial pressures below about 10^{-4} Torr, down to about 10^{-15} Torr. There are several forms of the instrument but the general principle of operation is common: the gas molecules are firstly ionized, then accelerated and finally separated into groups according to their masses.

The means of ionization are fairly standard, the gas molecules being bombarded by thermionically produced electrons. The acceleration is done by electric fields, while the separation is done by using magnetic deflection, resonance or time of flight techniques.

The methods used for partial pressure measurement have been reviewed by Huber (1963), Van Atta (1965), Redhead et al. (1968), Blauth (1968a, b), Greaves (1970), Van Oostrom (1972), Ramananda (1975), Huber (1977), Hoffman et al. (1980), Weston (1980), Lichtman (1984), Leckey and Boeckmann (1988). The calibration of mass spectrometers in the ultra-high vacuum range is discussed by Poulter (1973). A standard procedure for calibrating mass spectrometer gas analyzers was published by the American Vacuum Soc. (1972b). Details on the interpretation of mass spectra are given by Souchet (1972).

6.9.2. Magnetic deflection mass spectrometers

The 180° deflection instrument was first described by Dempster (1918), the sector instruments begun from that designed by Nier (1940).

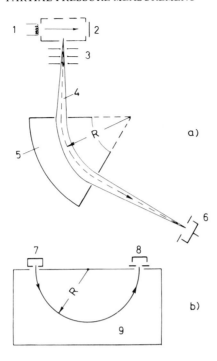

Fig. 6.41 Magnetic deflection mass spectrometers (schematic), (a) sector instrument; (b) 180°
deflection instrument. 1. Filament; 2. Electron collector; 3. Focusing electrodes; 4. Ion beam;
5. Magnetic field (normal to plane of figure); 6. Ion collector; 7. Ionization chamber; 8. Ion
collector; 9. Magnetic field.

In the magnetic deflection mass spectrometer the process starts from the ionization chamber which is a small box in which positive ions are formed by electron
impact as in the conventional ionization gauge. The positive ions formed are
drawn out of the ionization chamber through a narrow slit by means of an electric
field. The ions are deflected through $60-180°$ (fig. 6.41) by a magnetic field normal
to the direction of motion of the ions. For appropriate values of the applied
voltage and magnetic field all ions of a given charge-to-mass ratio ze/m are re-
focussed at the point of the collector slit (fig. 6.41), through which they pass to
impinge on the ion collector and be recorded. The kinetic energy of the ions issu-
ing from the ion source is

$$\tfrac{1}{2}mv^2 = 10^3 zeV/c \qquad (6.36)$$

where m – mass of the ion; v – velocity of the ion (cm/see); e – electron charge
(esu); z – number of electronic charges carried by the ion; V – applied voltage
(Volts); c – velocity of light (cm/sec).

The radius of curvature R (cm) of the orbit for an ion in a magnetic field is

$$R = cmv/zeB \tag{6.37}$$

where B is the magnetic flux density (Gauss). From eqs. (6.36) and (6.37), it results

$$R = \left(\frac{2c}{e} \, 10^8 \right)^{1/2} \left(\frac{mV}{z} \right)^{1/2} \frac{1}{B} = \left(\frac{2 \times 3 \times 10^{10}}{4.8 \times 10^{-10}} \, 10^8 \right)^{1/2} \left(\frac{mV}{z} \right)^{1/2} \frac{1}{B} =$$

$$= 1.12 \times 10^{14} \left(\frac{mV}{z} \right)^{1/2} \frac{1}{B} \tag{6.38}$$

Since the mass of an atom of unit atomic weight is 1.66×10^{-24} g, the mass of an atom of atomic weight M is $m = 1.66 \times 10^{-24} \, M$, thus eq. (6.38) can be written

$$R = (144/B) \, (MV/z)^{1/2} \tag{6.39}$$

As an example, for a singly charged atomic oxygen ion ($z = 1$, $M = 16$), and $V = 1560$ Volts, $B = 3600$ Gauss, R $(M = 16) = 6.35$ cm, so that if the instrument has a $180°$ deflection (fig. 6.41b) the collector slit should be located $2R = 12.7$ cm from the slit of the ion source. The radius of curvature of the orbit of an atomic hydrogen ion ($z = 1$; $M = 1.008$) will be $R_H = 6.35 \, (1/16)^{1/2} = 1.59$ cm, thus the collector slit for hydrogen should be located at $2R_H = 3.18$ cm from the slit of the ion source. The accelerating voltage required to record the hydrogen ion at $2R = 6.35$

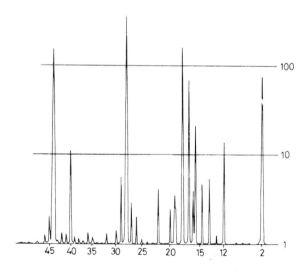

Fig. 6.42 Mass spectrum. Mass numbers of main constituents: $2-H_2$, $17-18-H_2O$; $28-N_2$ or CO $40-A$; $44-CO_2$.

cm is $V_H = 1560$ $(16/1.008) = 24\ 700$ V, which would be somewhat impractical. By providing one or two collector slits, and varying the voltage in a reasonable range the various parts of the mass ranges can be scanned. For each arrangement of collector slit, by imposing the accelerating voltage horizontally (on an oscilloscope) and the ion current received through the collector slit vertically, a trace as shown in fig. 6.42 is obtained. Such a trace shows current peaks which are roughly proportional to the partial pressures of the gases of the various mass numbers. For details see e.g. Souchet (1972).

It is obvious that the greatest resolution of ions is obtained by using a deflection of 180°. However this requires a relatively large system and a magnet of large-pole–piece area since the field must be uniform over the whole ion path. More compact and lighter mass spectrometers with deflections of 120°, 90°, and 60° were built, but the resolution of these instruments is not so high. Mass spectrometers with 180° deflection were described by Craig and Harden (1966), McKraken (1969), with 90° by Davis and Vanderslice (1961), with 60° by Reynolds (1956), Davis W.D. (1962).

The sensitivity of these instruments, when using a photomultiplier as the ion detector permits to measure partial pressures as low as 10^{-13} Torr at room temperature, and 10^{-15} Torr by taking the multiplier down to liquid nitrogen temperatures.

6.9.3. *The trochoidal (or cycloidal) mass spectrometer*

In these instruments (fig. 6.43) the positive ions are acted upon by crossed static electric and magnetic fields. In this situation the path of the ions are trochoidal, the

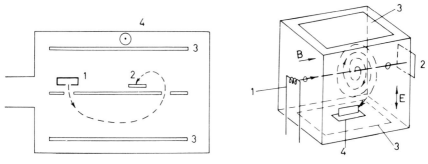

Fig. 6.43   Fig. 6.44

Fig. 6.43 Trochoidal mass spectrometer (schematic). 1. Ion source; 2. Collector; 3. Plates producing the electric field; 4. Magnetic field (normal to figure).

Fig. 6.44 Omegatron (principle). 1 Filament; 2. Electron collector; 3. R.F. plates; 4. Ion collector; B – Magnetic field; E – Electric field. After Alpert and Buritz (1954).

distance b between entry and collector slit being given by

$$b = \frac{2\pi EM}{eB^2} \quad \text{(metres)} \tag{6.40}$$

where E is the electric field (V/m), B the magnetic field (Wb/m^2), e the ionic charge (C) and M the ion mass (kg).

The instrument (fig. 6.43) has two fixed slits, and by adjusting the electric field, ions of a given value of M/e which originate behind one slit can be caused to pass through the other and fall on a collector. The mass spectrum can be scanned by varying the electric field over wide limits, keeping B constant.

The trochoidal mass spectrometer was originally described by Bleakney and Hipple (1938). Trochoidal (cycloidal) mass spectrometers are discussed by Robinson and Hall (1956), Huber and Trendelenburg (1962), Lange (1965), Andrew (1967).

6.9.4. The omegatron

The omegatron, originally developed by Sommer, Thomas and Hipple (1951), and adapted for residual gas analysis by Alpert and Buritz (1954), is a mass spectrometer using the principle of cyclotron resonance. The positive ions move perpendicular to a magnetic field and are accelerated along helical paths of ever increasing radii (Archimedes spiral) by a sinusoidally alternative electric field. This is somewhat similar to the cyclotron, where ions move in circular paths being accelerated with a sudden increase of radius twice per revolution. In the omegatron (fig. 6.44) a narrow beam of electrons passes from the filament to the electron collector parallel to a magnetic field B. Above and below the beam are the two plates which provide the r.f. field. Ions formed along the central axis by electron impact are accelerated by this field. If the resonant frequency of the ions in the magnetic field is the same as the field alternating frequency they will gain energy continuously and so move with ever increasing radius, until they strike the collector. All ions not in resonance have no continuous build-up of energy and hence remain in the vicinity of the central axis.

A singly charged ion moving in a direction perpendicular to a uniform magnetic field moves in a circular orbit of radius R according to eq. (6.37) with $z=1$. Thus the period of rotation is

$$\xi = 2\pi R/v = 2\pi mc/Be \quad \text{(sec)} \tag{6.41}$$

so that the rotational frequency is

$$\phi = \frac{1}{\xi} = \frac{eB}{2\pi cm} = 1.53 \times 10^3 \frac{B}{M} \quad \text{(cycles/sec)} \tag{6.42}$$

since $e = 4.8 \times 10^{-10}$ esu, $c = 3 \times 10^{10}$ cm/sec; and $m = 1.66 \times 10^{-24}$ M.

It can be seen that the frequency of the motion remains constant (eq. 6.42) even if the kinetic energy, thus the radius increases (eq. 6.37). If in the plane of the motion there is superimposed an alternating electric field of strength $E = E_0 \sin wt$ then provided $| w - 2\pi\phi | \ll 2\pi\phi$ the particles follow approximately spiral paths with a radius given by

$$r = \frac{E_0}{B \, | \, w - 2\pi\phi \, |} \sin \left(\frac{w - 2\pi\phi}{2} t \right) \tag{6.43}$$

The radius of the path thus passes through successive maxima and minima, except in the special case of "resonance" when $w = 2\pi\phi$, when the radius increases indefinitely. Therefore the ions having an M so that the frequency given by eq. (6.42) is in resonance, with that of frequency w of the applied field, will reach the collector. Ions which are not in resonance, oscillate in radius (eq. 6.43) but never get to the ion collector.

Commercial omegatrons are able to show a total pressure of max. 1×10^{-5} Torr, and detect partial pressures of the order of 10^{-11} Torr. Omegatrons are described and discussed by Klopfer and Schmidt (1960), Zdanuk et al. (1960), Combley and Milner (1960), Lawson (1962), Petley and Morris (1968), Culton and Peacock (1970), Bijma and Drijfholt (1972), Gentsch et al. (1974), Winkel and Hemmerich (1987).

6.9.5. The Farvitron

Linear high frequency spectrometers were described by Bennett (1950), Moddy (1957), Tretner (1960). The compact version, known as Farvitron, was described by Reich (1961).

The Farvitron is a linear resonance mass spectrometer, with a cylindrical symmetrical electrode system with a cathode K (fig. 6.45), an electron collector A and an ion collector S. Because of the geometry of the electrodes and the d.c. voltages applied, the axial potential distribution is approximately a parabola, that is $\phi = V - kx^2$, in which V is the voltage applied between the two end electrodes and the central ring electrode. An ion of charge-to-mass ratio e/m injected into such a field experiences an axial oscillation of frequency

$$\varphi = [4/(\pi L)] \, [(e/m)V]^{1/2} \tag{6.44}$$

where L is the distance between the end electrodes at which the electrical potential $\phi = 0$. If an alternating potential of frequency φ is superimposed upon the d.c. potential, an ion of e/m satisfying the above frequency relation will resonate and gain sufficient energy to escape from the potential pocket.

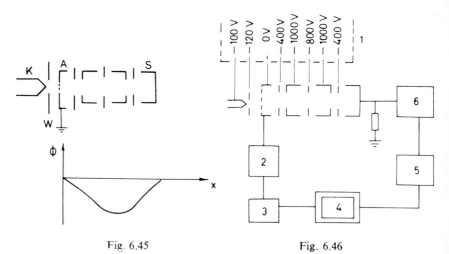

Fig. 6.45   Fig. 6.46

Fig. 6.45 Schematic diagram of the electrodes and potential distribution in the Farvitron. After Reich (1961).

Fig. 6.46 Circuit of the Farvitron. 1. D-c supply; 2. R.F. generator; 3. Wobbler; 4. Oscilloscope; 5. Demodulator; 6. R.F. amplifier. After Reich (1961).

In the Farvitron the ions are produced by accelerating a regulated current of electrons from a tungsten filament K (figs. 6.45, 6.46) axially into the electrode the end of which is a wire mesh. The electrons start from a cathode potential of -100 V (fig. 6.46), and will therefore penetrate the parabolic field to a depth of -100 V, producing positive ions by collisions with any molecules present. These ions oscillate in the parabolic field, most of them not having sufficient energy to reach the cup-shaped electrode S. However, when an r.f. voltage is applied to the electrode on the left, ions of the e/m corresponding to the frequency (eq. 6.44) escape to the collector S.

The Farvitron is a bakeable instrument; the sensitivity is apparently limited to partial pressures not less than about 10^{-8} Torr.

6.9.6. *Quadrupole and monopole mass spectrometers*

The quadrupole mass spectrometer does not require a magnetic field, and in consequence is much less bulky than magnetic types. The instrument (fig. 6.47) consists of four cylindrical rods to which are applied a combination of d.c. and a.c. potentials. For a given applied frequency only ions of a particular value of e/m pass through the spectrometer to the collector (Paul *et al.*, 1958). Ions of different e/m are collected by altering the a.c. frequency. The quadrupole is able to detect partial pressure of the order of 10^{-12} Torr.

The quadrupole mass filter was first described by Paul and Steinwendel (1953). Its quite sophisticated theory was analyzed by Paul *et al.* (1958), and summarized by Greaves (1970). Quadrupole mass spectrometers have been discussed by Bultemann and Delgmann (1965), Swingler (1968), Schwarz and Tourtellotte (1969), Arnold (1970), Kluge (1974), Fultz and Carver (1975), Dawson (1976), Buckingham and Holme (1977), Huber (1976, 1977), Reid (1978), Huber and Rettinghaus (1979), Komiya et al. (1979), Rosenthal et al. (1987).

A modification of this instrument is the *monopole* spectrometer. This consists of a single cylindrical rod and two plane electrodes, which act as reflectors giving three electrostatic images of the rod, thus completing the quadrupole arrangement as before.

The monopole mass spectrometer was originally described by Von Zahn (1963), and it was discussed by Hudson and Watters (1966), Dawson and Whetten (1969).

6.9.7. Time-of-flight mass spectrometers

The most straight forward time-of-flight spectrometer consists of a pulsed ion source and an ion collector at the opposite end of an evacuated tube. The ions are formed by electron bombardment and accelerated out of source towards the collector by either one, or a series of, electric fields.

The short pulse of ions traverses the tube of length L; the transit time t of ions of a particular m/e value, with velocity v is decided by

$$t = L/v \tag{6.45}$$

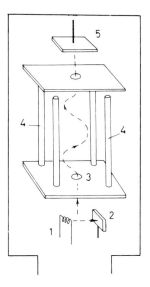

Fig. 6.47 Quadrupole mass spectrometer (principle). 1. Filament; 2. Electron collector; 3. Entrance aperture; 4. Quadrupole rods; 5. Ion collector.

If the potential difference through which the ions are accelerated initially is V, then

$$v = (2Ve/m)^{1/2} \tag{6.46}$$

and

$$t = [L/(2V)^{1/2}](m/e)^{1/2} \tag{6.47}$$

At the ion collector at the far end of the tube, ions will therefore arrive at times decided by m/e, the heavier ions requiring the longer times of travel. Thus each source pulse results in a mass spectrum which can be displayed on an oscilloscope.

Time-of-flight instruments are usually able to detect partial pressures of the order of 10^{-9} Torr. These instruments were described by Cameron and Eggers (1948), Kendall (1962), Wilson (1969). Calibration (under pulsed sensing) is discussed by Winterbottom (1973).

CHAPTER 7

High vacuum technology

7.1. Criteria for selection of materials

7.1.1. General

For the construction of vacuum systems or vacuum devices it is conventional to use metals, glasses, ceramics and some rubbers and plastics. The materials that become part of the vacuum system, forming the enclosure (vessels, pipes) must have *sufficient mechanical strength* to withstand the pressure difference, must be *impermeable enough* to gases, must have *low vapour pressures*, and good resistance to special working conditions (e.g. temperature).

Materials for vacuum technology are discussed in detail by Knoll (1959), Monch (1959), Kohl (1960, 1967), Rosebury (1965), Espe (1966/68), Weston (1975). Reviews of relevant properties of these materials are included in many of the books listed in §1.4.1.

7.1.2. Mechanical strength

Vacuum enclosures are made up from cylindrical, plane and hemispherical parts. All these parts tend to deform inward as a result of the difference between external (atmospheric) and internal (zero) pressures.

Cylindrical parts tend to collapse easier if their length is greater than the critical length L_c defined by

$$L_c = 1.11 D(D/h)^{1/2}$$

where D is the mean diameter and h the wall thickness (fig. 7.1). Table 7.1 lists the permissible values of D/h calculated for cylinders longer than the mentioned

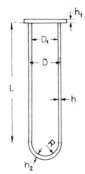

Fig. 7.1 Dimensions of cylindrical, plane and hemispherical parts of vacuum enclosures.

critical length L_c. These values are satisfactory also for shorter cylinders, but for such cases even larger values of D/t are satisfactory. Table 7.1 also lists the values of D_1/h_1 (fig. 7.1) for clamped circular plates, where a deflection δ at the center is permitted. For unclamped circular end plates D_1/h_1 values are greater by a factor of approximately 1.2. The thickness h_2 of hemispherical ends have to be at least that resulting from the ratios R/h_2 (fig. 7.1) mentioned in Table 7.1. For

Table 7.1.
Permissible dimension ratios* for parts of vacuum systems (Roth, 1966).

Material	Cylinders		End plates		Hemisphe-rical
	D/h	L_c/D	D_1/h_1	h_1/δ	R/h_2
Copper at 20°C	84	10	52	15	600
Copper at 500°C	58	8.5	—	—	—
Nickel at 20°C	100	11	73	8	780
Nickel at 500°C	90	10.5	—	—	—
Aluminum 20°C	70	9	37	57	470
Aluminum 500°C	62	8.7	—	—	—
Stainless steel 20°C	105	11.6	89	3	830
Stainless steel 500°C	89	10.5	—	—	—
Glass (hard) 20°C	70	9	16	117	470
Neoprene 20°C	2.5	1.7	10	0.2	30
Teflon 20°C	12	3.8	14	9	—
PVC (Tygon)	3.7	2.1	—	—	—
Perspex	—	—	30	—	—
Mica	—	—	58	15	—

*Notations, see fig. 7.1.

more detailed information refer to Timoshenko (1936), Morrison (1952), Brownell and Young (1959), Steinherz (1963), Roth (1966), Vollbrecht (1974), Envelopes (1983).

7.1.3. Permeability to gases

The metallic, glass or rubber walls of vacuum vessels or pipes are more or less permeable (§ 4.2) to gases. The quantity of gas which permeates the walls can be really large as in the case of porous ceramics or castings or low as for the case of gas diffusion through "non porous" walls. Porous, rough-textured, and loosely laminated materials are to be avoided. In the case of "non porous" materials, their permeability to gases (figs. 3.45; 4.10–4.15) must be taken into account in the evaluation of the gas load.

7.1.4. Vapour pressure and gas evolution

The materials used in vacuum systems should have a low vapour pressure at the maximum working temperature. Vapour pressure data are summarized in §4.1.3. Some metals (Zn, Cd, Pb) have at 400–500°C vapour pressures exceeding the pressures required in high vacuum systems and therefore these metals (or their alloys) cannot be used. For ultra-high vacuum work (fig. 1.1) the choice of metals is only stainless steels, high nickel alloys and oxygen-free high conductivity (OFHC) copper. The gas evolution from the metal surface should be low. To meet this requirement previously degassed materials are recommended, and the outgassing rate should be decreased (fig. 3.44) by cleaning (§7.2) and baking.

7.1.5. Working conditions

The main influence of the working conditions on high vacuum systems is the temperature. In some vacuum systems such influences as chemical corrosion, radiation damage, magnetic field, and others may be very important in selecting the appropriate materials.

The outgassing for high and ultra-high vacuum (figs. 3.44; 4.29–4.31) requires baking of the system to temperatures in the range of 400–500°C. Small bakeable vacuum systems can be constructed of glass, but any greased joint should be excluded. The metal bakeable systems are constructed using the various stainless steels.

7.1.6. Metal vessels and pipes

Metals are extensively used as building materials for vacuum plants (pumps, connexion pipes, valves or vacuum chambers). The metals and alloys used vary from brass and aluminum to stainless steel (table 7.2). The dimensions of the vessels vary from centimeters to kilometers, e.g. Karlsson and Siegbahn (1960), Neal (1965), Halliday and Trickett (1972), Fischer (1972), Lewin and Tenney (1974), Prevot (1974), Pustovoit (1974), Sledziewski *et al.* (1974), Bostic *et al.* (1975), Clausing *et al.* (1976), Cohen (1976), Aggus *et al.* (1977), Hartwig and Kouptsidis (1977c), Heiland (1977), Rees (1977), Trickett (1977, 1978), Ishimaru (1978), Bennet *et al.* (1978), Gomay (1979), Ishimaru and Horikoshi (1979), Clausing *et al.* (1980), Miller (1982), Reddan (1982), Ishimaru (1984).

Table 7.2.
Useful range of materials for vacuum vessels and pipes.

Material	Pressure (Torr)				
	760–1	$1-10^{-3}$	$10^{-3}-10^{-5}$	$10^{-5}-10^{-7}$	$10^{-7}-10^{-16}$
Iron, steels	good	good	good	only after degassing	only stainless steels
Cast iron, copper or aluminum	good	good	bad	bad	bad
Rolled copper or alloys	good	good	good	only after degassing	only OFHC copper
Nickel and alloys	good	good	good	good	good
Aluminum	good	good	only after degassing		not recommended
Glass, quartz	good	good	good	good with degassing	only thick-walled
Ceramics	good	good	only with vitreous coating		only special types
Mica	good	good	only after strong degassing		not recommended
Rubbers	good	good	only degassed	bad	bad
Plastics	good	only special types		only Teflon, Araldite	not recommended

As the chamber size increases, the weight of the chamber and the thickness of the wall increase rapidly (table 7.1). Above some critical size the baking of such a chamber becomes very difficult. A solution in such cases can be the *double chamber*. Here the outer chamber may remain cool and can be built from common metals (e.g. mild steel) with a wall thickness to withstand atmospheric pressure. The inner chamber (e.g. stainless steel) can be made quite thin since it supports practically no pressure difference, the space between the two chambers being evacuated by a separate pump to a "guard vacuum" level (see eq. 7.22). Double chambers have been described by Rivera and Le Riche (1960), Ehlers and Moll (1960), Kienel and Lorenz (1960b), Metcalfe and Trabert (1962).

7.1.7. *Glass vessels and pipes*

Glass is used as the envelope of many vacuum devices (lamps, electron tubes), as bell jars in small evaporation plants, as reaction vessels and connection pipes, and in the construction of some diffusion pumps and gauges. The vacuum range covered by glass is quite large (table 7.2).
Glass is a noncrystalline material that has no regular internal structure. It is rigid at ordinary temperatures and almost fluid at higher temperatures. It has no definite freezing point but becomes solid because its viscosity increases progressively to very large values.
Glass is a fragile material, and for this reason its main mechanical characteristics have to be listed according to the factors generally leading to its breaking: mechanical stresses (tension, bending, impact), thermal stresses and composition changes (weathering, devitrification).
Glass *fractures only as a result of tensile stresses* and not due to shear or compression. The useful strength of glass is but a small fraction of its intrinsic strength because of imperfections (small cracks) usually existing on its surface. The useful tensile stress of glass is about 0.7 Kg/mm².
When glass is *suddenly cooled*, tensile stresses are introduced in the cooled surfaces, and a compensating compression stress in the mass of the glass. *Sudden heating* leads to surface compression and internal tension. Since glass fails only in tension of the surface, *sudden cooling is much more dangerous than sudden heating*.
If the glass is exposed to *steady temperature differences between the two faces*, thermal gradients are developed through the glass. These are even *more dangerous* to the integrity of the glass than sudden cooling.
The weathering and devitrification are discussed in §7.2.2; for the thermal properties of glasses see §7.3.2. Glass, its properties and technology are treated by Springer (1950), Stanworth (1950), Morey (1954), Jones (1956), Marx (1957), Shand (1958), Monch (1959), Colnot and Gallet (1962), Kohl (1960, 1967), Espe (1966/68), Roth (1966), Jones (1972).

Machinable glass-ceramics are discussed by Mog (1976), Grossman (1978), Altemose and Kacyon (1979).

7.1.8. Elastomer and plastic pipes

Elastomers and plastics are used in vacuum technology as pipes and gaskets. The elastomers used are classified in table 7.3.

The field where elastomer and plastic pipes can be used is limited (table 7.2) to backing lines (fig. 3.35) because of the gas evolution of these materials (figs. 4.30; 3.44). The use of rubbers is also limited by their narrow temperature range,

Table 7.3.
Classification of elastomers.

Group	Name	Chemical comp.	Electr resistance	Flame resistance	Imper-meabi-lity	Heat resistance	Cold resistance
Non-oil resistant	Natural rubber	Isoprene	G	P	F	F	G
	S.B.R. Buna S	Styrene/butadiene	G	P	F	F	G
	Butyl I.I.R.	Isoprene/isobutylene	G	P	E	G	F
	Polybutadyene	Butadiene	G	P	F	F	G
Oil and Petroleum resistant	Thiokol	Organic polysulfide	F	P	E	G	F
	Nitrile, Phil-prene, Hycar, Buna N,	Acrylonitrile/buta-diene					
	Perbunan		P	P	E	G	F
	Polyurethane	Diisocyanate/polyes-ter of polyether	F	F	G	G	G
	Neoprene	Chloroprene	F	G	G	G	F
	Hypalon	Chlorosulfonated polyethylene	G	G	—	G	P
Heat resistant	Silicone, Silastic	Polysiloxane	E	F	F	E	E
	Fluocarbon	Vinylidene fluoride/					
	Viton	hexafluoropropylene	E	G	—	E	F
	Kalrez**	Perfluoroelastomer	E	G	—	E	F

*In comparison with the other elastomers: E=excellent, G=good, F=fair, P=poor.
** du Pont – ECD/006.

which extends to $+80°C$ and $-40°C$. At higher temperatures, or after long time rubbers present "ageing" effects in the form of hardening. At low temperatures rubbers become brittle. Silicone rubbers have a wider temperature range extending up to $180°C$ and for shorter heating periods even up to $250°C$.

Because of their mechanical properties the *wall thickness* of elastomer pipes must be *relatively large* (table 7.1) to prevent collapsing. Thinner walled elastomer pipes can be prevented from collapsing by a helical wire spring inserted in the pipe, which reduces the unsupported length of the pipe to the spacing between turns.

The plastics used in vacuum technology are listed in table 7.4. *Acrylics* have a relatively high outgassing rate and are not recommended for high vacuum, except for low temperatures or windows of small surface areas. *Fluocarbons* are suitable for use at operating temperatures ranging from $-100°C$ to $+300°C$. and have a

Table 7.4.
Plastics used in vacuum technique.

Group	Chemical composition	Common trade names	Remarks
Acrylics	Polymethyl metacrilate	Lucite, Perspex, Plexiglas	Transparent, but high vapour pressure.
Fluocarbons	Polytetra-fluorethylene Polytrifluor-chlorethylene	P.T.F.E., Teflon, Fluon P.T.F.C.E., Kel-F, Hostaflon	Inert, heat and cold resistant, low vapour pressure.
Polyethylene	—	Polythene, Alkathene, Hostalen	Chemical resistant
Polystyrene	—	Styron, Lustrex, Polystyrol, Styrofoam	Radiation and cold resistant
P.V.C.	Polyvinyl chloride and acetate-chloride vinyl copolymer	Tygon, Vinylite, Astralon	
Vinylidene chloride		Saran Velon	

relatively low outgassing rate, thus they can be used in high vacuum systems. *Polyethylene* has outgassing rates near to that of fluocarbons, but can be heated just up to 80–100°C. *Polystyrene* has a low outgassing rate but is tough and brittle, and is not suitable for construction of large parts. *Polyvinyl chloride* is used as transparent tubing in the backing line of high vacuum systems, where its outgassing rate is tolerable. *Vinylidene chlorine* is satisfactory for vacuum of the range 10^{-4} Torr.

For details on elastomers and plastics see e.g. Payne and Scott (1960), Simonds (1959/61), Espe (1966/68), Young J. F. (1966), Kraus and Zollinger (1974), Zabielski and Blaszuk (1976), Chernatony (1977a, b), Edwards *et al.* (1977).

Vacuum chambers of composite materials reinforced by carbon fibers are described by Engelmann *et al.* (1987).

7.2. Cleaning techniques

7.2.1. *Cleaning of metals*

Cleaning generally means the removal of undesirable materials lying on the surface. In vacuum technology, the cleaning must be regarded not only as the removal of the visible dirt from the surfaces, but including the subsequent removal of all the contaminants physically stuck on the surface (oil, grease, dust) or resulting from a chemical reaction (oxides, sulphides). The degree of cleanliness must be higher, for higher vacuum.

The oxides and other similar surface layers can be removed by mechanical and/ or chemical methods, as abrasive blasting, wire brushing or pickling and etching.

The cleaning of oils and greases depends on their nature, i.e. if they are soap-forming or not. The soap-forming oils and greases are those of animal or vegetable origin; the mineral oils do not form soaps. The soap-forming oils and greases can be removed by transforming them by hydrolysis in fatty acids and by reacting these acids with alkaline solutions to obtain water soluble soaps. The mineral oils can be removed by dissolving them in organic solvents, (Sheflan and Jacobs, 1953) and, in particular cases only, they can be washed with alkaline solutions containing detergents. Since the nature of the contaminants is usually unknown, a reliable cleaning must consist of two successive steps: (a) the degreasing with organic solvents, followed by (b) an alkaline degreasing (Doré, 1962).

The sequence of the cleaning operations begins generally with mechanical cleaning, followed by pickling, detergent cleaning and degreasing (Janecke, 1958; Espe, 1966/68). As a supplementary stage, very advanced cleaning uses ion bombardment (§4.5.2).

The *mechanical cleaning* methods are not specific for vacuum technology. For the purpose of cleaning from scale, rust, etc., abrasive blasting, or wire brushing is usually utilized.

Table 7.5.
Pickling solutions.

Metal (Alloy)	Pickling solution	Remarks
Aluminum	NaOH (10 % sol.) saturated with NaCl (if blackening appears, Al has Cu; in this case subsequent pickling in HNO_3 (20–30%) required; good washing)	At 80°C, 15–50 sec. Subsequent immersion in HCl (10%) for shining surface.
	$NiCl_2$ (25% sol.) diluted 5 : 1 with HCl (1.16)	
	Electrolytic etching in sol. of 100 g $H_3 BO_3$ in 1000 ml dist. water plus 0.5 g $Na_2B_4O_7$	50–100 V, up to 600 V
	Electrolytic etching in 5–10% sol. of chromic acid	3–15 mA/cm², 20–40 V about 30 min.
Beryllium	NaOH (or KOH) 50–100 g/1000 ml. dist. water at 20°C, electrolytic etching	2.5–7 A/dm²
Constantan (CuNi55/45)	H_2SO_4 (10% sol.) 50–60°C	
Copper	250 ml. HNO_3 (1.40) with 600 ml. H_2SO_4 (1.83) and 20 ml. HCl (1.16) in 130 ml. dist. water	Bright dip
	500 ml. HNO_3 (65%) with 500 ml. H_2SO_4 (conc.) and 10 ml. HCl (37%) and 5 g carbon black	Immersion (2–3 sec) immediate rinsing
	1000 ml. HNO_3 with 1000 ml. H_2SO_4 and 15 g NaCl and 20 g carbon black. To be diluted 1 : 1 with dist. water 24 hr before use	Immersion (1–5 sec)
	10% $Fe_2 (SO_4)_3$ in citric acid (0.1 −1.0%) or in acetic acid (0.3−0.5%)	Bright dip
	HCl dil. immersion for 5 min followed by immersion in sol. of 100 g Cr_2O_3, 7 ml. H_2SO_4 (conc.) in 1000 ml. dist. water	
	Sol. (1) : 40 ml. H_3PO_4, 15 ml. HNO_3, 1.5 ml. HCl, 20 g NH_4NO_3, 45 ml. dist. water. Immersion for 3–4 min. After rinsing immersion in Sol (2); 65 ml. glacial acetic acid, 30 ml. H_3PO_4(1.75), 5 ml. HNO_3 (1.42). Immersion about 1 min until gas evolves evenly all over	Sol. 1 at 35°C Sol. 2 at room temp. Bright dip
Invar (FeNi 64/36)	Cathodic etching in 1 vol. HCl (37%) with 1 vol. H_2SO_4 (96%) and 1 vol. HNO_3 (70%) in 1 vol. dist. water. 26 mA/cm². Carbon anode	

(contd.)

Table 7.5. (contd.)
Pickling solutions.

Metal (Alloy)	Pickling solution	Remarks
Iron	50% sol. of HCl or 5–15% sol. of H_2SO_4. Recommended to add hydrogen evolution inhibitor (e.g. Ferrocleanol.	
Iron-Chromium	500 g Cr_2O_3 with 5 ml. H_2SO_4 filled up with dist. water to 1000 ml.	
	Anodic etching in 335 ml. acetic acid with 240 ml. perchloric acid ($HClO_4$) in 100 ml. dist. water. 6 V, Cathode of graphite	8–10 min
Stainless steel (FeNiCr)	1–3% HNO_3 (conc) with 25% HCl (1.16) in dist. water. Temp. 65° C	Bright dip
	7% H_2SO_2 (1.83) with 3% HCl (1.16) in dist. water. Temp. 65°C	
	30 pbw $Fe_2 (SO_4)_3$ with 16 pbw HF (48–52%) in 380 pbw dist. water. Temp. 70°C	
	27 pbw HCl (1.16) with 23 pbw H_2SO_4 (1.83) in 50 pbw dist. water. Immersion 60 min at 45°C, followed by immersion for 20 min in a sol. of 11 pbw H_2SO_4 (1.83), 13 pbw HCl (1.16), 1 pbw NO_3H (1.40) in 75 pbw dist. water, at 60°C.	
Kovar (FeNiCo)	1 pbw HNO_3 (65%) with 1 pbw acetic acid (50%)	About 50 sec
	75 g $(NH_4)_2SO_4$ in 100 ml. H_2SO_4 (20%) at 50–100°C	3–5 min
	1 vol. HCl (10%) with 1 vol. HNO_3 (10%) at 70°C. Stirring required	2–5 min
	Electrolytic etching: 1% NaCl in HCl (10–15%) a.c. 10–12 V, 1.6 A/cm². Electrode graphite or Kovar	
Molybdenum	HF (40%) or HCl (5–8%)	
	2000 ml. H_2SO_4 with 37.5 g CrO_3, 100 ml. HF, 10 ml. HNO_3 (conc) at 90°C	10 sec
	10 g NaOH in 750 ml. dist. water with 250 ml. H_2O_2 (30–35%) at 40°C	2–5 min
	Electrolytic etching in 20% KOH sol. d.c. 7.5 V. Carbon electrode	

(contd.)

Table 7.5 (contd.)
Pickling solutions.

Metal (Alloy)	Pickling solution	Remarks
Nickel	HNO_3 sol. 10%, at 70°C 150 ml. H_2SO_4 (1.83), 225 ml. HNO_3 (1.3), 3 g NaCl in 100 ml. dist. water, at 20–40°C Electrolytic etching in sol. of 130 ml. H_2SO_4, 25 g $NiSO_4$ in 200 ml. dist. water 6–12 V, nickel electrode Acetic acid sol. 10%	1–2 min 5–20 sec
Tantalum	Hot HF Anodic etching in 75–98% H_2SO_4 (or HCl) with 2–7% HF, in dist. water. 40–160 mA/cm²	1–2 sec
Tungsten	50 ml. HNO_3 with 30 ml. H_2SO_4 in 20 ml. dist. water Boiling H_2O_2 (3% sol.) Anodic etching in 250 g KOH and 0.25 g $CuSO_4$ in 1000 ml. dist. water	

Pickling is the chemical removal of oxides and other surface layers, leaving the cleaned part with a metallic appearance with a smooth or rough finish, depending on the concentration of the solution and the pickling time. Pickling solutions to be used for various metals are listed in table 7.5. After pickling the part should be always thoroughly rinsed and subsequently neutralized in an alkaline bath, and dried with hot (oil-free) air (Bolz, 1958; Turnbull *et al.*, 1962; Roth, 1966; Espe, 1966/68; Sowell *et al.*, 1974; Mathewson *et al.*, 1977; Brand and Kaan, 1980).

Electrolytic etching and polishing is the anodic (or cathodic) treatment of metal surfaces (table 7.6) in appropriate etching solutions (Steyskal, 1955; Angerer and Ebert 1959; Young J.F., 1966; Roth, 1966; Espe 1966/68).

Alkaline detergent cleaning is performed either by immersion or as an electro-cleaning process. The *immersion cleaning* is used usually with hot (60–85°C) solutions. For ferrous metals and difficult cleaning operations stronger cleaners, containing sodium hydroxide (silicates or phosphates), soaps and wetting agents are used, in concentrations of 10–40 g of each component per liter of solution. Nonferrous metals, and especially aluminum, are cleaned with inhibited alkaline cleaning solutions, i.e. sodium silicates, phosphates, carbonates with soap or synthetic organic detergents, at a concentration of 15–45 g/l. *Electrocleaning* in alkaline solutions can be used with the metal to be cleaned as the cathode or as the anode, the tank being the second electrode. With anodic cleaning, oxygen is liberated on the surface of the metal being cleaned, and the process requires

Table 7.6.
Electrolytic polishing.

Metal (Alloy)	Bath composition (ml.)	Voltage (V)	Current density (mA/cm²)	Temp (°C)	Time (min)
Aluminum	40 ml. H_2SO_4, 40 ml. phosphoric acid 20 ml. dist. water	10–18	720	95	5
	165 ml. perchloric acid, 785 ml. acetic acid. 50 ml. dist. water	50–100	30–50	< 50	15
	45 ml. perchloric acid, 800 ml. ethanol, 155 ml. dist. water	100–200	2000–4000	< 35	30
Beryllium	100 ml. orthophosphoric acid, 30 ml. H_2SO_4, 30 ml. glyceryne, 30 ml. ethanol	—	2000–4000	—	—
Copper	670 orthophosphoric acid, 100 ml. H_2SO_4, 270 ml. dist. water	2–2.2	100	22	—
	7 g CrO_3, 22 g sodium dichromate, 7 ml. acetic acid, 6 ml. H_2SO_4, 58 ml. dist. water	20–60	—	—	—
Iron	530 ml. orthophosphoric acid, 470 ml. dist. water	0.5–0.2	6	20	10
Molybdenum	35 ml. H_2SO_4, 140 ml. dist. water	12	—	50	—
Monel	200 ml. HNO_3, 400 ml. methanol	2.4–2.6	125–150	20–30	10
Nickel	60 ml. orthophosphoric acid, 20 H_2SO_4 2 dist. water	10–18	900	60	5
	210 ml. perchloric acid, 790 ml. acetic acid	22	180	20	—
Carbon steel	50 ml. H_2SO_4, 40 ml. glycerol, 2 ml. HCl 8 ml. dist. water	10–18	50	10	60
	185 ml. perchloric acid, 765 ml. acetic acid, 50 ml. dist. water	50	40–70	< 30	5–10
Stainless steel	50 ml. H_2SO_4, 40 ml. glycerol, 10 ml. dist. water	10–18	300–1000	30–90	3–9
	133 ml. glacial acetic acid, 25 g CrO_3, 7 ml. dist. water	20	900–2500	18	4–6
Tantalum	90 ml. H_2SO_4 (conc), 10 ml. HF	—	100	34–45	9
Tungsten	100 g NaOH, 900 ml. dist. water	—	30–60	20	20–30

6–12 V d.c. at 50–100 mA/cm². With cathodic cleaning hydrogen is liberated on the cleaned surface, and the process requires 6–12 V d.c. for a maximum of 50 mA/cm². For steels the anodic cleaning is recommended but, for nonferrous alloys the cathodic should be used. A recommended solution for electrocleaning consists of : 1 parts by weight caustic soda, 1 pbw trisodium phosphate and 1 pbw soda ash, in water.

Solvent cleaning is done by using the solvent in a liquid or in a vapour·state. Liquid cleaning can use benzene, xylene (if the necessary safety precautions are provided) or inflammable solvents (dichlorethylene $C_2Cl_2H_2$, carbon tetra-chloride CCl_4, trichlorethylene C_2Cl_3H, perchlorethylene C_2Cl_4).

Vapour degreasing is much more effective than liquid solvent cleaning. The solvent is heated to boiling, the parts to be cleaned are hung in the chamber in the hot vapour, which condenses on the metal surfaces, dissolves the oil and grease, and flows back to the solvent container.

Stowers (1978) describes an advanced cleaning method using high-pressure liquid (solvent) spraying.

Grunze *et al.* (1988) discuss *cleaning* methods for metals (Fe, Ni, Cu, Ag, stainless steel) in vacuum *by exposure to reactive gases* (O_2, NO, H_2, NH_3).

7.2.2. Cleaning of glass

In order to obtain inside surfaces sufficiently clean for vacuum purposes, the glass or quartz vessels, pipes or other parts should be thoroughly cleaned, even if the glass is perfectly transparent. More careful cleaning is necessary if the glass has a rough appearance.

The glass may have this rough appearance due to *weathering* or to *devitrifica-tion*. *Weathering* is a result of the influence of the atmospheric vapours, and consists of the hydrolysis of the alkali silicates, forming alkali hydroxides and colloidal silicic acid. The alkali hydroxides react with the carbon dioxide from the air, forming a film of alkali carbonates, with separation of silica. To cause minimum weathering *the glass is best stored unwrapped* or packed in plastics. When the weathering is not too advanced the matt surface can be cleaned by acid washing. The *devitrification* is a result of recrystallization. Due to this process the glass loses its transparency and becomes brittle. To avoid devitrification glasses must not be cooled too slowly. The appearance of the devitrified glass cannot be changed by cleaning its surface.

Glass vessels and pipes (new or weathered) can be washed by immersing them in hydrochloric acid solution (1–5%) for 3–10 sec, followed by a subsequent rinsing in water (40–50°C) and drying. Tap water if quickly dried leaves salts on the surfaces. To remove these salts a second rinsing in distilled (or demineralized) water is recommended before drying. For the washing of soda–lime glasses an acetic acid solution (3–5%) is recommended instead of hydrochloric acid.

Chromic acid is satisfactory for cleaning glass (subsequent to a washing in water) provided that care is taken to make sure that the glass is free of mercury. If mercury is present a residue is precipitated which is difficult to remove. The usual cleaning solution known as chromic acid contains about 50 ml of saturated aqueous sodium dichromate in a litre of concentrated sulphuric acid. The chromic acid solution should be used only if it has its brown colour. If the colour is changed the solution is decomposed.

A solution which is much more effective than the chromic acid solution consists of 5% HF with 33% HNO_3 in 60% water. The solution should be used cold.

Weathered glass can be washed with HF solution (40% in volumes) by an immersion of 1–5 min. The glass is superficially attacked by this solution, but it remains smooth. A subsequent washing with (distilled) water, and neutralizing in NaOH solution is absolutely necessary. The final washing is made in distilled water (40°C) and alcohol.

Organic solvents are adequate to remove greases from the inner surfaces of glass parts. Silicone grease can be washed from glass surfaces by using dichlorethlene or kerosene with subsequent washing with a solution of 10 g NaOH and 5 g borax in 100 ml distilled water or a solution of 10–15 ml KOH (50%) in 100 ml ethyl alcohol (maximum immersion 10 min).

The *drying* of washed glass parts can be done by hot air (free of oil). Acetone or alcohol may also be useful for drying glass parts. In order to obtain extremely clean surfaces for vacuum coating ion bombardment (discharge) cleaning (§4.5.2) is used.

Cleaning techniques are discussed by Strong (1938), Angerer and Ebert (1959), Roth (1966), Espe (1966/68), Stowers (1978).

Quartz parts have to be cleaned (before heating) with alcohol. Any material (alkaline, sweat) left on the surface of quartz causes at high temperatures its reversion to the crystalline state, which show up as permanent marks on the surface.

7.2.3. Cleaning of ceramics

Suitable cleaning of ceramic parts is obtained by firing the ceramic parts in air at 800–1000°C. Alternatively an alkaline cleaning solution can be used, followed by immersion in dilute nitric acid (2–5 min.). Chromic acid or other glass cleaning solutions are also satisfactory.

7.2.4. Cleaning of rubber

Rubbers evolve a great amount of gases (figs. 4.30; 3.44), especially when they are new and untreated. In order to obtain lower outgassing rates the rubber must be cleaned with a solution of KOH (20%) at 70°C with subsequent washing with distilled water and drying with clean air, and/or degassing in vacuum at 70°C for 4–5 hours (Angerer and Ebert, 1959).

Rubbers that are to be used in contact with mercury must be treated for at least one hour with NaOH (20%) solution at 70°C. After this treatment, the mercury will remain uncontaminated in contact with rubber.

7.2.5. Baking

The most efficient method of reducing the outgassing rates of the parts of vacuum systems is their baking. The useful range of baking temperatures is 400–500°C (for metal or glass systems), and the efficiency is much reduced if only 150–200°C is used (see fig. 3.44).

Baking techniques are discussed by Espe (1966/68), Beijerinck and Verster (1973), Fischer (1974, 1977), Calder et al. (1977), Rees (1977), Blechschmidt (1978), Dean et al. (1978), Kubiak et al. (1983).

7.3. Sealing techniques

7.3.1. General, classifications

A vacuum system or even a vacuum chamber cannot be constructed as a single unit. One must use various components of various shapes and different materials and provide for the possibility to change the parts or to open and close the chambers. These various parts are joined together using various seals, which afford joining the parts but prevent leakage through the joint. A detailed treatment of sealing methods was published by Roth (1966).

A common requirement and permanent problem of all the vacuum seals is their *leak tightness*. Any vacuum seal must be leak tight but must not necessarily be hermetic. A *hermetic seal* is designed to permit no detectable leak through it (on a sensitive leak detector such as a helium mass spectrometer), (see §7.4.5), while a *leak tight* seal is just free of leaks according to a given specification.

Besides their function to prevent gas penetration, some vacuum seals must be capable of allowing the *transmission of an electric current* or of *motion*, the *transfer of material* or the passage of *radiation* into or from the system.

The classification of seal used in vacuum technology can be based on the purpose of the seal, the requirements, the materials or the construction techniques used (Roth, 1966). The selection of the appropriate seal is discussed by Peacock (1980).

For the present short description we divide the seals in: *permanent seals* (welded and brazed metal joints, glass-to-glass, glass-to-metal, and ceramic-to-metal seals); *demountable seals* (wax and resins, ground, liquid and gasket seals); *electri-*

cal lead-throughs; seals for *motion transmission*; seals for *transfer of materials* (cut-offs, valves, vacuum locks).

7.3.2. *Permanent seals*

Metal parts are joined permanently by *welding* or by *brazing*. *Glass-to-glass* is joined permanently by fusion; *glass-to-metal*, and *ceramic-to-metal* seals are constructed by using specific techniques.

Welded seals. Welding is the generic term to describe metal joining processes based on *localized melting* of the metal produced as a result of temperature and/or pressure (fig. 7.2). For detailed discussions of welding methods we refer to Jefferson (1955), Laughner and Hargan (1956), Espe (1966/68), Roth (1966), Young J.F. (1966), Sullivan (1966), Hartwig and Kouptsidis (1977b).

The *non-pressure welding* processes include the the techniques of metal joining the application of heat without the use of pressure. In these processes a mixture of molten metal is formed as a result of the local melting of the surfaces or edges. This liquid metal mixture (to which eventually a filler metal is also added) bridges the gap between the components to be welded to each other. After the source of welding heat has been removed, this liquid solidifies, thus "welding" the parts together. The sources of heat used are: the flame (gas welding), the electric arc or the electron beam.

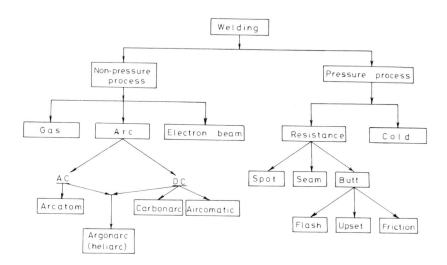

Fig. 7.2 Welding methods for vacuum sealing.

The *pressure welding* processes are techniques for joining metals using pressure with or without heating the components. The *resistance welding* heats the parts, but in some cases *cold welding* is possible.

In *gas* (*torch*) *welding* using acetylene or hydrogen flames the torch can be adjusted to produce a reducing, neutral or oxidizing flame. *Neutral* (or slightly reducing) flames are preferred in vacuum sealing. The use of a reducing flame can be the cause of *porous welds* due to the occlusion of hydrogen in the weld, a fact especially evident when copper is welded. The use of oxidizing flames affects the strength of the weld introducing an oxide layer between the parts. The necessity to use *fluxes* (high outgassing rate) limits the use of torch welding for vacuum sealing only to joints in very heavy copper or iron vessels.

Arc welding is based on the heat obtained from an electric arc formed between the work and an electrode or between two electrodes. From the many commercial arc welding systems those used for vacuum sealing are: the atomic hydrogen (arcatom) process, the carbon arc, the aircomatic and the argon arc (heliarc) process.

In the *arcatom* process the heat is obtained from the a.c. arc (60–100 V; 20–62 A) between two tungsten electrodes surrounded by hydrogen. The molecular hydrogen supplied through the electrode holder is dissociated in the arc to atomic hydrogen and recombines on contact with the cooler metal, producing temperatures up to 4000°C. The welds are homogeneous and clean. This method is suitable for iron, mild steels, aluminum and chromium, but is *unsuitable* for nickel and copper alloys. The hydrogen is soluble in molten nickel (stainless steels) and it escapes when the metal is setting, producing cracks and pores. Copper and copper alloys become brittle due to the hydrogen.

The *carbon arc welding* process uses a d.c. straight polarity arc (1–10 A; 100 V) between the carbon electrode and the work or between two carbon electrodes. The process may be used to weld iron, nickel, aluminum or copper.

The *aircomatic welding* uses a d.c. current with consumable electrodes and a shielding of hydrogen or argon. The process can be used for aluminum and stainless steel.

In the *inert gas arc welding* (argonarc, heliarc) d.c. or a.c. is used between the work and a tungsten electrode. The arc works in a shielded atmosphere of argon or helium. Aluminum and its alloys are usually welded with a.c. (100 V; 250–300 A), while d.c. (45–75 V; 15–200 A) is used for steels, stainless steels, nickel, copper, silver and titanium. *Inert gas arc welding is the most commonly used welding procedure for vacuum sealing for high and ultra-high vacuum.*

The *electron beam welding* process is carried out in a vacuum chamber (5×10^{-5} Torr) within which a stream of electrons is accelerated through a high potential and focused on the work piece. Very accurate and clean welds are obtained. Electron beam welding can be used for stainless steel, aluminum alloys, tungsten, molybdenum, titanium and tantalum.

Table 7.7.
Weldability of metals and alloys (Roth, 1966).

	Zr	W	Ti	Ta	Stainless steel	Pt	Ni	Mo	Kovar	FeNi
Ag	—	C	—	R_3	—	C R_3	C R_3	—	—	C R_3
Al	—	—	—	R_3	—	R_3	—	—	—	R_2
Au	—	—	—	—	—	—	—	—	—	—
Be	—	—	—	—	—	—	—	—	—	—
Brass	—	—	—	—	—	—	R_2	—	—	R_2
Co	—	—	—	—	—	—	—	—	—	—
Cr	—	—	—	—	—	—	—	—	—	—
CrFe	—	—	—	—	—	R_1 C	R_2 C	—	—	—
CrNi	—	C	—	C	R_2	R_1 C	R_2	R_2 C	—	C C
Cu	R_3	C	R_3	—	(7)	R_2	(5)	(6) C	(6)	R_2
Fe	R_3	R_3	—	R_3	—	R_1	R_1	R_3	R_3	R_3
FeNi	—	C	—	C	R_3	C	C	C	—	C, H P
Kovar	—	—	—	—	(7)	—	—	(6)	A, H(1) (4, 5)	
Mo	R_2 H	C,A C	—	—	R_2	C R_2 C	C R_2 C,H(5)	A,C E(3)		
Ni	R_2	R_2 C	R_3	R_2	R_1	R_1 C	M, R_1			
Pt	—	R_2	—	R_2 H	R_1 H,M(5)	R_1				
Stainless steel	—	R_2	—	R_2	E(3)					
Ta	—	—	—	—						
Ti	—	—	H,E(3) R_2 C,A							
W	R_2	E								
Zr	H R_2									

(contd

Table 7.7. (contd.)

Fe	Cu	CrNi	CrFe	Cr	Co	Brass	Be	Au	Al	Ag
		C								
R_3	C	R_3	—	—	—	—	—	—		C,H
									A,C,H	
R_2	—	—	—	—	—	—	—		M,R_3	
								T,P		
—	—	—	—	—	—	—	—	R_2		
—	—	—	—	—	—	—	$E(3)$			
						C				
R_2	R_2	R_2	—	—	—	R_2				
—	—	—	—	—	A					
				A,H						
				(1)						
—	—	—	R_2							
R_2	C	C								
R_2	R_2	R_3								
	$H(2),T$									
R_2	$P(5),R_3$									
C,H										
T,R_1										

T—torch weld. A—arcatom. C—carbonarc. H—heli (argon) arc. M—aircomatic. E—electron beam. P—cold weld. R—resistance weld. 1 very easy. 2 good. 3 difficult.

(1) After cleaning with phosphoric acid
(2) Only OFHC copper
(3) Limitation for large parts
(4) Use electrodes of CrNi-steel (18 Cr, 8 Ni)
(5) Vacuum-tight for many cycles up to 480°C
(6) Not recommended for vacuum sealing
(7) Danger of leakage (cracks)

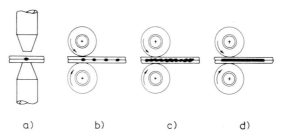

a)   b)   c)   d)

Fig. 7.3 Resistance welding. (a) spot weld; (b) roll spot weld; (c) overlapping spotweld; (d) seam weld.

In *resistance welding* the joining of the pieces is produced by the heat obtained from the resistance set up in the metal parts by the passage of high intensity currents, the parts being pressed against each other. Resistance welding is used as spot welding or seam welding (fig. 7.3).

Butt welding is the joining of two parts placed with their ends abutting each other. The fusion of the ends is accomplished by *flash or upset*. In *flash* welding the parts are placed slightly separated, the electric voltage applied produces an arc (flash) in the gap. At this moment the parts are pressed against each other. *Upset* welding is similar to flash welding, except that the areas to be joined are pressed into contact and then heated by an electric current which passes across the abutting surfaces.

Friction welding is a process limited to welding the ends of objects (tubes, caps) one of which is rotated about its axis (1500–6000 rpm). The heat is obtained due to the friction between the parts. The pressure of the parts on each other has to be e.g. for Al to stainless steel friction welding, about 5 kg/mm² (Hartwig and Kouptsidis, 1977b).

Cold welding. Certain metals can be welded together to give a vacuum tight seal, by applying sufficient pressure. The pressures required for cold welding are about 17–25 kg/mm² for aluminum, 50–75 kg/mm² for copper and about 200 kg/mm² for stainless steel. The surfaces must be *free of oxides* and well degreased. The fact that the whole procedure is cold, and practically no gases are released during the sealing, makes this procedure adequate for the sealing-off of evacuated metal tubes. Cold welding for ultra-high vacuum is discussed by Murko-Jezovsek (1973).

Weldability indicates the amount of precautions necessary for successful welding. As a guide to the first choice of the welding method which can be used, table 7.7 summarizes the recommended (or dangerous) solutions (Lander *et al.*, 1962; Von Ardenne, 1962; Roth, 1966).

In the design and construction of *welded seals for vacuum technology* the following points should be observed:

(1)   The joints must be designed and welded with *full penetration* avoiding trapped volumes in which contaminants may collect. Some recommended arrangements of welded joints for vacuum seals are shown in fig. 7.4, together with the incorrect constructions which are to be avoided.

(2)   Whenever possible, *single-pass* welds should be used. Double-pass welds create trapped volumes, and make impossible the leak detection.

(3)   Welds should be made from the vacuum side of the vessel.

(4)   If for strength reasons double weld is necessary, the inside weld should be the leak-tight one. For leak detection, drilled and plugged holes have to be provided on the outside weld.

(5)   If structural welds are necessary inside the vessel, they should be made discontinuous to allow easy flow of gases from any pocket. These structural welds should not cross the sealing ones.

(6)   The welded assembly should be designed so that a maximum number of welds could be tested separately in the construction stage and corrected prior to making the final assembly.

The maximum leak rate (air) permitted in welded seals is about 10^{-8} Torr·lit/sec· cm length of weld (see fig. 3.43). If *higher leak rates* are found *the weld should be*

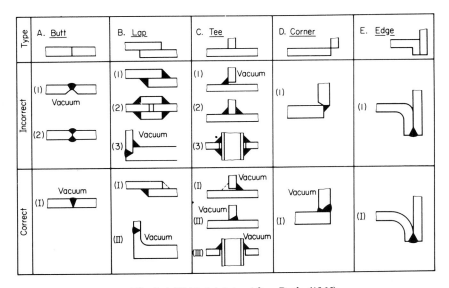

Fig. 7.4 Welded joints. After Roth (1966).

ground off to the base metal and a new weld must be made. The supposition that applying another layer of weld over the original (leaking) one will correct the situation, is wrong. Leaks are seldom corrected this way, since stresses are likely to be set up, which cause a new crack.

Brazed seals. The joining process of two metal parts with a third one having a lower melting point is known generally as *soldering*. When the solder has a melting point lower than 400°C the process is known as *soft soldering*; if the solder melts above 500°C the process is known as *hard soldering*.

Brazing is defined as the metal joining process in which molten filler metal is drawn by *capillary attraction* into the space between the closely adjacent surfaces of the parts to be joined. The temperatures required for brazing are above 500°C and they should be 50–200°C lower than the melting point of the brazed

Table 7.8.
Brazing metals and alloys*.

No.	Liquidus (°C)	Solidus (°C)	Type of alloy**	Composition (parts by weight)
1	3180	3180	P	Rhenium
2	2996	2996	P	Tantalum
3	2497	2497	P	Niobium
4	2427	2427	P	Ruthenium
5	2444	2444	P	Iridium
6	1966	1966	P	Rhodium
7	1950	1935	—	Rhodium (40), Platinum (60)
8	1852	1852	P	Zirconium
9	1770	1770	P	Platinum
10	1695	1645	—	Au (5)–Pd(20)–Pt(75)
11	1550	1550	P	Palladium
12	1452	1452	P	Nickel
13	1423	1423	L	Ni(36)–Fe(64)
14	1320	1320	E	Mo(46.5)–Ni(53.5)
15	1320	1290	—	Pd(30)–Ni(70)
16	1305	1260	—	Pd(13)–Au(87)
17	1300	—	—	Ni(51)–Mo(49)
18	1300	1230	H	Ni(45)–Cu(55)
19	1240	1190	—	Pd(8)–Au(92)
20	1238	1238	E	Ni(40)–Pd(60)
21	1232	1149	—	Mn(3)–Pd(33)–Ag(64)
22	1205	1150	H	Ni(25)–Cu(75)

(contd.)

Table 7.8. (contd.)
Brazing metals and alloys.*

No.	Liquidus (°C)	Solidus (°C)	Type of alloy**	Composition (parts by weight)
23	1160	995	—	Pt(27)–Ag(73)
24	1135	1080	—	Fe(3)–Si(10)–Cr(19)–Ni(68)
25	1084	1084	P	Copper (OFHC)
26	1083	1083	—	Ni(3)–Cu(35)–W(62)
27	1065	1000	—	Pd(10)–Ag(90)
28	1063	1063	P	Gold
29	1060	1000	—	Ag(5)–Cu(95)
30	1050	1030	L	Ni(30)–Mn(70)
31	1035	1015	H	Au(30)–Cu(70)
32	1030	975	—	Cu(62)–Au(35)–Ni(3)
33	1025	970	—	Cu(97)–Si(3)
34	1025	960	H	Cu(95)–Ag(5)
35	1020	—	—	Fe(33)–Ni(56)–P(11)
36	1018	1018	E	Ni(40)–Mn(60)
37	1015	990	H	Cu(63)–Au(37)
38	1015	970	—	Cu(77)–Au(20)–In(3)
39	1010	985	H	Cu(60)–Au(40)
40	1005	996	—	B(3.5)–Si(5)–Cr(16)–Ni(72.5)–Fe(3)
	993	976	—	B(2.9)–Si(4.5)–Ni(91)–Fe(1.6)
41	975	950	—	Cu(50)–Au(50)
42	971	960	—	Ag(85)–Mn(15)
43	962	—	—	Ni(3)–Ag(35)–W(62)
44	960	960	P	Silver
45	950	950	L	Au(82)–Ni(18)
46	—	946	—	Cu(85)–Sn(8)–Ag(7)
47	920	904	—	Au(58)–Cu(40)–Ag(2)
48	910	779	—	Cu(60)–Ag(40)
49	900	860	—	Au(60)–Cu(37)–In(3)
50	900	900	E	Cu(76)–Ti(24)
51	900	—	—	Cu(50)–Ni(10)–Mn(40)
52	900	714	—	Cu(95)–P(5)
53	896	885	—	Au(75) Cu(20)–Ag(5)
54	889	889	L	Au(80) Cu(20)
55	885	779	—	Ag(62) Cu(32)–Ni(6)
56	—	800	—	Cu(50) Ag(40)–Mn(10)
57	880	—	—	Ni(89)–P(11)
58	870	779	H	Ag(90)–Cu(10)
59	845	835	—	Au(60) Cu(20)–Ag(20)
60	830	779	—	Ag(77) Cu(21)–Ni(2)
61	821	794	—	Au(50) Ag(30)–Cu(20)

(contd.)

Table 7.8. (contd.)
Brazing metals and alloys.*

No.	Liquidus ($^\circ$C)	Solidus ($^\circ$C)	Type of alloy**	Composition (parts by weight)
62	779	779	E	Ag(72)–Cu(28)
63	770	714	—	Cu(93) P(7)
64	721	640	—	Cu(87) P(7.5)–Ag(5.5)
65	630	—	—	Ag(80)–In(20)
66	636	—	—	Ag(50) Cu(15.5)–Cd(18)–Zn(16.5)
67	625	(565)	—	Al(95)–Si(5)
68	570	(550)	—	Al(85)–Cu(3)–Si(12)
69	550	—	—	Ag(60) Cu(23)–Sn(17)

*Brazing possibilities, see table 7.9.
**P = pure metal, E = eutectic; L = lowest melting point alloy; H = alloy having higher liquidus temperatures if more of the main metal component is added.

parts. For detailed discussions of brazing methods we refer to Strong (1938), Knoll (1959), Monch (1959), Roth (1966), Kohl (1967), Sullivan (1966), Espe (1966/68).

Brazing is carried out by torch, furnace or induction.

Torch brazing uses flames of oxyacetylene, oxyhydrogen, oxygen–butane, etc. A neutral or reducing flame can be used, except for brazing of copper where an oxidizing flame is used to avoid embrittlement. Torch brazing needs fluxes which must be carefully cleaned from the joint, *the flux remaining on the vacuum side of the seal has a high outgassing rate.*

Furnace brazing consists in heating an assembly of metal parts to be brazed in a furnace with a protective atmosphere (vacuum, neutral gas).

Induction brazing utilizes a high frequency current (400–2000 kc) to heat the parts which are placed in a specially fitted coil of size and shape that match the assembly to be heated (Curtis, 1950; Kohl, 1967).

Brazing can also be achieved below the melting point of the brazing metal, by using the *diffusion process*. In this process a thin layer of suitable metal (gold, silver, copper) is placed between the parts. The layer may be formed by electrodeposition on one or both of the members, or may be interposed as a foil of about 0.01 mm thickness. The parts are assembled, pressed against each other, and heated to 400–700°C (about 15 min.).

The *brazing materials* for vacuum technology must have low vapour pressure, must be pure, have ability to wet and flow at brazing temperatures, and alloy with the joined metals. Based on these criteria the list of suitable elements to be

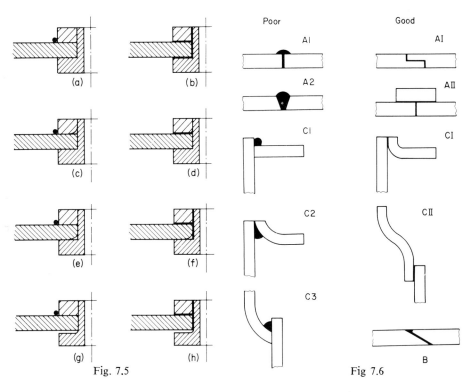

Fig. 7.5

Fig 7.6

Fig. 7.5 Control of brazing by means of the gap.
Fig. 7.6 Brazed joints. After Roth (1966).

used in brazing alloys for high vacuum seals is limited. These brazing alloys are listed in table 7.8 and the possibilities for using them with various metals are summarized in table 7.9 (Roth, 1966).

In some cases of physical incompatibility, some brazing alloys are excluded from use with particular metals. So, *Kovar*, (Fe Ni Co) cannot be brazed with *silver*, because the silver penetrates in the Kovar and produces its splintering. *Silver and gold* (and copper) brazing materials cannot be used in plants where *mercury* vapour will be present, except if they are protected by nickel electroplating.

To obtain a *leak-tight brazed joint* the following points should be observed:

(1) The smallest quantities of brazing alloy should be used. With *small*

Table 7.9.

	W	Ti	Ta	Stainless steel	Ni	Monel	Mo
Ag	—	—	—	—	62(43)	—	—
Au	—	—	—	—	—	—	—
Be	—	—	—	—	62	—	—
CrFe	—	—	—	—	—	—	—
CrNi	—	—	—	—	—	—	—
Cu	54, 45 44*[8], 62*[3]	50	—	62,40	62,45 32,39 49	32	45,28 64,32
Fe	22	—	—	—	37,39	26	22
FeNi	—	—	—	—	—	—	—
Inconel	45	—	—	20,21 24,42	45	—	45
Kovar	45,32 54*[10]	—	—	28	45,28, 38,32,39	26,28,32 62*[6]	45(44)
Mo	22,23 32,45	—	—	19,27	27,32 45	32	3–14,23 32,45
Monel	28,32	—	—	—	32	32	
Ni	— 32	50	—	27	17,31,32 28,40,41 62		
Stainless steel	19,20 21,24	—	—	15,19 24,30,36 40,45			
Ta	—	—	28*[6]				
Ti	—	44*[7]					
W	1–16 23,45						

(contd.)

clearances and clean surfaces better joints are obtained than when applying big quantities of brazing alloy.

(2) Wide or irregular spaces between parts are to be avoided.

(3) The overlap between the two parts brazed to each other must be minimum 3 mm, in order to allow the capillary forces to suck in the brazing alloy.

(4) If metals with different thermal expansions are to be brazed, the assembly must be arranged to compress the brazing alloy during the cooling, i.e. the *outer part must have the bigger thermal expansion.*

Brazing possibilities of metals* (Roth, 1966).

Kovar	Inconel	FeNi	Fe	Cu	CrNi	CrFe	Be	Au	Ag
—	—	62	—	62,64	62	—	—	—	54,62
									64
—	—	—	—	—	—	—	—	62	
—	—	—	25[5]	62[1]	—	—	—		
—	—	—	25[6]	—	—	25			
44,28	—	—	—	66[9]	66[9]				
				29,33,39					
54,31,32	62,45	62,54	31,39,48	34,46,47					
·39,62[4]			54,62[2]	54,62					
28,31,34	—	—	39,28,47						
37,39			44,54						
—	—	28							
		62[4]							
45	21,24								
	20,45								
28,39,54									
62[5]									

* Numbers refer to alloy number from table 7.8.

[1] Short h.f. heating; alternative technique: braze Be to Fe using Cu and Fe to Cu using Ag alloy (830°C)

[2] Cu plated Fe, sintered in H_2

[3] W first etched (HNO_3/HF 1 : 1)

[4] Kovar first Cu plated

[5] H.f. heating in H_2 (5 min)

[6] H.f. heating in vacuum

[7] In vacuum, or with acetylene flame and AgCe flux

[8] W first Ni plated

[9] Cd and Zn evaporates on heating in vacuo

[10] Heating in H_2, for 5 min at 920°C

(5) The flow of the brazing alloy can be controlled by the construction of the joint. The clearance at the corners determines how the brazing alloy will flow around these corners. *Square corners* (fig. 7.5a) will give a good flow of the brazing alloy through all the joint (fig. 7.5b), the resulting joint being strong and leak-tight. *Round corners* stop the flow. If the first corner, from the side where the brazing alloy is applied, is round (fig. 7.5c) the brazing alloy will not pass this corner (fig. 7.5d). When only the second corner is round (fig. 7.5e), the joint will be stronger and leak-tight (fig. 7.5f). A square edge pressed against

Table 7.10.
Glasses used in vacuum technique.

Type	Constituents % weight	Examples				
		Glass	Expansion coeff. $(10^{-7}/^{o}C)$	Strain point ^{o}C	Annealing point ^{o}C	Manufacturer**
Soft glasses	SiO_2;					
	$PbO=$	Minos 1650[111]	88	—	415*	Jena
Lead	40–50;					
glasses	alkali					
	<10	Lead W2	82	—	415	Moosbr.
		Iron 153	123	380	396	Sovirel
		Lead N (915a)	100	—	425*	Moosbr.
		FeCr L 14	98	360	430	GEC
		Lead K 1 A	95	—	425*	Philips
	SiO_2;	Lead 111	92	—	425*	Philips
Lead	$PbO=$	Soft lead L 1	91	340	430	GEC
alkali	20–35;	Copperclad C 12	91	380	435	BTH
silicate	alkali	Soft glass 0010	91	397	428	Corning
glasses	<10	Soft glass 0120	89	400	433	Corning
		123a M	88	—	425*	Osram
		Lead GWB				
		(GW2)	86	—	410	Chance
		Lead 3079[111]	81	—	480*	Jena
		B8	96	460	530	GEC
		X 4	96	465	500	GEC
Soda	SiO_2;	FeCr C 19	95	—	530	BTH
lime	$CaO=$	X 8	95	465	500	GEC
silicate	5–12;	Bulb 0080	92	478	510	Corning
glasses	alkali=	Magnezia 105	89	—	508*	Osram
	13–20	GWA (GW1)	87	—	530	Chance
Hard glasses	SiO_2;	Iron R L 114	114	—	500	GEC
	$Al_2O_3=$	Lime C 22	104	—	505	BTH
Alumino	3–10;	Apparate Glass				
lime	$CaO=$	584d	88	—	530*	Osram
silicate	6–12;	Thermometer				
glasses	alkali=	16[111]	80	495	537	Jena
	8–23					

(contd.

Table 7.10. (contd.)
Glasses used in vacuum technique.

Type	Constituents % weight	Examples				
		Glass	Expansion coeff. $(10^{-7}/^\circ C)$	Strain point $^\circ C$	Annealing point $^\circ C$	Manufacturer**
Hard glasses	SiO_2; $B_2O_3 =$					
Alumino boro- lime silicate glasses	3–8; $Al_2O_3 =$ 3–10; $CaO =$ 6–12; alkali= 8–23	Amber Ma 1	75	400	580	GEC
		Mo. B.B.	47	563	594	Russian
Alumino boro- lime zinc glass	As before but $CaO =$ 3–12; $ZnO = 3–7$ alkali= 8–14	Mo. 1447^{III}	50	483	529	Jena
Boro- silicate	SiO_2; $B_2O_3 > 10$ $Al_2O_3 < 3$	Kovar C.40	48	455	505	BTH
		W seal, W 1	38	540	580	GEC
		Duran 3891 III	37	516	567	Jena
Alkali- boro- silicate glasses	SiO_2; B_2O_3 > 10; Al_2O_3 < 6; alkali = 6–8	Thermometer 7520	61	530	566	Corning
		Mo 637h	48	—	550*	Osram
		Neutrohm E(Mo)	48	—	505*	Baccarat
		FeNiCo 756	48	—	500*	Osram
		Mo. H.H.	47	500	590	GEC
		3072(Gerätte 20)	46	—	558*	Jena
		Clear seal 7050	46	461	496	Corning
		Mo. C 11	45	500	575	BTH
		Uran 3320	41	497	535	Corning
		W glass C 9	36	480	525	BTH
		Hysil GH 1	33	513	556	Chance
		Pyrex 7740	33	515	555	Corning

(contd.)

Table 7.10. (contd.)
Glasses used in vacuum technique.

Type	Constituents % weight	Examples				
		Glass	Expansion coeff. $(10^{-7}/^{\circ}C)$	Strain point $^{\circ}C$	Annealing point $^{\circ}C$	Manufacturer**
Hard glasses Alumino borosilicate glasses	SiO_2; $B_2O_3 =$ 5–20; $Al_2O_3 =$ 3–20; alkali < 6	Kovar GS 3	50	400	450	Chance
		Mo. H 26 X	46	600	725	GEC
		Supremax 3058[III]	33	—	738*	Jena
Lead borosilicate glasses	SiO_2; $B_2O_3 =$ 15–18; $PbO = 4$–7	W seal 362a	39	—	522*	Osram
		Nonex 7720	36	484	518	Corning

*Transformation point, with viscosity $10^{13.3}$ poise.
**BTH — The British Thomson–Houston Co. Ltd., Rugby, England.
Chance — Chance Brothers Ltd. Glass Works, Birmingham, England.
Corning — Corning Glass Works, Corning, N.Y., U.S.A.
GEC — Osram–G.E.C Glass Works, East Lane, Wembley, Middlesex, England.
Sovirel — Sovirel Co., Bagneaux-sur-Loing, France.
Osram — Osram, Berlin, West Germany.
Moosbr. — Moosbrunner Glasfabrik, Vienna 4, Austria.
Jena — Jenauer Glaswerk Schott u. Gen., Mainz, West Germany.
Philips — Philips, Eindhoven, Holland.
Baccarat — Cristallerie de Baccarat, France.

a round corner (fig. 7.5g) would similarly stop the flow (fig. 7.5h).

(6) If the flow of the brazing alloy is to be avoided on a surface, the area must be coated with carbon or chromium.

(7) Lap joints, or step joints are to be preferred in vacuum sealing applications (fig. 7.6).

Glass-to-glass (quartz) seals. Although silica (SiO_2) is the principal constituent of most glasses, the addition of other melting agents and modifiers gives to glasses a wide range of properties. A general classification divides glasses into *soft and hard ones* (table 7.10) corresponding to the temperature range in which the glass is soft enough to be worked.

Among the points established to define the state of the glass the two most important are the *strain point* and the *annealing point*. A detailed list of the many characteristic points of glasses is given by Roth (1966).

The *strain point* is defined as the temperature at which the internal stress in the glass is substantially relieved in 4 hours, and "absolutely" relieved in 15 hours. *The annealing point* is the temperature at which the internal stress is substantially relieved in 15 min. The strain point corresponds to a viscosity of $10^{14 \cdot 5}$ poises, while the annealing point corresponds to 10^{13} poises. Sometimes an intermediate point, the *transformation point* (viscosity $10^{13 \cdot 3}$ poises), is given.

Soft glasses have their annealing point between 350° and 450°C, while the annealing point of *hard* glasses is higher than 500°C.

Glass-to-glass sealing techniques are performed by the glass blower using manual or mechanized tools. The description of the operation of glass blowing exceeds the scope of this section, and we refer for this subject to Strong (1938), Barr and Anhorn (1949), Reimann (1952), Parr and Hendley (1956), Robertson (1957), Wheeler (1958), Friedrichs (1960), Barbour (1968).

Nevertheless the following points are important enough to be mentioned:

(1)　Two glasses can be sealed together if their *thermal expansions do not differ more than about 10%*.

(2)　The contact of the hot glass with conducting materials (metals) produces stresses.

(3)　The free cooling of a glass assembly is generally too fast to form a stress-free joint. The *stresses may be released* by a suitable annealing, i.e. heating some degrees above the annealing point, holding it at this temperature for 5–10 min., and cooling slowly (at 1–3° C/min).

Glasses of *widely different expansion coefficients* can be joined by *graded seals*. These seals consist of a number of segments of glass, having progressively slightly different expansions; together the segments form a zone of gradual transition between high and low expansion.

Using graded seals, quartz can be sealed to hard or soft glasses.

Glass-to-metal seals. To obtain a reliable, leak-tight glass–metal seal, the following requirements should be fulfilled:

(1)　To achieve a good *bond* between the metal surface and the adjacent glass.

(2)　To base the seal either on the *matching of the expansion* characteristics of the metal and glass, or on the *plasticity of the metal*.

(3)　To control the cooling process in order to minimize the stresses in the seal.

(4)　To choose the geometry (shape) of the seal so as to obtain minimum and not dangerously oriented stresses.

Table 7.11
Metals and alloys for glass–metal seals.

Metal or alloy	Composition (%) Ni	Cr	Co	Fe	Specific gravity	Expansion coefficient α×10⁶(1/°C) from (20°C) to (°C) 100	200	300	400	500	600	Inflexion point (°C)	Electrical resistivity (Ohm·cm × 10⁶)	Thermal conductivity (cal/cm sec °C)	Manufacturer*	Remarks
Tungsten	—	—	—	—	19.2	—	—	45	—	46	46	—	5.5	0.5	—	—
Nilo 40	40	—	—	60	8.1	—	41	—	—	—	—	330–250	62–70	0.025	W	Curie Point 330 °C
Vacon 12	28	—	18	54	8.3	58	53	48	47	59	76	430	45		V	M.P. 1450 °C
Nilo 42															W	
Driver 42	42	—	—	58	8.2	52	48	49	56 (from 0 °C)	76	—	340	60	0.025	D	Curie Point 375 °C
Carpenter 42	42	—	—		8.2										CS	M.P. 1450 °C
A L 42	42	—	—	58	8.2	47	47	47	55	78	93	340	65	0.026	A	
Dilver P	29	—	17	54	8.5	56	52	48	47 (from 0 °C)	60	—		45	0.042	L	Curie Point 425 °C
Nilo K															W	
Sealvac A	29	—	17	54	8.3	58	51	47	46	60		430	44–50		VM	Curie Point 453 °C
Rodar															WD	M.P. 1450 °C
Therlo															D	
Vacon 10	28	—	18	54	8.3	60	56	51	50	61	78	425	45		V	
Fernico I	29	—	17.8	53.2	8.3	59	57	52	50	63	77	430	50		G	
Sivar 48		—	17.3	52.9				(from 30 °C) 45–51		57–		423	—		M	
Kovar A	{28.7 / 29.2}	—	17.8	53.4	8.2	43–53	44–52 57	45–51 54	53	62 65	79	435	49	0.044	St	Curie Point 453 °C
Fernico 11	31	—	15	54		—	57	54	53	65		420	44		—	M.P. 1450 °C
Molybdenum					10.3					55		—	58	0.3		
Vacodil 42	42	—	—	58	8.2	49	52	53	63	55	95	355	60–100		V	
Vacodil 43	43	—	—	57	8.3	52	61	60	65	81	95	370	60		V	
Vacon 20	28	—	21	51	8.3	64	68	65	63	81	76	480	45		V	
Vacodil 46		—	—			71				63			60–100		V	
Driver 46	46	—	—	54	8.2	74	74	73	75	88	99	400	45		D	
Vacon 70	28	—	23	49	8.3	85	80	77	74	71	81	515	50		V	
Nilo 48	48	—	—	52		87	83	83	83	88	—	435		0.030	W	Curie Point 450 °C
Vacovit 426															V	
Driver 14															D	
Sylvania HC-4	42	6	—	52	8.2	69	72	83	101	114	124	265	95	0.033	S	
Carpenter 426															CS	
Sealmet 4															A	

(contd.)

Table 7.11. (contd.)
Metals and alloys for glass-metal seals.

Metal or alloy	Composition (%)				Specific gravity	Expansion coefficient $\alpha \times 10^7 (1/°C)$ from (23 °C) to (°C)						Inflexion point (°C)	Electrical resistivity Ohm·cm × 10⁶	Thermal conductivity (cal/cm·sec·°C)	Manufacturer*	Remarks
	Ni	Cr	Co	Fe		100	200	300	400	500	600					
Platinite	49	—	i	51	8.2	87	88	88 (from 0 °C)	87	92	—	—	50	0.038	I	Curie Point 480 °C
Driver FeNi	48	—	—	52		79	86	88	88	93	—	425	50	0.037	D	
A L 4750	47	—	—	53		84	88	89	91	98	108				A	
Vacovit 501															V	
Carpenter 49	49	1	—	50	8.2	91	91	91	89	97	107	445	58		CS	
Platinum	—	—	—	—	21.4	89	91	92 (from 0 °C)		96		—	9.8	0.17	—	
Nilo 50	50	—	—	50	8.2	93				100		470	41–47	0.032	W	Curie Point 489 °C
Driver 52	52	—	—	48						105			43		D	M.P. 1450 °C
Sealmet 1		28	—	72	7.6	84	93	98	102	100	108			0.059	A	M.P. 1480 °C
Chrom Iron	<0.4	23.5 27		76 73	7.5	84		99	104	105				0.04	Ph	M.P. 1490 °C
Dilver 0		28–25	—	72–75	7.5	84	93	98	102	108	108	340	65	0.029	I	Curie Point 570 °C
FeNiCr	47	5	—	48	8.2			88		105	109	480	51		W	
Vacovit 511	51	1	—	48		101	101	101	101	102					V	
Driver 74/26		26	—	74		86	97	102	106	112					D	
Telemet		16–23	—	84–77			—	104 (from 0 °C)		110			60	0.057	A	
Dilver T	—	x	—	x 80 (Nb)	7.6	93	99	104	108	110	—		65	0.029	I	Curie Point 640 °C
Novar B	0.3	19	—	0.5–1	8.2	96	99	106	108	110	112		35		H	
Vacovit 540	54	—	—	46		106	106	107	107	108	113	550	70		V	
Vacovit 025		25	—	75		103	105	107	109	111	112		—		V	
Fernicochrom	30	8	28	37		88	91	94	101	115	—	380			—	

*A – Allegheny Ludlum Steel Corp. Brackerrnridge, Pa., U.S.A.; Ph – Phillips, Eindhoven, Holland; D – Driver Harris Co. Harrison, N.J., U.S.A.; V – Vakuumschmelze AG., Hanau, Germany; CS – Carpenter Steel Comp. Reading, Pa., U.S.A.; I – Aciéries d'Imphy, Nièvre France; W – Henry Wiggin & Co. Ltd., Birmingham, England; S – Sylvania Electric Prod. Inc., Hayside, New York, U.S.A.; St – Stupakoff Ceramic Mfg. Co. Latrobe. Par, U.S.A.; G – General Electric Co. U.S.A.; M – Metallwerk, Plansee, Reute-Tirol, Austria; H – Stahlwerk Hagen. Germany; VM–Vacuum Metals Corp; WD – Wilbur W. Driver Co.

The *bond* in glass–metal seals is based either on direct glass to metal adhesion or on an oxide–metal bond. In the direct glass to metal bond the metal surface adheres to the glass without any intermediate layer. This kind of seal can be vacuum-tight but the bond is not strong. The seal is mechanically stronger if between the metal and the glass an oxide layer is formed, containing a graded series of oxide mixtures from the oxide of the metal to those forming the glass. The kind of oxides formed is determined by the composition of the metal and of the glass, the kind of the atmosphere of the flame and the temperature during the sealing. *Platinum* can be sealed in glass only without oxides, since it does not form them; thus the platinum glass seal has a metallic appearance and a limited strength. *Copper* can give very adherent seals; if the oxide is Cu_2O and its thickness is the proper one. The correct colour of copper to glass seals is from gold–yellow to purple; grey or black colour seal is not recommended. *Nickel* can adhere to glass with a metal or oxide bond (grey–green colour). *Tungsten* and *molybdenum* can give metallic bonds, but the seals having an oxide layer have to be preferred. The colour of an adequate tungsten to glass seal is from golden yellow to brown if the glass contains sodium or potassium, blue if the glass contains lithium and grey–brown in lead glasses. The molybdenum seals are brown. *Chromium* forms an oxide bond and produces very strong seals which are dark green. The FeCr glass seals (table 7.11) are brown–green, those with FeNiCo grey or blue, and FeNiCr seals are brown.

The metals and alloys used for glass–metal seals are listed in table 7.11 (Roth 1966).

The name *matched seals* refers to those seals where an attempt is made to have partners of equal thermal expansion. These seals can be carried out with pairs of metals and glasses as shown in tables 7.12 and 7.13 (Monch, 1959; Kohl, 1967; Espe 1966/68). Each such seal includes a specific technology of cleaning, oxidizing, heating, cooling, etc. For details we refer to Partridge (1949), Reimann (1952), Zincke (1961), Roth (1966). From the various sealing techniques we summarize here only that concerning the Kovar–glass seals, which is one of the most extensively used techniques in vacuum technology.

FeNiCo alloys (table 7.11) should be oxidized either before or during the sealing. To ensure freedom of gas bubbles in the seal it is useful to decarbonize the surface of these alloys by heating the parts in wet hydrogen (about 4 hr at 900°C or 1 hr at 1100°C), or better still in vacuum, since the remaining hydrogen can produce bubbles as well. The oxide layer is produced by heating the Kovar in air for about 17 min. at 800°C, 3 min. at 900°C or 1 min. at 1000°C.

Figure 7.7 shows the typical steps in completing a Kovar–glass seal. A sleeve of appropriate glass (table 7.13) is slipped over the oxidized Kovar part (1. fig. 7.7) and the glass is fused to build up a ring on the outside of the tube (2). The

Table 7.12.
Soft glasses for glass–metal seals.

Metal*	Glass**
Platinum	C; 0280, 0041, (7550), 7570, 7560, 0050, 0080, 0010 Ch: GW2 (GW1), PWD, PWL BTH: C 12, C 19, C 94 J: 16, 2962 O: 301b GEC: X 4, L 1(L 15) P: DIAL 444
FeNi (50/50 or 46/54)	C: 0010, 0120, 0080 BTH: C 12 J: 16 III Ch: PWD, PWL GEC: L 1
Dumet	C: 0050 J: 2962 III Ch: GW2, PWD, PWL O: 352, 743g, 123a BTH: C 12, (C 19, C 94) GEC: L 1
FeCr (74/26 or 80/20)	C: 0050, 0060, 0080 (9019, 9010) Ch: GW2 (GW1), PWD BTH: C 31, (C 12) GEC: L 14, X 8, (L 1) O: 123a
FeNiCr	C: 8870, 0080, 0014, 0120, 0010, 0050 Ch: PWD, PWL BTH: C 12 GEC: L 1
FeNiCoCr (37/30/25/8)	C: 0050, 0080 Ch: GW2, PWD
Nickel	GEC: NSG 2
Iron	C: 7290, 1990 (1991) GEC: R 16, ISG 20, NSG 2 BTH: C 76, (C 41) J: 4210
Copper	C: 7295 GEC: CSG 3
Titanium	BTH: C 77, C 78

*See table 7.11.
**See footnotes table 7.10.

Table 7.13.
Hard glasses for glass–metal seals.

Metal*	Glass**
Tungsten	C: 3320, 7720, 7780 (7070, 5420, 7741, 7252, 7750, 7331, 7050) Ch: GS 1 (Intasil), (GH 1 Hysil) BTH: C 14, (C 9) J: 1646, 8212 (3891, 8330, 2955, 8409) O: 712b, 712h, 742c (362a) GEC: W 1 P: Bluesil, Dial 36
Molybdenum	C: 7040, 7052, 7050, 7042, 7510, 8830 (7750, 1720, 7731, 7055, 7720) Ch: GS 4, GSB BTH: C 14 (C 11, C 37, C 46) J: 1639, 2877, 2954 (8401, 1447) O: 637h, 637n, 637x, 906c, 632a GEC: HH (H 26) P: (Kodial)
FeNiCo	C: 7052, 7040, 8800, 7520, 7055, 7050, 7750, 7340, 7060, 1720 Ch: GS 3 BTH C 40 J: 1447, 8243, 8401, 8482 O: 756b, 911b GEC: FCN, SBN 124 P: Kodial
FeNiCoCr	C: 0080, 0050 Ch: PWD
Rhenium	GEC: HH, H 26 X
Zirconium	C: 7052 BTH: C 40
Tantalum	C: 7052, 7720 J: 1447 Ch : GS 4
Silver	O: 424dd

*See table 7.11.
**See footnotes table 7.10.

end of a glass tube is shaped as required and it is sealed to the glass ring (3). An alternative technique consists in sealing the ring so as to extend to the end of the Kovar tube (4. fig. 7.7) or even to cover the rounded edge of the Kovar tube (5. fig. 7.7). It is recommended to anneal the Kovar–glass seal by heating it to about 490°C for 20 min and decreasing the temperature at a rate of 1°C/min down to

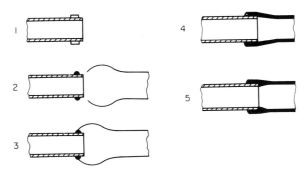

Fig. 7.7 Stages in completing a Kovar–glass seal.

450°C, followed by a cooling at 7–10°C/min to room temperature.

During the glass–metal sealing process, the Kovar parts near the seal are always oxidized. They can be cleaned by immersion (10–60 min) in a hot solution (60–80°C) of 50 g ferric ammonium sulphate, 125 cm³ sulphuric acid (1.84), 150 cm³ hydrochloric acid (1.16) made up with water to 1000 cm³.

The glass–metals known as "*unmatched seals*" are based either on the fact that the stresses developed in the glass are minimized by the elastic or plastic deformation of the metal or on the fact that the developed stresses are only compression. The first kind of seals are represented by those known as *Housekeeper seals*, (Housekeeper, 1923), and the second kind by the *compression seals* (Hull, 1946).

Housekeeper seals can be made with copper, platinum, stainless steel and molybdenum. Using one or the other of these metals, the seal can be made in various shapes (fig. 7.8): (1) wire seal; (2) ribbon seal; (3) feather-edge seal, or (4) disc seal. In principle these seals can be made with any glass having an expansion smaller than that of the metal part. Copper wire up to $d=0.05$ mm (fig. 7.8) and platinum wires up to $d=0.2$ mm can be sealed in glass by this technique. For copper ribbons (2. fig. 7.8) $a=4$ mm; $b=0.1$ mm is appropriate.

Molybdenum ribbons of $a=1$–3 mm, $b=0.01$–0.05 mm can be sealed even *into quartz*. Feather-edge seals (3. fig. 7.8) can be made with copper tubing of $d=$ 10–100 mm, $t=0.07$–0.1 mm (measured at a distance of 1 mm from the edge), $s=3$–4.5 mm and $\alpha=2$–3°. For platinum tubing $\alpha=1$–1.5°, while for stainless steel $\alpha=1°$ is necessary. Copper discs up to $t=0.4$ mm (4. fig. 7.8) can be sealed at the end of glass tubes.

A *compression seal* consists of a metal ring (1. fig. 7.9) surrounding a glass window (2), which may have metal rods (3) or pipes (4) sealed through it. In the window seal (fig. 7.9a, c) the expansion coefficient of the outside metal ring should be always greater than that of the glass. In the rod seal (b, e) or pipe seal (d, f) the expansion of the metal ring (1), glass (2) and rod (or pipe) (3) must be in such a ratio as to develop only compression stresses in the glass part, thus $\alpha_1=\alpha_2>\alpha_3$

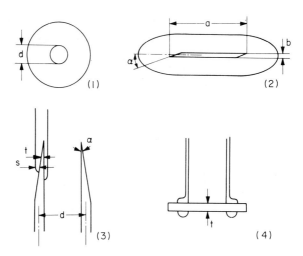

Fig. 7.8 Housekeeper seals.

Table 7.14.
Faults in glass–glass or glass–metal seals.

Nature	Cause
Mechanical	Excessive pressure difference on the glass walls (collapsing, bursting) Excessive bending of the glass tube Knocking with hard materials Scratching with hard materials Inclusions of heterogenous grains or bubbles in the glass Lack of adhesion (bond)
Thermal	Excessive thermal gradients Excessive thermal shock Expansion differential (between two glasses, or between glass and metal) Inadequate annealing Incorrect shape (of the seal)
Chemical	Weathering Devitrification Electrolytic effects Chemical attack (flame, solutions, vapours) Radiation damage

or $\alpha_1 > \alpha_2 = \alpha_3$ are possible solutions. In multiple rod seals (fig. 7.9e) the distance between rods or between the rods and the outer ring must be larger than the diameter of the rods.

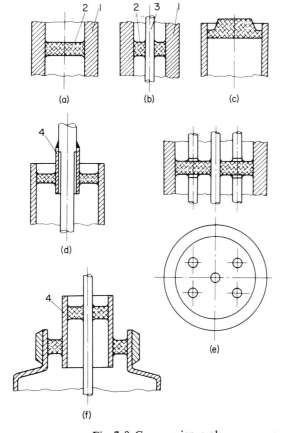

Fig. 7.9 Compression seals.

Unfortunately the *faults* occurring in glass–glass (or quartz) and glass–metal seals are numerous (table 7.14). The drastic result of these faults is obviously the breakdown (cracking) of the seal, but for vacuum seals any cause leading immediately or after some time to a leak rate higher than that specified (e.g. 8×10^{-10} Torr·lit/sec) must be considered as a fault.

Ceramic–metal seals. Ceramic–metal seals can be made by using glass as an intermediate material, or by using the "sintered metal" technique*. These techni-

*Other techniques are: the hydride process, the carbide process, active metal process (see Roth, 1966).

Table 7.15.
Ceramics used for seals with glass and metal.

Ceramic		Manufacturer*	α (10⁻⁷/°C)	Tensile	Compress	Bending
				strength (kg/mm²)		
Forsterites	Alsimag 243	AL	(25 − 700 °C) 112	7	60	14
	Frequenta M	SM	(20 − 800 °C) 106	4	85	11
	Rosalt 7	RI	(20 −1000 °C) 90	5	90	15
	Forsterite 352	HC	from 125 − 85	7	60	13
	BN 3054	GC	(20 − 400 °C) 105	7	60	14
	Frequentite S	SPP	(20 − 700 °C) 111	—	70	14
Steatites	Alsimag 196	AL	(25 − 700 °C) 86	7	63	14
	Almanox 13889	FP	(25 − 700 °C) 73	5	52	12
	Steatit	SM	(20 −1000 °C) 90	5	88	13
	Calit	H	(20 −1000 °C) 85	5	95	15
	Steatite	SPP	(20 − 700 °C) 85	5	85	13
Alumina		(Al₂O₃				
	Alsimag 576 (85%)	AL	(25 − 700 °C) 75	14	98	28
	AD-85 (85%)	CP	(25 −1000 °C) 79	12	140	—
	Stemag A 16 (90%)	SM	(20 − 800 °C) 85	15	170	—
	Almanox 4462 (94%)	FP	(25 − 700 °C) 73	11	131	25
	Alsimag 614 (96%)	AL	(25 − 700 °C) 79	18	280	54
	Alsimag 652 (98%)	AL	(25 − 700 °C) 80	18	294	44
	AD-99 (99%)	CP	(25 −1000 °C) 92	—	220	—
	Degussit Al 23 (99.5%)	D	(0 −1000 °C) 83	26	300	—
Zircon	Alsimag 475	Al.	(25 − 700 °C) 41	8	70	13
	Almanox 3569	FP	(25 − 700 °C) 45	5	44	12
	ZI-4	CP	(25 −1000 °C) 57	—	55	—
	Zr Porcellan	SM	(20 −1000 °C) 52	5	80	11
Al-silicates	Hartporcellan	SP	(20 − 700 °C) 38	3	42	8
	Porcelain	SPP	(20 − 700 °C) 39	3	50	8

(contd.)

*AL – American Lava Corp., Chattanooga 5, Tenn., U.S.A.; FP – Frenchtown Comp. Trenton, N.J., U.S.A.; CP – Coors Porcelain Comp., Golden, Colorado, U.S.A.; D – Degussa, Frankfurt a.M., Germany; SM – Steatit-Magnesia A.G., Lauf/Pegnitz, Germany; RI – Rosen-

Table 7.15. (contd.)
Ceramics used for seals with glass and metal.

Safe temperature (°C)	Softening temperature (°C)	Specific gravity	Thermal conductivity (cal/cm·sec·°C)	Matching Glass**	Matching Metal
1000	1440	2.8	0.008		
1100	—	2.8	0.006		FeNi (46–50%
—	1460	2.7	0.006	BTH C22, C 19,	Ni); FeCr
1000	—	2.9	—	GEC RL 16, X 4, B 8	(16% Cr); Fe;
1000	—	3.0	0.008		FeNiCr
—	—	2.9	—		
1000	1440	2.6	0.006	GEC L 1; Corning 0010,	FeNi (42–
—	1388	2.7	0.006	0080; Osram 562m,	46% Ni)
—	—	2.7	0.006	850, 584d; Jena 2954,	(FeNiCo)
—	1470	2.7	0.006	16 III; BTH C 12;	
—	—	2.6	—		
1100	1440	3.4	0.040		
1400	—	—	0.034		
1400	—	3.5	0.010		
1500	1920	3.5	0.018		Ni; FeNi,
1550	1650	3.7	0.045		(FeNiCo)
1600	1700	3.8	0.045		
1725	—	—	0.070		
1900	2030	3.8	0.012		
1100	1440	3.7	0.012	Mo seal glass; Corning	
—	1550	3.1	0.007	3320, 7050; GEC H26X;	Molybdenum,
1300	—	—	0.010	HH; BTH C 11;	Kovar
—	1500	3.0	—	Jena 3072	
				Corning 7740, 7720; Jena	
1100	—	2.5	0.004	2950, 8330; GEC W 1;	
—	—	2.4	—	BTH C40, C9; Osram	
				394b, 3891;	

thal-Isolatoren GmbH, Selb/Oberfranken, Germany; H – Hescho, Hermsdorf, Thuringen, Germany; SP – Staatlichen Porzellanmanufaktur, Berlin, Germany; HC – Hackney & Co. Ltd., England; GC – General Ceramic Corp., Keashey, N.J., U.S.A.; SSP – Steatite & Porcelain Prod. Ltd., Stourport on Severn, Worcs., England.
**For the significance of abbreviations see table 7.10.

Table 7.16.
Sintered ceramic–metal seals (Roth, 1966).

Technique	Powder mixture	Suspension liquid	Mixing
Moly–manganese	160 g Mo (200 mesh) 40 g Mn (150 mesh) particle size 3–10 μ	50 cm³ amylacetate 50 cm³ acetone 100 cm³ pyroxyline binder (Du Pont 5511)	Ball milling 24 hr
Moly–iron	40 g Mo 0.8 g Fe particle size 5–8 μ	10 g nitrocellulose sol in 150 cm³ lacquer thinner from: aromatic naphta 9 cm³ ethyl alcohol 27 cm³ ethyl acetate 24 cm³ normal buthyl acetate	amylacetate 90 cm³ ethyl (buthyl) (85–88%) (83–92%)
	40 g Mo 1.6 g Fe (carbonyl) particle size 3 μ	100 g nitrocellulose sol from 10 g nitrocellulose 90 g ethyl acetate	
Tungsten–iron	90 g W 10 g Fe particle size 1–4 μ	nitrocellulose in ethyl acetate	
	90 g W 10 g Fe	not specified (shellac 2% sol in alcohol)	
Moly–manganese-iron	200 g Mo (400 mesh) 40 g Mn (400 mesh) 10 g Fe (H$_2$ reduced) 2 g silicic acid powder 2 g calcium oxide (200 mesh)	55 cm³ acetone 25 cm³ methyl ethyl ketone 50 cm³ ethyl ether 45 cm³ nitrocellulose lacquer* (600–1000 sec)	Ball milling 100 hr
Activated Mo–Mn	172 g Mo (200 mesh) 44 g Mn (200 mesh) 9 g titanium hydride	same as above	as above
Activated Mo–Mn–Fe	200 g Mo (400 mesh) 40 g Mn (400 mesh) 10 g Fe (H$_2$ reduced) 2 g silicic acid powder 8 g alumina powder (90 mesh) 8 g titanium hydride	60 cm³ acetone 30 cm³ methyl ethyl ketone 40 cm³ ethyl ether 40 cm³ nitrocellulose lacquer* (600–1000 sec) 20 cm toluene	as above

(contd.)

Table 7.16. (contd.)
Sintered ceramic–metal seals (Roth, 1966).

Coating	Sintering	Plating	Brazing
Brushed, or sprayed 25–50 μ layer	1300–1400 °C, in hydrogen or dissociated ammonia 15–30 min	Cu, or Ni or Cu–Ni fired at 1000°C 10 min in reducing atmosphere	Cu–Ag eutectic
Brushed 15–20 μ (on Steatite)	1250 °C, 20 min in hydrogen–nitrogen (28/72)	Sprayed with Ni suspension (4 μ particle) 15 μ thick layer; fired 1100 °C, 15 min in wet hydrogen	Butt sealed to Kovar, plated with 1.5 mg/cm² Cu, by interposing between Kovar and metalized ceramic an Ag washer (25 μ) and heating to 1000°C for 10 min.
Brushed about 100 μ	1400 °C, 30 min hydrogen–nitrogen (30/70)	Brushed with 40 g Ni powder in 10 g nitro-cellulose in 90 g ethyl acetate; Fired H₂, 1000 °C, 15 min	
Thickness 25–50 μ	1340–1360 °C (Zircon) 1350–1400 °C (Alumina) 15–30 min, in hydrogen–nitrogen (15/85)		
—	1400 °C in hydrogen	After wire brushing plated Ni or Cu	Ag–Cu eutectic
—	For high alumina ceramics	—	—
Coated and fired in two layers	1525 °C	Ni plated to total thickness of 25μ	Ag–Cu eutectic, brazed to copper parts
Painted or sprayed	1500 °C 30 min or 1250°C 45 min in wet hydrogen	Ni plated 5–8 μ	Good seal to OFHC stacked Cu discs (0.25 mm thick) by brazing with Ni–Cu–Au alloy or Ag–Cu eutectic

*Nitrocellulose lacquer: 40 g nitrocellulose, 165 cm³ toluene, 75 cm³ ethyl alcohol, 60 cm² ethyl acetate

Table 7.17.
Sealing waxes.

Wax	Composition	Softening	Max. safe	Vapour pressure (Torr)	Solvents (in up to 24 hr)
		temperature (°C)			
Soft red wax	beeswax (5 pbw) Turpentine (1 pbw) dyestuff	55–60 (wetting)	25	10^{-5} (25°C)	Acetone, alcohol benzene, chloroform, ether, turpentine, xylene
Faraday wax	rosin (5 pbw) beeswax (1 pbw) Venetian red (1 pbw)	60–75 (75–95 wetting)	—	—	Acetone, alcohol, benzene, ether, xylene
Beeswax–rosin	Rosin (1 pbw) Beeswax (1 pbw)	47	40	5.10^{-6} (25°C)	Mixture of carbon tetrachloride and alcohol (1:1)
Celvacene heavy	dark yellow-brown transparent, greasy	130	—	10^{-6} (25°C)	Chloroform, acetone
Shellac	Insect and tree resin (India), mixture of polyhydroxy acids and esters. Used dissolved in warm spirit*	60–80 (100–125 wetting)	—	—	Acetone, alcohol chloroform, ether, butylphtalate
Red sealing wax	Shellac, Venice turpentine, Vermillon or Chinese red	60–80 (100–125 wetting)	—	10^{-5} (25°C)	Acetone, alcohol, benzene, chloroform, ether, xylene
De Khotinsky cement	Shellac and Caroline (wood) tar	85–100 (95–150 wetting)	40	10^{-3} (25°C)	Acetone, alcohol, benzene, chloroform, ether, xylene
Sealstix	De Khotinsky type	—	—	—	ethyl alcohol, acetone
W.E. wax	Shellac base, brown	80	—	—	alcohol
Picein	Hydrocarbons from rubber, shellac, bitumen, Black wax	80 (90 wetting)	50 60	1.10^{-6} ($-25°C$) 4.10^{-4} (25°C) 5.10^{-3} (50°C)	Benzene, benzine chloroform, ether, turpentine, xylene
Wax V	Solid high molecular weight hydrocarbons, fine inorganic powder, rubber	183 (drops)	30	10^{-4} (25°C)	—

(contd.)

Table 7.17. (contd.)
Sealing waxes.

Non-solvents (up to 60 hr)	Attacked (A) Resistant (R)		Remarks (Supplier)
	acids	alkalines	
water	as shellac	A by potassium sol, R to sodium sol.	Slightly harder than plasticine. Loses its plasticity by oxidation
water	as shellac	R (slight dis-colouration)	Characteristics may be changed by various mixtures
—.	—	—	Good adhesion to cold metals
—	—	—	Vacuum-tight bond for rubber-to-metal or-glass (CVC)
Benzene, turpentine, xylene, water, most oils	A by HNO_3 and H_2SO_4 R to HCl	A	Moderately tough resin. Disintegrates and sol. no more suitable for sealing. By heat (30 hr at 90°C, or 3 hr at 150°C) becomes harder due to poly-merization
oils, water	as shellac	A	Bends or prone to slow changes; brittle when dropped
Petroleum, turpentine oil, water	R	A	Tough but very slightly plastic. Poly-merizes at room temperature in 6 months (Cenco)
water	A by chromic acid	A	Polymerizes (6 months) Cleaning off with chromic acid (Cenco)
benzene, toluene	—	—	(Edwards)
acetone, alcohol water	A by H_2SO_4 R to HCl HNO_3 (and chromic)	R	Suitable for metal, glass sealing, Vibration resistant; not brittle; Avail-able in 2 grades; Separation of joints by heating (Edwards, Leybold)
—	—	—	For sealing ungrounded joints (Leybold)

(contd.)

Table 7.17. (cont'd.)
Sealing waxes.

Wax	Composition	Softening temperature	Max. safe (°C)	Vapour pressure (Torr)	Solvents (in up to 24 h)
White sealing wax	Shellac with resins and heat resistant minerals	106 (drops)	50	10^{-3} (25°C)	Petroleum, benzene alcohol (water)
Apiezon sealing compound Q	Graphite, grease or paraffin oil distillation products	45 60 (wetting)	30	10^{-4} (25°C) 2.10^{-4} (70°C)	—
Vacoplast	similar to wax Q	—	—	—	—
Apiezon Wax W-40 (soft)	Black wax in sticks	45	30	10^{-6} (25°C) 10^{-3} (180°C)	xylene
Apiezon Wax W-100 (medium)	Black wax in sticks	55	50	as W-40	xylene
Apiezon Wax W (hard)	Black wax in sticks	85 (100 wetting)	80	10^{-7} (25°C) 10^{-3} (180°C)	xylene (benzene, chloroform)

(contd.)

*Shellac used alone is brittle and tends to form hair-line cracks. It is therefore reinforced with other materials making it more readily fusile, more adhesive and stronger. A mixture which after melting

ques are used extensively in vacuum sealing, with ceramics such as Steatites, Forsterites, Alumina or Hard Porcelains (table 7.15).

The process of *sintered ceramic–metal seals* consists basically in covering the ceramic part with a layer of molybdenum (or tungsten) powder with a slight addition of manganese (iron or titanium), and in sintering the layer at high temperature. After an eventual coating with Ni or Cu, the ceramic part can be brazed to the metal part (details in table 7.16).

Ceramic-to-copper seals for low temperatures are discussed by Schauer (1977). Al-alloy ceramic cryogenic feedthroughs are described by Ishimaru (1982).

Table 7.17. (contd.)
Sealing waxes.

Non-solvents (up to 60 hr)	Attacked (A) Resistant (R)		Remarks (Supplier)
	acids	alkalines	
—	R	R	Adheres well to glass and metal (Leybold)
—	—	—	Consistency of plasticine; temporary sealing or blanking during leak detection (Edwards)
—	—	—	(Comp. Gen. Radiologie)
—	—	—	Application where it is required to flow the wax in or round the joint. For joints subjected to vibration, but not heated (Edwards)
—	—	—	Safe for cracks in joints subjected to vibration (Edwards)
acetone, alcohol (water)	A by H_2SO_4 R to HCl, HNO_3	R	High vacuum work where the parts tend to warm up in use. Brittle to shock. (Edwards)

can be poured into moulds (to form sticks) consists of: shellac (50 pbw), wood creosote (5 pbw), turpentineol (2 pbw) and ammonia 0.88 (1 pbw).

7.3.3. *Semi-permanent and demountable seals*

Vacuum seals which have to be opened from time-to-time (semi-permanent) or often (demountable) can be made by using waxes or adhesives, ground joints, liquids or gaskets.

Waxed seals are used especially in unique or temporary applications. They can be used to join *temporarily* metal, glass, quartz, ceramic (or plastic) parts or to seal temporarily pin-holes or leaky joints. *It is recommended to avoid their use in any long-term vacuum work.*

Waxes (table 7.17) are compounds, which when warmed are plastic but become rigid at room temperature, and this effect can be used in vacuum sealing. A good waxed seal is obtained only with clean (degreased) surfaces, assembled with a minimum amount of wax applied at temperatures high enough for the particular wax. Procedures consisting of applying hot wax on colder surfaces do not give reliable seals.

Adhesives are used for sealing pinholes or porous walls (sealing lacquers) or for proper sealing purposes (sealants). Table 7.18 summarizes the characteristics of lacquers and sealants (irreversible adhesives). This table does not contain the epoxy resins which are summarized in table 7.19. For details we refer to Skeits (1958, 1962), Schrade (1957), Lee and Neville (1967), Salomon and Houwink (1967), Potter (1971). Neuhauser (1979) uses indium (or aluminium) for bonding glass-to-glass (or metal).

The *sealing lacquers* (e.g. Glyptal) painted on the surfaces in order to seal pinholes are drawn into the orifices and as the volatile solvent is removed, the residue plugs the orifice.

The *epoxy adhesives* are available under various trade names from various suppliers*. The choice of the adhesive to be used in a particular application depends on the shape of the seal, the thickness of the joint and the desired or possible heating (table 7.19) for curing. Heat curing resins are preferred for sealing similar materials with small differences in their thermal expansion. For the sealing of materials with very different thermal expansions or for heat sensitive materials it is better to use adhesives cured at room temperature.

Epoxy adhesives were found to have many physical and chemical properties useful in applications for high vacuum seals, they are very stable and have tolerable outgassing rates.

In order to make an epoxy resin seal the procedure includes: the mixing, the applying and the curing of the resin. *Mixing* is needed since epoxy adhesives are usually supplied as a separate resin and a hardener. These have to be mixed immediately before use. The mixed adhesive may be *applied* on the (previously cleaned and dried) surfaces by painting, spraying or immersion. Usually it is sufficient to apply the adhesive on one of the surfaces to be bonded together. The adhesive should be applied on both surfaces if they are very rough or if one part should be inserted into the other (e.g. telescopic joint). Although curing can take place without pressure, the parts to be joined should nevertheless be fixed in such a way that the thickness of the resulting adhesive layer is that given in table 7.19. *Curing* is done at the temperature and for the time recommended for each adhe-

---

*e.g Araldite (Ciba Ltd., Basle, Switzerland); Epon (Shell Comp., New York ; Gen Epoxy (General Mills, Kankakee, U.S.A.); Torr Seal (Varian, Palo Alto, U.S.A.).

sive (Table 7.19). If heating is necessary this can be done in an oven, with infrared lamps, induction heating, etc.

Dismantling of an epoxy joint can be done by immersing the joint (for several days) in trichlorethylene or in warm dimethyl–formamine. Sometimes the joint can be heated to 130–150°C and at this temperature the components can be pulled apart. Then the adhesive residue left on the surfaces may be scraped off after softening by immersion for a few hours in dimethyl–formamine, nitrobenzene, phenol or cresol.

Silver chloride may be used for vacuum seals that must sustain higher temperatures than those to which wax or adhesive seal can resist. Silver chloride seals can be used generally up to 300°C, where its vapour pressure is only 10^{-7} Torr (e.g. Greenblatt, 1958; Roth, 1966; Aslam, 1972).

The technique of sealing with silver chloride consists in heating the assembled parts to about 500°C, and melting the silver chloride by placing it on the hot joint, or by placing it on the cold assembly and heating it. On cooling the silver chloride expands, the design of the joint must allow for this effect. Silver chloride can be used to seal metals, glass, mica, etc. To obtain a coating on glass (polished surfaces) it is necessary to cover it first with a film of platinum or silver.

Silver chloride is extensively used for sealing of windows, which is carried out by using a thin metal part (fig. 7.10a), providing a channel (fig. 7.10b) around the window, in a step (fig. 7.10c) or on a tapered edge (fig. 7.10d) of the walls. Mulder (1977) describes a seal with lead fluoride which can be heated to 300°C (for short time to 400°C).

Ground glass (quartz) seals or *lapped metal seals* consist of two parts connected to each other on their ground or lapped surfaces. For vacuum-tight seals the surfaces have to be *greased* before placing them together. The ground surface may be plane, conical or spherical.

Plane ground joints are used in applications where the joint must be closed and opened without moving the parts axially, or where the diameter is too large to use conical or spherical joints. Metal lapped plane seals are used in slide valves of exhaust machines for pumping of electric lamps, electron tubes, etc. (Matheson, 1955; Roth, 1955, 1966; Espe 1966/68). Such seals (fig. 7.11) consist of two discs, one of which is stationary and the other rotating. The holes of the stationary disc (the lower in fig. 7.11) are connected to the pumps, while those of the rotating disc (the upper part in fig. 7.11) are connected to the vessels (lamps, tubes) to be evacuated. When the holes in the two lapped parts correspond to each other, a vacuum-tight connexion is established between vessel and pump. The joint is lubricated and sealed by the oil circulated in the concentric grooves on the plates (fig. 7.11b).

Conical ground seals consist of the inner and outer parts which fit together, having the same taper. The dimensions of a conical ground joint are expressed

Table 7.18.
Irreversible adhesives.

Lacquer or sealant	Composition	Max. safe temperature (or range °C)	Characteristics	Remarks (Supplier)
Sealing lacquer	Mixture of high polymers in solvents	−20 80	After drying insoluble in oils, petroleum, water	(Leybold)
Glyptal lacquer	Alkyd resin, formed by condensation of phtalic anhydride, glycol	100	Vapour pressure 10^{-8} Torr ($-25°C$), 2×10^{-4} ($25°C$), 10^{-1} ($70°C$). Soluble in acetone, xylol or benzine. Insoluble in mineral oils, alcohol, water.	Good fluidity and wetting, also Al, and Pexiglass; Drying 8 hr at room temp. Polymerizes 1-2 hr at 140°C. (Comp. Gen. Radiologie)
Helmitin sealing compound	Polyester to be mixed with hardener (600 °C) (Hardener irritant of mucose!)	100	Soluble in methylene chloride, methanol, water. Insoluble organic solvents. Tensile strength 5 kg/mm² (25°C)	Hardens: 1 hr (18°C), 15–20 (20°C) (Leybold)
Desmodur	Ester of diysocyan acid. Two solutions are to be mixed max. 3 hr before use.	110		Seals glass, metals ceramics, rubber. Needs pressure 3–4 atm for 2 hr at 90–130°C or 8–10 hr a room temperature. (I.G. Farben)
Anaerobic permafil	Polymer of polyglycol dimethacrylate	—	Sets by exclusion of oxygen (vacuum)	Applied by brushing on the outside o porous walls. (Gen. Electric)
PVA solution sealant	Polyvinil acetate (1pbw) solution in toluol (10 pbw) or PVA in acetone	65	Vapour pressure 1×10^{-5} Torr (25°C)	After drying th coated surface i heated for 30 min a 150°C.

(contd

Table 7.18. (contd.)
Irreversible adhesives.

Lacquer or sealant	Composition	Max. safe temperature (or range °C)	Characteristics	Remarks (Supplier)
Loctite sealant	Thin liquid	−40 150	Liquid soluble in trichlorethylene. Sets by exclusion of air.	Bonds metal, glass ceramics, 4–12 hr (25°C), 10 min (100°C) 5 min (180°C) (Amer. Sealants)
Natural rubber sealant	Natural rubber (1 pbw), in benzol (2 pbw)	70	Applied on rough surfaces.	8 hr drying, then pressed
Silicone rubber sealant	RTV 102 silicone rubber adhesive	−60 150	Air cure at room temp; first forms a surface skin; completely cured 24 hr (3 mm layer) (Gen. Electric)	

Anaerobic polymers used as sealants are discussed by Kendall (1982a); sealing of leaks in an ultra-high vacuum system by using a G.E. silicon resin is described by Egelhoff (1988).

by the diameters (d, D) at the small and large ends, and the length L of the ground zone. The taper defined as $(D-d)/L$ is $\frac{1}{10}$ in the standard joints, and $\frac{1}{5}$ in special joints.

Spherical or ball and socket *ground joints* were developed to be used in applications where the alignment of the parts to be joined is difficult or where angular motion of the parts with respect to each other is required. Spherical joints are designated by the ratio between the diameter of the ball and the inside diameter of the tubing.

Ground seals can be used in vacuum technology only if they are properly *selected, assembled* and *maintained*. If small-bore ground seals are necessary, spherical joints should be used; for medium size (3–100 mm diameter), conical joints may be used, and for large diameters flat joints are useful. For radial openings only flat joints are useful; if a small axial displacement is possible, spherical joints may be used. The opening of standard conical joints requires an axial displacement equal at least to the length of the joint.

Ground joints may be *assembled* in any position. It is recommended that *the parts be clamped together* even if it does not appear imperative.

Table 7.19.
Epoxy adhesives.

Resin	Hardener	Parts by weight of hardener to 100 pbw of resin	Recommended joint thickness (mm)	Curing time (min) at temp. (°C)						Recommended uses and remarks
				20	40	70	100	150	180	
Araldite 101	951	5–6	0.1–0.2	24hr	—	—	10	—	—	For bonding small metal surface
	930	6–7	max. 0.5	24hr	—	—	—	—	3–5	More heat resistant
	936	6.5	—	—	—	—	—	—	—	Curing only at 60°C for 2.5 hr
Araldite 102	951	6–7	0.05–0.15	24hr	—	—	—	5	3	Bonding porous materials
	936	6	—	—	—	—	—	—	—	Curing only at 60°C for 2.5 hr
Araldite 103	951	7–8	0.05–0.2	36hr	14hr	3hr	60	20	10	For bonding metals, ceramics, rubber; large surfaces. Hard joints
	930	6–10	0.2–0.5	24hr	14hr	2hr	60	20	10	
Araldite 105	960	6–10	3	24hr	—	—	—	10	—	Filling up cracks
Araldite 106	953 U	80	—	7hr	3hr	1hr	10	5	—	Resistant to vibration
Araldite 121	951	4–4.5	0.1–3	36hr	14hr	2hr	30	10	5	Bonding ceramics, synthetic resins to themselves and to metals; low expansion or heat sensitive materials
	930	2.5–5	0.1–0.5	36hr	14hr	2hr	30	10	5	
Araldite 123	951	5.5–6	0.1–3	36hr	14hr	1hr	15	5	3	Bonding large, metal ceramic surfaces
	930	3–6	0.1 1.5							
Araldite 1 (Natural)	one part, rods	—	0.05–0.2	—	—	—	—	3hr	55 10(200°C)	Bonding nonferrous metals. ceramics

(contd.)

Table 7.19. (contd.)
Epoxy adhesives

Resin	Hardener	Parts by weight of hardener to 100 pbw of resin	Recommended joint thickness (mm)	Curing time (min) at temp. (°C)						Recommended uses and remarks
				20	40	70	100	150	180	
Araldite VIII (Natural)	paste	—	0.05–0.5	—	—	—	—	4hr	1hr 30(200°C)	Bonding large badly fitting parts
Epon 1X	one part paste	—	—	—	—	—	—	—	90	Bonding metal to metal; Max. service temp. 150°C
Epon 901	B–1	23	—	24hr	—	—	1hr	—	—	Max. service temp. 120°C
	B–3	11	—	—	—	—	—	—	1hr	Max. service temp. 160°C
Epon 907	B	80	—	24hr	—	1hr	—	—	—	Max. service temp. 80°C
Epon 929	one part paste	—	—	—	—	—	—	—	2hr 15(200°C)	Max. service temp. 250°C
Gen Epoxy M 180	Versamid 140	100	—	—	—	3hr	—	20	—	Bonding metals
	Versamid 125	65–40	—	—	—	—	2hr	10	—	Bonding metals, rubber, PVC, Polyester
Gen Epoxy 190	Versamid 125	100–60	—	24hr	—	2hr	—	10	—	Bonding metals, rubber, PVC, Polyester
Torr Seal	Hardener	equal squeezed length	—	24hr	—	80	—	—	—	Bonding metal, ceramic, glass, or sealing up leaks; service up to 10^{-8} Torr and 100°C

Bonding, metal, glass, rubber, plastics

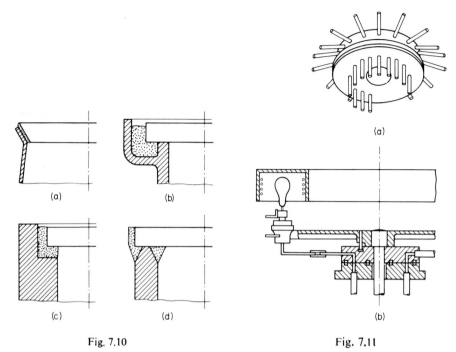

Fig. 7.10   Fig. 7.11

Fig. 7.10 Silver chloride window seals.

Fig. 7.11 Lapped seal.

The maintenance of ground joints includes their *greasing* and *cleaning*. For the greasing of ground joints the proper grease should be used, corresponding to the range of temperatures in which the seal will work and to the vapour pressure requirements of the system (table 7.20). The grease is usually applied as strips on two diametrically opposed sides along the taper of the inner part. It is recommended that grease be applied only on the portion which is on the atmospheric side of the joint (fig. 7.12). By rotating the members on each other, the joint should become transparent. If after rotation, the surface of the joint presents any lines which do not disappear, cleaning and regreasing is required.

The washing-off of Ramsay and Apiezon greases may be done with benzine, benzol or carbon tetrachloride. For cleaning of silicone greases see §7.2.2.

Liquid seals are joints in which the gap between the connected parts is sealed by a material in the liquid phase. If the liquid seal separates spaces having a pressure difference of an atmosphere, the sealing action may be based on: *hydrostatic pressure*, on a *high impedance* to the flow of the liquid or on the *surface tension*

Table 7.20.
Vacuum greases.

Designation	Max. service temp. (°C)	Dropping (melting) point (°C)	Vapour pressure (Torr)		Remarks	Supplier
			at 25°C	at higher temp.		
Vacuum grease P	25	55	10^{-5}	10^{-4} (100°C)	10^{-8} (25°C) degassed 2 hr at 90°C	Leybold
Graise PB 1	—	—	—	—	light	Comp. Gen. Radiologie
Ramsay grease	25–30	56	10^{-7}	10^{-4}(38°C)	—	Leybold
Apiezon L	30	—	10^{-9}	10^{-6}(135°C)	only temporarily in wellfitting joints	
Apiezon M	30	—	10^{-8}	10^{-6}(30°C)	—	Edwards Shell
Apiezon N	30	—	10^{-7}		for tapered joints	
Vacuum grease R	30	65	$5 \cdot 10^{-8}$	—	10^{-8}(25°C) degassed	Leybold
Graise PB 2	(30)	—	—	—	for large joints	Comp. Gen.
Graise PB 3	(30)	—	—	—	heavy	Radiologie
Lubriseal	30	40	—	—	for glass/metal	CENCO
Vacuseal light	50	—	10^{-5}	—	—	CENCO
Joint grease DD	58	120	—	—	for rotary seals	Leybold
Vacuseal heavy	60	—	10^{-5}	—	—	CENCO
Celvacene light	—	90	10^{-6}	—	—	CVC
Celloseal	—	100	10^{-8}		soluble in chloroform	Fischer
Apiezon T	110	—	10^{-8}	—	light m.p. grease	Edwards Shell
Celvacene medium	—	120	$<10^{-6}$	—	—	CVC
Cello grease	—	120	$<10^{-6}$	—	—	Fischer
Lithelen	150	210	very low	—	lithium soap	Leybold
Silicone stopcock grease	200	—	10^{-7}	10^{-5}(170°C)	—	Dow Corning
Silicone high vacuum grease	200	250	10^{-7}	10^{-5}(170°C)	useful down to −40°C	Edwards

of the sealing liquid. In order to minimize the height of the liquid column needed for the seal, in some seals a *guard vacuum* is used (see fig. 7.41) or the sealing liquid is *frozen*.

In seals where the sealing material is permanently liquid, mercury or oil is used, e.g. Gaunt and Redford (1959), Brueschke (1961), Roth (1966). In frozen seals molten indium or Wood's metal can be used, e.g. Reynolds (1955), Toby and Kutsche (1957). Magnetic-liquid seals are discussed by Raj and Grayson (1981).

Oil seals are used especially where the lubrication is needed simultaneously with the sealing action. The most extensively used seal of this type is the cylinder seal in the rotary pumps (fig. 7.13). For a reliable seal the clearance between rotor

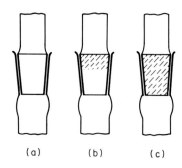

(a)   (b)   (c)

Fig. 7.12 Conical joints; (a) before greasing; (b) recommended greasing; (c) not recommended greasing. Dotted part is greased.

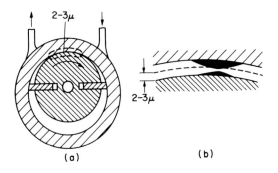

Fig. 7.13 Oil seal in rotary pumps.

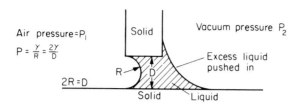

Fig. 7.14 Surface tension seal. After Milleron (1959).

and stator must be 2–3 μ. If the two materials of the stator and rotor are not properly chosen the wear increases the clearance until the oil cannot seal (fig. 7.13b).

Hablanian (1986) mentions that a spacing of 0.02 mm between rotor and stator is sufficient for oil sealed rotary pumps (of 200–300 liter/min).

The sealing action due to surface tension (Milleron, 1959) is shown in fig. 7.14. The liquid which fills the gap between the parts bears on one side the atmospheric (or high) pressure P_1 and on the other the vacuum (P_2), and the pressure difference pushes the liquid toward the low pressure. The liquid will withstand the pressure difference if the gap D is small enough and the surface tension of the liquid γ is high enough. The equilibrium condition (fig. 7.14) is given by

$$P_1 - P_2 = \gamma[(1/R_1) + (1/R_2)] \tag{7.1}$$

Since $P_2 = 0$, and $R_1 \ll R_2$, and $2\,R_1 = D$, this gives

$$D = 2\gamma/P_1 \tag{7.2}$$

Table 7.21.
Surface tension and clearance for liquid seals.

Liquid	Temp. (°C)	Surface tension dyn/cm	Maximum clearance (μ) for a pressure difference of	
			1 atm	100 Torr
Gallium	40	735	14.7	112
Tin	300	520	10.4	78
Mercury	15	487	9.5	72
Lead	350	420	8.4	64
Bismuth	300	370	7.4	56
Silver chloride	803	114	2.3	18
Water	20	73	1.4	11
Organic liquids	20	25–30	0.5–0.6	3.8–4.5

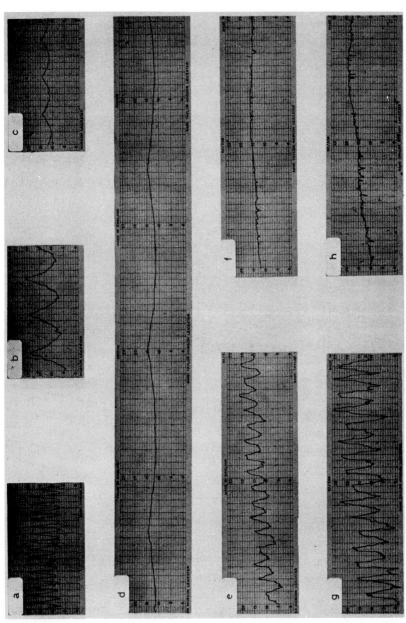

Fig. 7.15 Records of surface roughness. a–d. Same surface recorded at (a) vertical magnification (v.m.) 2000, horizontal magnification (h.m.) 20; (b) v.m. 2000, h.m. 100; (c) v.m. 500, h.m. 100; (d) v.m. 500, h.m. 500. (e–h) Concentrically machined surface; (e) radial profile, v.m. 5000, h.m. 100; (f) tangential profile, v.m. 5000, h.m. 100; (g) radial profile, v.m. 10 000, h.m. 100; (h) tangential profile, v.m. 10 000, h.m. 100 (Roth, 1971).

i.e. a liquid metal with $\gamma = 500$ dyn/cm (table 7.21) seals $P_1 = 1$ atm $= 10^6$ dyn/cm^2, if the gap is less than $D = 1000/10^6 = 10^{-3}$ cm.

7.3.4. Gasket seals

Sealing mechanism. Two flanges with a very good surface finish compressed against each other leave between them micron-size channels which constitute the leakage paths through such a seal. These leak paths are determined by the profiles of the surfaces in contact.

We are used to the image of the surface profile in which the peaks have considerable slopes (fig. 7.15a,b) since these profiles are recorded by instruments which have a much larger vertical than horizontal magnification. In order to obtain the image of the profile with the real slopes existing on the surface, the record should be taken at equal vertical and horizontal magnification (fig. 7.15d). On lapped surfaces the profile is the same in any direction, while concentrically machined surfaces present very different surface profiles in radial and tangential (circumferential) directions (fig. 7.15e–h).

It was established that on machined surfaces more than 90% of the peaks have slopes of $1°$–$4°$. From these values, the typical form of the cross section of the leak paths formed at the interface–contact was determined (Roth, 1967, 1971) to be a triangle with $\alpha = 4°$ (fig. 7.16b). The length of the typical leak path is practically equal to the width w of the seal, as it results from the difference between the outside and inside radii (r_o and r_i fig. 7.16a) of the sealing annulus.

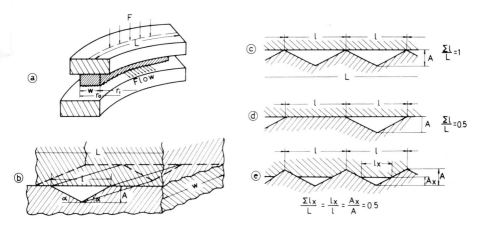

Fig. 7.16 Dimensions of a seal; (a) the interface–contact annulus; (b) a typical leak path; (c) the surface contact (machined surface); (d) single path on a surface; (e) loaded interface–contact. After Roth (1971).

According to eq. (3.118), the conductance (molecular flow) of an individual leak path (fig. 7.16b) was expressed by Roth (1971) as

$$C_1 = 19.3\,(T/M)^{1/2}\frac{A^3}{2\,(1+1/\cos\alpha)\,\mathrm{tg}\,\alpha}\;\frac{K}{w} \tag{7.3}$$

where C_1 is in liter/sec, A (cm) and K the correction factor (table 3.5). For helium ($M=4$) at $T=298°$K, and $\alpha=4°$ (thus $K=1.7$; extrapolated according to table 3.5 for the A/l value corresponding to $\alpha=4°$), eq. (7.3) is written

$$C_1 = 1000\,A^3/w \quad \text{liter/sec} \tag{7.4}$$

This equation is plotted in fig. 7.17 for the range of usual seal widths w, together with the throughput $Q_1 = C_1\Delta P$ (eq. 3.26). Since the peak-to-valley height of the roughness of machined surfaces is in the range 10^{-3}–10^{-5} cm, and the lowest

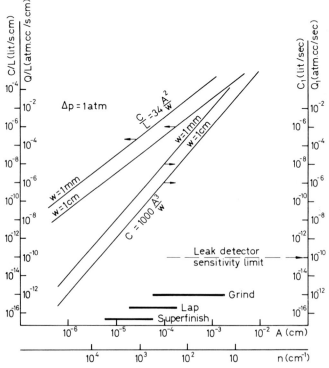

Fig. 7.17 Conductance and throughput of individual leak path (right scale), and of unloaded interface–contact (left scale). After Roth (1971).

sensitivity of helium leak detectors (see §7.4) is about 10^{-10} atm·cm^3/sec, at $\Delta P = 1$ atm the leakage through one single leak path can still be detected (fig. 7.17).

If the density of the leak paths on the surface is $\Sigma l/L$, their number per unit length (L, cm, fig. 7.16a) is $n = (\text{tg } \alpha/2A)(\Sigma l/L)$. A machined surface is usually completely covered by peaks and valleys, thus $\Sigma l/L = 1$ (fig. 7.16c). Hence the conductance of the unit length of *an unloaded interface contact* will be (for He, and $\alpha = 4°$):

$$C/L = 34 \ (A^2/w) \ (\Sigma l/L) = 34 \ A^2/w \ \text{liter/sec·cm} \tag{7.5}$$

This equation is also plotted on fig. 7.17, and it shows that the leak rate of one cm length of seal made just bringing into contact surfaces having the best achievable finish ($A = 6 \times 10^{-6}$ cm) is easily detectable.

In order to reach *lower leak-rates*, the cross section of the leak path has to be reduced by *applying a load* onto the contacting surfaces. This decreases the height A of the paths (fig. 7.16c) to A_x (fig. 7.16e). Since in this case $\Sigma l/L = A_x/A$, the conductance per unit length of seal is given by

$$C/L = 34(A_x^2/w) \ (\Sigma l/L) = 34 \ (A^2/w) \ (A_x/A)^3 \tag{7.6}$$

By using the integrated result (eq. 3.176)

$$[\ln(r_o/r_i)]/(2\pi) = (r_o - r_i)/[\pi(r_o + r_i)] = w/L \tag{7.7}$$

eq. (7.6) is written

$$C = 34[2\pi/\ln(r_o/r_i)]A^2(A_x/A)^3 \tag{7.8}$$

which is an equation similar to eq. (3.177). A_x/A is the factor changing under the effect of the load, according to eq. (3.178) as

$$A_x/A = \exp \ (-P/R) = \exp \ [-F/(LwR)] \tag{7.9}$$

where P is the pressure exerted by the tightening force F on the contact area Lw, and R is the *sealing factor* expressing the sealing ability of the material (see fig. 7.18).

From eqs. (7.6) and (7.9) the sealing process (for He at 25°C) is expressed by

$$C = 34 \ A^2(L/w) \exp \ [-3F/(LwR)] \tag{7.10}$$

or the throughput (leak rate) for any gas is expressed by

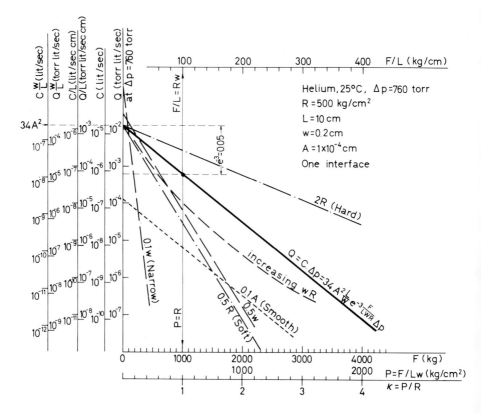

Fig. 7.18 Plot of the equations of the sealing process in various systems of scales. After Roth (1971).

$$Q = C \, \Delta p \approx 4 \left(\frac{T}{M} \right)^{1/2} A^2 \frac{L}{w} \exp\left(-3F/LwR \right) \Delta p \ \text{Torr} \cdot \text{liter/sec} \quad (7.11)$$

where Δp is the pressure difference of the gas across the seal.

Equation (7.10) can be written in the normalized form

$$C \, \frac{w}{L} = 34 \, A^2 \exp - 3\kappa \qquad\qquad (7.12)$$

by using the concept of the *tightening index* $\kappa = P/R = F/(L \, wR)$.

Equations (7.5–7.12) refer to *one* interface contact. Gasket seals are constituted of two such surface contacts (fig. 7.16a), thus the *leakage* in a gasket seal is *double* that given by these equations. Equations (7.10, 7.11, 7.12) are represented on fig. 7.18. The value of the sealing factor R is obtained on this graph as the $P = F/(wL)$ value which produces a decrease of the conductance by a factor $e^{-3} \approx 0.05$. The value of Cw/L which results for $F=0$, represents the initial surface roughness, according to $Cw/L = 34\ A^2$.

Figure 7.18. (eq. 7.10) shows that the influence of the various factors on the sealing process is:

– The curves for seals using hard gasket materials (2 R) will have lower slopes than those for soft materials (0.5 R), i.e. for the same decrease of the leak rate soft materials require less sealing force (figs. 7.18, 7.19).

– The slope of the curve is also influenced by the width w; the curves of narrow seals (0.5 w; 0.1 w) have higher slopes, i.e. for the same decrease of the leak rate narrow seals require less sealing force.

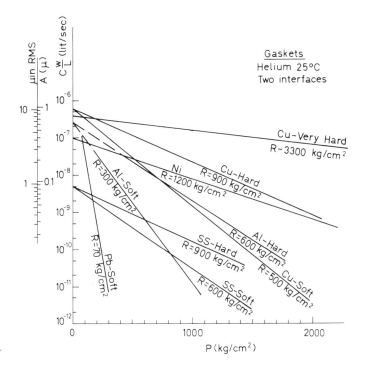

Fig. 7.19 Sealing curves of various metals. After Roth (1971).

– The width w of the seal also influences the position of the curve; curves of narrow seals (0.1 w) are shifted toward higher C/L values.
– The initial roughness A of the surface determines the position of the curve relative to the C/L scale. If the surface is smoother (0.1 A) the curve is shifted toward lower C/L values; a change in the roughness by one order of magnitude shifts the curve by two orders of magnitude of C/L.
– The straight lines (fig. 7.18) represent cases in which R and w are constant during the sealing process. If the material hardens and/or the width increases during the sealing process (increasing wR) the curves will be bent toward higher C/L values (fig. 7.18).

These conclusions are in good agreement with the measurements of Armand *et al.* (1962, 1964), Gitzendanner and Rathbun (1965), Roth and Amilani (1965), Roth (1967, 1970a, 1971).
Figure 7.19 shows a comparison of the sealing factors of various metals. The positions of the curves show the ranges of the initial surface roughness; the roughness (A) scale refers to zero load ($P=0$).
While copper has a sealing factor R from 500 kg/cm² (soft) to 3300 kg/cm² (very hard), and lead $R=70$ kg/cm²; Teflon has $R=150$ kg/cm² and neoprene $R=5$–30 kg/cm².

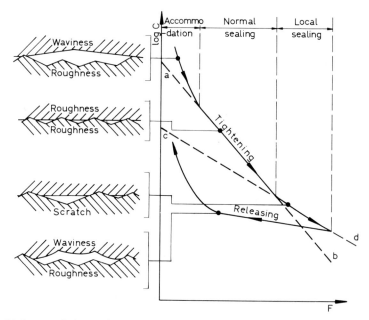

Fig. 7.20 Stages of the sealing process. After Roth and Inbar (1967a, b), Roth (1971).

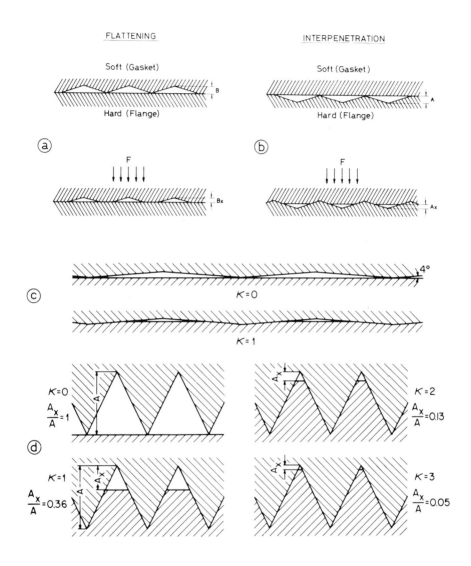

Fig. 7.21 The normal sealing process. After Roth and Amilani (1965), Roth (1967, 1971).

Experimentally obtained sealing curves always include a part which corresponds to eq. (7.10 or 7.11), but also present various deviations from this equation due to superimposed effects and imperfections. These appear both on the *tightening curve* obtained by increasing the load and on the *releasing curve* obtained when F is reduced (fig. 7.20). These two curves together were termed a *force cycle* (Roth and Inbar, 1967b).

The *tightening curve* (fig. 7.20) may contain three different sealing stages: *accommodation, normal sealing* and *local sealing*. The accommodation stage appears at the beginning of the tightening process where the force is mainly used to overcome the waviness of the surfaces, to crush foreign particles or to push local protrusions into the softer sealing surface. This is the stage in which the interface contact is gradually established along the entire circumference. The accommodation stage extends usually up to a tightening index $\kappa = 0.5$-0.6.

Upon further tightening, the process corresponds to eq. (7.10); this is the *normal sealing* stage (figs. 7.20; 7.21). In this stage the leak paths existing at the interface – contact are gradually throttled by *interpenetration* or *flattening*. If the smooth sealing part is the harder one, the sealing occurs by flattening (fig. 7.21a), while if the smooth sealing surface is softer than the rough one, sealing is a result of interpenetration (fig. 7.21b). In both cases eq. (7.10) is valid, but the value of the sealing factor R is different: sealing factors for flattening are smaller than for interpenetration. Figure 7.21 shows steps of the normal sealing process. Early steps can be represented (fig. 7.21c) by real slopes, but advanced steps (fig. 7.21d) must be represented by using an expanded vertical scale (the slopes becoming schematic).

Upon further increasing of F, the effect of *local sealing* (fig. 7.20) may appear. This stage is due to the conductance of local deeper grooves which is initially (line c–d, fig. 7.20) much less than that of the whole seal (line a–b) and decreases very slowly with the applied force. Thus the sum of the two conductances appears to be equal to the normal sealing at smaller loads and to the local sealing, when the force is higher than that corresponding to the intersection of the lines a–b and c–d (fig. 7.20). Local grooves can be radial scratches, eccentrically mated concentric machining marks, or grooves of helical machining (Roth and Inbar, 1968).

The *releasing curve* (fig. 7.20) reflects the kind of deformation which took place during the tightening curve. If this deformation was entirely elastic the releasing curve coincides exactly with the tightening curve. If the deformation during tightening was entirely plastic then during the release the conductance remains constant, i.e. the releasing curve is a horizontal line. For partly elastic deformation the releasing curve is located between the two curves mentioned. Even when the releasing curve is a horizontal, it continues only up to forces corresponding to the limit between the accommodation and the normal sealing stage, where the conductance begins to rise, since in this region the elastic deformation of the waviness begins to separate again the parts in contact (Roth and Inbar, 1967a, b).

The bottom of the leak paths is round (fig. 7.22), rather than notched (as it was schematically represented in fig. 7.16). When the opposite sealing surface approaches the rounded surface of the path, a *sudden closure of the path* occurs.

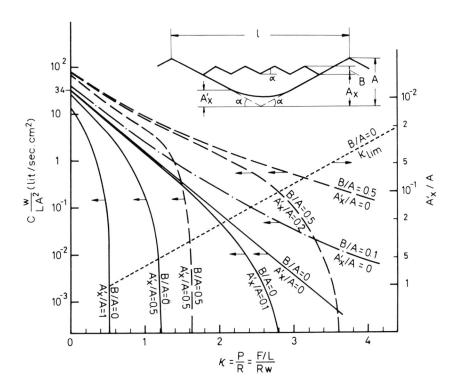

Fig. 7.22 The sudden drop in conductance due to rounded bottom of the leak paths. After Roth (1971).

Supposing that the curvature begins from a height A'_x (fig. 7.22), the conductance is given by

$$Cw/L = 34 A^2 e^{-3\kappa} (1 + Be^{\kappa}/A)^2$$
$$\times [1 - (0.35 A'_x/A) \{e^{2\kappa}/ (1 + Be^{\kappa}/A)\}]^2$$
$$\times [1 - 0.012 (A'_x/A) e^{\kappa}]^{-1} \tag{7.13}$$

which is plotted on fig. 7.22. It can be seen that the tightening index κ at which the sudden drop appears is not a function of the kind of material but of the shape of the surface roughness (value of A'_x/A). Since the tightening index depends on the sealing factor R, the force at which the drop appears is different at various materials even if the shape (A'_x/A) is the same. The value of κ at which the conductance drops to zero, results from eq. (7.13) as being (for $B/A=0$)

$$\kappa_{\lim} = \ln 1.7/(A'_x/A) \tag{7.14}$$

function which is also plotted on fig. 7.22. This phenomenon suddenly bends the sealing curves toward low values of the conductance (fig. 7.22).

The sealing curve is bent toward higher values of C if the width of the seal increases during sealing. Such phenomena are almost inevitable in seals, since the increase of w is either due to plastic deformation of the softer sealing surface, or to elastic straightening of the cylindrical contacting surfaces (fig. 7.23). The effective contact width is given in this later case by

$$w' = k(F/L)^{1/2} \tag{7.15}$$

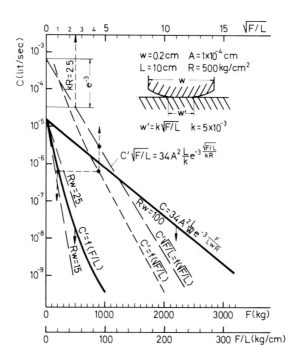

Fig. 7.23 Conductance of seal with cylindrical surface. After Roth (1971).

where κ depends on the elasticity moduli of the two materials in contact, and the radii of curvature of the surfaces. From eqs. (7.10) and (7.15) it results that in this case the sealing is described by

$$C' = 34 \, A^2 \, \frac{L}{k \, (F/L)^{1/2}} \, \exp\left[-3 \, (F/L)^{1/2}/kR\right] \qquad (7.16)$$

which is plotted on fig. 7.23 in terms of $(F/L)^{1/2}$ as well as of F/L. It can be seen that $C' = f(F/L)$ is a curve, which has a much higher slope than the line of the normal sealing C for the nominal width w, while

$$C'(F/L)^{1/2} = f(F/L)^{1/2} \text{ is a straight line.}$$

Equation (7.11) shows that in order *to obtain a leak rate as small as possible* or to use a tightening force as small as possible, the various possibilities used separately or combined are:

– To use gasket materials with a low sealing factor R; this is done practically by using neoprene, Viton or indium as the gasket material.

– To have a seal width w as small as possible; this is done by using *0–rings* (figs. 7.24 and 7.34), *shear seals* (fig. 7.37) or *knife edges* (figs. 7.38–7.39).

– To have surfaces as smooth as possible (low A value); this is done by using polishing, lapping, etc.

– To have Δp as low as possible; this is done by using a *guard vacuum* (fig. 7.41).

An analysis of the mechanisms by which the leak rate of demountable bakeable seals is maintained at the required low level is presented by Roth (1983).

0–ring seals. An 0–ring seal is a demountable joint which uses a gasket with circular cross section. The 0–ring, made of an elastomer or a metal is compressed between the sealing parts. If the main compression force is exerted axially the seal is known as *flange seal;* if the force works radially the connection is a *shaft seal.* When the joint closes, the sealing parts touch the cross section of the 0–ring in two, three or more points (fig. 7.24), but in all these cases the *contact surface is just a thin band,* thus the seal is based on a small width (eqs. 7.11, 7.16).

The various shapes shown in fig. 7.24 are obtained by : (1) placing the 0–ring in a *groove* machined in one of the sealing parts; (2) closing the 0–ring between the sealing parts and keeping it in place with *spacers;* (3) compressing the 0–ring in a *conical seal;* (4) keeping the 0–ring on a *step* and compressing it between the flanges; or (5) compressing the 0–ring between *flat flanges* (other seals and details, see Roth, 1966).

In *groove seals* one of the sealing parts (flange, shaft) has a machined recess designed to receive the 0–ring. Basically a groove seal may be designed either

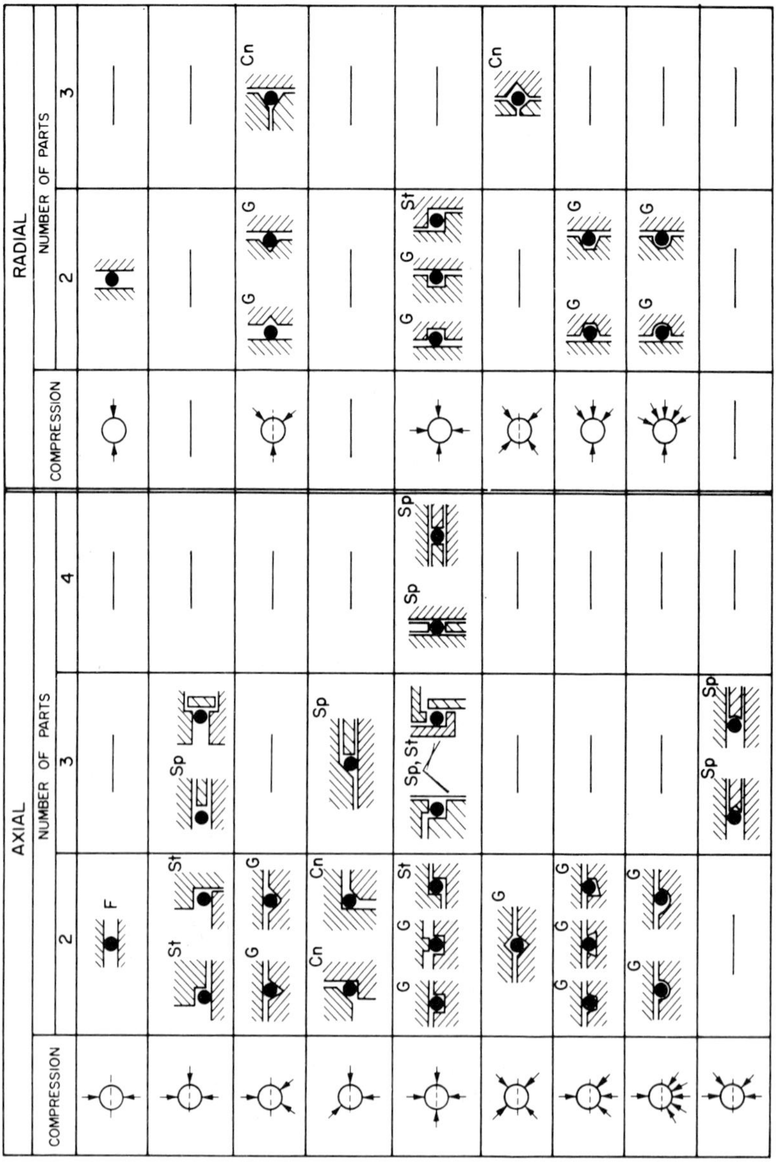

Fig. 7.24 0–ring seals. The axis of the seal is vertical and on the right side of the seal. F – flange seal; G – groove seal; Sp – spacer seal; Cn – conical seal; St – step seal. After Roth (1966).

for constant deflection of the 0–ring or for constant load on the 0–ring. The *constant deflection* seals (known also as seals with *limited compression*) are designed so that when the required compression ratio is reached the sealing parts meet in a *metal-to-metal* contact enclosing the 0–ring in the groove. The limited compression seals are preferred for elastomer gaskets, while the designs with unlimited compression are used with Teflon and metal gaskets (Brown, 1959; Von Ardenne, 1962; Roth, 1966; Auerbach *et al.*, 1978; Ishimaru and Horikoshi, 1979).

The cross section of the grooves can be rectangular, trapezoidal or rounded. The dimensions of the *rectangular groove are* determined (fig. 7.25) by

$$AB = \kappa \pi \ d^2/4 \qquad\qquad (7.17)$$
$$B/d = \kappa_r$$

where κ is a factor determining the dead volume, and κ_r is the compression ratio required for the gasket material. With the usual value of $\kappa_r = 0.72$ (for rubber gaskets having a Shore hardness of 40–60) and a dead volume of 5% ($\kappa = 1.05$) it results that the dimensions of the groove are $B = 0.72 \ d$; and $A = 1.15 \ d$ (fig. 7.25a). The edges of the groove should have a radius of $0.15 \ d \leqslant R \leqslant 0.22d$. For better retention of the 0–ring when the seal is not closed (vertical flange), the width of the groove can be made equal to the cross sectional diameter of the 0–ring (fig. 7.25b). In this case $C = 2 \ L = 0.32 \ d$. The sides of the grooves can be bevelled at $45°$ (fig. 7.25c).

If the rectangular cross section groove is intended to be used with metal 0–rings

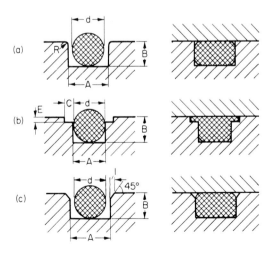

Fig. 7.25 Rectangular grooves for 0–rings.

(e.g. indium) the groove should be made with its width equal to the cross sectional diameter of the 0–ring (fig. 7.26a) and having a depth 15–30% greater than the width. In order to overcome the difficulty in removing the 0–ring such seals can be constructed with demountable parts (fig. 7.26b, c).

Sliding or rotating *shafts* (fig. 7.27) may be sealed with 0–rings. The groove may be machined on the shaft (fig. 7.27a) or in the cylinder bore (fig. 7.27b). The dimensions are listed in table 7.22.

Three types of *trapezium* grooves are used to receive 0–rings: the closed trapezium or dovetail groove (fig. 7.28a), the open trapezium groove (fig. 7.28b) and the trapezium groove with parallel side walls (fig. 7.28c). The dovetail groove is difficult to machine and always has some trapped volume in the seal, but has a very good retention of the gasket. The dimensions should be (fig. 7.28a) $C/d = 0.75$–0.80; $A/d = 0.9$.

To avoid the difficult machining needed for the dovetail groove the trapezium groove with parallel side (fig. 7.28c) can be used. The dimensions of this groove are given (Barton, 1953) by

$$W_{max} = d_{min} - 0.001 \text{ (in)} = d_{min} - 0.025 \text{ (mm)}$$
$$W_{min} = W_{max} - 0.03d$$
$$X_{max} = X_{min} + 0.03d$$
$$Y_{max} = Y_{min} + 0.03d$$
$$X_{min} = A_{max}/W_{min} - 0.312W_{min} \tag{7.18}$$
$$Y_{min} = A_{max}/W_{min} + 0.312W_{min}$$
$$A_{max} = \pi d^2_{max}/4$$

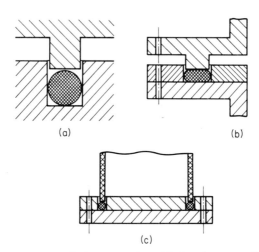

(a)   (b)

(c)

Fig. 7.26 0–ring seals with unlimited compression; (a) closed groove; (b) demountable groove; (c) pipe edge seal.

Table 7.22.
Dimensions* of 0-ring shaft seals (mm).

0-ring cross sectional diameter			B–A	C	G	R₁	R₂
nominal	in	minimum					
mm	in	maximum					
1.78	0.070	1.71	2.80	1.92			
		1.85	2.90	2.02	0.13	0.5	0.1
2.62	0.103	2.55	4.10	2.76	0.13	0.5	0.1
		2.69	4.25	2.93			
3.53	0.139	3.43	5.60	3.70			
					0.15	0.8	0.2
		3.63	5.80	3.90			
5.33	0.210	5.21	8.50	5.50			
					0.18	0.8	0.2
		5.45	8.70	5.80			
7.0	0.275	6.85	11.2	6.90			
					0.20	0.8	0.25
		7.15	11.5	7.25			

*Notations, see fig. 7.27.

d_{min} and d_{max} are the minimum and maximum cross section diameters as specified for a given type of 0-ring.

Rectangular flanges may be equipped with 0-ring gaskets by cutting the groove of one of the cross sections used for circular 0-rings (figs. 7.25, 7.28) but follow-

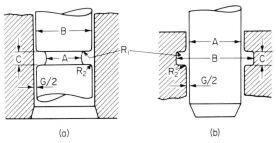

(a)   (b)
Fig. 7.27 0-ring shaft seals.

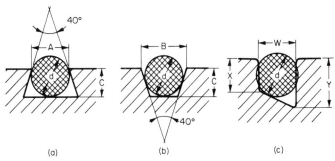

Fig. 7.28 Trapezium grooves.

ing the rectangular outline of the flange. The straight parts of the groove should be connected by circular bends whose radius R is not less than that specified in fig. 7.29. Bakeable seals on noncircular ports are discussed by Flemming *et al.* (1980).

The typical *spacer seal* is shown in fig. 7.30. The spacer seals use a retaining ring 3 (fig. 7.30a) which has a V or rounded V groove machined on its circumference. The retaining ring is closed between identical flanges 1.2 by the action of the external clamping ring 4 (fig. 7.30b) made of two half circles.

The spacer seal can also be used without specially machined retaining rings (fig. 7.31). The 0-ring can be enclosed between two concentric cylindrical rings (1, 2 fig. 7.31a) and the ends of the pipes (3, 4). In the specific seal shown, the components are pulled together by the clamping flanges 5 pulling against the

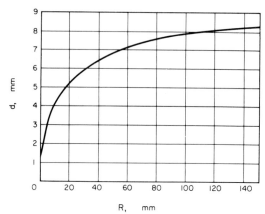

Fig. 7.29 Minimum radius R of bend as a function of the cross section diameter d of the 0-ring. After Roth (1966).

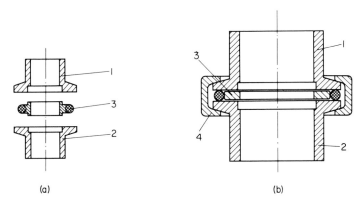

Fig. 7.30 Spacer seal with retaining ring (Leybold–Heraeus).

external spring steel retaining rings 6. Such seals can also be based on a single spacer (fig. 7.31b).

A pipe 1 (fig. 7.32) can be jointed to a flange 2 of a vacuum vessel or pipe, by surrounding it with an 0–ring, and compressing the 0–ring toward the pipe and the flange with the *conically shaped* sealing ring 3. The conical surface should form an angle of 45° with the surface of the pipe 1 and the flange 2. In this case the optimum dimension will be $A = 1.32\ d$. If A is smaller a seal with metal-to-metal contact is not possible (fig. 7.33a); if A is larger, the compression ratio is insufficient for a tight seal (fig. 7.33b) and the seal will enclose large trapped volumes.

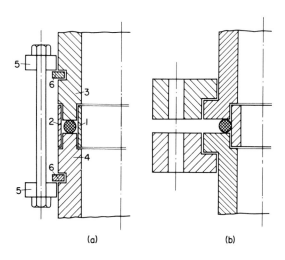

Fig. 7.31 0–ring seals with spacers; (a) with concentric tubular spacers; (b) with single spacer.

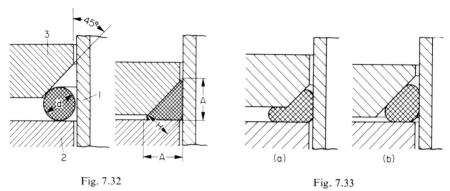

Fig. 7.32   Fig. 7.33

Fig. 7.32 Principle of conical 0-ring seals.

Fig. 7.33 Conical seal having dimension A (fig. 7.32); (a) too small; (b) too large.

The arrangement shown in Fig. 7.32 is useful for elastomer 0-rings. Figure 7.34 shows the *Wheeler seal*, a conical bakeable seal with metal (copper) 0-rings (Wheeler, 1963). This seal is constructed for diameters up to 18 inches, where $A = 21\ 5/8$ in; $B = 20$ in and $C = 18$ in. The seal uses OFHC copper wire gasket which is captured between the flanges (fig. 7.34a). The sealing surfaces of the flanges are at $20°$ to the horizontal and the wire is supported by a vertical face. To ensure the location of the gasket against this vertical face, the gasket is made somewhat undersized and is slightly stretched as it is snapped into place. The sealing force is applied directly over the gasket by means of clamps.

Techniques to be used in large seals are discussed by Wheeler (1977).

The 0-ring can be placed in the *step* machined in one of the sealing parts and is compressed against the other (fig. 7.35a) flange or against a second step machined on the other sealing part (fig. 7.35b).

If metal (gold) gaskets are used, and the steps are constructed with appropriate clearances (fig. 7.36) very reliable bakeable seals are obtained (Caswell, 1960; Mark and Dreyer, 1960; Hawrylak, 1967). The location of the gasket in the joint should permit the crushing of the gasket in the shape shown in fig. 7.36b, the ratio H/d being 0.5–0.4. The radial clearance should be $T = 0.06\ d$ to $T = 0.12\ d$, and the surface finishes of the step about 16 μ in r.m.s. (Grove, 1959).

Large metal gasket seals are discussed by: Head *et al.* (1982) discuss seals with hollow metal gaskets; Grössi *et al.* (1987) describe reusable Ag gaskets; Wikberg and Poncet (1987) mention Ag-plated Cu wire gaskets, or Au wire gaskets.

Assembly and maintenance of 0-ring seals. The main suggestions for the correct handling regard: the *0-ring*, the *sealing surfaces* and the *tightening process.*

The proper size of 0-ring should be used. When using 0-rings with *smaller cross-sectional diameters* than that required, the needed compression ratio cannot be reached and the seal may leak immediately or after a short time. With an 0-ring having a *larger cross-sectional diameter* than designed, the seal cannot be brought

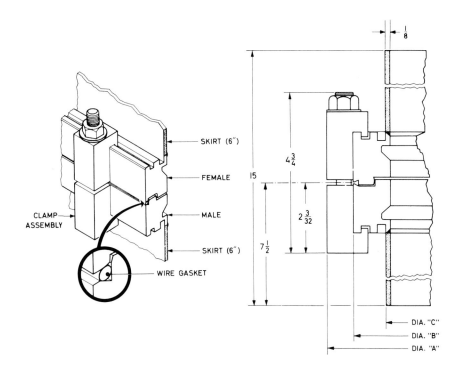

Fig. 7.34 Wheeler wire gasket seal. After Wheeler (1963).

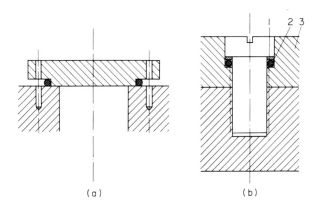

Fig. 7.35 Step seals with O-rings; (a) single step: (b) double step; 1. Washer; 2. O-ring; 3. Flange.

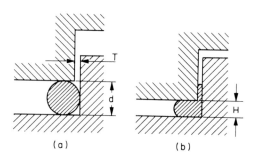

Fig. 7.36 Corner seal; (a) before and (b) after compression

to metal–metal contact or the 0–ring is sheared in the seal. The resulting seal can be leak-tight, but the alignment of the sealing parts is difficult and a larger surface area of the 0–ring is exposed to the evacuated space.

It is impossible to use 0–rings with *larger diameters* (circumference) than that of the groove, since the rubber is not compressible. The opposite solution, i.e. the use of 0–rings with a *smaller diameter* (circumference) than that required, *stretched* over the inside diameter of the groove, step or spacer is *the most common mistake* made in 0–ring seals. Such a solution is equivalent to the use of 0–ring with a smaller cross-sectional diameter than that provided for the seal but has an *additional danger.* Due to the stretching, all the small irregularities on the surface of the 0–ring are enlarged to real channels which make the seal leaky. Very often the 0–ring is damaged during the mounting operation itself by the cutting action of the edges over which it is stretched, especially when these edges are not radiused (e.g. end of a glass pipe not fired previously).

To make a tight seal the surface of the 0–ring must be free of dust or any particles which would prevent the direct contact between the 0–ring and the sealing surfaces (see §7.2.4).

0–rings may be greased. *The grease may seal for the moment* small scratches on the surface of the flange or of the 0–ring itself, but it cannot be recommended as a sealing material. Greasing is particularly not recommended in seals where the gasket is not enclosed in a groove and may slide laterally.

0–rings cannot be reused if they present any permanent deformation.

The seal is restricted to a very small area of the parts, the *surface finish* of these areas is a basic criterion for a reliable seal. The required surface finish is about 60 μin r.m.s. for elastomer gasket seals, and 4–20 ηin r.m.s. for metal gasket seals (see also fig. 3.43). Small machining marks remaining on the sealing surfaces are incomparably less dangerous *if they are parallel to the 0–ring.* Radial scratches, even if they are very small are dangerous (see fig. 7.20) for the leak rate obtained.

The tightening should be done using the required torques. It is no use in tightening with heavy tools, seals constructed for hand tightening. One of the most important facts in assembling 0–ring seals is to prevent and to *avoid drag* (rotation) on the 0–ring during tightening. This is done either by keeping the two sealing parts without rotation (bolt seals) during the tightening or by the use of compression washers interposed between the 0–ring and the (rotating) tightening coupling.

Seals, tightened by remote control, are described by Fischer (1974).

Shear seals. Shear seals or step seals, based on the shearing of the gasket were developed by Lange and Alpert (1957). They may be designed either with a clearance between the opposite shearing steps (fig. 7.37a) or with overlapping steps (fig. 7.37b). The gasket used should be 1 mm thick OFHC copper, hydrogen annealed at 950°C after cutting. For details refer to Wheeler and Carlson (1962), Roth (1966).

Knife edge seals. The most extensively used technique to concentrate the tightening force onto a small width (as required by eq. 7.11) is to provide a "knife edge" on one of the sealing parts. This technique was initiated by Van Heerden (1955), and discussed by Hees *et al.* (1956), Wheeler and Carlson (1962), Unterlerchner (1977), Edwards *et al.* (1979). The gaskets for knife edge seals are usually of OFHC copper, but nickel, silver, silver-plated copper or indium-plated copper were also used. The knife edge must be of a harder material than the gasket. It is in principle a ridge having a V-shape cross section

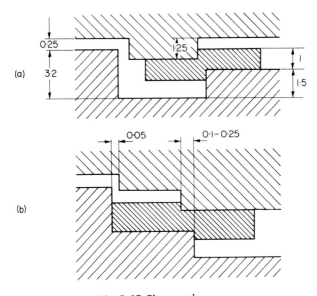

Fig. 7.37 Shear seals.

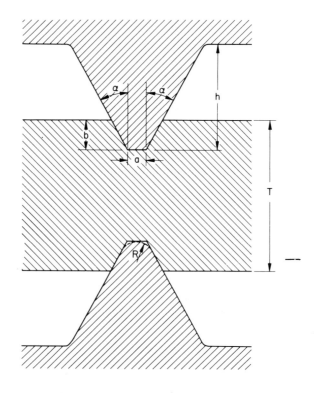

Fig. 7.38 Knife edge seal (detail).

(fig. 7.38), with $\alpha = 30-45°$. The apex of the knife edge may have a flat edge $a = 0.1-0.2$ mm, or a rounded one $R = 0.1-2.5$ mm.

The height of the knife edge is $h = 0.7-2$ mm, the gasket thickness is usually $T = 1-1.5$ mm and the depth of bite $b = 0.2-0.4$ mm.

A widely used application of the knife edge principle is the *Conflat seal* manufactured by *Varian* for ultra-high vacuum systems. The sealing ridge is made (fig. 7.39) with a vertical edge (normal to the gasket) and with the second side of the ridge inclined at 70°; the angle is not critical (Wheeler and Carlson, 1962).

Large size Conflat seals are discussed by Unterlerchner (1987). A tool for removing ultra-high vacuum gaskets is described by Waclawski (1983).

Another type of edge seal uses the "H-gasket", an elastic conical washer inserted between the sealing parts (Ullman, 1962; Horikoshi and Miyahara, 1977). *Guard vacuum in the seals.* According to eq. (7.11) the leak rate of gasket seals can be reduced, by reducing the pressure difference across the seal, i.e. using the technique known as *guard vacuum* or differential pumping. This technique consists of providing an enclosed evacuated space intermediate between the vacuum

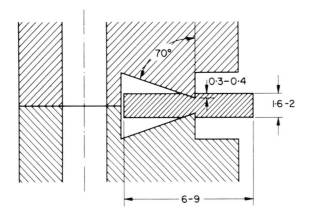

Fig. 7.39 The Conflat seal (Varian). After Wheeler and Carlson (1962).

system (gasket) and the atmosphere. The guard vacuum is generally built as a double gasket system, but also double chambers are constructed (see §7.1.6).

A gasket seal (fig. 7.40a) with a conductance C, or a leak rate $Q = C(P_o - P)$ seals against the external pressure (usually atmospheric) P_o. If a pumping speed S is used, the lowest pressure which can be reached is given by

$$SP = C(P_o - P)$$

thus it is

$$P = [C/(C+S)]P_o \tag{7.19}$$

If this gasket seal is doubled by a second one of conductance C_1 (fig. 7.40b), the limiting pressure obtained with the same pumping speed S will be

$$P' = \frac{CC_1}{(C + C_1) S + CC_1} P_o \tag{7.20}$$

or with $C_1 = C$

$$P'' = \frac{C}{2S + C} P_o = \frac{S + C}{2S + C} P \tag{7.21}$$

i.e. always $P > P' > P/2$, thus the gain by this double gasket is not of great importance.

If a guard vacuum is used (fig. 7.40c) and the pumping speed between the two gaskets is S_1 (and the pressure P_1), the limiting pressure in the main vessel will be

$$P'' = \frac{C}{C+S} P_1 = \frac{C}{C+S} \frac{C_1}{C_1+S} P_o = \frac{C_1}{C_1+S_1} P \qquad (7.22)$$

or

$$P'' = (P_1/P_o)P$$

Considering $P_o = 760$ Torr, and $P_1 = 10^{-1}$ Torr (easily obtained with a rotary pump), the pressure inside the vessel will be

$$P'' = 1.3 \times 10^{-4}P$$

thus a considerable gain is obtained.

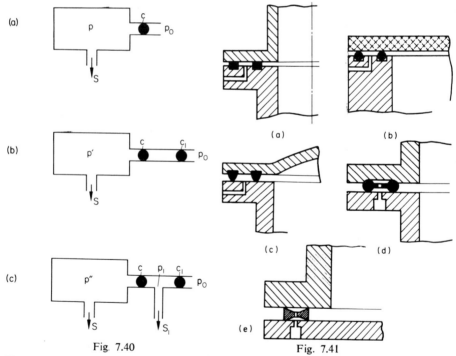

Fig. 7.40

Fig. 7.41

Fig. 7.40 Single gasket, double gasket and guard vacuum seal (principle).
Fig. 7.41 Guard vacuum in gasket seals.

The guard vacuum may be used between concentric seals (fig. 7.41a,b,c). A pump-out connexion is provided in one of the flanges, having one or more connexions to the space between the two gaskets. For ease of assembly, the double 0-ring can be made as a single moulding (fig. 7.41d). Seals with guard vacuum can also be made by using concentric knife edges, or double-edged copper gaskets (fig. 7.41e). Motion seals using guard vacuum have been described by Milleron (1959), Ehlers and Moll (1960), Brueschke (1961), Armstrong and Blais (1963), Merrill and Smith (1971), Harra (1978).

Guard vacuum sealed windows are described by Lucatorto *et al.* (1979), Hollins and Pritchard (1980).

7.3.5. *Electrical lead-throughs*

To transmit an electric current into a vacuum system or envelope lead-throughs should be used. These leads should be electrically insulated from the envelope (and other leads) and joined to the system by vacuum tight seals. The selection of the materials forming the electrical lead-through and its vacuum seal is determined by the service requirements, i.e. the voltage, current, frequency, temperature, etc. According to the requirements, *permanent* or *demountable* electrical lead-throughs may be used.

Permanent lead-throughs are based on glass–metal or ceramic–metal seals

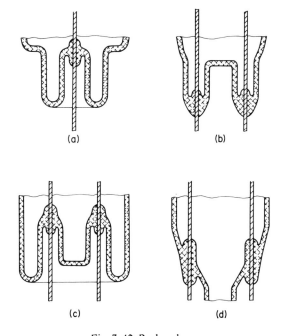

(a)   (b)

(c)   (d)

Fig. 7.42 Rod seals.

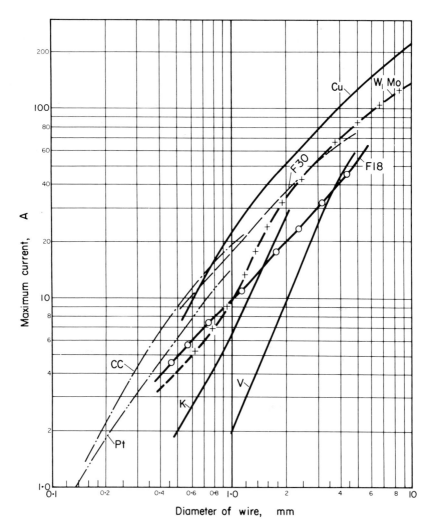

Fig. 7.43 Admissible current in wire (rod) seals. CC – copper clad wire; K – Kovar; V – Vacon; F18–Fe Cr(18% Cr); F30 – Fe Cr (30% Cr). Plotted after data from Steyskal (1955), Roth (1955), Mönch (1959), Von Ardenne (1962), Espe (1966/68).

(§7.3.2). According to their shapes, permanent lead-throughs can be rod seals, stem seals, pin seals, ribbon seals, disc or cup seals (Roth, 1966; Ishimaru, 1976, 1977).

In *rod seals*, metal rods (W, Mo, FeNiCo) are sealed in a glass (table 7.13) part having a specially designed shape (fig. 7.42a) to avoid the build-up of stresses in the glass. The rod seals are usually used as double (multiple) lead-throughs on extensions (fig. 7.42b), on parts protruding into vacuum vessel (fig. 7.42c) or crossing the wall of the system (fig. 7.42d). The lead should carry only currents which do not heat it sensibly, i.e. less than the values in fig. 7.43.

Metal rods sealed in ceramics are used as assemblies connected to metal flanges which are then joined to vacuum vessels by brazing, elastomer or metal gasket seals.

The *stems* are lead-throughs obtained by flattening a glass tube over the lead-wires. Such stems used in electric lamps are based on lead-glasses and copper-clad (or Dumet) wire. The copper-cladded Fe Ni (58/42) wire has a radial expansion which matches that of the lead-glass, while the much higher axial expansion difference is compensated by the elasticity of the thin copper layer. The stem lead-throughs may have various shapes (fig. 7.44) and number of wires.

A *pin seal* is a glass disc in which metal pins are sealed perpendicular to the faces of the disc. These seals are also called *horizontal stems* due to the position in which they are placed in the electron tubes, constituting their bottom. Pin seals can be

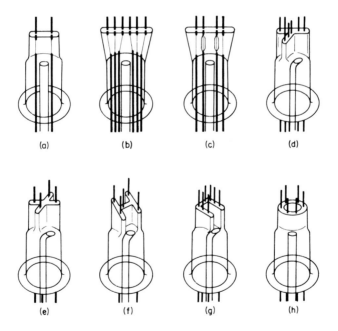

Fig. 7.44 Stem seals of various shapes; (a) flat stem with 2 wires; (b) flat stem with 6 wires; (c) flat stem with 2 single and 2 multiple wire seals; (d) T-stem; (e) X-stem; (f) H-stem; (g) U-stem; (h) 0-stem.

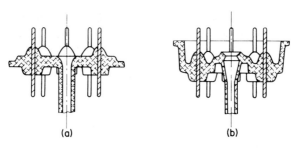

Fig. 7.45 Pin seals.

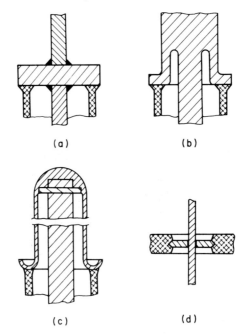

Fig. 7.46 Lead-throughs with disc seals.

in the form of a disc (fig. 7.45a) or a cup (fig. 7.45b) according to the requirements of the subsequent sealing into the vacuum device.

Ribbon seals are represented especially by molybdenum ribbons sealed in quartz as electrical lead-throughs of quartz envelopes (lamps).

Electrical lead-throughs carrying heavy currents are constructed so as to separate the seal from the electrical lead. This is done by connecting the lead through a metal *disc or cup* which is sealed (on its circumference) to the glass (or ceramic). The electric lead is brazed (fig. 7.46a) to the two faces of the disc, or it is machined as one piece (fig. 7.46b) with an annular cut which assures the required temperature gradient to the glass part. The same result is obtained by using a metal cup (fig. 7.46c). For lead-throughs sealed in quartz, a molybdenum foil (disc) may be used (fig. 7.46d).

For various applications, *demountable lead-throughs* are required. In such cases, the electrical lead-through may be either demountable by separating the electrical lead from the insulating part and this part from the vacuum vessel, or being able to separate the electrical lead-through as a whole from the vacuum vessel. In both cases the sealing and insulation, or just the sealing is done either by wax (resin) seals (§7.3.3), or by gasket seals (§7.3.4). Such lead-throughs are supplied by most vacuum equipment manufacturers.

Sliding electrical contacts in ultrahigh vacuum are discussed by Newstead *et al.* (1984); O-ring sealed rotary feed-throughs for ultra-high vacuum are described by Pararas *et al.* (1982); A bakeable all-metal feed-through is described by Edmonds and Shipley (1988).

7.3.6. *Motion transmission*

The transmission of motion from the outside into the vacuum chamber is a very frequent requirement in the various vacuum technologies: closing systems of valves, motion of shutters, targets, tilting of crucibles. The techniques used to transmit motion into the evacuated space are: (a) the tilting of the vacuum device; (b) the bending of elastic pipes; (c) the deformation of bellows; (d) the deformation of diaphragms; (e) the relative motion of ground seals; (f) the motion of gasket seals; (g) the use of magnetic fields; (h) the use of heat transfer or electric

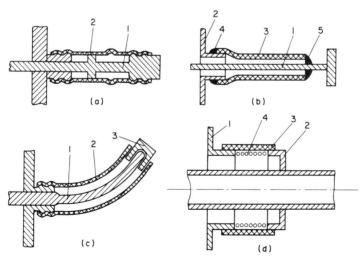

Fig. 7.47 Transmission of motion by elastic pipes.

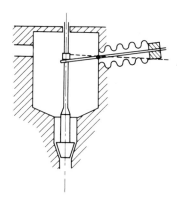

Fig. 7.48 Bellows used for tilting motion. After Della Porta (1960).

current (Monch, 1959; Von Ardenne, 1962; Roth; 1966, Maurice, 1974a; Mulder, 1976; Pearce and Barker, 1977; Head and Griffiths, 1978; Homan, 1980).

Limited translational or rotational motion may be transmitted using a shaft (1 fig. 7.47a) sealed by a rubber sleeve 2, slipped over (fig. 7.47a) or joined by wax (fig. 7.47b) to the parts. Rubber tubing may be used to obtain *unlimited* rotary motion by using a crank (1 fig. 7.47c), the handle of which is enclosed in the rubber tubing 2 in such a way that the rubber flexes on the crank handle. The end of the

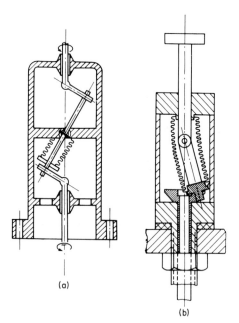

(a)

(b)

Fig. 7.49 Rotation transmission using: (a) tilted bellows; (b) bent bellows.

crank rotates freely in the stopcock 3 which seals the end of the rubber. For tilting the shaft, a large diameter tube (1 fig. 7.47d) and a flange 2 are joined by a rubber tube 3, supported from the inside by a helical wire spring 4 (Bachman, 1948; Fremlin, 1952; Roth, 1966). Rotation motion by using thin-walled metal tubes is described by Hunter (1963).

Bellows are pipes having their walls bent to form consecutive rings. This corrugated shape allows limited axial compression or bending. The bellows are made from rubber, Teflon, metals, and in special cases from glass (e.g. Boll, 1961). Metal bellows allow an axial compression of about 20% of their net length, and are extensively used in all-metal valves and opening devices (Grady, 1973). Bellows can also be used to transmit tilting motion (fig. 7.48) or rotary motion (fig. 7.49). Such devices were described by Della Porta (1960), Snyder and Steel (1962), Tasman (1963).

Rotary motion can also be transmitted by using *diaphragms* (fig. 7.50). The driving shaft 1, in alignment with the driven shaft 2, rotates this latter by means of a wobble shaft 3, sealed through the centre of the diaphragm 4.

Ground and lapped seals may be used to transmit motion into vacuum chambers, if the vapour pressure of the oil or grease used with these seals may be tolerated in the system. The lapped part (1 fig. 7.51a) is sealed to the system, while the shaft 4 is connected to the second lapped part 2. The sealing and lubrication is done by the oil flowing through the circular channel 3. Lapped seals are able to transmit rotation up to 3–4 r.p.m. Conical ground seals (fig. 7.51b) or spherical

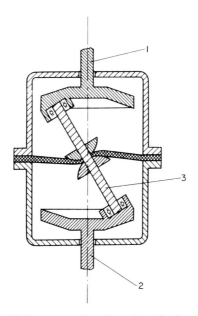

Fig. 7.50 Rotary motion through a diaphragm.

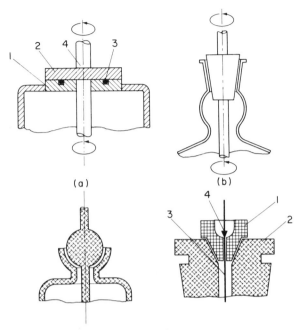

Fig. 7.51 Rotary motion through ground seals. (a) lapped flat seal; (b) conical ground seal; (c) spherical ground seal; (d) seal with Teflon plug (1); 2. Adapter; 3. Shaft; 4. Bearing.

ones (fig. 7.51c) are also frequently used to transmit slow rotation (e.g. Howe, 1955; Nester, 1956b). The gas evolution from rotary seals is discussed by Gorodetsky and Skurat (1975).

Sliding and/or rotary motion can be transmitted in principle through *shaft seals* using one 0–ring (fig. 7.52a) but in current practice seals using two 0–rings are preferred (fig. 7.52b, c) since they ensure better centering of the shaft. It is easier to machine the groove on the shaft (fig. 7.52a) but when the shaft must have a long stroke the groove should be made in the wall of the port (fig. 7.52 b, c). A sliding or rotary shaft seal with a double 0–ring can be constructed without grooves, by placing a cylindrical spacer (fig. 7.52d) between the 0–rings and compressing the assembly with a closing ring (e.g. Gore, 1957; Lake *et al.*. 1963; Armstrong and Blais, 1963).

The *lip seals* known as *Wilson seals* are based (Wilson, 1941) on the sealing action of a rubber sheet toward a shaft crossing it through a hole, cut with a diameter considerably smaller than that of the rod. The periphery of the rubber sheet is held tightly by a circular metal part (4 fig. 7.53), and the rubber close to the shaft is distorted and bent out from the plane of the sheet, its lip sealing against the shaft. The sliding shaft does not necessarily have to be particularly straight, but its surface must be smooth. The gasket is usually 1.4–1.6 mm thick and made

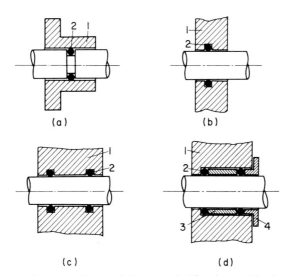

Fig. 7.52 Sliding and rotary 0-ring shaft seals 1. Chamber wall; 2. 0-rings; 3. Spacer; 4. Sealing ring.

of a rubber of Shore hardness 50–60. The hole should be 0.65–0.8 of the shaft diameter, thus (fig. 7.53) $d=0.65$ D–0.8 D; $d_2=3D$–$4D$ for small D values, and $d_2=2D$ for large D.

Wilson seals can be used with shafts 1.5 mm up to very large (e.g. 70 mm) diameters, but at larger sizes (over about 20 mm) it must be ensured that the pressure difference will not force the shaft into the chamber.

Wilson seals are very often constructed using a double gasket the space between the gaskets being evacuated (fig. 7.54a) or filled with grease (fig. 7.54b) or with liquids (Brueschke, 1961).

Shaft seals for the transmission of rotary and/or translational motion are manufactured by the various firms constructing vacuum equipment.

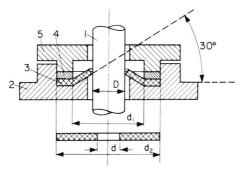

Fig. 7.53 Wilson seal; 1. Shaft; 2. Base plate; 3. Rubber washer; 4. Metal ring; 5. Lock nut.

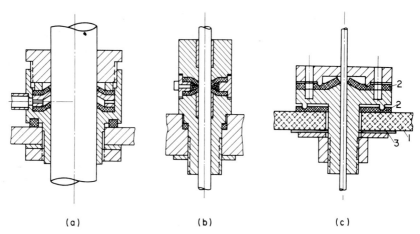

(a)   (b)   (c)

Fig. 7.54 Wilson seals with double gasket; (a) with parallel gaskets and guard vacuum; (b) with opposite gaskets; (c) Wilson seal on a glass plate; 1. Glass; 2. Neoprene; 3. Mica.

Magnetic fields may be used to transmit motion inside the evacuated space by placing a magnet (or electromagnet) outside the chamber and transmitting the field through the wall, which should be of non-magnetic material. The electromagnet may be placed in the chamber as well and energized from the outside, but this solution presents the need of electrical lead-throughs and degassing or breakdown difficulties.

Vacuum-tight, bakeable magnetic feed-throughs for rotary motion (up to 700 rpm) are available from vacuum equipment manufacturers (e.g. Varian). Non-bakeable seals may achieve very high speeds (Coenraads and Lavelle, 1962).

Lubrication in vacuum. At pressures higher than 10^{-4} Torr, the water vapour on the surfaces is enough to permit motion of components in contact. At lower pressures the surfaces desorb the various gases (and vapours) as discussed in §4.4, and the "bare" contacting surfaces present a gradually higher friction up to adherence. In order to separate the contacting surfaces, a low shear material – a *lubricant* – is inserted between the moving parts. The lubricant may be fluid (wet) or solid (dry) and must have a *low vapour pressure* in the required temperature ranges. The lubricant has to conform to the lubricating and wear conditions as well as to the conditions required from any vacuum material (see §7.1).

Fluid lubricants (silicones, fluorocarbons, etc.) have vapour pressures at room temperature extending from 10^{-9} to 10^{-12} Torr, corresponding to rates of evaporation (losses) of 10^{-9} to 10^{-13} $g \cdot cm^{-2} \cdot s^{-1}$ (eq. 4.7). These liquid lubricants are useful in the temperature range from 225 to 375 K (Roller, 1988).

Solid (dry) lubricants used are PTFE, MoS_2, WS_2, $NbSe_2$ and soft metals as Au, Ag (Pb, Sn) (Spalvins, 1987). These lubricants have vapour pressures of $10^{-12}-10^{-14}$ Torr at room temperature, corresponding to rates of evaporation (loss rates) of 10^{-10} to 10^{-14} $g \cdot cm^{-2} \cdot s^{-1}$. The solid lubricants are useful in the

4–700 K temperature range (with substrates permitting these temperatures). Solid lubricants may be applied to the surfaces: by rubbing, coating by tumbling in powdered lubricant or by high-velocity impingement of the lubricant entrained in a gas. Lubricants applied by these techniques have limited wear life (tens of sliding cycles; thousands of rolling cycles).

Bonded lubricants are applied by spraying in a binder and curing the parts. Such coatings have better wear life (7000 to 15 000 sliding cycles). The best wear life is obtained by ion plated, sputtered coatings. MoS₂ (often co-sputtered with Ni), Au, PTFE have been sputter coated as solid lubricants for parts moving in vacuum (Roller, 1988).

The friction in vacuum and the problems of lubrication were discussed by Buckley (1974), Matsunaga and Hoshimoto (1974), Friebel and Hinricks (1975), Kirby et al. (1982), Marmy (1983), Spalvins (1987), Panitz et al. (1988), Roller (1988).

7.3.7. Material transfer into vacuum systems

In every vacuum application, gases (liquids) or solids are to be transferred from the vacuum system to the surrounding space, from outside into the vacuum

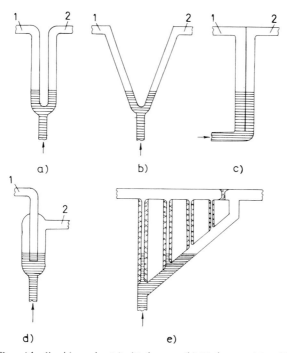

Fig. 7.55 Cut-offs with liquid seal; (a) U-shape; (b) V-shape; (c) with separating wall; (d) concentric pipes; (f) for variable flow. After Roth (1966).

system or from one part of the system to another. The seals used in connexion with material transfer must allow a port of adequate dimensions to be open during the transfer, while still keeping the rest of the vacuum system vacuum tight. The port itself should be vacuum-tight when closed.

For the transfer of gases from one part of the system to another (exhaust) it is always desirable to have means of shutting-off various parts: cut-offs can be used when the pressure difference is not too large; *stopcocks* or *valves* for larger pressure differences.

When gases are to be introduced into the vacuum system, *stopcocks* or *valves* are used for large throughputs and *controlled leaks* for small and very small throughputs.

To achieve the transfer of solids (specimens, photo plates, etc.) into or from the evacuated chamber *vacuum locks* are used.

The transfer of radiation (through windows) into or from vacuum chambers is discussed e.g. by Roth (1966), Heasley (1976), Fisher (1988).

A *cut-off* is a device in which a liquid surface is used to separate two parts of the vacuum system. A cut-off consists of a system for raising the liquid and a closing system. The raising systems are those described in connection with the McLeod gauge (figs. 6.8–6.10) which always includes a cut-off.

The closing action may be achieved by the liquid itself or by floats moved on the level of the liquid. The liquid may close the connection between two parts of the system, either by entering into a Y-shaped part of the connexion (fig. 7.55) or by sealing on a porous (sintered) glass surface (fig. 7.56) (e.g. Glaister, 1956; Almond, 1958).

Cut-offs using floats (fig. 7.57) have the advantage that when closed the vapours of the liquid are excluded from the evacuated space. They also permit positive closing (Hammond, 1954; Neville, 1955; Wyllie 1957; Knor, 1960).

The liquid used is usually mercury, but oil or molten metals (indium, gallium) are also used (e.g. Paty and Schurer, 1957; Axelrod, 1959; Beynon, 1960).

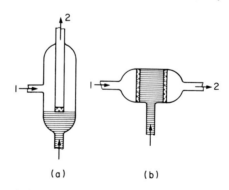

Fig. 7.56 Cut-offs with sintered glass; (a) with single glass frit (unlimited raising of the liquid); (b) with double glass frit (limited raising).

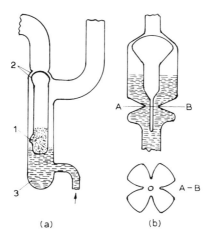

Fig. 7.57 Cut-offs with float; (a) cylindrical; 1. Guiding bulge; 2. Closing taper; 3. Mercury; (b) guided mushroom float.

Glass stopcocks consist of a plug and a body. The plug can be solid (fig. 7.58a) or hollow (fig. 7.58b). Sometimes hollow plugs are provided with an inside connecting pipe (fig. 7.58c). In order to guarantee the leak tightness of the stopcock, the design with the body closed over the plug (fig. 7.58b) is recommended. This arrangement avoids the atmospheric pressure on the small end of the plug and thus ensures that the plug will be pressed into the shell, greatly reducing the leak rate of the stopcock.

The main characteristic in defining a stopcock is the *number of ways* (connexion

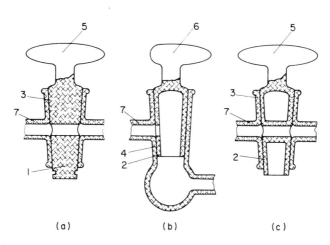

Fig. 7.58 The parts of stopcocks: 1. Solid plug; 2. Hollow plug; 3. Open shell; 4. Closed shell; 5, 6. Handles; 7. Outlets.

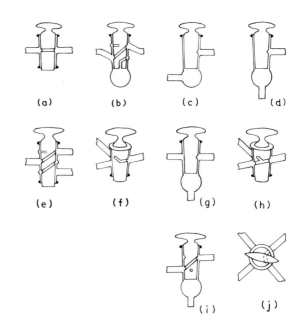

Fig. 7.59 Shapes of glass stopcocks.

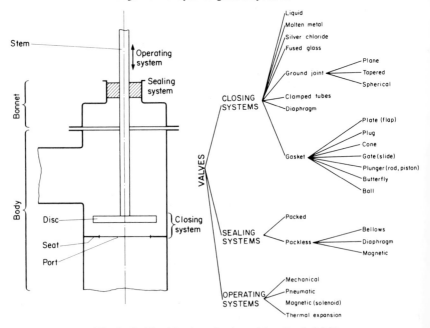

Fig. 7.60 Classification of valves. After Roth (1966).

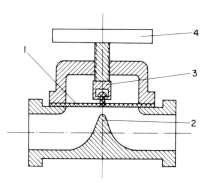

Fig. 7.61 Diaphragm valve.

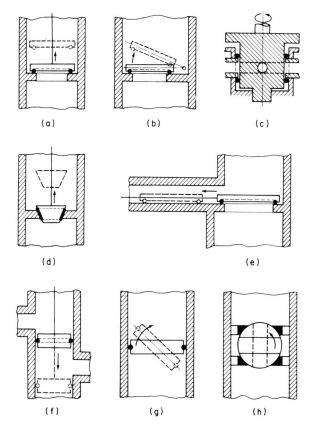

Fig. 7.62 Closing systems of vacuum valves using gasket seals. (a) plate valve; (b) flap valve; (c) plug valve; (d) cone valve; (e) gate valve; (f) plunger valve; (g) butterfly valve; (h) ball valve. After Roth (1966).

possibilities), which *is not necessarily equal to the number of outlets*. Thus the stopcocks in fig. 7.59a–d have one way and two outlets, those in fig. 7.59e–g have two ways and three outlets; those in fig. 7.59h–i have three ways and three outlets, and that in fig. 7.59j four ways and four outlets.

The usual glass stopcocks must be lubricated using vacuum grease (table 7.20). The gas evolution from the greased stopcock can be decreased by degassing the grease before use after being applied onto the stopcock or by maintaining the stopcock at lower temperatures.

In vacuum systems where gases are handled which dissolve the grease, *greaseless stopcocks* must be used. These are glass stopcocks lubricated with graphite and sealed with mercury, or stopcocks having the plug of Teflon.

A *valve* consists of: the *body*, the *bonnet* and the *stem* (fig. 7.60). The function of a valve is to adjust or stop the flow and is achieved by the *closing system*. This system needs motion inside the valve (*operation system*) and the moving parts should be sealed (*sealing system*). From the combination of these systems (fig. 7.60) a great number of valve designs is possible, the number of types constructed being really very large. The most frequently used are: the *diaphragm valves, plate valves* and *gate valves*, using *elastomer or metal gaskets, bellows seals* and *mechanical or magnetic operation*. (For a detailed discussion see Roth, 1966.)

In *diaphragm valves* a rubber membrane (1. fig. 7.61) is forced against the seat 2, by the external shaft 3 operated from the head 4. The diaphragm functions here as both the closing and the sealing system of the valve. These valves have a good conductance but the large area of exposed elastomer leads to outgassing properties which make the valve useful only in backing vacuum lines.

If the valve is to be used on the high vacuum side it must be of high conductance (large opening) and the amount of elastomer exposed must be kept to a minimum to reduce the gas load from this source. This is done by using one of the various *gasket* seals (fig. 7.62). The gasket may be of Neoprene, Viton, Teflon

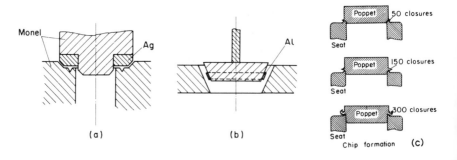

Fig. 7.63 Closing systems of all-metal valves; (a) with flat silver ring (Bills and Allen, 1955); (b) with aluminum conical ring (Kienel and Lorenz, 1960); (c) with copper poppet (Parker and Mark, 1961).

or a metal, and may have rectangular, trapezium, or circular cross sections. It may be placed on the closing disc or on the seat. The sealing system of the valve may be an 0–ring shaft seal, a Wilson seal or metal bellows. Bellows sealed valves have leak rates less than 1×10^{-4} lusec, i.e. by a factor of 100 less than 0–ring (or Wilson) sealed ones. Bellows seals are absolutely necessary in bakeable (all-metal) valves. The performance of Viton, bellows sealed valves was studied by Wheeler (1971).

All-metal valves are based on closing systems consisting of *edge seals*. In these constructions a silver part is compressed against a harder (monel) part (fig. 7.63a), an Al ring is closed in a conical joint (fig. 7.63b) or a copper poppet (fig. 7.63c) is pressed into a stainless steel cutter seat. Cross sections of bakeable ultra-high vacuum valves are shown in figs. 7.64 and 7.65. The valve in fig. 7.64 uses a bellows which is sealed at the upper part by a gold gasket seal. The closing system is based on a copper nose against a stainless steel seat. The valve shown in fig. 7.65 uses a stainless steel knife edge against a copper disc.

The all-metal valves for ultra-high vacuum have used many solutions, and a large number of such valves were described. On this matter we refer to Bills and Allen (1955), Lange (1959), Kienel and Lorenz (1960a), Parker and Mark (1961), Baker (1962), Ullman (1962), Lewin and Mullaney (1964), Scheffield (1965), Roberts and Bahn (1965), Bultemann (1965), Hauff *et al.* (1965), Comsa and Simionescu (1966), Bannenberg (1966), Bernard (1970), Wikberg (1971), Teutenberg (1972), Basta *et al.* (1976), Gilmour (1976), Wheeler (1976), Harra (1978), Stone and Scully (1979), Pathak (1982), Singleton (1984), Weston (1984), Schrag and Colgate (1987).

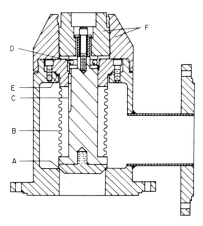

Fig. 7.64 Ultra-high vacuum valve. (A) copper nose; (B) bellows; (C) guide; (D) ball bearing; (E) gold wire seal; (F) driving mechanism. After Lange (1959).

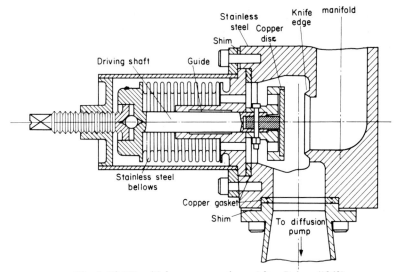

Fig. 7.65 Ultra-high vacuum valve. After Baker (1962).

The various valves should fit the kind of process achieved in the plant and the place where they are connected into the vacuum system. Valves whose only purpose is separating two spaces at different pressures are known as *isolation valves*. Valves that have also other purposes or that should correspond to special requirements are known as: seal-off valves, throttling valves; air admittance valves; baffle valves. *Seal-off valves* are used to close an evacuated vessel, the sealing system being taken away, leaving the vessel sealed by the closing system of the valve. *Throttling valves* are used to adjust the rate of flow. *Baffle valves* are so designed that the disc of the valve remains *in line* with the port, thus acting as a baffle. These latter valves are installed above the diffusion pumps (fig. 3.35). Valves for specific purposes in the vacuum system are described e.g. by Fuchs (1976), Peters and Pingel (1977), Baker *et al.* (1978), Fischer *et al.* (1978), Weston (1984), Singleton (1984).

Controlled leaks are used to introduce metered quantities of gases in the evacuated vacuum systems. The basic features used to construct such leaks are listed in table 7.23, the throughput ranges being those mentioned in table 7.24.

Systems for calibrating standard leaks are discussed by Miller (1973b), Iverson and Hartley (1982), Rubet (1983), Scuotto (1986), Thornberg (1988). Solomon (1986) discusses the calibration of permeation (diffusion) leaks used for calibrating leak detectors (§4.2.3) and stresses the subject of the temperature corrections.

Diffusion (permeation) leaks are well suited as secondary standards as their stability over several years is within a few percent (Grosse *et al.* 1987b).

To achieve *the transfer of solids* into or from the vacuum chamber, *vacuum locks* are used. These consist of a space (chamber) adjacent to the vacuum system, and which can be opened either to the atmosphere or to the vacuum system. The lock is first opened to the atmosphere (being sealed towards the vacuum chamber) and the object is introduced into the lock. The port to the atmosphere is closed and the vacuum lock is evacuated (through its own pumping line). The port to the vacuum chamber is opened and the object is transferred from the lock into the vacuum chamber, and the port to the vacuum chamber is closed. Vacuum locks may be based on sliding rods, rotating plugs or chambers with double port.

Vacuum locks were described by Stevens (1953), Zovac (1954), Leisegang (1956), Brunee (1960), Von Ardenne (1962), Kofoid and Zieske (1962), Colombani and Rano (1963), Porteous (1963), Boerboom et al. (1964), Roth (1966), Salle (1970), Ono et al. (1973), Polaschegg and Schirk (1975), Miller et al. (1976), Clausing et al. (1979), Hobson and Kornelsen (1979), Polizzotti and Schwarz (1980), Schlier (1982), Heinemann and Poppa (1986).

Vacuum locks using *sliding rods* (fig. 7.66a) consist of a rod 1, which may be pushed into the vacuum chamber 2 through a seal 3. The rod carries the objects to be transferred in a pit (or pits) 4 of appropriate shape. A more sophisticated construction (fig. 7.66b) uses a long rod made of three sections; the two end sections 1, are carriages and the middle section 2, is a blank rod. The sample is mounted inside a carriage, and the rod is moved through the channel 3. Three pumping lines (P_1, P_2, P_3) assure the pressure gradient (guard vacuum) from the atmosphere to the vacuum chamber 4. At the two ends of the rod, double 0-ring seals 5, were provided.

In *vacuum locks using plugs* (fig. 7.67) the plug receives the object from one side, and by its rotation allows the transfer of the object to the other side. In the technique shown in fig. 7.67a the object is pushed into the plug 4, through the

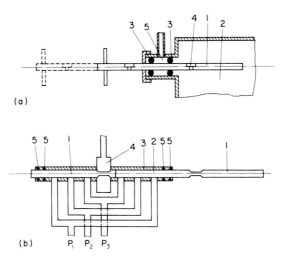

Fig. 7.66 Vacuum locks using a sliding rod; (a) with single pumping; (b) with guard vacuum.

opening 1. By the rotation of the plug its opening reaches the evacuation outlet 2, and by further rotation the object is transferred to the chamber 3. In an alternative construction (fig. 7.67b) the plug 1 may be withdrawn from the shell, since a second rotating shell 3 is included in the lock. After inserting the object into the pit 2, the plug is placed again in the shell, and by rotation the pit is brought to the evacuation outlet 4 and then to chamber outlet 5.

The principle of a *double port* chamber used as a vacuum lock in electron microscopes is shown in fig. 7.68. Here the probe 1 is introduced in the chamber 4 between the ports P_1 and P_2, and then by opening the port P_1 into the chamber 2.

Table 7.23.
Gas leaks. (Roth, 1966).

Basic feature of the leak	Characteristics	References
Pinhole	through thin walled glass, quartz or metal	Munson (1955) Marks (1957)
Orifice	at the end of pipes	Gordon (1958) Beynon (1960)
Crack	in glass walls; variable by twisting the tube or change of mercury level	Hopfield (1950) Amoignon (1957)
Capillary	constant leak	Laufer (1962)
Flattened tube	variable by bending, twisting, squeezing	Ochert (1951)
Knife edge	forced into soft metal	Winkelman, Davidson (1975
Porous plug	ceramic, glass frit or metal; variable by changing the mercury level	Morrison (1953) Jenkins (1958), Lawson (197
Annular impedance	concentric pipes, cone and seat, washers, O–rings	Amariglio (1958) Allensworth (1963)
Needle valves		Review: Roth (1966)
Temperature actuation	longitudinal expansion bimetal radial expansion temperature of the gas	Flinta (1954) Nester (1956a) Smither (1956)
Diffusion	for helium, hydrogen, and oxygen	see fig. 3.45
Pulsed leaks	bubbling vibrated cap

rotating pits

vibrated needle | Littman (1961)

Gorowitz et al. (1960) Ishimaru Fukomoto (1974) Chavet (1976) |

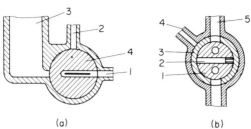

(a)   (b)

Fig. 7.67 Vacuum locks, using plugs; (a) plug with pit; (b) plug with pit and double shell.

7.4. Leak detection

7.4.1. *Leak rate and detection*

An ideal vacuum chamber should maintain forever the vacuum (pressure) reached at the moment of its separation from the pumps. Any real chamber presents a rise in pressure after being isolated from the pumping system. The pressure rise is produced by the gas which penetrates through leaks, that which evolves from the walls (outgassing) and that entering by permeation (see figs. 3.43–3.45).

The leak rate (or real leak rate) is the quantity of gas (in PV units) flowing per unit time into the system. * Obviously a perfectly tight vacuum system or chamber is one having a zero real leak rate, but to achieve this is as impossible as it is to reach zero pressure. The leak rate is expressed in throughput units (Torr · lit/sec, lusec, table 3.3). Indirectly the leak rate of a given system or chamber is sometimes expressed as the pressure rise in a given time, and for a specific volume, or as the time required for a given quantity of gas to flow into the system. Table 7.25 compares these specifications.

In approaching the problem of a supposed leak greater than the admissible value, it is necessary to determine first if such a leak actually exists. Plotting of pressure vs. time curves (fig. 7.69) will assist in determining the actual leak rate. First the system is evacuated to a stable minimum pressure. To minimize the effect of the vapours present in the system it is useful to employ liquid nitrogen traps. When no further improvement in the pressure is evident during further pumping, the pumps are valved off from the system. The recorded pressure rise vs. time curve during a long period of isolation shows if the rise of pressure is produced only by real leaks (straight line 1 fig. 7.69), only by outgassing (curve tends to a limiting maximum value as 2 fig. 7.69) or by a combined effect of these two (curve 3 fig. 7.69). The recording should be continued until the shape of the curve becomes evident. The curve of the pressure drop is analyzed by Hamacher (1974a); the separation of the effect of real leaks from outgassing, in the pressure rise curve, is discussed by Schalla (1975).

* It is recommended (Ehrlich 1986; Solomon, 1986) to specify the temperature or to express the leak rate in units of mol/s.

Table 7.24.
Throughput ranges of gas leaks.

Leak	Throughput range (lusec)[*]	Lowest pressure (Torr)	Remarks	References
Orifice through platinum disc.	$6 \times 10^{-1} - 30$	—	5–15 diameter	Nief (1952)
Orifice on the end of tapered capillary	$1-9 \times 10^{-4}$	—	1.2 μ bore taper 39 μ/cm	
	1.6	—	24 μ bore taper 30 μ/cm	Gordon (1958)
Crack on capillary glass tube	$2 \times 10^{-3} - 1.8$		varied by twisting the tube	Hopfield (1950)
Slit on glass tube	$0 - 3.8 \times 10^2$	10^{-8}	varied by mercury level	Kunzl and Slavik (1935)
Circular cross section capillary	$10^{-7} - 10^{-2}$	—	0.2 to 10 μ dia. 1 cm long	see §3.3.3 and 3.6.7
Flattened copper tube	$3 \times 10^{-3} - 3$	—	varied by rolling the tube	Nier (1947)
Glass needle in capillary tube	$2 \times 10^{-2} - 5$	—		Hopfield (1950)
Porous metal plug	$5 \times 10^{-5} - 10^{-2}$	—	compressed at various loads	Jenkins (1958)
Porous material	$>7 \times 10^{-1}$	—	varied by the level of the mercury	Rose (1950)
Porous ceramic rod	$1 \times 10^{-3} - 10$	10^{-6} 10^{-2}		Morrison (1953)
Permeation through silicone rubber sheet	$<1 \times 10^{-4}$	—		—
O–ring seals with variable compression	$>2 \times 10^{-3}$	10^{-5}		Amoignon ('957)
Conical, spring washer seal	5×10^{-4} -7×10^{-2}	—		Wishart (̇9 ̇)
Steel ball on seat	7×10^{-1} -7×10^{-3}	—	differential screw, belows valve	Allensworth (1963)
Needle valves	$1 \times 10^{-2} - 1.0$	—		Riggs (1962)
Pin valve	$>6 \times 10^{-5}$	10^{-4} 10^{-6}	bellows sealed	Bouyer et al (1960)
Needle valve	$>10^{-4}$	—	PTFE seat	Cope (1958)

(contd.)

Table 7.24. (contd.)
Throughput ranges of gas leaks.

Leak	Throughput range (lusec)*	Lowest pressure (Torr)	Remarks	References
Bakeable all-metal valve	$>7 \times 10^{-8}$	—	continuously variable	Alpert (1955)
Platinum wire expanding in glass capillary	$3 \times 10^{-2} - 25$	10^{-8}		Martin (1948)
Tungsten rod expanding in stainless steel body	1×10^{-2} -9.10^{-1}	— —		Green (1953)
Heated capillary	$4-40$	5×10^{-2}	throughput varied as a result of gas temperature	Smither (1956)
Solid state components in "hermetic package"	$10^{-3}-10^{-4}$	—	—	Davy (1975)

*Conversion factors to other throughput units, see table 3.3.

Table 7.25.
Leak rate specifications.

Leak rate* Torr·l/sec	Pressure rise in 1 litre volume	Time for 1 micron pressure rise/litre	Time for 1 cm³ STP gas inflow	Equivalent opening
10^{-3}	$1~\mu/\text{sec}$	1 sec	12.7 min	Rectangular slit with 1 cm width, 0.1 mm height and 1 cm depth
10^{-4}	$6~\mu/\text{min}$	10 sec	2.1 hr	Rectangular slit with 1 cm width, 30μ height and 1 cm depth
10^{-5}	$36~\mu/\text{hr}$	1.66 min	21 hr	Capillary 1 cm long and 7μ dia
10^{-6}	$3.6~\mu/\text{hr}$	16.6 min	8.7 days	Capillary 1 cm long μ4 dia
10^{-7}	$8.6~\mu/\text{day}$ (24hr)	2.77 hr	87 days	Capillary 1 cm long 1.8μ dia
10^{-8}	$0.86~\mu/\text{day}$	27.7 hr	2.4 yr	Capillary 1 cm long 0.8μ dia
10^{-9}	$31~\mu/\text{yr}$	11.6 days	24 yr	Capillary 1 cm long 0.4μ dia
10^{-10}	$3~\mu/\text{yr}$	116 days	240 yr	Capillary 1 cm long 0.2μ dia

*Conversion factors to other units, see table 3.3.

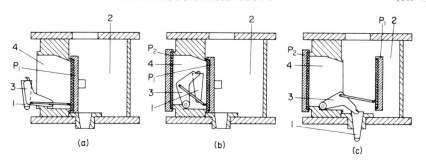

Fig. 7.68 Vacuum lock with double port; (a) the object in its holder; (b) the object in the closed vacuum lock; (c) the object transferred into the vacuum chamber.

The use of the variation of pressure as the leak detection method in large systems is discussed by Callis (1985). Settina *et al.* (1987) used the pressure rise method and measured leak rates down to 10^{-15} Torr · liter/sec, with a spinning rotor gauge.

In order to obtain higher sensitivities and to locate the exact position of the leaks a series of leak testing methods were developed using various *test gases* (or liquids). In principle in all these tests the gas (or liquid) is spread over the outside of the part and the presence of the gas is tested inside, or the vacuum system is filled with the test gas and its presence detected outside. Table 7.26 lists most of these methods.

Reviews on the various methods of leak detection were published by Blears and Leck (1951), Raible (1955), Briggs (1959), Pirani and Yarwood (1961), Turnbull (1965), Holkeboer *et al.* (1967), Winkelman and Davidson (1975, 1979). For a detailed treatment of the subject we refer to Marr (1968), Maurice (1971), Wilson and Beavis (1976), Santeler (1984).

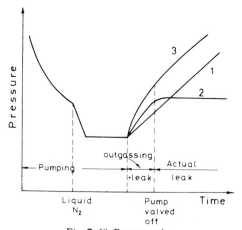

Fig. 7.69 Pressure-time curves.

7.4.2. *Leakage measurement*

The leakage measurement gives the value of the *total leakage*, of the whole system being measured. For this purpose the test gas should enclose the whole contact surface and the detector should measure the test gas concentration on the opposite side.

Leakage measurement techniques fall into two major classes: *static* and *dynamic*.

In *static testing* the device to be tested is pressurized with the test gas and placed in the test chamber. The concentration of test gas in this chamber is then monitored as a function of time; since the concentration increases during the testing period, this method is also called "accumulation testing". The measurement is carried out and interpreted according to

$$Q = V(\mathrm{d}P/\mathrm{d}t) \tag{7.23}$$

where Q is the leak rate, V the volume of the collecting system, P the test gas pressure in the collecting volume, and t the accumulation time.

In *dynamic testing*, the system or envelope being tested is continuously evacuated. The test gas flowing into the pump passes through a detector section where its concentration is measured. The leakage measurement can be done (fig. 7.70a) by continuously *pressurizing the system* to be tested and placing it in an envelope connected to the leak detector or by introducing the *tracer gas into the envelope* and connecting the system to the detector (fig. 7.70b).

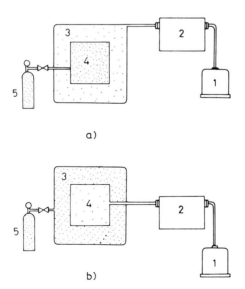

a)

b)

Fig. 7.70 Dynamic leakage measurement methods. (a) pressurized system; (b) pressurized envelope. 1. Pump; 2. Leak detector; 3. Envelope; 4. System under test; 5. Test gas container. After Marr (1968).

Table 7.26.
Leak detection methods.

Method	Test gas (liquid)	Test principle	Pressure range	Minimum detectable air leak (Torr·l/sec)	Remarks	References
Flame wavering	air, nitrogen	The gas stream is heard by its hissing or is seen by the wavering of a flame	up to 3 atm	4×10^{-2}	Draught free or quiet room required	
Electric discharge	acetone, methanol, CO_2, hydrogen	Color change of the discharge		1×10^{-2}		
Wet outside surface	pressurized	The inside is filled with liquid (observe the points where outside will be wet)	up to 3 atm	4×10^{-3}	Involves wetting the inside of the vessel and subsequent cleaning	Minter (1960)
Bubbles in liquid	air, nitrogen	The leak is indicated by a bubble appearing where the gas pressurized inside can pass to the outside	up to 3 atm	1×10^{-4}		Biram and Burrows (1964)
Bubbles on soap film	air, nitrogen		up to 3 atm	4×10^{-5} $*(8 \times 10^{-6})$	*If soap film maintained 5 min	Bloomer (1973) Ratcliffe (1964)
Ammonia fumes	ammonia, CO_2 SO_2	Ammonia inside; detected outside with CO_2 or HCl or CO_2 resp. SO_2 inside; detected outside with ammonia	up to 3 atm	4×10^{-5}		

(contd.)

Table 7.26. (contd.)
Leak detection methods.

Method	Test gas (liquid)	Test principle	Pressure range	Minimum detectable air leak (Torr·l/sec)	Remarks	References
Ammonia sensitive paper	ammonia	Ammonia inside; where leaks to outside produces black spots on wet ammonia sensitive paper rolled over the parts	about 2 atm	1×10^{-6} $*(1 \times 10^{-8})$	*Detectable in about 30 hr	Delafosse et al. (1960)
Single Pirani gauge	CO_2 hydrogen butane	The test gas changes the thermal conduction inside the gauge	1 to 1×10^{-3} Torr	2×10^{-5} 1×10^{-5} 5×10^{-6}		Minter (1958)
Differential Pirani	CO_2 butane	A pair of gauges, one sensitive to both air and test gas – the other connected via a trap is sensitive only to air	1 to 10^{-6} Torr	1×10^{-6} 5×10^{-7}		Steckelmacher and Tinsley (1962)
Charcoal Pirani	hydrogen	Cooled charcoal trap is inserted in the gauge line to reduce the effect of pressure fluctuations	1 to 10^{-6} Torr	4×10^{-7}		Kent (1955)
Single ionization gauge	hydrogen CO_2 butane	Test gas changes gauge reading	10^{-3} to 10^{-8} Torr	5×10^{-6} 1×10^{-6} 1×10^{-7}		Varicac (1956) Barrington (1965)
Differential ionization gauge	CO_2 butane	Pair of ionization gauges arranged as in diff. Pirani test	10^{-3} to 10^{-8}	3×10^{-9} 5×10^{-10}		Benvenuti and Decroux (1972)

(contd.)

Table 7.26. (contd.)
Leak detection methods

Method	Test gas (liquid)	Test principle	Pressure range	Minimum detectable air leak (Torr·l/sec)	Remarks	References
Palladium barrier	hydrogen	The gauge (Pirani, ionization) separated from vacuum system by a Pd barrier which when hot is permeable to H_2 only	as by Pirani or ionization	5×10^{-8}	Care not to poison the gauge by impurities	Ochert and Steckelmacher (1952)
Halogen detector	freon (CCl_2F_2) trichloroethylene tetrachlor-carbon	Based on the positive ion emission from hot Pt anode when exposed to traces of halides	2×10^{-1} to 7×10^{-2} Torr (optimum range)	10^{-5} to 10^{-6}	Sudden loss of sensitivity after exposure to high concentration of halide vapours. Advisable to check frequently the sensitivity	White and Hickey (1948) Drawin (1959)
Mass spectrometer (cold cathode)	helium	Based on separation of ions produced by the test gas from those formed by the residual gases	10^{-2} to 10^{-4} Torr	10^{-8}	Can be used only at high vacuum. A separate leak testing pumping system is required in addition to the pumping system of the mass spectrometer	Nier (1947) Daly (1960) Moody (1957)
Mass spectrometer (hot cathode ion source)	hydrogen argon helium		5×10^{-4} to 10^{-8} Torr	5×10^{-9} 5×10^{-9} 5×10^{-11}		
RF mass spectrometer	helium		10^{-5} to 10^{-10} Torr	10^{-10}		

(contd.)

Table 7.26 (contd.)
Leak detection methods.

Method	Test gas (liquid)	Test principle	Pressure range	Minimum detectable air leak (Torr·1/sec)	Remarks	References
Omegatron	hydrogen, argon	Acceleration of ions	10^{-6} to 10^{-10} Torr	4×10^{-11}		Bell (1956) Nicollian (1961)
Ion pump	hydrogen, air	Ionization	10^{-6} to 10^{-10} Torr	1×10^{-11}		Young (1961)
Radioactive	Kr 85	Detection of gamma radiation	2 Torr	5×10^{-12}		Cartois and Gasnier (1963) Waldschmidt (1973)

The magnitude of the response in the dynamic method is dependent on the sensitivity of the detector to the test gas used and the speed at which the gas is removed by the pump. Thus the leak rate Q is given by

$$Q = SKI = SP \qquad (7.24)$$

where S is the pumping speed, P the pressure, KI is the pressure reading, K is the amplification factor of the detector (gauge) and I the detector current.

When test gas is introduced into a system through leaks at rate Q, the rate of pressure build-up is

$$dP/dt = Q/V \qquad (7.25)$$

where V is the volume of the system. However if simultaneously the test gas is continuously pumped out, the balance is

$$V(dP/dt) = Q - PS \qquad (7.26)$$

By integrating (and for the case that the initial test gas pressure is zero)

$$P = (Q/S) \; [1 - \exp(-St/V)] \qquad (7.27)$$

or

$$t = -(V/S) \; \ln[1 - (PS/Q)] \qquad (7.28)$$

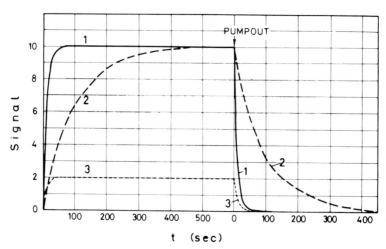

Fig. 7.71 System response time. After Marr (1968).

At long exposure time, the exponential term in eq. (7.27) approaches zero, and the equation reduces to the equilibrium value given by eq. (7.24). The time required to reach equilibrium is shown in fig. 7.71 for three examples. In case of curve 1 ($V = 10$ l; $S = 1$ l/sec; $V/S = 10$ sec) the detector requires 10 seconds to reach $1 - e^{-1} = 0.63$ of the ultimate signal, and the equilibrium signal is attained in approximately one minute. In case 2 ($V = 100$ l; $S = 1$ l/sec; $V/S = 100$ sec), in 10 seconds the signal reaches ($1 - e^{0.1} = 0.095$) only about 10% of the ultimate value, and about 10 minutes are required to reach the ultimate signal.

By increasing the pumping speed as in case 3 ($V = 100$ l; $S = 5$ l/sec; $V/S = 20$ sec) the time to reach the ultimate signal is about the same as in case 1, but the value of this signal is lowered to 20% of its value in case 1.

When removing the test gas from the system (clean-up, pumpdown, pumpout) the process is expressed by

$$dP/dt = PS/V \tag{7.29}$$

thus

$$t = (V/S) \ln (P_t/P_o) \tag{7.30}$$

and

$$P_t/P_o = \exp (-St/V) \tag{7.31}$$

P_t and P_o being the pressures at time t and time zero.

Equation (7.31) is plotted in the right side of fig. 7.71 for the three cases considered. It may be seen that in cases 1 and 3 the *clean-up* is done rapidly, while in case 2 the pump-out takes considerable time.

The response time and the clean-up time are determined by the time constant of the system ($\tau = V/S$), thus the detector response and detector cleaning can be expressed as values valid for any system (table 7.27, fig. 3.37). The *response time* is usually understood as the time required to reach $1 - e^{-1} = 0.63$ of the equilibrium signal, while the *clean-up* time is the time required to reduce (by pumping) the signal to $e^{-1} = 0.37$ of its initial value. The factors which influence the response time of the system have been discussed by Blears and Leck (1951), Goldbach (1956), Florescu (1962a), Lee (1963), Westgaard (1970), Beavis (1970), Young (1970).

To increase the sensitivity it is often suggested that the leak detector be connected between the diffusion pump and the fore pump, thus *on the forepump side* instead of *on high vacuum side* (fig. 7.72). The subject was discussed by Florescu (1962a), Santeler (1963).

The gain in sensitivity depends on the factor which limits the ultimate sensitivity

Table 7.27.
Detector response time and cleaning time.

Fraction of time constant	Detector response Fraction of ultimate signal	Detector cleaning Fraction of starting signal
0.001	0.0009	0.999
0.002	0.0019	0.998
0.004	0.0039	0.996
0.006	0.0059	0.994
0.008	0.0079	0.992
0.01	0.0099	0.990
0.02	0.0198	0.980
0.04	0.0392	0.961
0.06	0.0582	0.942
0.08	0.0769	0.923
0.1	0.095	0.905
0.2	0.181	0.819
0.3	0.259	0.741
0.4	0.329	0.670
0.5	0.393	0.607
0.6	0.451	0.549
0.7	0.503	0.496
0.8	0.550	0.449
0.9	0.593	0.407
1	0.6321	0.3679
2	0.8647	0.1353
3	0.9502	0.0498
4	0.9816	0.0183
5	0.9932	0.0067
6	0.9975	0.0025
7	0.9991	0.0009
8	0.9997	0.0003
9	0.9999	0.0001
10	1.0000	4×10^{-5}

of the test. In a *clean system*, the ultimate sensitivity may be limited by the partial pressure sensitivity of the detector. In this event, the pressure amplification obtained on the forepressuie side will result in a sensitivity gain. In a *contaminated system* or when searching for extremely small leaks in the presence of large leaks, the sensitivity is frequently limited by the resolution of the detector in distinguishing test gas partial pressure from the background. In this event, forepressure leak detection results in amplification of both test signal and background, and unless selective pumping means are employed (e.g. Pd barrier for H_2) no gain in the concentration ratio is realized.

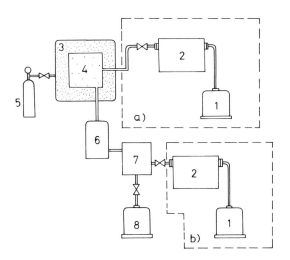

Fig. 7.72 Alternative sites for leak detectors; (a) on high vacuum side; (b) on forepump side. 1. Pump of leak detector; 2. Leak detector; 3. Envelope; 4. System under test; 5. Test gas container; 6. Diffusion pump; 7. Ballast tank; 8. Fore pump. After Marr (1968).

If the leak detector is connected on the forepump side of the diffusion pump, the forepump removes gas from the system by a batch process. The pressure of its inlet side tends to fluctuate, and these fluctuations would contribute to the noise level of the detecting element which is directly connected with the mechanical pump. By placing the detecting element between two diffusion pumps, this effect is greatly reduced. An alternative approach is to place a ballast tank (fig. 7.72) or a throttling valve between the pumps and the detector.

Leak testing at relatively low vacuum (e.g. 0.2 Torr) is carried out by a method in which the system under test is connected to the discharge side of the diffusion pump (between diffusion and fore pump) while the helium leak detector is connected to the high vacuum side of the diffusion pump. This method is described and discussed by Hablanian and Briggs (1977), Hablanian (1980), Reich (1987) and is known as the "counterflow" method. It is based on the fact that the compression ratio of diffusion pumps and turbomolecular pumps is much higher for heavy gases than for light ones. At a compression ratio of 100 for He, compression ratios of 1.75×10^4 for H_2O and 2×10^5 for N_2 are obtained. Hence the enrichment for He in H_2O is $1.75 \times 10^4/100 = 175$, and that of He in N_2 is $2 \times 10^5/100 = 2000$.

7.4.3. Leak location

There are two distinct *techniques of locating leaks:* the use of a *tracer probe* and the use of a *detector probe* (fig. 7.73). In the *tracer probe* technique a stream of test gas is spread on the suspected area, and the gas penetrating into the system is pumped via the detector. In the *detector probe* technique the test gas is filled into the system under test and a detector probe (sniffer) connected to the leak detector is passed over the suspected area to receive the test gas escaping through the leaks.

The tracer probe technique is usually more sensitive than the detector probe technique, nevertheless this latter should be used if (a) the system has to be tested in a pressurized condition; (b) the test gas is one which may be readily absorbed on the leak surfaces; (c) the detector has a sensing element that may be operated under atmospheric pressure. The *tracer probe* technique should be used if (a) the detector sensing head has to be evacuated for use; (b) leak location is performed after dynamic leakage measurement.

A coaxial helium leak detection probe which permits to pinpoint the helium jet is described by Fowler (1987). It consists of a delivery tube for He and a coaxial outer tube which pumps away the excess He, preventing drift to any other leaks in close proximity of the test area.

The problems arising in leak detection of large vacuum systems were discussed by Garrod and Nankivel (1961), King (1962), Lee (1963), Guilbard and Guihery (1967), Westgaard (1970), Moore and Camarillo (1973), Wilson (1975), Hopkins and Valania (1977), Moraw and Prasol (1978), Moore and Walker (1979), Falland (1981), Blanchard *et al.* (1982), Jackson (1982), Holme (1983), Kozman (1983), Katheder and Lennermann (1984), Sänger and Franz (1984), Callis (1985), Wilson (1985), Winkel and Hemmerich (1987), Brooks *et al.* (1988).

Remote leak testing is discussed by Blanchard *et al.* (1982; 1988).

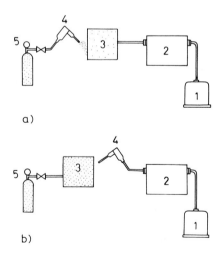

Fig. 7.73 Leak location techniques; (a) tracer probe technique; (b) detector probe technique; 1. Pump; 2. Leak detector; 3. System under test; 4. Probe; 5. Test gas container. After Marr (1968).

7.4.4. *Sealed unit testing*

Sealed units are tested by back-pressurizing. This technique consists of three stages: (a) The application to the external surface of the test specimen of test gas at a high pressure; the gas is flowing in through the leaks. (b) The period between the release of the external test gas pressure and the leak test. (c) The leak test.

If it is considered that the flow through the leaks is molecular, the leak rate is described (Howl and Mann, 1965), by

$$Q = C_A P_E [1 - \exp(-C_A t_E/V)] \exp(-C_A t_R/V) \qquad (7.32)$$

where Q is the measured gas leakage, C_A is the conductance of the leaks for the test gas; P_E is the test gas pressurizing pressure; t_E the pressurizing time; V internal free volume of the system; t_R residence time at one atmosphere after pressurizing.

Fig. 7.74 shows a plot of Q vs. C_A for typical values of P_E, V, t_E and t_R.

For very small leaks (less than 10^{-6} Torr·lit/sec) it is possible to reduce eq. (7.32) to a simple form

$$Q = (P_E t_E/V) C_A^2$$

or

$$P_E t_E = QV/C_A^2 \qquad (7.33)$$

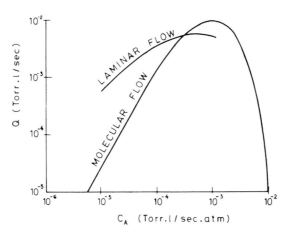

Fig. 7.74 Back-pressurizing. Computed values of leak rate for $P_E=8$ atm; $V=2$ cm³; $t_E=2$h; $t_R=10$ min. After Howl and Mann (1965).

The smallest detectable leak C_A depends upon the minimum leak rate Q, which gives a signal on the detector appreciably above the background. Equation (7.33) defined the value of the product $P_E t_E$ necessary to detect leaks as small as C_A in a system with a minimum detectable signal Q_{min}. The plot fig. 7.75 is based on $Q_{min} = 2.5 \times 10^{-8}$ Torr·lit/sec. The background level of the detector results mainly from the presence of test gas which has been adsorbed on the surface of the system during pressurizing. The amount of this adsorption and of the subsequent desorbtion during the leak test is dependent on the gas used, the material and finish of the surface. As a first step in any back-pressurizing test, the signal from desorbed test gas should be measured experimentally, using a specimen of the material which will be used. The effect of significant adsorption background can be reduced by heating the specimen.

Figure 7.74 also shows a part of the curve calculated (Howl and Mann, 1965) for laminar (viscous) flow, according to

$$Q = \frac{P_E - P_O}{P_E} C_A P_O \left[\frac{1 + \dfrac{P_E - P_O}{P_E + P_O} \exp\left(-2 C_A t_R/V\right)}{1 - \dfrac{P_E - P_O}{P_E + P_O} \exp\left(-2 C_A t_R/V\right)} \right]^2 \qquad (7.34)$$

where P_o is the atmospheric pressure, and other notations are as in eq. (7.32). The problems of leak detection of sealed units were discussed by Thorpe (1968), Westgaard (1970), Beeck (1974), Beeck and Reich (1974), Davy (1975), Briggs and Blumle (1977), Selhofer and Wagner (1977), Rosenthal et al. (1987).

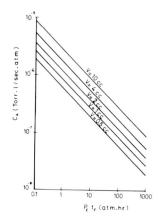

Fig. 7.75 $P_E t_E$ values for detecting minimum leak C_A. After Howl and Mann (1965).

7.4.5. *Sensitive leak detection methods*

Halogen leak detector. This detector was first described by White and Hickey (1948), and makes use of a red-hot (~900°C) platinum filament which emits positive ions. The presence of small traces of halogen vapours (Cl, F, Br, I) increases the emission of positive ions markedly. It is this increase in emission that is measured to indicate the presence of a leak. The detector consists (fig. 7.76) of a platinum cylinder mounted on a ceramic-clad heating element placed centrally within an outer cylinder. The heated inner cylinder is made positive (100–500 V) relative to the outer cylinder, and the ion current is read on a microammeter. The halogen detector is most effectively used as a leak detector by placing it inside the vacuum system and probing the system with a fine jet of Freon 12 or other halogen containing gas.

One feature of the halogen leak detector which can cause difficulty is the relatively long "memory" of the detector once it has been exposed to a surge of halogen gas. To reduce the memory period, the detector head has to be purged with a gas free of halogens. The sensitivity of the halogen leak detector (table 7.26) is appropriate for detecting medium leaks (Drawin, 1959). Halogen leak detection of small parts is discussed by Mennenga (1980).

Detectors using vacuum gauges. These procedures are based on the fact that most vacuum gauges (ionization, thermal conductivity) have a pressure response

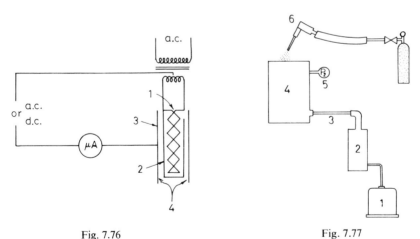

Fig. 7.76   Fig. 7.77

Fig. 7.76 Schematic diagram of the halogen leak detector. 1. Heating element; 2. Platinum cylinder; 3. Outer cylinder; 4. Air flow. After White and Hickey (1948).

Fig. 7.77 Idealized system for vacuum gauge response testing. 1. Backing pump; 2. Diffusion pump; 3. Conductance; 4. System being tested; 5. Gauge; 6. Tracer gas probe. After Marr (1968).

dependent on gas composition (e.g. figs. 6.25, 6.30). If the composition of the gas in a system changes, the reading on the gauge (detector) reflects this change. Leak location therefore consists of spraying a test gas on the suspected leak and observing any response of the gauge to the test gas that enters the system through the leak.

The procedure is very popular for leak location on vacuum systems, because a gauge is usually present on the system. Its major limitation is, that it is directly applicable only to the major leaks in the system. It would be very difficult (if not impossible) to locate a leak 1/100 the size of the total system leakage. The procedure is dependent on a constant pressure in the system. If the system pressure varies for reasons unrelated to testing, leak location using this procedure is impossible. The sensitivity of detection using single Pirani or ionization gauges (table 7.26) is relatively low.

The *principles of operation* (Blears and Leck, 1951) of this technique are the following, considering a system being tested as shown in fig. 7.77. If a leak is present, the system pressure will be P_2

$$Q = P_2 C \tag{7.35}$$

where C is the conductance of the pipes leading to the pumps.

If the flow through the *leak is viscous* the leakage will be inversely proportional (eq. 3.52) to the viscosity of the gas η, thus

$$Q = \frac{C_L}{\eta_a}(P_1^2 - P_2^2) \approx \frac{C_L}{\eta_a}P_1^2 \tag{7.36}$$

where C_L is the conductance of the leak, P_1 is the pressure outside the system (one atmosphere) and η_a is the viscosity of air; $P_1 \gg P_2$. If the concentration of the test gas is x atmospheres $(x<1)$, the concentration of air is $1-x$. The leak rate of the test gas will be

$$Q_x = C_x P_{2x} = \frac{C_L P_1^2}{\eta_x} x \tag{7.37}$$

while that of air will be

$$Q_a = C_a P_{2a} = \frac{C_L P_1^2}{\eta_a}(1-x) \tag{7.38}$$

The total resulting pressure is

$$P_{2a} + P_{2x} = C_L P_1^2 \left(\frac{1-x}{\eta_a C_a} + \frac{x}{\eta_x C_x} \right) \tag{7.39}$$

while the pressure difference due to the presence of the test gas is (from eqs. 7.39 and 7.36)

$$\Delta P_2 = C_L P_1{}^2 x \left(\frac{1}{\eta_x C_x} - \frac{1}{\eta_a C_a} \right) \tag{7.40}$$

If the sensitivity of the gauge to air is K_a and to the test gas K_x, the change of gauge reading ΔG will be

$$\Delta G = C_L P_1{}^2 x \left(\frac{K_x}{\eta_x C_x} - \frac{K_a}{\eta_a C_a} \right) \tag{7.41}$$

From eq. (7.41) it results that the maximum sensitivity is obtained when the test includes:

(a) Complete coverage of the leak by the test gas ($x = 1$);
(b) High sensitivity of the gauge to the test gas (large K_x);
(c) Low value of viscosity (η_x) of the test gas;
(d) A small value of C_x. Since the conductance is inversely proportional to the square root of the molecular weight (eq. 3.92), the test gas should have a high molecular weight.
  The use of search gases in leak detection is discussed by Jansen (1980).

If the leaks are small (10^{-6} atm·cc/sec) *the flow is molecular,* and the leakage Q_x is inversely proportional to the square root of the molecular weight (eq. 3.92) of the gas. The same relationship applies to the conductance C_x which determines the pumping speed of the tubulation. In this case the equation corresponding to eq. (7.40) shows that *the pressure* in the system is *independent of the property of the leaking gas.* The gauge response is then dependent only on the relative sensitivity of the gauge to the test gas as compared to air.
  Since there are a variety of factors involved in choosing a proper gas and gauge combination it is often easier to determine the sensitivity factor as

$$\phi = \left(\frac{\text{pressure caused by tracer gas on the leak}}{\text{pressure on system with air on leak}} \right)$$

which determines the minimum detectable leak

$$Q_{min} = \Delta P_{2a} C_a / \phi \tag{7.42}$$

where ΔP_{2a} is the smallest measurable air pressure variation. Some experimental values of ϕ are listed in table 7.28.

Table 7.28.
Leak testing substitution factors ϕ (Blears and Leck, 1951).

Test gas	Hot cathode ionization gauge	Pirani gauge
Butane	10	1
Diethyl ether	5	0.7
Carbon dioxide	1	0.3
Carbon tetrachloride	1	0.05
Benzine	0.3	0.1
Hydrogen	−0.4	0.4
Coal gas	0.25	0.25

The gases are preferred to the liquids (diethyl ether, carbon tetrachloride and benzine) as these liquids may block the leak or may enter the vacuum system in considerable concentration through a large leak. Hence the *most suitable gas is butane* (table 7.28) which has low viscosity, high molecular weight, good thermal conductivity and high ionization probability. *For an ionization gauge,* the *second best* choice is carbon dioxide, while for a Pirani gauge it is hydrogen (table 7.28). Carbon dioxide has the practical advantage, compared with butane and hydrogen, of being non-inflammable.

If in a certain leak detection plant the conductance (pumping speed) for air is 5 lit/sec, the air pressure fluctuation is 2×10^{-6} Torr, and an ionization gauge is used with butane ($\phi = 10$), the minimum detectable leak is (eq. 7.42)

$$Q_{min} = \frac{2 \times 10^{-6} \times 5}{10} = 10^{-6} \text{ Torr lit/sec}$$

The leak detection with vacuum gauges can use: (*a*) a *single gauge*; (*b*) a single gauge with a barrier that admits only the test gas; (*c*) two identical vacuum gauges in a *differential mounting*.

Single gauge leak detection. The response of the *Pirani gauge* head to a test gas is caused by the change in thermal conductivity (see §§2.7.3; 6.6) compared with that of air. Since the pressure indication of a Pirani gauge is the out-of-balance current of a Wheatstone bridge it is most sensitive to pressure changes when the out-of-balance current is zero. For this reason the control unit is a modification (fig. 7.78) of that used in low pressure measurement, in which the balancing arm resistance can be readily varied. For maximum sensitivity a more sensitive meter is used, and the head is protected from temperature changes by thermal insulation or by use of a compensator head.

Both *hot and cold cathode ion gauges* may be used for leak detection (table 7.26). Their response to a test gas is due to the change in ionization potential.

Barrier leak detection. A method to overcome the difficulty due to pressure fluctuations produced by outgassing is to arrange between the gauge and the vacuum system a barrier which passes only the test gas. Two techniques of achieving this both use hydrogen as the test gas: the *charcoal–Pirani method* and the *palladium barrier method.*

The *charcoal–Pirani* technique depends on the fact that degassed, activated charcoal cooled with liquid nitrogen will adsorb (fig. 4.24) readily atmospheric gases, but will adsorb much less hydrogen. A cooled charcoal trap is therefore inserted in front of the Pirani gauge (table 7.26).

The *palladium–barrier detector* is based on the fact that in the cold state palladium is impermeable to all gases, but when red hot it is highly permeable (fig. 4.12) to hydrogen. The detector head is a hot cathode ion gauge (fig. 7.79) which is evacuated, sealed-off and gettered. In operation the palladium barrier is heated by electron bombardment, the electrons being produced thermally and accelerated toward the positively charged palladium. Hydrogen test gas which enters the system via leaks diffuses through the hot palladium and is ionized by collision with the electron stream. The resulting positive ions are collected by an electrode which is at a negative potential with respect to the cathode. The current flowing in the collector circuit is amplified as in the hot cathode ionization gauge (see table 7.26).

If water or hydrocarbon vapours reach the hot palladium, there is a high probability that they will be dissociated to produce hydrogen. This will cause an erroneous indication of a leak or a high background current. To avoid this effect

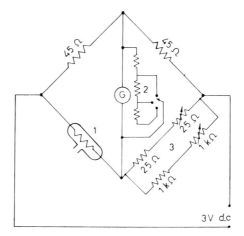

Fig. 7.78 Control circuit for Pirani leak detector; 1. Pirani gauge head; 2. Meter sensitivity control; 3. Balance control.

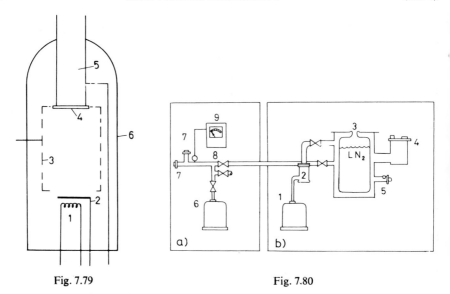

Fig. 7.79   Fig. 7.80

Fig. 7.79 Palladium barrier leak detector head. 1. Heater; 2. Cathode (-100 V); 3. Ion collector (-150 V); 4. Palladium barrier-cap anode (zero potential); 5. Kovar tube; 6. Glass envelope.

Fig. 7.80 Helium mass spectrometer leak detector system. (a) test station; (b) leak detector; 1. Fore pump; 2. Diffusion pump; 3. Liquid nitrogen trap; 4. Mass spectrometer; 5. Gauge; 6. Roughing pump; 7. Test or calibrated leak inlet; 8. Gauge for roughing manifold; 9. Gauge controls. After Marr (1968).

it is essential to use a liquid nitrogen cold trap between the detector and the vapour pumps. At the completion of leak detection the hydrogen is pumped out of the head until the ion current indication is a minimum.

Differential leak detection uses two similar gauges connected in an electrical circuit so that the difference between their readings is registered. Both these gauges (Pirani, ionization) record the residual permanent gas pressure in the system, but one of them has a trap which is selective for the test gas. Liquid nitrogen trap can be used with hydrogen as the test gas, while calcium hydroxide trap at room temperature is suggested for CO_2 as a test gas.

Mass spectrometer leak detectors. In principle any of the types of mass spectrometer described in §6.9 together with any probe gas may be used for leak detection, since the device can be adjusted to respond only to that gas. Although mass spectrometer leak detectors tuned to argon (or hydrogen) are also constructed, the widely used gas is *helium*, for the following reasons: (a) Helium's molecular weight is low thus it gives a high leak rate through small leaks; (b) Helium occurs in the atmosphere at an extent of only $5 \times 10^{-4}\%$ per volume (table 1.3); (c) There is little possibility that an ion from another gas will give an indication that can be mistaken for helium (except deuterium).

A helium mass spectrometer leak detection system (fig. 7.80) consists of: (a) A vacuum pumping system for pumping the spectrometer tube and associated lines; (b) a cold trap for pumping condensable vapours; (c) an appropriate test inlet (vacuum coupling) for connecting the vessel to be tested or the calibrated leak; (d) the mass spectrometer; (e) vacuum gauges controlling that the filament of the mass spectrometer is not "on" at too high pressures; (f) a pumping system to evacuate the vessel under test.

The problems of mass spectrometer leak detectors were discussed by Varadi and Sebestyen (1956), Roberts (1957), Moody (1957), Daly (1960), Cossuta and Stechelmacher (1960), Young (1970), Hain (1972), Hug (1973), Moore and Camarillo (1973), Winkelman and Davidson (1975, 1979), Briggs and Blumle (1977), Hablanian and Briggs (1977), Powell and Mullan (1978), Edwards (1978a), Frunzetti (1980), Falland (1980), Sinharoy and Lange (1982), Solomon (1984), Longsworth and Lahav (1987).

The mass spectrometer helium leak detector can be used for dynamic leakage measurement (§7.4.2), for leak location (§7.4.3) or for sealed unit testing (§7.4.4). Its sensitivity is very high (table 7.26). Calibration of leak detectors was discussed by Moller (1955), and standards are given by the American Vacuum Soc. (1966, 1973b), and the International Standard Organization (ISO, 1978a).

Ion pump as leak detector. Ion pumps can be successfully used as leak detectors in the range of ultra-high vacuum (Young, 1961; Barrington, 1962; Ackley, 1962). For a given type of gas the current drawn by the pump is proportional to the throughput. Thus for gas of type x

$$I_x = Q_x = S_x P_x \tag{7.43}$$

where I is the current drawn for a given throughput Q_x, P_x the resulting partial pressure, and S_x the pumping speed of the pump.

If a leak exists, through which the leak rate of gas 1 is Q_1, the ion-pump current is

$$I = I_0 + I_1 = \left(\frac{I}{SP}\right)_0 Q_0 + \left(\frac{I}{SP}\right)_1 Q_1 \tag{7.44}$$

where Q_0 represents the outgassing load. If at time $t = 0$, gas of type 1 is replaced by gas of type 2, then after a time t the change in the ion pump current is

$$\Delta I(t) = I(t) - I = \left(\frac{I}{SP}\right)_2 Q_2 (1 - \exp[-S_2 t/V])$$

$$= \left(\frac{I}{SP}\right)_1 Q_1 (1 - \exp[-S_1 t/V]) \tag{7.45}$$

Table 7.29.
Parameters for determing the change in getter-ion pump current. After Ackley (1962).

Probe gas	$(I/P)_2/(I/P)_1$	S_2/S_1	Q_2/Q_1	$\Delta I/I$ experimental
Helium	0.167	0.30	2.7	+0.5
Argon	1.25	0.834	0.85	+0.5
Hydrogen	0.50	1.73	3.8	+0.1
Oxygen	1.0	1.25	0.95	−0.5
Carbon dioxide	—	—	—	−0.5

since presumably Q_0 remains constant. After a sufficient time the exponential factors approach zero, and the fractional change in the current is

$$\frac{\Delta I}{I} (t = \infty) = \frac{(I/P)_2}{(I/P)_1} \frac{S_1 Q_2}{S_2 Q_1} - 1 \qquad (7.46)$$

The data for the parameters and the observed values for $\Delta I/I$ are given in table 7.29 (Ackley, 1962).

The value of I/SP for air was found to be about 20. Using the data from table 7.29, and this parameter for air, from eq. (7.46) it results that by substituting helium for air at a leak in the system, the change in current drawn by the ion pump is $\Delta I \simeq 10 \ Q_{air}$ (ΔI is in Amperes; Q_{air} is in Torr·lit/sec). Since electrometer circuits can measure currents of 10^{-12} A, the minimum detectable leak will be theoretically about 10^{-13} Torr·lit/sec. Practical values are in the range $10^{-11}–10^{-12}$ Torr·lit/sec.

Becker (1977) describes a method of using a turbomolecular pump as a leak detector.

7.5. Rules for operating vacuum systems

(1) When starting a mechanical pump, insure first that the rotor moves in the correct direction, and that the level of the oil is correct.

(2) Always vent a mechanical pump to atmospheric pressure when the power is turned off. The presence of a residual vacuum in the pump frequently will cause oil to be drawn into the casing or the system.

(3) Do not permit a mechanical pump to exhaust a high-vacuum system below a pressure of a few hundred microns unless the pump is separated from the high-vacuum chamber with a trap stopping the pump oil vapours from entering the chamber. At lower pressures, the flow is molecular, thus the oil vapour expands towards the chamber, yielding an excessively high vapour pressure hydrocarbon

contamination which may require many hours to remove.

(4) Do not run a mechanical pump at excessively high pressures for continuous periods. The motors are usually not sized for such runs. The pump will eject oil together with the gas.

(5) A diffusion pump should be cooled to a safe intermediate temperature before it is vented to atmosphere. Venting at too high a temperature results in oxidation of the pump fluid and an excessive carryover of the fluid into the mechanical pump.

(6) Check that the cooling water supply for the diffusion pump is turned on prior to the heating. Diffusion pumps should be provided with a thermal protection device to turn the diffusion pump heating off in the event of loss or failure of the cooling water supply.

(7) In liquid-nitrogen trapped systems, the trap should be cool enough to condense the diffusion pump oil prior to turning on the diffusion pump. Maximum backstreaming of the pump oil occurs during the startup and shutdown of the pump.

(8) Do not vent a liquid nitrogen trap to air while cold. Remove all liquid water condensate from the reservoir of a liquid nitrogen trap prior to use, to avoid water freeze-up in the trap. On the initial pumpdown, it is advisable to partially fill the liquid nitrogen trap, so that the principal vacuum system condensabies can be trapped on the lower portion of the trap. After high vacuum has been reached, the trap should be filled to a higher level. A liquid nitrogen trap which has elastomer seals at the top should not be overfilled, as the freezing of the gasket may cause leakage.

(9) When the chamber is vented to atmospheric pressure for a short period of time, it is advisable to use a dry, inert gas such as nitrogen to minimize the moisture adsorption on the surfaces of the vacuum system. Venting should always be on the chamber side, never vent the system from the foreline of the diffusion pump.

(10) Ionization gauges (hot cathode) should not be turned on until it is reasonably certain that the pressure is below 10^{-3} Torr.

Recommended practices for pumping *hazardous gases* (toxic, flammable, corrosive, pyrophoric, etc.) are given in the extensive paper of O'Hanlon and Fraser (1988).

References

A

Abbel, K., J. Henning and H. Lotz, 1982, New turbomolecular pump for radioactive gases, e.g. tritium. Vacuum **32**(10/11), 623.

Abel, B.D., 1978, TFTR (Tokamak Fusion Test Reactor) vacuum system. Proc. 24th Amer. Vacuum Symp., J. Vac. Sci. & Technol. **15**, 726.

Ackley, J.W., 1962, Leak detection using current changes in ionization gauges and sputter-ion pumps. Trans. 9th American Vacuum Symposium (Macmillan, New York) p. 380.

Adam, H., et al., 1980, Vacuum pumps in chemical processing. Vakuum-Technik **29**, 69, 72, 98, 131, 163, 169.

Aggus, J.R., et al., 1977, Characteristics of the ISABELLE vacuum system. IEEE Trans. Nucl. Sci. **NS-24**, 1287.

Akaishi, K., et al., 1987, New partial pressure gauge for helium and deuterium using the ion backscattering principle. J. Vac. Sci. & Technol. A **5**(4), 2444.

Akiyama, Y., K. Nakayama and M. Saito, 1971, Vacuum **21**, 167.

Aldao, C.M., and D.G. Löffler, 1984, Design of chromel–alumel thermocouple gauges. J. Vac. Sci. & Technol. A **2**(4), 1601.

Allensworth, D.I., 1963, Rev. Sci. Instrum. **34**, 448.

Almond, I., 1958, J. Sci. Instrum. **35**, 70.

Alpert, D., 1953, J. Appl. Phys. **24**, 860.

Alpert, D., 1955, Ultra-high vacuum technology. Trans. 2nd American Vacuum Symposium (Pergamon Press, Oxford) p. 69.

Alpert, D., and R.S. Buritz, 1954, J. Appl. Phys. **25**, 202.

Alpert, D., C.G. Matland and A.O. McCourbrey, 1951, Rev. Sci. Instrum. **22**, 370.

Altemose, V.O., 1961, J. Appl. Phys. **32**, 1309.

Altemose, V.O., and A.R. Kacyon, 1979, Vacuum compatibility of machinable glass ceramics. J. Vac. Sci. & Technol. **16**, 951.

Amariglio, I., and M.M. Benarie, 1958, J. Sci. Instrum. **35**, 385.

Amdur, I., and L.A. Guildner, 1957, J. Am. Chem. Soc. **79**, 311.

American Vacuum Society, 1956, Standards for performance ratings of vapour pumps. Trans. 2nd American Vacuum Symposium (Pergamon Press, Oxford) p. 91.

American Vacuum Society, 1958, Glossary of Terms Used in Vacuum Technology (Pergamon Press, New York).

American Vacuum Society, 1966, Standard for helium mass-spectrometer leak detector calibration. J. Vac. Sci. & Technol. **3**, 229.

American Vacuum Society, 1972a, Procedures for calibration of hot filament ionization gauge controls. J. Vac. Sci. & Technol. **9**, 1112.

American Vacuum Society, 1972b, Standard procedure for calibrating gas analyzers of the mass spectrometer type. J. Vac. Sci. & Technol. **9**, 1260.

American Vacuum Society, 1973a, Experimental Vacuum Science and Technology (Dekker, New York) p. 288.

American Vacuum Society, 1973b, Standard for calibration of mass spectrometer type leak detectors. J. Vac. Sci. & Technol. **10**, 568.

Ames, E.E., et al., 1978, Self maintaining high-voltage a.c. vacuum system. J. Vac. Sci. & Technol. **15**, 1568.

Ames, E.E., et al., 1979, Alternating and direct current ion pumping up to 30 kV. J. Vac. Sci. & Technol. **16**, 966.

Amoignon, J., 1957, Le Vide **12**, 176.

Amoignon, I., and I.P. Couillaud, 1969, Le Vide **24**, 181.

Anderson, R.M., 1978, Fundamentals of vacuum technology, Video tape course (Genesys Systems, Palo Alto, CA).

Andrew, D., 1967, A cycloidal path mass spectrometer with wirewound electric field structure. Trans. 3rd Int. Vacuum Congress (Pergamon Press, Oxford) p. 527.

Andrew, D., 1968, The development of sputter-ion pumps. Proc. 4th Int. Vacuum Congress (Institute of Physics, London) p. 325.

Angerer, E., and H. Ebert, 1959, Technische Kunstgriffe bei physikalischen Untersuchungen (Vieweg, Braunschweig) 464pp.

Angerth, B., and Z. Hulek, 1974, The tungsten evaporation limit of hot cathode ionization gauges. Proc. 20th American Vacuum Symposium, J. Vac. Sci. & Technol. **11**, 461.

Arakava, I., and Y. Tuzi, 1986, Temperature dependence of pumping speed for H_2 of a cryosorption pump with a condensed gas layer sorbent. J. Vac. Sci. & Technol. A **4**(3), 293.

Archard, M.H., R. Calder and A. Mathewson, 1979, The effect of bakeout temperature on the electron and ion induced gas desorption coefficient of some technological materials. Vacuum **29**, 53.

Armand, G., and Y. Lejay, 1967, Déformations des micro-géométries des surfaces planes en contact. Report C.E.A. PA-PIEL/RT No. 229 (Saclay, France).

Armand, G., J. Lapujoulade and I. Paigne, 1962, Correlations between leak rate and some phenomena observed in metal-to-metal contact. Trans. 2nd Int. Vacuum Congress (Pergamon Press, Oxford) p. 1091.

Armand, G., Y. Lejay and I. Paigne, 1964, Le Vide 19, 436.

Armstrong, D.E., and N. Blais, 1963, Rev. Sci. Instrum. 34, 440.

Arnold, W., 1970, J. Vac. Sci. & Technol. 7, 191.

Arnold, P.C., and D.G. Bills, 1984, Causes of unstable and nonreproducible sensitivities in Bayard–Alpert ionization gauges. J. Vac. Sci. & Technol. A 2(2), 159.

Aslam, M., 1972, J. Vac. Sci. & Technol. 9, 1191.

Audi, M., and M. DeSimon, 1988, The influence of heavier gases in pumping helium and hydrogen in an ion pump. J. Vac. Sci. & Technol. A 6(3), 1205.

Audi, M., and M. Pierini, 1985, High pressure pumping speed measurement on sputter-ion pumps. J. Vac. Sci. & Technol. A 3(2), 479.

Audi, M., and M. Pierini, 1986, Surface structure and composition profile of sputter-ion pump cathode and anode. J. Vac. Sci. & Technol. A 4(3), 303.

Audi, M., et al., 1987, A new ultrahigh vacuum combination pump. J. Vac. Sci. & Technol. A 5(4), 2587.

Auerbach, D.J., et al., 1978, Uhv application of spring-loaded Teflon seals. Rev. Sci. Instrum. 49, 1518.

Auger, P., 1925, J. Phys. Radium 6, 205.

Austin, W.E., 1969, Vacuum 19, 319.

Austin, W.E., and J.H. Leck, 1972, Vacuum 22, 331.

Autin, B., et al., 1977, CERN 400 GeV proton storage rings with superconducting magnets. IEEE Trans. Nucl. Sci. NS-24, 1187.

Auwärter, M., 1957, Ergebnisse der Hochvakuumtechnik und der Physik dünner Schichten (Wissenschaftliche Verlag, Stuttgart) 282pp.

Axelrod, N.N., 1959, Rev. Sci. Instrum. 30, 944.

B

Bächler, W., 1965, Optimale Ausnutzung des Magnetfeldes bei Ionen-Zerstauberpumpen. Trans. 3rd Int. Vacuum Congress (Pergamon Press, Oxford) p. 609.

Bächler, W.G., and D. Knobloch, 1972, J. Vac. Sci. & Technol. 9, 402.

Bächler, W., R. Frank and E. Usselmann, 1974, A new series of turbomolecular pumps. Trans. 6th Int. Vacuum Congress (Kyoto) p. 13.

Bachman, C.H., 1948, Techniques in Experimental Electronics (Wiley, New York) p. 1–67, 89–140.

Bailitis, E., 1975, On a method for the determination of residence time (in German). Vakuum-Technik 24, 194.

Baker, D., 1962, Vacuum 12, 99.

Baker, P.N., and L. Laurenson, 1972, J. Vac. Sci. & Technol. **9**, 375.

Baker, M.A., L. Holland and D.A.G. Stanton, 1972, The design of rotary pumps and systems to provide clean vacua. Trans. 5th Int. Vacuum Congress, J. Vac. Sci. & Technol. **9**, 412.

Baker, B.G., J.P. Hobson and A.W. Pye, 1973, Further measurements on physical factors influencing accomodation pumps. J. Vac. Sci. & Technol. **10**, 241.

Baker, F., D. Gibson and M. Petersen, 1978, Vacuum **28**, 75.

Bannenberg, J.B., 1966, Operating characteristics of a fast gas valve. Trans. 3rd Int. Vacuum Congress (Pergamon Press, Oxford) p. 193.

Barbour, R., 1968, Glassblowing for Laboratory Technicians (Pergamon Press, Oxford).

Barosi, A., and T.A. Giorgi, 1973, A non-evaporable getter for low temperatures. Vacuum **23**, 15.

Barosi, A., and E. Rabusin, 1974, Zr–Al alloy as a getter for high intensity discharge lamps. Proc. 6th Int. Vacuum Congress, Jpn. J. Appl. Phys., Suppl. 2, Pt. 1, p. 49.

Barr, W.E., and V.J. Anhorn, 1949, Scientific and Industrial Glass Blowing (Instrum. Publ., Pittsburg).

Barre, R.R., and G. Mongodin, 1957, Le Vide **12**, 195.

Barrer, R.M., 1951, Diffusion in and through Solids (Cambridge University Press, New York).

Barrington, A.E., 1962, Rev. Sci. Instrum. **33**, 1045.

Barrington, A.E., 1963, High Vacuum Engineering (Prentice-Hall, Englewood Cliffs, New York) 212pp.

Barrington, A.E., 1965, J. Vac. Sci. & Technol. **2**, 198.

Bartelson, C.L., et al., 1983, The Fermilab Tevatron, Vacuum for a superconductor storage ring. J. Vac. Sci. & Technol. A **1**(2), 187.

Bartness, J.E., and R.M. Georgiadis, 1983, Empirical methods for determination of ionization gauge relative sensitivities for different gases. Vacuum **33**(3), 149.

Barton, D.M., 1953, Vacuum **3**, 51.

Barton, R.S., and R.P. Govier, 1965, J. Vac. Sci. & Technol. **2**, 113.

Barton, R.S., and R.P. Govier, 1968, Some observations on the outgassing of stainless steel after cleaning by various methods. Proc. 4th Int. Vacuum Congress (Institute of Physics, London) p. 775.

Barton, R.S., and R.P. Govier, 1970, Vacuum **20**, 1.

Barz, A., and P. Kocian, 1970, J. Vac. Sci. & Technol. **7**, 200; Le Vide **25**, 172.

Basalaeva, N.I., 1958, Sov. Tech. Phys. **3**, 1027.

Basta, M., et al., 1976, A uhv bakeable diverter valve. J. Phys. E (Sci. Instrum.) **9**, 6.

Bauer, E., 1972, Vacuum **22**, 539.

Bauer, S.H., and P. Jeffers, 1965, J. Phys. Chem. **69**, 3317.

Bayard, R.T., and D. Alpert, 1950, Rev. Sci. Instrum. **21**, 571.

Beams, J.W., D.M. Spitzer and J.P. Wade, 1962, Rev. Sci. Instrum. **33**, 151.

Beaufils, P., and R. Geller, 1968, Le Vide **23**, 67.

Beavis, L.C., 1970, Vacuum **20**, 233.

Beavis, L., 1982, Thermal desorption measurements for estimating bakeout characteristics of vacuum devices. J. Vac. Sci. & Technol. **20**(4), 972.

Beck, A.H., ed., 1964, Handbook of Vacuum Physics. Vol. I. Gases and Vacua (1964, 1966) 419pp.; Vol. II. Physical Electronics (1965, 1968) 598pp.; Vol. III. Technology (1964) 270pp. (Pergamon Press, Oxford).

Becker, J.A., 1958, Solid State Phys. **7**, 379.

Becker, W., 1958, Vakuum-Technik **7**, 149.

Becker, W., 1961, Vakuum-Technik **10**, 223.

Becker, W., 1962, Vacuum **11**, 195.

Becker, W., 1966, Vacuum **16**, 625.

Becker, G.E., 1977, Operation of a cryopumped uhv system. Proc. 23rd American Vacuum Symposium, J. Vac. Sci. & Technol. **14**, 640.

Becker, W., 1977, A novel leak detection with turbomolecular pump. Proc. 7th Int. Vacuum Congress (Dobrozemsky, et al., Vienna) p. 203.

Becker, J.A., and C.D. Hartman, 1953, J. Chem. Phys. **57**, 157.

Becker, W., and J. Henning, 1978a, Problems with turbomolecular pumps in magnetic fields. Proc. 24th American Vacuum Symposium. J. Vac. Sci. & Technol. **15**, 768.

Becker, W., and J. Henning, 1978b, Vakuum-Technik **27**, 6.

Becker, W., and W. Nesseldreher, 1974, Vakuum-Technik **23**, 12.

Becker, W., and W. Nesseldreher, 1976a, Le Vide **31**, 59.

Becker, W., and W. Nesseldreher, 1976b, Vacuum **26**, 277.

Becker, J.A., C.B. Green and G.L. Pearson, 1947, Bell Syst. Tech. J. **26**, 170.

Beckmann, W., 1963, Vacuum **13**, 349.

Beeck, U., 1974, Possibilities and limitations of automatic leak detection below leak rates of 10^{-8} mbar·l/s (in German). Vakuum Technik **23**, 77.

Beeck, U., and G. Reich, 1972, J. Vac. Sci. & Technol. **9**, 126.

Beeck, U., and G. Reich, 1974, Vacuum **24**, 27.

Beeck, U., and G. Reich, 1974, Applicability and limits of automatic leak detection below 10^{-8} Torr l/s. Proc. 6th Int. Vacuum Congress, Jpn. J. Appl. Phys., Suppl. 2, Pt. 1, p. 257.

Behrisch, R., 1964, Ergeb. der exact Naturwissenschaften **35**, 295.

Beijerinck, H.C.W., and N.F. Verster, 1973, Vacuum **23**, 133.

Belk, J.A., 1963, Vacuum Techniques in Metallurgy (Pergamon Press, Oxford) 231pp.

Bell, R.L., 1956, J. Sci. Instrum. **33**, 269.

Bell, P.R., E.C. Moore and J. Wyrick, 1967, Vacuum **17**, 87.

Bennett, W.H., 1950, J. Appl. Phys. **21**, 143.

Bennett, J.R.J., 1987, The vacuum system for a British long baseline gravitational wave detector. J. Vac. Sci. & Technol. A **5**(4), 2363.

Bennett, J.R.J., et al., 1978, Vacuum system for an intense pulsed neutron source at the Rutherford Laboratory. Vacuum **28**, 507.

Benson, J.M., 1957, Thermopile vacuum gauges having transient temperature compensation and direct reading over extended ranges. Trans. 4th American Vacuum Symposium (Pergamon Press, Oxford) p. 87.

Bentley, P.D., 1980, The modern cryopump. Vacuum **30**(4/5), 145.

Bentley, P.D., and B.A. Hands, 1977, The initiation of deposition of gases on cryopumps, Proc. 7th Int. Vacuum Congress (Dobrozemsky, et al., Vienna) p. 73.

Benvenuti, C., 1973, Le Vide **28**, 235.

Benvenuti, C., 1974, Characteristics, advantages and possible applications of condensation cryopumping. J. Vac. Sci. & Technol. **11**, 591.

Benvenuti, C., 1977, Production and measurement of extreme vacua. Proc. 7th Int. Vacuum Congress (Dobrozemsky, et al., Vienna) p. 1.

Benvenuti, C., 1979, Cryopumping for cyclotrons. Trans. 8th Int. Conf. on Cyclotrons. IEEE Trans. Nucl. Sci. **NS-26**, 2128.

Benvenuti, C., and D. Blechschmidt, 1974, The optimization of a tubular condensation cryopump for pressures below 10^{-13} Torr. Proc. 6th Int. Vacuum Congress, Jpn. J. Appl. Phys., Suppl. 2, Pt. 1, p. 77.

Benvenuti, C., and R.S. Calder, 1972, Le Vide, Suppl. No. 157, 29.

Benvenuti, C., and J.C. Decroux, 1972, Le Vide **27**, 243.

Benvenuti, C., and J.C. Decroux, 1977, A linear pump for conductance limited vacuum systems. Proc. 7th Int. Vacuum Congress (Dobrozemsky, et al., Vienna) p. 85.

Benvenuti, C., and M. Firth, 1979, Improved version of the CERN condensation cryopump. Vacuum **29**(11/12), 427.

Benvenuti, C., and F. Francia, 1988, Room temperature pumping characteristics of a Zr–Al non evaporable getter for individual gases. J. Vac. Sci. & Technol. A **6**(4), 2528.

Benvenuti, C., and N. Hilleret, 1979, Cold bore experiments at CERN ISR. Trans. 8th Particle Accel. Conf. IEEE Trans. Nucl. Sci. **NS-26**, 4086.

Benvenuti, C., R.S. Calder and G. Passardi, 1976, Influence of thermal radiation on the vapour pressure of condensed hydrogen between 2 and 4.5 K. J. Vac. Sci. & Technol. **13**, 1172.

Benvenuti, C., R. Calder and N. Hilleret, 1977, A vacuum cold bore test section at the CERN ISR. IEEE Trans. Nucl. Sci. **NS-24**, 1373.

Berceanu, I., and V.K. Ignatovich, 1973, Molecular flow of ultracold neutrons through long tubes, Vacuum **23**, 441.

Bergandt, E., and H. Henning, 1976, Methods for producing uhv (in German). Vakuum-Technik **25**, 131.

Berman, A.S., 1965, Free molecule transmission probabilities. J. Appl. Phys. **36**, 3356.

Berman, A.S., 1969, Free molecule flow in an annulus. J. Appl. Phys. **40**, 4991.

Berman, A., 1974, Vacuum **24**, 241.

Berman, A., 1975, A comprehensive study of ionization gauges using radioactive isotopes. Vacuum **25**, 51.

Berman, A., 1979, Vacuum gauge calibration by the static method. Vacuum **29**(11/12), 417.

Berman, A., 1982, Pumping speed measurement in ionization gauge heads. Vacuum **32**(8), 497.

Berman, A., 1985, Total Pressure Measurements in Vacuum Technology (Academic Press, New York).

Berman, A., and J.K. Fremerey, 1987, Precision calibration of a static pressure divider by means of a spinning rotor gauge. J. Vac. Sci. & Technol. A **5**(4), 2436.

Bernard, G.R., 1970, J. Vac. Sci. & Technol. **7**, 267.

Bernardet, H., and P.S. Choumoff, 1970, Le Vide **25**, 167.

Bernhardt, K.H., 1983, Calculation of the pumping speed of turbomolecular pumps. J. Vac. Sci. & Technol. A **1**(2/1), 136.

Berry, R.W., P.M. Hall and M.T. Harris, 1968, Thin Film Technology (Van Nostrand, Princeton).

Beynon, I.H., 1960, Mass Spectrometry and its Applications to Organic Chemistry (Elsevier, Amsterdam).

Bhatia, M.V., and P.N. Chèremisinov, 1981, Air Movement in Vacuum Devices (Technomic Publishers, Westport, CN).

Bieger, W., et al., 1979, On the influence of magnetic fields on turbomolecular pumps. Vakuum-Technik **28**, 34.

Biguenet, C., A.M. Shroff, P. Mathiez and J. Borossay, 1968, Le Vide **23**, 15.

Bijma, J., and M.L. Drijfholt, 1972, A simple omegatron with high resolving power and small magnet (in German). Vakuum-Technik **21**, 48.

Bills, D.G., 1967, J. Vac. Sci. & Technol. **4**, 149.

Bills, D.G., 1973, J. Vac. Sci. & Technol. **10**, 65.

Bills, D.G., 1979, Sensitivity to various gases of Granville–Phillips Convectron. J. Vac. Sci. & Technol. **16**(6), 2109.

Bills, D.G., and F.C. Allen, 1955, Rev. Sci. Instrum. **26**, 654.

Bills, D.G., *et al.*, 1984, New ionization gauge geometries providing stable and reproducible sensitivities. J. Vac. Sci. & Technol. A **2**(2), 163.

Biram, J.G.S., and G. Burrows, 1964, Vacuum **14**, 221.

Blanc, D., and R. Dagnac, 1964, Vacuum **14**, 145.

Blanchard, W.R., *et al.*, 1982, He leak detection in the presence of deuterium background in Tokamaks. J. Vac. Sci. & Technol. **20**(4), 1162.

Blanchard, W.R., *et al.*, 1988, Proposal for a sensitive leak test telescope. J. Vac. Sci. & Technol. A **6**(2), 235.

Blauth, E.W., 1968a, Vakuum-Technik **17**, 90.

Blauth, E.W., 1968b, Mass spectrometry. Proc. 4th Int. Vacuum Congress (Institute of Physics, London) p. 21.

Blauth, A.W., H.G. Schaffer and W. Heiland, 1970, Z. Angew. Phys. **29**, 70.

Blazek, J., and Z. Hulek, 1977, On the modulation characteristics of Bayard–Alpert gauges. Proc. 7th Int. Vacuum Congress (Dobrozemsky, *et al.*, Vienna) p. 137.

Bleakney, W., and J.A. Hipple, 1938, Phys. Rev. **53**, 521.

Blears, J., 1947, Proc. R. Soc. London A **188**, 62.

Blears, J., and J.H. Leck, 1951, J. Sci. Instrum., Suppl. 1, **28**, 20.

Blears, J., E.I. Greer and I. Nightingale, 1960, Factors determining the ultimate pressure in large high vacuum systems. Trans. 1st. Int. Vacuum Congress (Pergamon Press, Oxford) p. 473.

Blechschmidt, D., 1973, J. Vac. Sci. & Technol. **10**, 376.

Blechschmidt, D., 1974a, A miniature extractor gauge for uhv. J. Vac. Sci. & Technol. **11**, 1160.

Blechschmidt, D., 1974b, Low pressure limitation of the orbitron ionization gauge. Proc. 6th Int. Vacuum Congress, Jpn. J. Appl. Phys., Suppl. 2, Pt. 1, p. 105.

Blechschmidt, D., 1975, New deflected beam gauge for pressures down to 10^{-14} Torr. J. Vac. Sci. & Technol. **12**, 1072.

Blechschmidt, D., 1977, New vacuum techniques for small aperture proton storage rings. IEEE Trans. Nucl. Sci. **NS-24**, 1379.

Blechschmidt, D., 1978, In situ conditioning for proton storage ring vacuum systems. J. Vac. Sci. & Technol. **15**, 1175.

Blechschmidt, D., and W. Unterlerchner, 1977, Prototype vacuum test section for the CERN large storage ring project. Proc. 7th Int. Vacuum Congress (Dobrozemsky, *et al.*, Vienna) p. 351.

Blinov, I.G., and V.E. Minaichev, 1974, Condensation cryopump with independent cryogenerator. Proc. 6th Int. Vacuum Congress, Jpn. J. Appl. Phys., Suppl. 2, Pt. 1, p. 101.

Bloomer, R.N., 1973, Vacuum **23**, 239.

Boerboom, A.J.H., A.P. Jongh and J. Kistemaker, 1964, Rev. Sci. Instrum. **35**, 301.

Boers, A.L., 1968, Pumping speed of molecular sieves in the forevacuum region. Proc. 4th Int. Vacuum Congress (Institute of Physics, London) p. 393.

Boffito, C., et al., 1987, Gettering in cryogenic applications. J. Vac. Sci. & Technol. A **5**(6), 3442.

Boissin, J.C., and J.J. Thibault, 1971, Le Vide **26**, 101.

Boissin, J.C., J.J. Thibault and A. Richardt, 1972, Le Vide, Suppl. No. 157, 103.

Boll, H.J., 1961, Rev. Sci. Instrum. **32**, 1415.

Bolz, R.W., ed., 1958, Metals Engineering Processes (McGraw-Hill, New York).

Borghi, M., and B. Ferrario, 1977, Use of nonevaporable getter pumps in experimental fusion reactors. Proc. 23rd American Vacuum Symposium, J. Vac. Sci. & Technol. **14**, 570.

Borghi, M., and L. Rosai, 1979, Gas evolution and gettering in high pressure discharge lamps. Vacuum **29**, 67.

Bostic, D., et al., 1975, Vacuum system for the Stanford-LBL storage ring (PEP). IEEE Trans. Nucl. Sci. **NS-22**, 1540.

Bottiglioni, M.F., et al., 1977, Semi automatic cryogenic pump for a particle accelerator (in French). Le Vide **32**, 81.

Boulassier, J.C., 1959, Le Vide **14**, 39.

Boulloud, J.P., and J. Schweitzer, 1959, Le Vide **14**, 241.

Bourgeois, M., et al., 1987, The large electron positron storage ring preinjector vacuum systems. J. Vac. Sci. & Technol. A **5**(4), 2346.

Boutry, G.A., 1962, Physique Appliquee aux Industries du Vide et de l'Electronique (Masson, Paris) 388pp.

Bouwman, R., et al., 1978, Surface cleaning low temperature bombardment with hydrogen particles. J. Vac. Sci. & Technol. **15**, 91.

Bouyer, P., C. Cassignol and P. Lazeyras, 1960, Le Vide **15**, 297.

Brand, J.F.J. van den, and A.P. Kaan, 1980, Cleaning of Al-alloys. Vacuum **30**(7), 249.

Breakwell, P.R., and P.J. Nash, 1987, Spinning rotor gauge carousel. J. Vac. Sci. & Technol. A **5**(5), 2979.

Breckenridge, R.A., and R.A. Russel, 1986, Space station technology experiments and uses. J. Vac. Sci. & Technol. A **4**(3), 281.

Brennan, D., and M.J. Graham, 1965, Philos. Trans. R. Soc. A **258**, 325.

Brennan, D., and F.H. Hayes, 1965, Philos. Trans. R. Soc. A **258**, 347.

Bretschi, J., 1975, A single crystal semiconductor vacuum gauge with extended range. J. Phys. E (Sci. Instrum.) **9**, 261.

Bridge, H., 1960, Some vacuum problems at low temperatures. Adv. Vacuum Science and Technology (Pergamon Press, Oxford) p. 481.

Bridwell, M.C., and J.G. Rodes, 1985, History of the modern cryopump. J. Vac. Sci. & Technol. A **3**(2), 472.

Briggs, W.E., 1959, Leak detection techniques. Trans. 5th American Vacuum Symposium (Pergamon Press, Oxford) p. 129.

Briggs, W.E., and L.J. Blumle, 1977, Frontiers of leak detection. Proc. 23rd American Vacuum Symposium, J. Vac. Sci. & Technol. **14**, 611.

Briggs, D., et al., 1980, Ultrahigh vacuum system for ISABELLE full cell. Proc. 26th American Vacuum Symposium, J. Vac. Sci. & Technol. **17**, 342.

Brombacher, W.G., 1961, Bibliography and index on Vacuum and Low Pressure Measurement (Nat. Bur. Standards) Monography No. 35.

Brombacher, W.G., 1967, Bibliography on low pressure measurement (Nat. Bur. Standards) Techn. Note No. 298.

Brooks, N.H., C. Baxi and P. Anderson, 1988, Leak detection on the DIII-D Tokamak using helium entrainment techniques. J. Vac. Sci. & Technol. A **6**(3), 1222.

Brown, F.C., 1959, Basic techniques in design and construction of vacuum plant. Trans. 5th American Vacuum Symposium (Pergamon Press, Oxford) p. 93.

Brown, R.D., 1967, Vacuum **17**, 505.

Brownell, L.E., and E.H. Young, 1959, Process Equipment Design (Wiley, New York).

Brueschke, E.E., 1961, Vacuum **11**, 255; Rev. Sci. Instrum. **32**, 732.

Brunauer, S., 1945, The Adsorption of Gases and Vapours (Princeton University Press).

Brunauer, S., P.H. Emmett and E. Teller, 1938, J. Am. Chem. Soc. **60**, 309.

Brunee, C., 1960, Ztschr. Instrkde **68**(5), 97.

Brunner, W.F., and T.H. Batzer, 1965, Practical Vacuum Techniques (Reinhold, New York) 197pp.

Bryant, P.J., and C.M. Gosselin, 1966, J. Vac. Sci. & Technol. **3**, 350.

Bryant, P.J., W.W. Longley and C.M. Gosselin, 1966, J. Vac. Sci. & Technol. **3**, 62.

Buch, S., 1962, Einführung in die Allgemeine Vakuumtechnik (Wissenschaftliche Verl., Stuttgart) 207pp.

Buck, T.M., and J.M. Poate, 1974, J. Vac. Sci. & Technol. **11**, 289.

Buckingham, J.D., 1976, The calibration of vacuum measuring instruments for industrial processes. Vacuum **26**, 143.

Buckingham, J.D., and A. Holme, 1977, The design and performance of a medical/industrial gas analyser system. Proc. 7th Int. Vacuum Congress (Dobrozemsky, et al., Vienna) p. 181.

Buckley, O.E., 1916, Proc. Natl. Acad. Sci. (USA) **2**, 683.

Buckley, D.H., 1974, Adhesion, friction, wear and lubrication in vacuum. Proc.

6th Int. Vacuum Congress, Jpn. J. Appl. Phys., Suppl. 2, Pt. 1, p. 297.

Buckman, J.S., et al., 1984, The operation of a capacitance manometer at 300°C. J. Vac. Sci. & Technol. A 2(4), 1599.

Budgen, L.J., 1982, Mechanical booster for pumping radioactive or dangerous gases. Vacuum 32(10/11), 627.

Budgen, L.J., 1983, Developments in transmission for mechanical booster pumps. J. Vac. Sci. & Technol. A 1(2/1), 147.

Buhl, R., 1981, Catalyzer trap. Vakuum-Technik 30(6), 166.

Bültemann, H., 1965, Vakuum-Technik 14, 132.

Bültemann, H.J., and L. Delgmann, 1965, Vacuum 15, 301.

Bunshah, R.F., 1958, Vacuum Metallurgy (Reinhold, New York) 472pp.

Bürger, H.D., 1983, Advanced operation of Roots pumps. Vakuum-Technik 32(5), 140.

Burrows, G., 1960, Molecular Distillation (Oxford University Press).

Burrows, G., 1973, Vacuum 23, 353.

Butler, R.S., B.R.F. Kendall and S.M. Rossnagel, 1977, The Brownian-motion pressure gauge. Vacuum 27, 589.

C

Cable, J.W., 1960, Vacuum Processing in Metalworking (Reinhold, New York) 202pp.

Calcatelli, A., et al., 1974a, A comparative study of the metrological characteristics of a fixed and variable conductance dynamic system for vacuum gauge calibration. Proc. 6th Int. Vacuum Congress, Jpn. J. Appl. Phys., Suppl. 2, Pt. 1, p. 127.

Calcatelli, A., et al., 1974b, Standards systems in vacuum and development trends. Proc. 6th Int. Vacuum Congress, Jpn. J. Appl. Phys., Suppl. 2, Pt. 1, p. 131.

Calder, R., 1974, Ion induced gas desorption problems in the ISR. Vacuum 24, 437.

Calder, R., and G. Lewin, 1967, Br. J. Appl. Phys. 18, 1459.

Calder, R., et al., 1977, Cleaning and surface analysis of stainless steel uhv chambers by argon gas discharge. Proc. 7th Int. Vacuum Congress (Dobrozemsky, et al., Vienna) p. 231.

Callis, R.W., 1985, Depressurization as a means of leak checking large vacuum vessels. J. Vac. Sci. & Technol. A 3(2), 538.

Cameron, A.E., and D.F. Eggers, 1948, Rev. Sci. Instrum. **19**, 605.

Caporiccio, G., and R.A. Steenrod Jr, 1978, Properties and use of perfluoro-polyethers for vacuum applications. Proc. 24th American Vacuum Symposium. J. Vac. Sci. & Technol. **15**, 775.

Carette, J.D., et al., 1983, Calculation of the molecular flow conductance of a straight cylinder. J. Vac. Sci. & Technol. A **1**(2/1), 143.

Carlson, G.L. Weissler and R.W., eds.., 1979, Vacuum Physics and Technology (Academic Press, New York).

Carpenter, L.G., and M.J. Watts, 1982, A note on the design of vapour traps for diffusion pumps. Vacuum **32**(5), 307.

Carter, G., D.G. Armour and L. Chernatony, 1973, Vacuum **23**, 85.

Carter, G., D.G. Armour and U.Z. Funuki, 1975, The solid state diffusion pump. Vacuum **25**, 315.

Casaro, F., 1988, Analysis of frequency response for different rotor configurations of small turbomolecular pumps. J. Vac. Sci. & Technol. A **6**(3), 1192.

Castaner, B.G., 1966, Introduction a la Tecnica de Alto Vacio (in Spanish) (Public. Junta Energia Nuclear Madrid) 230pp.

Caswell, H.L., 1960, An oil-free ultra-high vacuum system. Trans. 6th American Vacuum Symposium (Pergamon Press, Oxford) p. 66.

Cecchi, J.L., 1979, Tritium permeation and wall loading in the TFTR (Tokamak) vacuum vessel. J. Vac. Sci. & Technol. **16**, 58.

Cecchi, J.L., and R.J. Knize, 1984, Gettering in fusion devices. J. Vac. Sci. & Technol. A **2**, 1214.

Cecchi, J.L., et al., 1985, Technique for "in vacuo" passivization of Zr–Al alloy bulk getters. J. Vac. Sci. & Technol. A **3**(2), 487.

Cespiro, Z., 1973, Vacuum **23**, 277.

Champeix, R., 1958, Eléments de Technique du Vide (Dunod, Paris) 214pp.

Champeix, R., 1965, Le Vide (Hachette, Paris).

Champion, K.S.W., 1969, Review of the properties of the lower thermosphere in Space Research, IX (North-Holland, Amsterdam) p. 461.

Chang, C.C., 1971, Surf. Sci. **25**, 53.

Chapman, R., and J.P. Hobson, 1979, Sensitivity to various gases of Granville–Phillips vacuum gauge, series 275, Convectron. J. Vac. Sci. & Technol. **16**, 965.

Chapman, R., and J.P. Hobson, 1983, Solar powered magnetron pump/gauge. J. Vac. Sci. & Technol. A **1**(2/1), 133.

Chapman, R., and J.P. Hobson, 1985, Pumping of methane by a low power hot cathode ion pump/gauge, J. Vac. Sci. & Technol. A **3**(2), 476.

Chavet, I., 1976, Rotating variable leak valve. Vacuum **26**, 203.

Chen, J.R., 1987a, A comparison of outgassing rate of 304 stainless steel and

A6063-Ex aluminum alloy vacuum chamber after filling with water. J. Vac. Sci. & Technol. A **5**(2), 262.

Chen, J.Z., and C.D. Suen, 1982, An axial emission ultrahigh vacuum gauge. J. Vac. Sci. & Technol. **20**(1), 88.

Chen, J.R., et al., 1985, Thermal outgassing from Al alloy vacuum chambers. J. Vac. Sci. & Technol. A **3**(6), 2188.

Chen, J.Z., C.D. Suen and Y.H. Kuo, 1987a, An axial-emission magnetron suppressor gauge. J. Vac. Sci. & Technol. A **5**(4), 2373.

Chen, J.R., et al., 1987b, Outgassing behavior of A6063-Ex Al alloy and SUS 304 stainless steel, J. Vac. Sci. & Technol. A **5**(6), 3422.

Chernatony, L., 1977a, Recent advances in elastomer technology for uhv applications. Vacuum **27**, 605.

Chernatony, L., 1977b, Some characteristics of a novel uhv compatible elastomer. Proc. 7th Int. Vacuum Congress (Dobrozemsky, et al., Vienna) p. 255.

Cho, Y., et al., 1977, A cold-bore vacuum system design for POPAE. IEEE Trans. Nucl. Sci. **NS-24**, 1293.

Chou, T.S., 1987, Design studies of distributed ion pumps. J. Vac. Sci. & Technol. A **5**(6), 3446.

Chou, T.S., and H.J. Halama, 1977, Cryopumping of deuterium, hydrogen and helium mixtures on smooth 4.2 K surfaces. Proc. 7th Int. Vacuum Congress (Dobrozemsky, et al., Vienna) p. 65.

Chou, T.S., and H.J. Halama, 1979, Effect of neutron radiation of 4.2 K cryo-condensation panels. J. Vac. Sci. & Technol. **16**, 81.

Chou, T.S., and Z.Q. Tang, 1986, Investigation on the low pressure limit of the Bayard–Alpert gauge. J. Vac. Sci. & Technol. A **4**(5), 2280.

Choumoff, P.S., and H. Bernardet, 1970, J. Vac. Sci. & Technol. **7**, 270.

Choumoff, P.S., and B. Japteff, 1974, High pressure ionization gauge for calibration in the 10^{-7} to 10^{-1} Torr range. Proc. 6th Int. Vacuum Congress, Jpn. J. Appl. Phys., Suppl. 2, Pt. 1, p. 143.

Christian, R.G., 1966, Vacuum **16**, 175.

Christian, R.G., and J.H. Leck, 1966, J. Sci. Instrum. **43**, 229.

Chu, J.G., 1988, A new Hybrid molecular pump with large throughput. J. Vac. Sci. & Technol. A **6**(3), 1202.

Chubb, J.N., 1970, Vacuum **20**, 477.

Clausing, P., 1932, Ann. Physik **12**, 961, translated: The flow of highly rarefied gases through tubes of arbitrary length. J. Vac. Sci. & Technol. **8**, 636 (1971).

Clausing, R.E., 1962, A large scale getter pumping equipment using vapour deposited titanium films. Trans. 2nd Int. Vacuum Congress (Pergamon Press, Oxford) p. 345.

Clausing, R.E., et al., 1976, J. Vac. Sci. & Technol. **13**, 437.

Clausing, R.E., et al., 1979, Versatile uhv sample transfer system. J. Vac. Sci. & Technol. **16**, 708.

Clausing, R.E., et al., 1980, 15 cm diameter ultrahigh vacuum transfer system for remote plasma-wall interaction experiments. J. Vac. Sci. & Technol. **17**(3), 709.

Clay, F.P., and L.T. Melfi, 1966, J. Vac. Sci. & Technol. **3**, 167.

Cleaver, J.S., 1967, J. Sci. Instrum. **44**, 969.

Cleaver, J.S., and W.H. Zakrzewski, 1968, Vacuum **18**, 73.

Close, K.J., and R.S. Vaughan-Watkins, 1976, Reducing the effect of ion pumping on very low pressure measurements by a hot cathode ionization gauge. Vacuum **26**, 23.

Close, K., and J. Yarwood, 1970, Vacuum **20**, 56.

Close, K.J., R.S. Vaughan-Watkins and J. Yarwood, 1977, The design and study of a compact bakeable uhv system for calibrating absolutely low pressure gauges. Vacuum **27**, 511.

Close, K.J., D. Lane and J. Yarwood, 1979, Thermal effects on hot-cathode ionization gauges. Vacuum **29**, 249.

Coates, D.J.G., et al., 1977, Vacuum **27**, 531.

Coenraads, C.N., and I.E. Lavelle, 1962, Rev. Sci. Instrum. **33**, 879.

Cohen, S.A., 1976, Vacuum and wall problems in precursor reactor Tokamaks. Proc. 22nd American Vacuum Symposium. J. Vac. Sci. & Technol. **13**, 449.

Cohen, A., 1986, Space technology today. J. Vac. Sci. & Technol. A **4**(3), 263.

Cole, M., 1987, A new type of vacuum pump for corrosive gases and vapors. J. Vac. Sci. & Technol. A **5**(4), 2620.

Colgate, S.O., and P.A. Genre, 1968, Vacuum **18**, 553.

Colligon, J.S., 1961, Vacuum **11**, 272.

Colnot, P., and G. Gallet, 1962, Le Verre et la Céramique dans la Technique du Vide (Gauthier-Villars, Paris).

Colombani, A., and G. Ranc, 1963, CR. Acad. Sci. (France) **257**(10), 1682.

Combley, L.A., and C.J. Milner, 1960, Rev. Sci. Instrum. **31**, 776.

Comsa, G., and C. Simionescu, 1966, Bakeable metal valve with copper gasket. Trans. 3rd Int. Vacuum Congress (Pergamon Press, Oxford) p. 199.

Comsa, G., J.K. Fremerey and B. Lindenau, 1977, Tangential momentum transfer in spinning rotor molecular gauges. Proc. 7th Int. Vacuum Congress (Dobrozemsky, et al., Vienna) p. 157.

Comsa, G., et al., 1980, Calibration of a spinning-rotor gas friction gauge. J. Vac. Sci. & Technol. **17**(2), 642.

Cope, J.O., 1958, Rev. Sci. Instrum. **29**, 232.

Cossuta, D., and W. Steckelmacher, 1960, J. Sci. Instrum. **37**, 404.

Cost, J.R., and R.G. Hickman, 1975, Helium release from various metals. Proc.

21st American Vacuum Symposium, J. Vac. Sci. & Technol. **12**, 516.

Coupland, J.R., *et al.*, 1982, Experimental performance of an open structure cryopump. Vacuum **32**(10/11), 613.

Coupland, J.R., *et al.*, 1987, Experimental performance of a large-scale cryosorption pump. J. Vac. Sci. & Technol. A **5**(4), 2563.

Courtois, G., and M. Gasnier, 1963, Microtechnic **17**, 27, 67.

Craig, R.D., and E.H. Harden, 1966, Vacuum **16**, 67.

Crawley, D.J., and L. de Csernatony, 1964, Vacuum **14**, 7.

Crawley, D.J., E.D. Tolmie and A.R. Huntress, 1963, Evaluation of new silicone fluid for diffusion pumps. Trans. 9th American Vacuum Symposium (Macmillan, New York) p. 399.

Cruz, G.E., *et al.*, 1987, A vacuum-to-air interface for the advanced test accelerator. J. Vac. Sci. & Technol. A **5**(4), 2352.

Csernatony, L., 1966, Vacuum **16**, 427.

Cullingford, H.S., and J.W. Beal, 1977, Pumping and vacuum requirements for the magnetic fusion energy program. Proc. 23rd American Vacuum Symposium, J. Vac. Sci. & Technol. **14**, 567.

Culton, J.W., and R.N. Peacock, 1970, J. Vac. Sci. & Technol. **7**, 188.

Cummings, U., *et al.*, 1971, Vacuum system for Stanford storage ring SPEAR. Proc. 17th American Vacuum Symposium, J. Vac. Sci. & Technol. **8**, 348.

Cunningham, S.L., and W.H. Weinberg, 1978, Rev. Sci. Instrum. **49**, 752.

Curien, H., and A. Rolfo, 1983, Vacuum and space projects. 9th Int. Vacuum Congress (Madrid) p. 258.

Currington, I., *et al.*, 1982, Mechanical vacuum pumping: corrosive gases. J. Vac. Sci. & Technol. **20**(4), 1019.

Curtis, F.W., 1950, High frequency induction heating (McGraw-Hill, New York).

Cyranski, R., and J.H. Leck, 1976, Pumping characteristics of a vacuum system using a rotary pump and sublimation pump. Vacuum **26**, 371.

Czanderna, A.W., and T.M. Thomas, 1987, A quartz-crystal microbalance apparatus for water sorption by polymers. J. Vac. Sci. & Technol. A **5**(4), 2412.

D

Da, D., and X. Da, 1987, Kinetic theory of gas molecules in an extrahigh vacuum. J. Vac. Sci. & Technol. A **5**(4), 2484.

Dadyburjor, D.B., and S.I. Sandler, 1976, Effect of gas-surface interaction on thermal transpiration. J. Vac. Sci. & Technol. **13**, 985.

Dallos, A., and F. Steinrisser, 1967, J. Vac. Sci. & Technol. **4**, 6.

Daly, N.R., 1960, Rev. Sci. Instrum. **31**, 720.

Danielson, P.M., 1970, J. Vac. Sci. & Technol. **7**, 527.

Danilin, B.S., 1959, Construction of Vacuum Systems (in Russian) (Gosenergoizdat, Moscow).

D'Anna, E., et al., 1987, Fragmentation spectra of vacuum pump fluids. J. Vac. Sci. & Technol. A **5**(6), 3436.

Danziger, S., et al., 1982, Pressure pulsations above turbomolecular pumps. J. Vac. Sci. & Technol. **21**(3), 893.

Das, D.K., 1962, Outgassing characteristics of various materials in an ultra-high vacuum environment. Report AEDC, TDR-62-19.

Dauphin, J., 1982, Materials in space. Vacuum **32**(10/11), 669.

Davey, G., 1976, Cryopumping in the transition and continuum pressure regions. Vacuum **26**, 17.

Davis, D.H., 1960, J. Appl. Phys. **31**, 1169.

Davis, W.D., 1962, Sputter-ion pumping and partial pressure measurements below 10^{-11} Torr. Trans. 9th American Vacuum Symposium (Macmillan, New York) p. 363.

Davis, W.D., 1968, J. Vac. Sci. & Technol. **5**, 23.

Davis, R.H., and A.S. Divatia, 1954, Rev. Sci. Instrum. **25**, 1193.

Davis, W.D., and T.A. Vanderslice, 1961, A sensitive high-speed mass spectrometer for ultra-high vacuum work. Trans. 7th American Vacuum Symposium (Pergamon Press, Oxford) p. 417.

Davison, S.G., ed., 1971, Progress in Surface Science (Pergamon Press, Oxford).

Davy, J.R., 1951, Industrial High Vacuum (Pitman and Sons, London) 243pp.

Davy, J.G., 1975, Model calculations for maximum allowable leak rates of hermetic packages. Proc. 21st American Vacuum Symposium, J. Vac. Sci. & Technol. **12**, 423.

Dawe, R.A., 1973, A slip correction in accurate viscosity. Rev. Sci. Instrum. **44**, 1271.

Dawson, P.H., ed., 1976, Quadrupole Mass Spectrometry and its Applications (Elsevier, Amsterdam).

Dawson, P.H., 1977, Limiting factors in quadrupole mass filter design. Proc. 7th Int. Vacuum Congress (Dobrozemsky, et al., Vienna) p. 173.

Dawson, J.P., and J.D. Haygood, 1965, Cryogenics **5**, 57.

Dawson, P.H., and N.R. Whetten, 1969, J. Vac. Sci. & Technol. **6**, 97.

Dawton, R.H., 1957, Br. J. Appl. Phys. **8**, 414.

Dayton, B.B., 1960, Relations between size of vacuum chamber, outgassing rate and required pumping speed. Trans. 6th American Vacuum Symposium (Pergamon Press, Oxford) p. 101.

Dayton, B.B., 1962, Outgassing rate of contaminated metal surfaces. Trans. 2nd Int. Vacuum Congress (Pergamon Press, Oxford) p. 42.

Dayton, B.B., 1963, The effect of bake-out on the degassing of metals. Trans. 9th American Vacuum Symposium (Macmillan, New York) p. 293.

Dayton, B.B., 1968, Problems in vacuum physics influencing the development of standard measuring techniques. Proc. 4th Int. Vacuum Congress (Institute of Physics, London) p. 57.

De Luca, P.P., 1977, Freeze drying of pharmaceuticals. Proc. 23rd American Vacuum Symposium. J. Vac. Sci. & Technol. 14, 620.

De Marcus, W.C., and E.H. Hopper, 1955, Knudsen flow through a circular capillary. J. Chem. Phys. 23, 1344.

De Poorter, G.L., and A.W. Searcy, 1963, J. Chem. Phys. 39, 925.

Dean, N.R., et al., 1978, Glow discharge processing versus bakeout for aluminum storage ring vacuum chambers. Proc. 24th American Vacuum Symposium, J. Vac. Sci. & Technol. 15, 758.

Debe, M.K., 1986, Industrial materials processing on board the Space Shuttle Orbiter. J. Vac. Sci. & Technol. A 4(3), 273.

Debe, M.K., et al., 1987, Vacuum outgassing and gas phase thermal conduction of a microgravity physical vapor transport experiment. J. Vac. Sci. & Technol. A 5(4), 2406.

Degras, D.A., 1968, La chemisorption sur les solides. Proc. 4th Int. Vacuum Congress (Institute of Physics, London) p. 89.

Deichelbohrer, P.R., 1973, Four models of electron distribution in the Orbitron. J. Vac. Sci. & Technol. 10, 875.

Dekker, A.J., 1958, Solid State Phys. 6, 251.

Delafosse, J., and G. Mongodin, 1961, Les Calculs de la Technique du Vide (Le Vide, Paris) p. 107.

Delafosse, D., P. Noe and G. Troadec, 1960, Le Vide 15, 442.

Delbart, R., 1967, Un micromanomètre à haute sensibilité. Trans. 3rd Int. Vacuum Congress (Pergamon Press, Oxford) p. 255.

Della Porta, P., 1960, Vacuum 10, 182.

Della Porta, P., 1972, J. Vac. Sci. & Technol. 9, 532.

Della Porta, P., and E. Rabusin, 1974, Mercury dispensing and gettering in fluorescent lamps. Proc. 6th Int. Vacuum Congress, Jpn. J. Appl. Phys., Suppl. 2, Pt. 1, p. 45.

Dempster, A.J., 1918, Phys. Rev. 3, 316.

Denison, D.R., 1967, J. Vac. Sci. & Technol. 4, 156.

Denison, D.R., 1974, A Monte-Carlo study of the 3-gauge pumping speed test dome. Proc. 6th Int. Vacuum Congress, Jpn. J. Appl. Phys., Suppl. 2, Pt. 1, p. 155.

Denison, D.R., 1975, J. Vac. Sci. & Technol. 12, 548.

Denison, D.R., 1977a, Comparison of diode and triode sputter-ion pumps. Proc.

23rd American Vacuum Symposium, J. Vac. Sci. & Technol. **14**, 633.

Denison, D.R., 1977b, Characteristics of cryo- and cryo/ion pumped vacuum systems. Proc. 7th Int. Vacuum Congress (Dobrozemsky, et al., Vienna) p. 69.

Denison, D.R., and S. McKee, 1974, J. Vac. Sci. & Technol. **11**, 337.

Dennis, N.T.M., and T.A. Heppel, 1968, Vacuum System Design (Chapman and Hall, London) 223pp.

Dennis, N.T.M., et al., 1982a, Factors influencing the ultimate vacuum of single structure pumping groups. J. Vac. Sci. & Technol. **20**(4), 996.

Dennis, N.T.M., et al., 1982b, The effect of inlet valve on the ultimate vacuum above integrated pumping groups. Vacuum **32**(10/11), 631.

Dennison, R.W., and G.R. Gray, 1979, Cryogenic versus turbomolecular pumping in a sputtering application. Proc. 25th American Vacuum Symposium, J. Vac. Sci. & Technol. **16**, 728.

Deters, L., et al., 1987, Applications of turbomolecular pumps in manufacturing plants. J. Vac. Sci. & Technol. A **5**(4), 2367.

Diels, K., and R.J. Jaeckel, 1966, Leybold Vacuum Handbook (Pergamon Press, Oxford) 360pp.

Dillow, C.F., and J. Palacios, 1979, Cryogenic pumping of He, He_2, and a 90% H_2–10% He mixture. J. Vac. Sci. & Technol. **16**, 731.

Dobrott, J.R., and R.M. Oman, 1970, J. Vac. Sci. & Technol. **7**, 214.

Dobrozemsky, R., 1973, An improved cryosorption pump with high pumping speed, low final pressure and small liquified gas consumption (in German). Vakuum-Technik **22**, 41.

Dobrozemsky, R., and R. Moraw, 1971, Vacuum **21**, 587.

Dobrozemsky, R., and G. Moraw, 1974, Cryosorption pumping at 77K from atmospheric pressure to 10^{-7} Torr. Proc. 6th Int. Vacuum Congress, Jpn. J. Appl. Phys., Suppl. 2, Pt. 1, p. 97.

Doré, R., 1962, Le Vide **17**, 208.

Douglas, R.A., J. Zabritski and R.G. Herb, 1965, Rev. Sci. Instrum. **36**, 1.

Downing, J.R., and G. Mellen, 1946, Rev. Sci. Instrum. **17**, 218.

Drawin, H.W., 1958, Vakuum-Technik **7**, 177.

Drawin, H.W., 1959, Vakuum-Technik **8**, 215.

Drawin, H.W., 1965, Vacuum **15**, 99.

Dubois, L.H., 1988, Diffusion pumped ultrahigh vacuum systems. J. Vac. Sci. & Technol. A **6**(1), 162.

Dubrovin, J., 1933, Instruments **6**, 194.

Ducros, P., 1968, L'étude des surfaces par la diffraction des electrons lents. Proc. 4th Int. Vacuum Congress (Institute of Physics, London) p. 198.

Dunkel, M., 1975, Pumps in the early vacuum technology (in German). Vakuum-Technik **24**, 133.

Dunlap, G.C., and H.G. Trump, 1937, Rev. Sci. Instrum. **8**, 37.

Dunoyer, L., 1951, Le Vide et ses Applications (Presses Universitaires de France, Paris).

Durm, M., and K. Starke, 1972, Vakuum-Technik **21**, 11.

Durm, M., G. Stark and K. Starke, 1972, Vakuum-Technik **21**, 111.

Dushman, S., 1922, High Vacuum (G.E.C. Review, New York).

Dushman, S., 1949, Scientific Foundations of Vacuum Technique (Wiley, New York) 882pp.

Dushman, S., and J.M. Lafferty, 1962, Scientific Foundations of Vacuum Technique, 2nd Ed. (Wiley, New York) 806pp.

Duval, P., 1969, Le Vide **24**, 83.

Duval, P., 1975, Le Vide et ses Applications (Presses Universitaires de France, Paris).

Duval, P., 1982, Diffusion, turbo, or cryo-pumps. Vakuum-Technik **31**(4), 97.

Duval, P., et al., 1988, The molecular drag pump. J. Vac. Sci. & Technol. A **6**(3), 1187.

Dylla, H.F., 1978, Turbomolecular pump vacuum system for the Princeton Large Torus. Proc. 24th American Vacuum Symposium. J. Vac. Sci. & Technol. **15**, 734.

Dylla, H.F., 1988, Glow discharge techniques for conditioning high-vacuum systems. J. Vac. Sci. & Technol. A **6**(3), 1276.

Dylla, H.F., and T.J. Provost, 1982, Sensitivity of capacitance manometers to static magnetic fields. J. Vac. Sci. & Technol. **21**(2), 707.

Dylla, H.F., et al., 1979, Observations of changes in residual gas and surface composition with discharge cleaning of PLT. Proc. 25th American Vacuum Symposium, J. Vac. Sci. & Technol. **16**, 752.

Dylla, H.F., et al., 1984, Initial conditioning of the TFTR vacuum vessel. J. Vac. Sci. & Technol. A **2**, 1188.

E

East, H.G., and H.J. Kuhn, 1946, J. Sci. Instrum. **23**, 185.

Ebdale, B., 1978, Capabilities and limitations of steam jet ejectors and liquid ring pumps. Vacuum **28**, 337.

Ebsuzaki, Y., W.J. Kass and M. O'Keefe, 1967, J. Chem. Phys. **46**, 1378.

Eckertova, L., 1977, Physics of Thin Films (Plenum Press, New York [SNTL Publishing of Technical Literature, Prague]).

Edelmann, Chr., and P. Engelmann, 1982, Range extension of hot cathode ionization gauges. Vakuum-Technik **31**(1), 2.

Edenburn, N.W., 1972, Radiative energy transfer in an annular channel. J. Appl. Phys. **43**, 3868.

Eder, F.X., 1972, Entropy aspects during gas condensation in cryopumps (in German). Vakuum-Technik **21**, 76.

Edmonds, T., and J.P. Hobson, 1965, J. Vac. Sci. & Technol. **2**, 182.

Edmonds, P.H., and W.D. Shipley, 1988, A reentrant bakeable all-metal feed-through vacuum seal. J. Vac. Sci. & Technol. A **6**(1), 165.

Edwards Jr, D., 1977a, Upper bound to the pressure in an elementary vacuum system. Proc. 23rd American Vacuum Symposium, J. Vac. Sci. & Technol. **14**, 606.

Edwards Jr, D., 1977b, Vacuum **27**, 631.

Edwards Jr, D., 1977c, An upper bound to the outgassing rate of metal surfaces. J. Vac. Sci. & Technol. **14**, 1030.

Edwards Jr, D., 1978a, The nitrogen pressure in a vacuum system. Proc. 24th American Vacuum Symposium. J. Vac. Sci. & Technol. **15**, 755.

Edwards Jr, D., 1978b, Rate at which molecules strike a surface or orifice in a vacuum system. J. Vac. Sci. & Technol. **15**, 1182; **16**, 1573, 1979.

Edwards Jr, D., 1979a, An upper bound to the pressure (outgassing rate) in a dynamic vacuum system with uniform wall bombardment. J. Vac. Sci. & Technol. **16**, 82. .

Edwards Jr, D., 1979b, Ion and electron desorption of neutral molecules from stainless steel 304. Proc. 25th American Vacuum Symposium, J. Vac. Sci. & Technol. **16**, 758.

Edwards Jr, D., 1980, Methane outgassing from a Ti-sublimation pump. Proc. 26th American Vacuum Symposium, J. Vac. Sci. & Technol. **17**, 279.

Edwards Jr, D., and P. Limon, 1978, Pressure measurements in a cryogenic environment. J. Vac. Sci. & Technol. **15**, 1186.

Edwards, T.J., J.B. Rudge and W. Hauptli, 1977, Thin-film polyimide gasket seal for uhv. J. Vac. Sci. & Technol. **14**, 740.

Edwards, D., et al., 1979, Sealing of knife-edge flanges after a high temperature vacuum firing. J. Vac. Sci. & Technol. **16**(6), 2114.

Egelhoff Jr, W.F., 1988, Ultrahigh vacuum leak sealing with a silicon resin product. J. Vac. Sci. & Technol. A **6**(4), 2584.

Eggleton, A.E., and F.C. Tompkins, 1952, Trans. Faraday Soc. **48**, 738.

Ehlers, H., and J. Moll, 1960, Results with ultra-high vacuum systems. Trans. 6th American Vacuum Symposium (Pergamon Press, Oxford) p. 261.

Ehrlich, G., 1961, J. Chem. Phys. **34**, 29, 39; J. Appl. Phys. **32**, 4.

Ehrlich, G., 1962, Molecular processes in adsorption on metals. Trans. 2nd Int. Vacuum Congress (Pergamon Press, Oxford) p. 127.

Ehrlich, C.D., 1986, A note on flow rate and leak rate. J. Vac. Sci. & Technol. A 4(5), 2384.

Eisinger, J., 1957, J. Chem. Phys. 27, 1206.

Eisinger, J., 1959, J. Chem. Phys. 30, 410.

Elliott, K.W.T., D. Woodman and R. Dadson, 1967, Expansion system for calibration. Vacuum 17, 439.

Elo, D., et al., 1979, Cryogenic vacuum pumping at the LBL 88-inch cyclotron. Proc. 8th Int. Conf. on Cyclotrons, IEEE Trans. Nucl. Sci. NS-26, 2179.

Elsey, R.J., 1975, Outgassing of vacuum materials. Vacuum 25, 299; 25, 347.

Emerson, L.C., et al., 1986, Dissociative pumping of alkanes using non-evaporable getters. J. Vac. Sci. & Technol. A 4(3), 297.

Emerson, L.C., et al., 1987, Pumping of hydrocarbons using nonevaporable getters. J. Vac. Sci. & Technol. A 5(4), 2584.

Engelmann, G., et al., 1987, Vacuum chambers in composite material. J. Vac. Sci. & Technol. A 5(4), 2337.

Envelopes, 1983, Calculation of envelopes (cylindrical, spherical, etc.) under an external pressure (in French). Le Vide 38(217), 333.

Erikson, E.D., et al., 1984, Vacuum outgassing of various materials. J. Vac. Sci. & Technol. A 2(2), 206.

Eschbach, H.L., 1962, Praktikum der Hochvakuumtechnik (Akademische Verlag, Leipzig) 243pp.

Eschbach, H.L., and R. Werz, 1976, A generalized correction factor for the Knudsen radiometer gauge with arbitrary vane and heater shape. Vacuum 26, 67.

Eschbach, H.L., F. Gross and S. Schulien, 1963, Vacuum 13, 543.

Espe, W., 1955, Vakuum-Technik 4, 34.

Espe, W., 1966, Materials of High Vacuum Technology. Vol. I. Metals and Metalloids (1966) 912pp.; II. Silicates (1968) 660pp.; III. Auxiliary Materials (1968) 530pp. (Pergamon Press, Oxford).

Espe, W., and C. Hybl, 1965, Vakuum-Technik 14, 108.

Espe, W., and M. Knoll, 1936, Werkstoffe der Hochvakuumtechnik (Springer, Berlin).

Evrard, R., and G.A. Boutry, 1969, J. Vac. Sci. & Technol. 6, 279.

F

Fäber, W., 1976, Vakuum-Technik 25, 104.

Falland, Chr., 1980, Leaktest with air cooled turbo-pump and He. Vakuum-Technik 29(7), 205.

Falland, Chr., 1981, He leak detection of large accelerators and storage rings (PETRA). Vakuum-Technik 30(2), 41.

Falland, C., et al., 1974, The uhv system for the Desy Electron–Positron Double Storage Ring "DORIS". Proc. 6th Int. Vacuum Congress, Jpn. J. Appl. Phys., Suppl. 2, Pt. 1, p. 209.

Feakes, F., and F.L. Torney, 1963, The performance characteristics of three types of extreme high-vacuum gauges. Trans. 10th American Vacuum Symposium (Macmillan, New York) p. 257.

Feidt, M.L., and D. Paulmeir, 1980, Influence of end effects on optimum injection conditions for electrons in an orbitron device. Proc. 26th American Vacuum Symposium, J. Vac. Sci. & Technol. 17, 345.

Feidt, M.L., and D.F. Paulmeir, 1982, A model for optimum ionizing characteristics of an orbitron device. Vacuum 32(8), 491.

Ferrario, B., 1983, Introduzione alla Technologia del Vuoto (AIV Assoc. Italia, Vuoto).

Ferrario, B., and L. Rosai, 1977, New types of volume gettering panels for vacuum problems in plasma machines. Proc. 7th Int. Vacuum Congress (Dobrozemsky, et al., Vienna) p. 359.

Filippelli, A.R., 1987a, Operation of a Bayard–Alpert gauge in a uniform 0-0.16 tesla magnetic field. J. Vac. Sci. & Technol. A 5(2), 249.

Filippelli, A.R., 1987b, Residual currents in several commercial ultrahigh vacuum Bayard–Alpert gauges. J. Vac. Sci. & Technol. A 5(5), 3234.

Fischer, E., 1972, J. Vac. Sci. & Technol. 9, 1203.

Fischer, E., 1974, Vacuum and uhv for particle accelerators and storage rings. Proc. 6th Int. Vacuum Congress, Jpn. J. Appl. Phys., Suppl. 2, Pt. 1, p. 199.

Fischer, E., 1977, Uhv technology for storage rings. IEEE Trans. Nucl. Sci. NS-24, 1227.

Fischer, E., and H. Mommsen, 1967, Vacuum 17, 309.

Fischer, K., et al., 1982, Pumping of corrosive or hazardous gases with turbo- and rotary pumps. Vacuum 32(10/11), 619.

Fischhoff, E., 1962, Le Vide 17, 195.

Fisher, W.G., 1988, Large inexpensive quartz viewports for high-vacuum systems. J. Vac. Sci. & Technol. A 6(2), 246.

Fisher, P.W., and J.S. Watson, 1978, Cryosorption pumping of deuterium by MS-5A at temperatures above 4.2K for fusion reactor applications. Proc. 24th American Vacuum Symposium, J. Vac. Sci. & Technol. 15, 741.

Fisher, P.W., and J.S. Watson, 1979, Helium pumping at 4.2K by molecular sieve 5A. J. Vac. Sci. & Technol. 16, 75.

Fisher, A., et al., 1978, Rev. Sci. Instrum. 49, 872.

Fitch, R.K., T.N. Norris and W.J. Thatcher, 1969, Vacuum 19, 227.

Flécher, P., 1977, Recent state in turbomolecular pump development. Proc. 7th Int. Vacuum Congress (Dobrozemsky, et al., Vienna) p. 25.

Flécher, P., 1982, Vacuum outgassing of metal powders (in German), Vakuum-Technik **31**(3), 81.

Fleming, R.B., R.W. Brocker and D.H. Mullaney, 1980, Development of bakeable seals for large noncircular ports on TFTR. Proc. 26th American Vacuum Symposium, J. Vac. Sci. & Technol. **17**, 337.

Flesch, G., C. Henriot and G. Rommel, 1975, Performance of a cryogenic pump at 20 K (in French). Le Vide **30**, 182.

Fletcher, B., 1970, Vacuum **20**, 381.

Flinta, I., 1954, J. Sci. Instrum. **31**, 388.

Florescu, N.A., 1962a, Vacuum **12**, 227.

Florescu, N.A., 1962b, Reproducible low pressures and their application to gauge calibration. Trans. 8th American Vacuum Symposium (Pergamon Press, Oxford) p. 504.

Forth, H.J., 1965, Le Vide **20**, 343.

Forth, H.J., and R. Frank, 1977, Installation of cryopumps in the ESTEC Space Simulation chambers. Proc. 7th Int. Vacuum Congress (Dobrozemsky, et al., Vienna) p. 61.

Forth, H.J., et al., 1972, Application and design of cryopumps. Vakuum-Technik **21**, 81.

Fortucci, P.L., and V.D. Meyer, 1979, Apparatus for rapid and accurate measurement of pressure and volume. J. Vac. Sci. & Technol. **16**, 963.

Foster, C.A., 1987, High-throughput continuous cryopump. J. Vac. Sci. & Technol. A **5**(4), 2558.

Fouletier, J., and M. Kleitz, 1975, Electrically renewable and controllable oxygen getter. Vacuum **25**, 307.

Fowler, G.L., 1987, Coaxial helium leak detection probe. J. Vac. Sci. & Technol. A **5**(3), 390.

Fowler, P., and F.J. Brock, 1970, J. Vac. Sci. & Technol. **7**, 507.

Fox, M., 1970, Vacuum **20**, 97.

Frank, R., 1974, Turbomolecular pumps with a vertically arranged compressor (in German). Vakuum-Technik **23**, 109.

Frank, R., and E. Usselmann, 1976a, Production of hydrocarbon-free vacua using TURBOVAC turbomolecular pumps (in German). Vakuum-Technik **25**, 48.

Frank, R., and E. Usselmann, 1976b, Magnetomotive turbomolecular pump of the TURBOVAC type (in German). Vakuum-Technik **25**, 141.

Frank, R., and E. Usselmann, 1977, Magneto-bearing turbomolecular pump of the TURBOVAC type. Proc. 7th Int. Vacuum Congress (Dobrozemsky, et al., Vienna) p. 49.

Frank, R., E. Usselmann and W. Bächler, 1975, Performance data of vertical TURBOVAC turbomolecular pumps (in German). Vakuum-Technik **24**, 78.

Fraser, D.B., and H.D. Cook, 1977, Film deposition with the sputter gun. Proc. 23rd American Vacuum Symposium, J. Vac. Sci. & Technol. **14**, 147.

Frauenfelder, R., 1968, J. Chem. Phys. **48**, 3966.

Fremerey, J.K., 1972, J. Vac. Sci. & Technol. **9**, 108.

Fremerey, J.K., 1982, Spinning rotor vacuum gauges. Vacuum **32**, 685.

Fremlin, Y.H., 1952, J. Sci. Instrum. **29**, 267.

Frenkel, J., 1924, Zs. Phys. **26**, 117.

Friebel, V.R., and J.T. Hinricks, 1975, Lubrication for vacuum applications. Proc. 21st American Vacuum Symposium, J. Vac. Sci. & Technol. **12**, 551.

Friedrichs, F., 1960, Das Glas im chemischen Laboratorium (Springer, Berlin).

Fuchs, W., 1976, Vakuum-Technik **25**, 52.

Fujii, Y., et al., 1983, Modified extractor gauge. J. Vac. Sci. & Technol. A **1**(1), 90.

Fukutome, R., J. Aikawa and C. Hayashi, 1976, Three-stage Roots high vacuum pump. J. Vac. Sci. & Technol. **13**, 630.

Fulker, M.J., M.A. Baker and L. Laurenson, 1969, Vacuum **19**, 556.

Fuller, G.M., and J.R. Haines, 1984, Design considerations for achieving high vacuum integrity in fusion devices. J. Vac. Sci. & Technol. A **2**, 1162.

Fultz, C.R., and D.W. Carver, 1975, Uhv system for metering gas flow (leaks) and computing flow rate. J. Vac. Sci. & Technol. **12**, 1088.

Füstöss, L., 1977, Monte-Carlo calculations for free molecule and near-free molecule flow through cylindrical and conical tubes. Proc. 7th Int. Vacuum Congress (Dobrozemsky, et al., Vienna) p. 101.

Füstöss, L., 1981, Monte-Carlo calculations for molecular flow through axially symmetric tubes. Vacuum **31**(6), 243.

Füstöss, L., 1983, Evaluation and calculation of gas flow through axially symmetric tubes. Vacuum **33**(1/2), 13.

G

Gajewski, P., and A. Wisniewski, 1977, Two-dimensional flow through the channel in the presence of the condensation process. Proc. 7th Int. Vacuum Congress (Dobrozemsky, et al., Vienna) p. 109.

Gale, R.F., and C.F. Machin, 1953, J. Sci. Instrum. **30**, 97.

Garbe, S., and K. Christians, 1962, Vakuum-Technik **11**, 9.

Gareis, P.J., and G.F. Hagenbach, 1965, Ind. Eng. Chem. **57**, 27.

Garrod, R.I., and I.F. Nankivel, 1961, Vacuum **11**, 139.

Gaunt, A.J., and R.A. Redford, 1959, J. Sci. Instrum. **36**, 377.

Gear, P.E., 1976, The choice of cathode material in a hot cathode ionization gauge. Vacuum **26**, 3.

Geller, R., 1958, Le Vide **13**, 71.

Genot, B., 1975, Some remarks on the mathematical nature of the BET equation, and a new method of determination of V_m. J. Colloid Interface Sci. **3**, 413.

Gentsch, H., et al., 1974, Partial pressure measurements (in German). Vakuum-Technik **23**, 230.

Germer, L.H., 1965, Sci. Am. **212**(3), 32.

Gilmour, A.S., 1976, Bakeable uhv gate valve for microwave tube experimentation. J. Vac. Sci. & Technol. **13**, 1199.

Giorgi, T.A., 1974, Getters and gettering. Trans. 6th Int. Vacuum Congress (Kyoto) p. 53.

Giorgi, T.A., et al., 1985, An updated review of getters and gettering. J. Vac. Sci. & Technol. A **3**(2), 417.

Gitzendanner, L.G., and F.O. Rathbun, 1965, Statistical interface-leakage analysis, Report General Electric 65-GL-73.

Glaister, R.M., 1956, J. Sci. Instrum. **33**, 34.

Glaros, S.S., et al., 1979, Shiva and Argus target diagnostics vacuum systems. Proc. 25th American Vacuum Symposium, J. Vac. Sci. & Technol. **16**, 674.

Glassford, A.P.M., and C.K. Liu, 1980, Outgassing rates of multilayer insulation materials at ambient temperature. J. Vac. Sci. & Technol. **17**(3), 696.

Goetz, A., 1926, Physik und Technik des Hochvakuums (Vieweg, Braunschweig).

Goetz, G., 1982, Large turbomolecular pumps. Vacuum **32**(10/11), 703.

Goetz, G., and H.H. Henning, 1983, Heavy duty turbo-pump. Vakuum-Technik **32**(5), 130.

Goetz, D.G., et al., 1984, Advanced turbomolecular pumps used to pump reactive and abrasive media. J. Vac. Sci. & Technol. A **2**(2), 182.

Goetz, D., et al., 1987, The use of turbomolecular pumps in television tube production. J. Vac. Sci. & Technol. A **5**(4), 2421.

Goff, R.F., 1973, J. Vac. Sci. & Technol. **10**, 355.

Goldbach, G., 1956, Vakuum-Technik **5**, 7.

Göllnitz, H., H.G. Schneider and H. Rössler, 1978, Vakuumelektronik (Akademie Verl., Berlin) (in German).

Gomay, Y., T. Tazima and N. Fujisawa, 1978, Discharge cleaning experiment in the JFT-2 Tokamak with surface observation by AES. J. Vac. Sci. & Technol. **15**, 103.

Gomay, Y., et al., 1979, Wall conditioning by low power discharge in the ISX-A Tokamak. J. Vac. Sci. & Technol. **16**, 918.

Gomer, R., 1959, J. Phys. Chem. **63**, 468.

Gopalaraman, C.P., R.A. Armstrong and P.A. Redhead, 1970, J. Vac. Sci. & Technol. **7**, 195.

Gordon, S.A., 1958, Rev. Sci. Instrum. **29**, 501.

Gore, G.W., 1957, J. Sci. Instrum. **34**, 459.

Gorinas, G., 1977, Advantages of hybrid turbomolecular pumps for totally oil-free vacuum. Proc. 7th Int. Vacuum Congress (Dobrozemsky, et al., Vienna) p. 45.

Gorman, J.K., and W.R. Nardella, 1962, Vacuum **12**, 19.

Gorodetsky, I.G., and V.E. Skurat, 1975, Mass-spectrometric observation of gas evolution in some devices for transmission of rotational motion into vacuum. Vacuum **25**, 427.

Gorovitz, B., K. Moses and P. Gloersen, 1960, Rev. Sci. Instrum. **31**, 146.

Górsky, W., K. Litynski and A. Smiech, 1977, The compensation of ambient temperature influence on Pirani gauge indication. Proc. 7th Int. Vacuum Congress (Dobrozemsky, et al., Vienna) p. 165.

Gosselin, C.M., G.A. Beitel and A. Smith, 1970, J. Vac. Sci. & Technol. **7**, 233.

Gottwald, B.A., 1973, Beam formation in molecular flow (in German). Vakuum-Technik **22**, 106, 141.

Govier, R.P., and G.M. McCracken, 1970, J. Vac. Sci. & Technol. **7**, 552.

Grady, B.I., 1973, J. Vac. Sci. & Technol. **10**, 208.

Graham, W.G., and L. Ruby, 1979, Cryopumping measurements relating to safety, pumping speed and radiation outgassing. J. Vac. Sci. & Technol. **16**, 927.

Grant, W.A., and G. Carter, 1965, Vacuum **15**, 477.

Grant, W.A., and G. Carter, 1967, Br. J. Appl. Phys. **18**, 527.

Greaves, C., 1970, Vacuum **20**, 65.

Green, G.M., 1953, J. Sci. Instrum. **31**, 473.

Green, W., 1968, The Design and Construction of Small Vacuum Systems (Chapman and Hall, London) 181pp.

Greenblatt, M.H., 1958, Rev. Sci. Instrum. **29**, 738.

Grigorov, G., 1973, Le Vide **28**, 14.

Grigorov, G., and V. Kanev, 1965, High Vacuum (in Bulgarian) (Jusantor, Sofia); adapted in French by E. Thomas, 1970, Le Vide Poussé (Masson, Paris) 482pp.

Grigorov, G.I., and K.K. Tzatzov, 1977, Sticking coefficient of gases on continuously deposited getter films. Proc. 7th Int. Vacuum Congress (Dobrozemsky, et al., Vienna) p. 97.

Grigorov, G.I., and K.K. Tzatzov, 1983, Theory of getter pump evaluation. Vacuum **33**(3), 139.

Gröbner, O., et al., 1983, Studies of photon induced gas desorption using synchrotron radiation. Vacuum **33**(7), 397.

Groskowski, J., 1955, High Vacuum Technology (in Polish) (Panstwowe Wydawnictwa Techn., Warszawa).

Groskowski, J., 1967, Electrode dimensions of the Bayard–Alpert ionization gauge and its sensitivity. Trans. 3rd Int. Vacuum Congress (Pergamon Press, Oxford) p. 241.

Groskowski, J., 1968, The development of ionization gauges for very low pressures. Proc. 4th Int. Vacuum Congress (Institute of Physics, London) p. 631.

Groskowski, J., 1969, Le Vide **24**, 226.

Groszkowski, J., S. Pytkowski and W. Trzoch, 1977, Continuous modulation method of uhv measurements. Proc. 7th Int. Vacuum Congress (Dobrozemsky, et al., Vienna) p. 121.

Grosse, G., and G. Messer, 1981, Calibration of vacuum gauges below 10^{-9}mbar with a molecular beam method. Vakuum-Technik **30**(8), 226.

Grosse, G., et al., 1987a, Secondary electrons in ion gauges. J. Vac. Sci. & Technol. A **5**(5), 3242.

Grosse, G., et al., 1987b, Summary abstract: Calibration and long-term characteristics of helium reference leaks. J. Vac. Sci. & Technol. A **5**(4), 2661.

Grössi, M., et al., 1987, Summary abstract: Reusable metal seal for high- and ultra-high vacuum applications. J. Vac. Sci. & Technol. A **5**(4), 2629.

Grossman, D.G., 1978, Machining a machinable glass-ceramic. Vacuum **28**, 55.

Grove, D.J., 1959, Applications of ultra-high vacuum techniques. Trans. 5th American Vacuum Symposium (Pergamon Press, Oxford) p. 10.

Grunze, M., et al., 1988, Chemical cleaning of metal surfaces in vacuum systems by exposure to reactive gases. J. Vac. Sci. & Technol. A **6**(3), 1266.

Guilbard, C., and A. Guihery, 1967, Les controles d'étancheité sur grands ensembles. Trans. 3rd Int. Vacuum Congress (Pergamon Press, Oxford) p. 271.

Gupta, A.K., and J.H. Leck, 1975, An evaluation of the titanium sublimation pump. Vacuum **25**, 362.

Gupta, A.C., and J.K.N. Sharma, 1980, Pirani and Penning control circuits for system automation. Vacuum **30**(7), 291.

Gureswitch, A.M., and W.F. Westendrop, 1954, Rev. Sci. Instrum. **25**, 389.

Guthrie, A., 1963, Vacuum Technology (Wiley, New York) 532pp.

Guthrie, A., and R.K. Wakerling, 1949, Vacuum Equipment and Techniques (McGraw-Hill, New York).

H

Habets, A.H.M., et al., 1975, Simple device for measurements of the sticking coefficient of cryopumps. Rev. Sci. Instrum. **46**, 613.

Hablanian, M.H., 1974, Diffusion pump technology. Trans. 6th Int. Vacuum Congress (Kyoto) p. 25.

Hablanian, M.H., 1984, Comments on the history of vacuum pumps. J. Vac. Sci. & Technol. A **2**, 118.

Hablanian, M.H., 1986, Performance characteristics of displacement-type vacuum pumps. J. Vac. Sci. & Technol. A **4**(3), 286.

Hablanian, M.H., 1987, Recommended procedure for measuring pumping speeds. J. Vac. Sci. & Technol. A **5**(4), 2552.

Hablanian, M.H., 1988, The emerging technologies of oil-free vacuum pumps. J. Vac. Sci. & Technol. A **6**(3), 1177.

Hablanian, M.H., and W.E. Briggs, 1977, New developments in helium leak detection. Proc. 7th Int. Vacuum Congress (Dobrozemsky, et al., Vienna) p. 199.

Hablanian, M.H., and A.A. Landfors, 1974, J. Vac. Sci. & Technol. **11**, 344.

Hablanian, M.H., and A.A. Landfors, 1976, Stable pumping with small diffusion pumps. Proc. 22nd American Vacuum Symposium, J. Vac. Sci. & Technol. **13**, 494.

Hablanian, M.H., and J.C. Maliakal, 1973, J. Vac. Sci. & Technol. **10**, 58.

Hablanian, M.H., et al., 1987, Elimination of backstreaming from mechanical vacuum pumps. J. Vac. Sci. & Technol. A **5**(4), 2612.

Hachenberg, O., and W. Brauer, 1959, Adv. in Electronics and Electr. Phys. **11**, 413.

Haefer, R.A., 1953/54, Acta Phys. Austriaca **7**, 251; **8**, 213.

Haefer, R.A., 1980a, On the pumping speed of large area cryopumps. Vacuum **30**(1), 19.

Haefer, R.A., 1980b, Ionization gauges in vacuum chambers with cryopumping surfaces. Vacuum **30**(4/5), 193.

Haefer, R.A., 1980c, Addition theorem for the resistance to flow of composite systems. Vacuum **30**(6), 217.

Haefer, R.A., 1981, Kryo-Vakuumtechnik (in German) (Springer, Berlin).

Haefer, R.A., and J. Hengevoss, 1961, Studies on the Blears effect at pressure measurements in the ultra-high vacuum range. Trans. 7th American Vacuum Symposium (Pergamon Press, Oxford) p. 67.

Haefer, R.A., and P. Kleber, 1976, Measurement of molecular flow patterns in vacuum chambers with cryo-surfaces. Le Vide **31**, 19.

Haefer, R., and O. Winkler, 1956, Vakuum-Technik **5**, 149.

Hagstrum, H.D., and C. D'Amigo, 1960, J. Appl. Phys. **31**, 715.

Hain, K., 1972, Nomographs for evaluation of detection efficiency in the determination of leaks with the He mass spectrometer (in German). Vakuum-Technik **21**, 133.

Halama, H.J., 1977, Behavior of titanium sublimation and sputter-ion pumps in the 10^{-11} Torr range. Proc. 23rd American Vacuum Symposium, J. Vac. Sci. & Technol. **14**, 524.

Halama, H.J., 1979, Performance of vacuum components operating at 1×10^{-11} Torr. Proc. 25th American Vacuum Symposium, J. Vac. Sci. & Technol. **16**, 717.

Halama, H.J., 1983, Vacuum systems for large colliders using superconducting magnets. Proc. 9th Int. Vacuum Congress (Madrid), p. 283.

Halama, H.J., and J.R. Aggus, 1974, Measurements of adsorption isotherms and pumping speed of helium on molecular sieve in the 10^{-11}–10^{-7} Torr range at 4.2K. Proc. 20th American Vacuum Symposium, J. Vac. Sci. & Technol. **11**, 333.

Halama, H.J., and J.R. Aggus, 1975, Cryosorption pumping for intersecting storage rings. Proc. 21st American Vacuum Symposium, J. Vac. Sci. & Technol. **12**, 532.

Halama, H.J., and C.L. Foerster, 1987, Vacuum performance of the uv and X-ray rings at the National Synchrotron Light Source. J. Vac. Sci. & Technol. A **5**(4), 2342.

Halama, H.J., and J.C. Herrera, 1975, Comparison of cold and warm vacuum systems for intersecting storage rings. IEEE Trans. Nucl. Sci. **NS-22**, 1492.

Halama, H.J., and J.C. Herrera, 1976, Thermal desorption of gases from Al 6061, their rates and activation energies. Proc. 22nd American Vacuum Symposium, J. Vac. Sci. & Technol. **13**, 463.

Halama, H.J., and C.H. Hseuh, 1987, Summary abstract: Gauging requirements and practices for storage rings and accelerators. J. Vac. Sci. & Technol. A **5**(5), 3232.

Hall, L.D., 1958, Rev. Sci. Instrum. **29**, 367.

Hall, L.D., 1969, J. Vac. Sci. & Technol. **6**, 44.

Halliday, B.S., and B.A. Trickett, 1972, Vacuum problems on a 5 GeV electron synchrotron. Trans. 5th Int. Vacuum Congress (American Vacuum Society, New York) p. 42.

Hamacher, H., 1974a, Computing the pressure profile in a volume when exposed to an exponentially dropping lower pressure through an orifice (in German). Vakuum-Technik **23**, 1.

Hamacher, H., 1974b, Experimental investigations with after coolers of Roots type pumps (in German). Vakuum-Technik **23**, 129.

Hamacher, H., 1976, Computation of the pumping speed in space simulation chambers (in German). Vakuum-Technik **25**, 33.

Hamacher, H., 1977, The development of a vacuum system for Spacelab. Proc. 7th Int. Vacuum Congress (Dobrozemsky, et al., Vienna) p. 207.

Hamilton, A.R., 1957, Rev. Sci. Instrum. **28**, 693.

Hammond, V.J., 1954, J. Sci. Instrum. **31**, 258.

Hands, B.A., 1976, Introduction to cryopump design. Vacuum **26**, 11.

Hands, B.A., 1982, Developments in cryopumping. Vacuum **32**(10/11), 603.

Hands, B.A., and G. Davey, 1977, Cryogenic aspects of cryopump design. Proc. 7th Int. Vacuum Congress (Dobrozemsky, et al., Vienna) p. 53.

Hansen, N., and W. Littman, 1967, Uber die Bestimmung der Haftwarscheinlichkeit von Gasen an reinen Metalloberflachen. Trans. 3rd Int. Vacuum Congress (Pergamon Press, Oxford) p. 465.

Haque, C.A., 1976, Simple Ti and Ni sublimation pump. J. Vac. Sci. & Technol. 13, 1088.

Harding, G.L., 1977, Improvements in a dc reactive sputtering system for coating tubes. J. Vac. Sci. & Technol. 14, 1313.

Harra, D.J., 1974, J. Vac. Sci. & Technol. 11, 331.

Harra, D.J., 1974b, Predicting and evaluating titanium sublimation pump performance. Proc. 6th Int. Vacuum Congress, Jpn. J. Appl. Phys., Suppl. 2, Pt. 1, p. 41.

Harra, D.J., 1975, Improved Ti sublimation pumping, for long-term, clean, applications. Proc. 21st American Vacuum Symposium, J. Vac. Sci. & Technol. 12, 539.

Harra, D.J., 1976, Review of sticking coefficients and sorption capacities of gases on titanium films. Proc. 22nd American Vacuum Symposium, J. Vac. Sci. & Technol. 13, 471.

Harra, D.J., 1978, Differentially pumped six-inch metal sealed gate valve. Proc. 24th American Vacuum Symposium, J. Vac. Sci. & Technol. 15, 779.

Harra, D.J., and T.W. Snouse, 1972, J. Vac. Sci. & Technol. 9, 552.

Harra, D.J., D.R. Nichols and K.M. Welch, 1980, Description and application of fully integrated cryopumped systems. Proc. 26th American Vacuum Symposium, J. Vac. Sci. & Technol. 17.

Harries, W., 1949, Chem. Ing. Tech. 21, 139.

Harris, L.A., 1968, J. Appl. Phys. 39, 1419, 1428.

Harris, N.S., 1977, Diffusion pump back-streaming. Vacuum 27, 519.

Harris, N.S., 1978, Rotary pump back-migration. Vacuum 28, 261.

Harris, N.S., 1979, Vacuum Engineering (London Caledonian, Watford).

Harris, N.S., 1980, Modern Diffusion pumps vs. turbomolecular pump systems. Vacuum 30(4/5), 175.

Harris, N.S., and L. Budgen, 1976, Design and manufacture of modern vacuum pumps. Vacuum 26, 525.

Harrower, G.A., 1959, Phys. Rev. 104, 52.

Hartwig, H., and J.S. Kouptsidis, 1974, A new approach for computing diode sputter-ion pump characteristics. J. Vac. Sci. & Technol. 11, 1154.

Hartwig, H., and J.S. Kouptsidis, 1977a, Design and performance of integrated sputter-ion pumps for particle accelerators. Proc. 7th Int. Vacuum Congress (Dobrozemsky, et al., Vienna) p. 93.

Hartwig, H., and J.S. Kouptsidis, 1977b, Two novel techniques for uhv joints between Al and stainless steel. Proc. 7th Int. Vacuum Congress (Dobrozemsky, et al., Vienna) p. 259.

Hartwig, H., and J. Kouptsidis, 1977c, New techniques for the PETRA vacuum system. IEEE Trans. Nucl. Sci. **NS-24**, 1248.

Hasse, T., G. Klages and H. Klumb, 1936, Physik. Zs. **37**, 440.

Hauff, A., G. Kienel and A. Wamser, 1965, Vakuum-Technik **14**, 139.

Hauser-Gaswindt, I., and H. Rukop, 1920, Telefunken-Zeitung **19**, 21.

Hawrylak, A., 1967, J. Vac. Sci. & Technol. **4**, 364.

Hayashi, C., 1966, J. Vac. Sci. & Technol. **3**, 286.

Hayward, W.H., and R.L. Jepsen, 1962, A simple high-vacuum gauge calibration system. Trans. 9th American Vacuum Symposium (Macmillan, New York) p. 459.

Head, P.V., and B.W. Griffiths, 1978, Vacuum **28**, 421.

Head, P.V., et al., 1982, Sealing large ultrahigh vacuum systems. Vacuum **32**(10/11), 639.

Heap, R.D., 1973, Evacuating to 0.5–10 Torr through long tubes. Vacuum **23**, 317.

Heasley, J.H., 1976, J. Vac. Sci. & Technol. **13**, 649.

Hees, G.W., W. Eaton and J. Lech, 1956, The knife-edge seal. Trans. 2nd American Vacuum Symposium (Pergamon Press, Oxford) p. 75.

Heiland, W., 1977, Special aspects of vacuum systems for fusion experiments. Proc. 23rd American Vacuum Symposium, J. Vac. Sci. & Technol. **14**, 576.

Heinemann, K., and H. Poppa, 1986, An ultrahigh vacuum multipurpose specimen chamber with sample introduction system for in situ transmission electron microscopy investigations. J. Vac. Sci. & Technol. A **4**(1), 127.

Heinze, W., 1955, Einfuhrung in die Vakuumtechnik (Verlag Technik, Berlin) 451pp.

Helmer, J.C., 1967a, Applications of an approximation to molecular flow in cylindrical tubes. J. Vac. Sci. & Technol. **4**, 179.

Helmer, J.C., 1967b, Solution of Clausing's integral equation for molecular flow. J. Vac. Sci. & Technol. **4**, 360.

Helmer, J.C., and H. Hayward, 1966, Rev. Sci. Instrum. **37**, 1652.

Hemmerich, J.L., 1988, Primary vacuum pumps for the fusion reactor fuel cycle. J. Vac. Sci. & Technol. A **6**(1), 144.

Hengevoss, J., and E.A. Trendelenburg, 1963, Continuous cryotrapping of hydrogen and helium by argon at 4.2°K. Trans. 10th American Vacuum Symposium (Macmillan, New York) p. 101.

Hengevoss, J., and E.A. Trendelenburg, 1967, Vacuum **14**, 495.

Hengevoss, J., H. Reisinger and H. Voessner, 1970, J. Vac. Sci. & Technol. **7**, 251.

Henning, J., 1974, New developments in the field of high performance turbo-molecular pumps. Trans. 6th Int. Vacuum Congress (Kyoto) p. 5.

Henning, J., 1974b, Turbomolecular pumps for oil-free uhv (in German). Vakuum-Technik **23**, 65.

Henning, H., 1975, The memory effect for argon in triode type sputter-ion pumps (in German). Vakuum-Technik **24**, 37.

Henning, J., 1977, New turbomolecular pumps and their application in physics (in German). Vakuum-Technik **26**, 177.

Henning, H., 1978a, The approximate calculation of transmission probabilities for the conductance of tubulations in the molecular flow regime. Vacuum **28**, 151; Comments: W. Steckelmacher and H. Henning (1979), Vacuum **29**, 31.

Henning, J., 1978b, Trends in the development and use of turbomolecular pumps. Vacuum **28**, 391.

Henning, J., 1979, A miniaturized turbolmolecular pump. Vacuum **29**(11/12), 447.

Henning, J., 1980a, A new line of air-cooled turbomolecular pumps. Proc. 26th American Vacuum Symposium, J. Vac. Sci. & Technol. **17**.

Henning, J., 1980b, Comparison of the construction of old and new turbomolecular pumps. Vacuum **30**(4/5), 183.

Henning, J., 1988, Thirty years of turbomolecular pumps: A review and recent developments. J. Vac. Sci. & Technol. A **6**(3), 1196.

Henning, J., and H. Lang, 1976, Roots pumps for high pressure difference with cooling by gas circulation. Vacuum **26**, 273.

Henning, J., et al., 1982, High throughput pumping of dangerous gases with a multistage roots pump. J. Vac. Sci. & Technol. **20**(4), 1023.

Henry, R.P., 1969, Le Vide **24**, 316.

Henry, R.P., 1971, Cours de Science et de la Technique du Vide (Soc. Franc., Ing. Techn. Vide, Paris).

Heppel, T.A., 1967, J. Sci. Instrum. **44**, 686.

Herb, R.G., T. Pauly and H.J. Fischer, 1963, Bull. Am. Phys. Soc. **8**, 336.

Heydemann, P.L.M., C.R. Tilford and R.W. Hyland, 1977, Ultrasonic mano-meters for low and medium vacua under development at NBS. Proc. 23rd American Vacuum Symposium, J. Vac. Sci. & Technol. **14**, 597.

Hickman, K.C.D., 1962, High vacuum with the Polyphenyl ethers, a self containing technology. Trans. 2nd Int. Vacuum Congress (Pergamon Press, Oxford) p. 307.

Hilleret, N., and R. Calder, 1977, Ion desorption of condensed gases. Proc. 7th Int. Vacuum Congress (Dobrozemsky, et al., Vienna) p. 227.

Himmelblau, D.M., 1974, Basic Principles and Calculations in Chemical Engineering, 3rd Ed. (Prentice-Hall, Englewood Cliffs, NJ).

Hirata, M., *et al.*, 1982, Calibration of ionization gauges. J. Vac. Sci. & Technol. **20**(4), 1159.

Hirata, M., *et al.*, 1987, Design and testing of a quartz friction vacuum gauge, using a self-oscillating circuit. J. Vac. Sci. & Technol. A **5**(4), 2393.

Hirsch, E.H., 1964, Br. J. Appl. Phys. **15**, 1535.

Hirsch, E.H., and J. Richards, 1974, Pressure fluctuations in a diffusion pump using polyphenyl ether. Vacuum **24**, 123.

Ho, T.L., 1932, Physics **2**, 386.

Ho, W., *et al.*, 1982, Calculation of sputter-ion pump speed. J. Vac. Sci. & Technol. **20**(4), 1010.

Hobson, J.P., 1963, Br. J. Appl. Phys. **14**, 544.

Hobson, J.P., 1964, J. Vac. Sci. & Technol. **1**, 1.

Hobson, J.P., 1970, J. Vac. Sci. & Technol. **7**, 351.

Hobson, J.P., 1973, J. Vac. Sci. & Technol. **10**, 73.

Hobson, J.P., 1974, The relationship of solid surfaces to vacuum science and technology. Proc. 6th Int. Vacuum Congress, Jpn. J. Appl. Phys., Suppl. 2, Pt. 1, p. 317.

Hobson, J.P., 1977, Methods of improving vacuum in space. J. Vac. Sci. & Technol. **14**, 1279.

Hobson, J.P., 1979, On the difference between true and net outgassing rates. J. Vac. Sci. & Technol. **16**, 84.

Hobson, J.P., 1984, The future of vacuum technology. J. Vac. Sci. & Technol. A **2**(2), 144.

Hobson, J.P., and E.V. Kornelsen, 1979, Uhv technique for intervacuum sample transfer. Proc. 25th American Vacuum Symposium, J. Vac. Sci. & Technol. **16**, 701.

Hobson, J.P., and P.A. Redhead, 1958, Can. J. Phys. **36**, 271.

Hobson, J.P., B.G. Baker and A.W. Pye, 1973, J. Vac. Sci. & Technol. **10**, 241.

Hoch, H., 1961, Vakuum-Technik **10**, 235.

Hoffman, D.M., 1979, Operation and maintenance of a diffusion-pumped vacuum system. J. Vac. Sci. & Technol. **16**, 71.

Hogg, J.L., 1906, Proc. Am. Acad. Arts Sci. **42**, 115.

Hojo, H., M. Ono and K. Nakayama, 1973, A gauge calibration system for 10^{-2}–10^{-7}Pa range. Proc. 7th Int. Vacuum Congress (Dobrozemsky, *et al.*, Vienna) p. 117.

Holanda, R., 1973, J. Vac. Sci. & Technol. **10**, 1133.

Holden, I., L. Holland and L. Laurenson, 1959, J. Sci. Instrum. **36**, 281.

Holkeboer, D.H., D.W. Jones, F. Pagano and D.J. Santeler, 1967, Vacuum Engineering (Boston Technical Publishers Cambridge, Mass.) 333pp.

Holland, L., 1956, Vacuum Deposition of Thin Films (Wiley, New York) 542pp.

Holland, L., 1970, Vacuum **20**, 175.

Holland, L., 1978, Design and operation characteristics of low pressure plasma systems. Vacuum **28**, 437.

Holland, L., and R.E.L. Cox, 1974, Getter sputtering – a review. Vacuum **24**, 107.

Holland, L., W. Steckelmacher and J. Yarwood, 1974, Vacuum Manual (Spon, London) 350pp.

Holland-Merten, E.L., 1953, Handbuch der Vakuumtechnik (Knapp, Halle).

Holleck, G.L., 1970, J. Phys. Chem. **74**, 503.

Hollins, P., and J. Pritchard, 1980, Guard vacuum sealed ultrahigh vacuum window. J. Vac. Sci. & Technol. **17**(2), 665.

Holm, R., and S. Storp, 1976, Vakuum-Technik **25**, 41, 73.

Holme, A.E., 1983, Leak testing using a He mass spectrometer on industrial scales. Proc. 9th Int. Vacuum Congress (Madrid) p. 227.

Homan, F., 1980, Externally operable variable aperture for ultrahigh vacuum. J. Vac. Sci. & Technol. **17**(2), 664.

Honig, R.E., and H.O. Hook, 1960, RCA Rev. **21**, 360.

Honig, R.E., and D.A. Kramer, 1969, RCA Rev. **30**, 285.

Hood, C.B., 1984, Cryopump speed measurement with the AVS/ASTM and ISO domes. J. Vac. Sci. & Technol. A **2**(1), 74.

Hooverman, R.H., 1963, J. Appl. Phys. **34**, 3505.

Hopfield, J.J., 1950, Rev. Sci. Instrum. **21**, 671.

Hopkins, J.S., and J.J. Valania, 1977, Helium leak testing of large pressure vessels of subassemblies. Proc. 23rd American Vacuum Symposium, J. Vac. Sci. & Technol. **14**, 617.

Hora, H., and H. Schwarz, 1974, Theory of quadrupole ion pump. Proc. 6th Int. Vacuum Congress, Jpn. J. Appl. Phys., Suppl. 2, Pt. 1, p. 69.

Horikoshi, G., 1987, Physical understanding of gas desorption mechanisms. J. Vac. Sci. & Technol. A **5**(4), 2501.

Horikoshi, G., and A. Miyahara, 1977, The H type metal gasket and its applications. Proc. 7th Int. Vacuum Congress (Dobrozemsky, et al., Vienna) p. 239.

Horikoshi, G., and H. Mizuno, 1977, A new type of uhv gauge by means of molecular density modulation principle. Proc. 7th Int. Vacuum Congress (Dobrozemsky, et al., Vienna) p. 125.

Horikoshi, G., K. Satoh and H. Mizuno, 1974, A scheme of initial start of a large vacuum system with sputter-ion pump, without baking procedure. Proc. 6th Int. Vacuum Congress, Jpn. J. Appl. Phys., Suppl. 2, Pt. 1, p. 205.

Housekeeper, W.G., 1923, J. Am. Inst. Electr. Eng. **42**, 954.

Howe, P.G., 1955, Rev. Sci. Instrum. **26**, 625.

Howl, D.A., and C.A. Mann, 1965, Vacuum **15**, 347.

Hseuh, H.C., 1982, The effect of magnetic field on the performance of Bayard–Alpert gauges. J. Vac. Sci. & Technol. **20**(2), 237.

Hseuh, H.C., and C. Lanni, 1987, Summary abstract: A thin-collector Bayard–Alpert gauge for 10^{-12} Torr vacuum. J. Vac. Sci. & Technol. A **5**(5), 3244.

Hseuh, H.C., et al., 1985, Glow discharge cleaning of stainless steel accelerator beam tubes. J. Vac. Sci. & Technol. A **3**(2), 518.

Hu, Y.Z., 1987, An evaluation of outgassing from titanium–molybdenum alloy wire. J. Vac. Sci. & Technol. A **5**(4), 2497.

Hua, Z.Y., 1987, Ultrahigh vacuum measurements in China. J. Vac. Sci. & Technol. A **5**(5), 3226.

Hua, Z.Y., et al., 1982, A group of terminal-flow ultrahigh vacuum gauges. J. Vac. Sci. & Technol. **20**(4), 1144.

Huang, Z.B., et al., 1987, Vacuum gauge calibration by combination of expansion and pumpdown methods. J. Vac. Sci. & Technol. A **5**(4), 2380.

Huber, W.K., 1963, Vacuum **13**, 399, 469.

Huber, W.K., 1977, Measurement of total and partial pressures in vacuum systems. Proc. 7th Int. Vacuum Congress (Dobrozemsky, et al., Vienna) p. 169.

Huber, W.K., and G. Rettinghaus, 1979, Validity of mass spectrometric measurements at pressures exceeding 10^{-4} mbar. Proc. 25th American Vacuum Symposium, J. Vac. Sci. & Technol. **16**, 681.

Huber, W.K., and E.A. Trendelenburg, 1962, Mass-spectrometer investigations in ultra-high vacuum systems. Trans. 8th American Vacuum Symposium (Pergamon Press, Oxford) p. 592.

Huber, H., and M. Warnecke, 1958, Le Vide **13**(74), 84.

Hudson, J.B., and R.L. Watters, 1966, Trans. IEEE **IM-15**, 94.

Hug, P., 1973, Test of vacuum tightness of metal-ceramic bond (in German). Vakuum-Technik **22**, 214.

Hughes, F.L., 1959, Phys. Rev. **113**, 1036.

Hulek, Z., 1977, On the X-ray limit measurement by variation of electron energy method. Proc. 7th Int. Vacuum Congress (Dobrozemsky, et al., Vienna) p. 145.

Hull, A.W., 1946, J. Appl. Phys. **17**, 685.

Hunter, W.R., 1963, Vacuum **13**, 193.

Hurrle, K., F.M. Jablonski and H. Roth, 1973, Technical Dictionary of Vacuum Physics and Vacuum Technology (Pergamon Press, Oxford).

I

Ichimura, K., et al., 1988, Absorption/desorption of hydrogen isotopes and isotopic waters by Zr-alloy getters. J. Vac. Sci. & Technol. A **6**(4), 2541.

Inanananda, S., 1947, High Vacua (Van Nostrand, New York).

Ishii, H., and K. Nakayama, 1962, A serious error caused by mercury vapour stream in the measurement with a McLeod gauge. Trans. 2nd Int. Vacuum Congress (Pergamon Press, Oxford) p. 519.

Ishimaru, H., 1976, Vacuum 26, 243.

Ishimaru, H., 1977, Vacuum 27, 465.

Ishimaru, H., 1978, Bakeable Al vacuum chamber, bellows and metal seal for uhv. J. Vac. Sci. & Technol. 15, 1853.

Ishimaru, H., 1982, Al-alloy ceramic ultrahigh vacuum cryogenic feedthrough. Vacuum 32(12), 753.

Ishimaru, H., 1984, All Al-alloy ultrahigh vacuum system for large scale electron-positron collider. J. Vac. Sci. & Technol. A 2, 1170.

Ishimaru, H., and G. Horikoshi, 1979, Bakeable Al vacuum chamber and bellows with an Al flange and metal seal for uhv. IEEE Trans. Nucl. Sci. NS-26, 4000.

Ishimaru, H., S. Fukumoto and N. Yagi, 1974, Pulsed gas shutter. Proc. 6th Int. Vacuum Congress, Jpn. J. Appl. Phys., Suppl. 2, Pt. 1, p. 191.

ISO (International Organization for Standardization), 1974a, Methods of measurement of the performance characteristics of sputter-ion pumps, DIS-3556/I/1974.

ISO, 1974b, Vacuum gauges – Calibration by direct comparison with a reference gauge, DIS-3567/1974.

ISO, 1974c, Ionization vacuum gauges – Calibration by direct comparison with a reference gauge, DIS-3568/1974.

ISO, 1975, Vacuum gauges – Standard methods for calibration. Pressure reduction by continuous flow in the pressure range 10^{-1}–10^{-5} Pa, DIS-3570/I/1975.

ISO, 1976, Vacuum gauges of the thermal conductivity type – Calibration by direct comparison with a reference gauge, DIS-5300/1976.

ISO, 1978a, Mass-spectrometer-type leak detector calibration, DIS-3530.2/1978.

ISO, 1978b, Methods of measurement of the performance characteristics of positive displacement vacuum pumps, ISO-1607/I/1970 (Measurement of pumping speed), ISO-1607/II/1978 (Measurement of ultimate pressure).

ISO, 1978c, Methods of measurement of the performance characteristics of vapour vacuum pumps, ISO-1608/I/1970 (Measurement of pumping speed), ISO-1608/II/1978 (Measurement of critical backing pressure).

ISO/DIS, 1981, Vacuum Technology Vocabulary, Part I; General Terms, Part II; Vacuum pumps and related terms.

Itoh, A., et al., 1988, Reduction of outgassing from stainless steel surfaces by glow discharge. J. Vac. Sci. & Technol. A 6(4), 2421.

IUVSTA, 1978, Visual Aid Project (American Vacuum Society, New York, L.C. Beavis).

Iverson, M.V., and J.L. Hartley, 1982, Calibration of standard leaks. J. Vac. Sci. & Technol. **20**(4), 982.

J

Jackson, G.L., 1982, The leak testing program of doublet III Tokamak. J. Vac. Sci. & Technol. **20**(4), 1182.

Jackson, A.G., and T.W. Haas, 1967, J. Vac. Sci. & Technol. **4**, 42.

Jaeckel, R., 1950, Kleinste Drucke, Ihre Messung und Erzeugung (Springer, Berlin) 302pp.

Jaeckel, R., 1962, Entgasung. Trans. 2nd Int. Vacuum Congress (Pergamon Press, Oxford) p. 17.

Jahrreiss, H., 1987, Otto von Guericke (1602–1686) in memoriam. J. Vac. Sci. & Technol. A **5**(4), 2466.

Janecke, D., 1958, Vakuum-Technik **7**, 52.

Jansen, W., 1980, Leak-testing with search gases. Vakuum-Technik **29**(4), 105.

Jefferson, T.B., 1955, The Welding Encyclopedia (McGraw-Hill, New York).

Jenkins, R.O., 1958, J. Sci. Instrum. **35**, 428.

Jenkins, L.H., and M.F. Chung, 1971, Surf. Sci. **26**, 151; **28**, 409.

Jepsen, R.L., 1961, J. Appl. Phys. **32**, 2619.

Jepsen, R.L., 1968, The physics of sputter-ion pumps. Proc. 4th Int. Vacuum Congress (Institute of Physics, London) p. 317.

Jepsen, R.L., S.L. Mercer and M.J. Callaghan, 1959, Rev. Sci. Instrum. **30**, 377.

Jepsen, R.L., A.B. Francis, S.L. Rutherford and B.E. Kietzmann, 1960, Stabilized air pumping with diode type getter-ion pumps. Trans. 7th American Vacuum Symposium (Pergamon Press, Oxford) p. 45.

Johnson, K.H., and R.P. Messmer, 1974, J. Vac. Sci. & Technol. **11**, 236.

Johnson, F.S., J.M. Carroll and D.E. Evans, 1972, J. Vac. Sci. & Technol. **9**, 450.

Jones, G.O., 1956, Glass (Wiley, New York).

Jones, G.O., 1972, Glass (Chapman and Hall, London).

Jones, D.W., and C.A. Tsonis, 1964, J. Vac. Sci. & Technol. **1**, 19.

Jones, A.W., E. Jones and E.M. Williams, 1973, Investigation, by techniques of electron stimulated desorption, of the merits of glow discharge cleaning of the surfaces of vacuum chambers at the CERN intersecting storage rings. Vacuum **23**, 227.

Jordan, I.R., 1962, New developments in vacuum sealing. Trans. 2nd Int. Vacuum Congress (Pergamon Press, Oxford) p. 1302.

Juillet, F., and F. van Nieuwenhuyze, 1969, Techniques et Applications de l'Ultra-Vide (Masson, Paris) 204pp.

K

Kageyama, K., et al., 1974, A pulsed discharge vacuum gauge. Proc. 6th Int. Vacuum Congress, Jpn. J. Appl. Phys., Suppl. 2, Pt. 1, p. 109.

Kaminsky, M., 1965, Atomic and Ionic Impact on Metal Surfaces (Academic Press, New York).

Kanaji, T., et al., 1977, Three-grid modulated Bayard–Alpert gauge. Proc. 7th Int. Vacuum Congress (Dobrozemsky, et al., Vienna) p. 133.

Kanaji, T., et al., 1987, Curious zig-zag in modulation characteristics, curves of a Bayard–Alpert gauge in the extrahigh vacuum region. J. Vac. Sci. & Technol. A 5(4), 2397.

Kanellopoulos, N.K., 1979, Double-acting mercury-piston gas pump. J. Phys. E (Sci. Instrum.) 12(8), 680.

Karlsson, E., and K. Siegbahn, 1960, Nucl. Instrum. & Methods 7, 113.

Katheder, H., and K. Lennermann, 1984, Leak testing of very long narrow channels (in German). Vakuum-Technik 33(2), 34.

Kato, S., M. Nojiri and H. Oikawa, 1974, Characteristics of Orbitron pump. Trans. 6th Int. Vacuum Congress (Kyoto).

Katz, H., 1974, Technologische Grundprozesse der Vakuumelektronik (Springer, Berlin).

Kay, E., 1962, Adv. Electr. & Electron. Phys. 17, 245.

Kaye, G.W.C., 1927, High Vacua (Longmans, Green & Co., London).

Kellner, E., et al., 1983, Use of turbomolecular pumps in rough/medium vacuum during pumpdown. Vakuum-Technik 32(5), 136.

Kendall, B.R.F., 1962, J. Sci. Instrum. 39, 267.

Kendall, B.R.F., 1974, Apparatus for measuring the internal volumes of vacuum systems. J. Vac. Sci. & Technol. 11, 610.

Kendall, B.R.F., 1982a, Anaerobic polymers as high vacuum leak sealants. J. Vac. Sci. & Technol. 20(2), 248.

Kendall, B.R.F., 1982b, Obtaining 10^{-5} Pa with oil-sealed rotary pumps. J. Vac. Sci. & Technol. 21(3), 886.

Kendall, B.R.F., 1983a, The pressure multiplier revisited. J. Vac. Sci. & Technol. A 1(4), 1875.

Kendall, B.R.F., 1983b, Comments on resistor network simulation for flow in vacuum systems (Ohta et al., 1983). J. Vac. Sci. & Technol. A 1(4), 1881.

Kendall, B.R.F., et al., 1987a, Passive levitation of small particles in vacuum:

Possible application to vacuum gauging. J. Vac. Sci. & Technol. A 5(4), 2458.

Kendall, B.R.F., et al., 1987b, Summary abstract: Electrostatic levitation in vacuum and possible gauge applications. J. Vac. Sci. & Technol. A 5(5), 3224.

Kent, T.B., 1955, J. Sci. Instrum. 32, 132.

Kidnay, A.J., and M.J. Hiza, 1970, Cryogenics 10, 271.

Kienel, G., and A. Lorenz, 1960a, Vakuum-Technik 9, 3.

Kienel, G., and A. Lorenz, 1960b, Vakuum-Technik 9, 217.

Kienel, G., and M. Wutz, 1966, Vakuum-Technik 15, 40.

Kindall, S.M., and E.S.J. Wang, 1962, Vacuum pumping by cryosorption. Trans. 9th American Vacuum Symposium (Macmillan, New York) p. 243.

King, J.G., 1962, Production leak testing of large pressure vessels. Trans. 8th American Vacuum Symposium (Pergamon Press, Oxford) p. 1100.

Kirby, R.E., et al., 1982, Materials and lubrication for bearings/gears in ultrahigh vacuum. Vakuum-Technik 31(3), 67.

Kisliuk, P., 1959, J. Chem. Phys. 31, 1605.

Kistemaker, J., D. Onderdelinden, F.W. Saris and W.F. van Derweg, 1968, The interaction of heavy particles with metal surfaces. Proc. 4th Int. Vacuum Congress (Institute of Physics, London) p. 164.

Kleber, P., 1975, Pressure, what does it mean in vacuum chambers with cryosurfaces. Vacuum 25, 191.

Kleber, P., 1977, SPACELAB – A European contribution to manned space activities in the eighties. What does it mean for a vacuum scientist. Proc. 7th Int. Vacuum Congress (Dobrozemsky, et al., Vienna) p. 333.

Kleber, P., 1983, Vacuum research in space. Proc. 9th Int. Vacuum Congress (Madrid) p. 264.

Klein, H.H., et al., 1984, Use of refrigerator-cooled cryopumps in sputtering plants. J. Vac. Sci. & Technol. A 2(2), 187.

Klemperer, O., and J.P.G. Shepherd, 1963, Adv. Phys. 12, 355.

Klipping, G., 1974, Relation between cryogenics and vacuum technology up to now and in the future. Proc. 6th Int. Vacuum Congress, Jpn. J. Appl. Phys., Suppl. 2, Pt. 1, p. 81.

Klopfer, A., 1962, An ionization gauge for measurement of ultra-high vacua. Trans. 2nd Int. Vacuum Congress (Pergamon Press, Oxford) p. 439.

Klopfer, A., and W. Ermrich, 1959, Properties of small titanium-ion pump. Trans. 6th American Vacuum Symposium (Pergamon Press, Oxford) p. 297.

Klopfer, A., and W. Schmidt, 1960, Vacuum 10, 363.

Kluge, A., 1974, A new quadrupole mass spectrometer of mass independent sensitivity (in German) Vakuum-Technik 23, 168.

Knoll, M., 1959, Materials and Processes of Electron Devices (Springer, Berlin) 480pp.

Knor, Z., 1960, Rev. Sci. Instrum. **31**, 351.

Knudsen, M., 1910, Ann. Phys. **31**, 205; **32**, 890; **33**, 1435.

Kobari, T., and H.J. Halama, 1987, Photon stimulated desorption from a vacuum chamber at the National Synchrotron Light Source. J. Vac. Sci. & Technol. A **5**(4), 2355.

Kobayashi, S., and K. Yada, 1960, Leak test of rubber gaskets. Adv. Vacuum Sci. & Technol. (Pergamon Press, Oxford) p. 248.

Kobayashi, M., et al., 1987, Angular distribution of photon stimulated desorption in a vacuum duct observed by using unidirectional detector. J. Vac. Sci. & Technol. A **5**(4), 2417.

Kofoid, M.J., and P. Zieske, 1962, Rev. Sci. Instrum. **33**, 1115.

Kohl, W.H., 1960, Materials and Techniques for Electron Tubes (Reinhold, New York) 638pp.

Kohl, W.H., 1967, Handbook of Materials and Techniques for Vacuum Devices (Reinhold, New York) 623pp.

Kokubun, K., et al., 1987, Unified formula describing the impedance of a quartz oscillator on gas pressure. J. Vac. Sci. & Technol. A **5**(4), 2450.

Komiya, S., et al., 1979, Direct-molecular-beam method for mass selective outgassing rate measurement. Proc. 25th American Vacuum Symposium, J. Vac. Sci. & Technol. **16**, 689.

Konishi, A., and S. Mizumachi, 1974, Starting characteristics of triode and diode sputter-ion pumps. Proc. 6th Int. Vacuum Congress, Jpn. J. Appl. Phys., Suppl. 2, Pt. 1, p. 65.

Kornelsen, E.V., 1960, A small ionic pump employing metal evaporation. Trans. 7th American Vacuum Symposium (Pergamon Press, Oxford) p. 29.

Kornelsen, E.V., 1967, Developments in the measurement of low pressures. Trans. 3rd Int. Vacuum Congress (Pergamon Press, Oxford) p. 65.

Korolev, B., 1950, Fundamentals of Vacuum Techniques (in Russian) (Goss-energoizdat, Moscow).

Kouptsidis, J.S., 1977, Vacuum problems of electron storage rings. Proc. 7th Int. Vacuum Congress (Dobrozemsky, et al., Vienna) p. 341.

Kozman, T.A., 1983, Leak testing of fusion devices. Proc. 9th Int. Vacuum Congress (Madrid) p. 233.

Kraus, T., 1982, Approximation of pressure curves during evacuation. Vakuum-Technik **31**(8), 226.

Kraus, T., and E. Zollinger, 1974, Vakuum-Technik **23**, 40.

Kreisel, W., 1976, A modified Knudsen manometer. Vacuum **26**, 339.

Kruger, C.H., and A.H. Shapiro, 1961, Vacuum pumping with a bladed axial-flow turbomachine. Trans. 7th American Vacuum Symposium (Pergamon Press, Oxford) p. 6.

Kubiak, R.A.A., et al., 1983, On baking a cryopumped ultrahigh vacuum system. J. Vac. Sci. & Technol. A 1(4), 1872.

Kudzia, J., and W. Stowko, 1981, Numerical method of calculating ion current in a high pressure ionization gauge. Vacuum 31(8/9), 359.

Kuhn, M., and P. Bachmann, 1987, Selection and analytical monitoring of backing pump fluids in semiconductor processes. J. Vac. Sci. & Technol. A 5(4), 2534.

Kunzl, V., and J.B. Slavik, 1935, Zs. Techn. Phys. 16, 272.

Kuo, Y.H., 1981, Non-liniarity of ionization gauges at high pressure side. Vacuum 31(7), 303.

Kuo, T.C., et al., 1988, Thin-film thermocouple gauge. J. Vac. Sci. & Technol. A 6(3), 1150.

Kutzner, K., and I. Wietzke, 1972, Method for evaluating the degassing rates (in German). Vakuum-Technik 21, 54.

Kuus, G., and W. Martens, 1980, Zr–Ni selective H_2 getters for high pressure discharge lamps. Vacuum 30(6), 213.

Kuypers, N.R., 1977, A new high vacuum (diffusion) pump and its application in the design of a gas chromatograph/quadrupole mass spectrometer. Proc. 7th Int. Vacuum Congress (Dobrozemsky, et al., Vienna) p. 9.

Kuznetzov, V.K., 1959, Mechanical Pumps (in Russian) (Gossenergoizdat, Moscow).

Kuznetzov, M.V., A.S. Nazarov and G.F. Ivanovsky, 1969, J. Vac. Sci. & Technol. 6, 34.

L

Lafferty, J.M., 1961, A hot–cathode magnetron ionization gauge for the measurement of ultra-high vacua. Trans. 7th American Vacuum Symposium (Pergamon Press, Oxford) p. 97.

Lafferty, J.M., 1962a, Further developments in the hot-cathode magnetron ionization gauge. Trans. 2nd Int. Vacuum Congress (Pergamon Press, Oxford) p. 460.

Lafferty, J.M., 1962b, The hot-cathode magnetron ionization gauge with an electron multiplier detector. Trans. 9th American Vacuum Symposium (Macmillan, New York) p. 438.

Lafferty, J.M., 1963, Rev. Sci. Instrum. 34, 467.

Lafferty, J.M., 1964, Techniques of high vacuum. Report 64–RL–3791 G (General Electric, Schenectady).

Lafferty, J.M., 1972, Review of pressure measurement techniques for ultra-high vacua. Trans. 5th Int. Vacuum Congress (American Vacuum Society, New York) p. 101 [J. Vac. Sci. & Technol. **9**, 101].

Lafferty, J.M., 1987, History of the International Union for Vacuum Science Technique and Applications. J. Vac. Sci. & Technol. A **5**(4), 405.

Lafferty, M.S. Kaminsky and J., eds.., 1980, Dictionary of Terms for Vacuum Science and Technology (American Institute of Physics, New York).

Lake, G.J., P.P. Lindley and A.G. Thomas, 1963, J. Sci. Instrum. **40**, 40.

Lambert, R.M., and C.M. Comrie, 1974, J. Vac. Sci. & Technol. **11**, 530.

Lander, J.J., 1953, Phys. Rev. **91**, 1382.

Lander, H.J., W.T. Hess and S.S. White, 1962, Electron beam welding of high strength alloys. Trans. 2nd Int. Vacuum Congress (Pergamon Press, Oxford) p. 685.

Landfors, A.A., and M.H. Hablanian, 1959, Diffusion pump speed measurements at very low pressures. Trans. 5th American Vacuum Symposium (Pergamon Press, Oxford) p. 22.

Landfors, A.A., and M.H. Hablanian, 1983, Pumping speed measurement for water vapour. J. Vac. Sci. & Technol. A **1**(2/1), 150.

Lang, H., 1977, Gas cooled Roots pumps (in French). Le Vide **32**, 60.

Lange, W.J., 1957, Ultrahigh vacuum techniques, Report AECU-3889/1957 (Westinghouse, Pittsburgh).

Lange, W.J., 1959, Rev. Sci. Instrum. **30**, 602.

Lange, W.J., 1965, J. Vac. Sci. & Technol. **2**, 74.

Lange, W.J., 1975, Methane production and sorption capacities of flashed barium getters. Proc. 21st American Vacuum Symposium. J. Vac. Sci. & Technol. **12**, 543.

Lange, W.J., 1977, Experimental studies of Zr/Al getter pumps. Proc. 23rd American Vacuum Symposium, J. Vac. Sci. & Technol. **14**, 582.

Lange, W.J., and D. Alpert, 1957, Rev. Sci. Instrum. **28**, 726.

Lange, W.J., and D.P. Eriksen, 1966, J. Vac. Sci. & Technol. **3**, 303.

Lange, W.J., and J.H. Singleton, 1978, Evaluation of a turbomolecular pump. J. Vac. Sci. & Technol. **15**, 1189.

Lange, W.J., J. Singleton and D.P. Eriksen, 1966, J. Vac. Sci. & Technol. **3**, 338.

Langmuir, I., 1913, J. Am. Chem. Soc. **35**, 105.

Langmuir, I., 1915, J. Am. Chem. Soc. **37**, 1139.

Langmuir, I., 1918, J. Am. Chem. Soc. **40**, 1361.

Lapalle, R.R., 1972, Practical Vacuum Systems (McGraw-Hill, New York) 238pp.

Laporte, H., 1957, Hochvakuum, Seine Erzeugung, Messung und Anwendung im Laboratorium (Knapp, Halle) 175pp.

Latham, R.V., 1981, High Voltage Vacuum Insulation – The Physical Basis (Academic Press, London).

Laufer, M.K., 1962, Standard leaks and their calibration. Trans. 8th American Vacuum Symposium (Pergamon Press, Oxford) p. 1086.

Laughner, V.H., and A.D. Hargan, 1956, Handbook of fastening and joining metal parts (McGraw-Hill, New York).

Laurenson, L., 1980, Vacuum fluids. Vacuum 30(7), 275.

Laurenson, L., 1982, Technological applications of pumping fluids. J. Vac. Sci. & Technol. 20(4), 989.

Laurenson, L., and G. Caporiccio, 1977, Perfluoropolyethers – Universal vacuum fluid. Proc. 7th Int. Vacuum Congress (Dobrozemsky, et al., Vienna) p. 263.

Laurenson, L., et al., 1979, The vacuum use of perfluoropolyether. Vacuum 29(11/12), 433.

Laurenson, L., et al., 1988, Rotary pump backstreaming. J. Vac. Sci. & Technol. A 6(2), 238.

Laurent, J.M., and O. Gröbner, 1979, Distributed sputter-ion pumps for use in low magnetic fields. IEEE Trans. Nucl. Sci. NS-26, 3997.

Laurent, J.M., C. Benvenuti and F. Scalambrin, 1977, Pressure measurments for the ISR at CERN. Proc. 7th Int. Vacuum Congress (Dobrozemsky, et al., Vienna) p. 113.

Laville Saint-Martin, B., 1973, Le Vide 28, 255.

Lawson, R.W., 1962, J. Sci. Instrum. 39, 281.

Lawson, R.W., 1967, Br. J. Appl. Phys. 18, 1763.

Lawson, R.P.W., 1975, Temperature dependence of gas conductance through porous silicon carbide plugs. Vacuum 25, 377.

Lawson, R.W., and J.W. Woodward, 1967, Vacuum 17, 205.

Leblanc, M., 1951, La Technique du Vide (Colin, Paris).

Leck, J.H., 1964, Pressure Measurement in Vacuum Systems, 2nd Ed. (Chapman and Hall, London).

Leck, J.H., 1970, Vacuum 20, 369.

Leck, J.H., and A. Riddoch, 1956, Br. J. Appl. Phys. 7, 153.

Leck, J.H., and B.P. Stimpson, 1972, J. Vac. Sci. & Technol. 9, 293.

Leckey, J.H., and M.D. Boeckmann, 1988, Static gas analysis by transient flow technique. J. Vac. Sci. & Technol. A 6(4), 2514.

Lee, T.H., 1963, Vacuum 13, 167.

Lee, H.L., and K.O. Neville, 1967, Handbook of Epoxy Resins (McGraw-Hill, New York).

Lee, C.O., and J.H. Peavey, 1976, Simple method for determining the internal volume of a vacuum system. J. Vac. Sci. & Technol. 13, 1108.

Lee, D., H. Tomasche and D. Alpert, 1962, Adsorption of molecular gases on surfaces and its effect on pressure measurement. Trans 2nd Int. Vacuum Congress (Pergamon Press, Oxford) p. 153.

Leiby, C.C., and C.L. Chen, 1960, J. Appl. Phys. 31, 268.

Leisegang, S., 1956, Elektronenmikroskope, in: Handbuch der Physik, Vol. 33 (Springer, Berlin) p. 416.

Levenson, L.L., N. Milleron and D.H. Davis, 1960, The optimization of molecular flow conductance. Trans. 7th American Vacuum Symposium (Pergamon Press, Oxford) p. 372.

Levenson, L.L., N. Milleron and D.H. Davis, 1963, Le Vide 18, 42.

Lewin, G., 1965, Fundamentals of Vacuum Science and Technology (McGraw-Hill, New York) 248pp.

Lewin, G., and D. Mullaney, 1964, A fast acting ultra-high vacuum gas inlet valve, Trans. 10th American Vacuum Symposium (Macmillan, New York) p. 176.

Lewin, G., and F.H. Tenney, 1974, Vacuum problems associated with the conceptual design of Tokamak Fusion Reactors. Proc. 6th Int. Vacuum Congress, Jpn. J. Appl. Phys., Suppl. 2, Pt. 1, p. 221.

Li, W., and Z. Zhang, 1987, A miniature deflected-ion-beam ionization gauge with the channel electron multiplier. J. Vac. Sci. & Technol. A 5(4), 2447.

Lichtman, D., 1984, Perspectives on residual gas analysis. J. Vac. Sci. & Technol. A 2(2), 200.

Littman, C., 1961, Rev. Sci. Instrum. 32, 1154.

Liu, B.K., et al., 1982, Cryosorption pumping of hydrogen. J. Vac. Sci. & Technol. 20(4), 1000.

Liu, B., et al., 1987, The thermodynamic treatment of the cryosorption process in a cryopump. J. Vac. Sci. & Technol. A 5(4), 2577.

Liu, Y.C., et al., 1988, Pumping mechanism for N_2 gas in a triode ion pump with a 1100 Al cathode. J. Vac. Sci. & Technol. A 6(1), 139.

Liversey, R.G., 1982, Pumpdown calculations for small volume rapid cycle duties. Vacuum 32(10/11), 651.

Liversey, R.G., and L.J. Budgen, 1982, Computation of the performance of mechanical boosters. J. Vac. Sci. & Technol. 20(4), 1014.

Livshitz, A.I., 1979, Superpermeability of solid membranes and gas evacuation. Vacuum 29, 103.

Livshitz, A.I., and M.E. Notkin, 1979, Permeation of hydrogen through a Pd membrane. Vacuum 29, 113.

Logan, J.S., J.H. Keller and R.G. Simmons, 1977, The rf glow discharge sputtering model. Proc. 23rd American Vacuum Symposium, J. Vac. Sci. & Technol. 14, 92.

Longsworth, R.C., and G.E. Bonney, 1982, Cryopump regeneration studies. J. Vac. Sci. & Technol. 21(4), 1020.

Longsworth, R.C., and Y. Lahav, 1987, A cryopumped leak detector. J. Vac. Sci. & Technol. A 5(4), 2646.

Loriot, G., and T. Moran, 1975, Reliability of a capacitance manometer in the range 2 × 10^{-4}–5 × 10^{-6} Torr. Rev. Sci. Instrum. **46**, 140.

Loyalka, S.K., T.S. Storvich and H.S. Park, 1976, Poiseuille flow and thermal creep flow in long, rectangular channels. J. Vac. Sci. & Technol. **13**, 1188.

Lu, M., and P.W. Fang, 1987, New cathode material for sputter ion pumps: Aluminum mixed rare earth alloy. J. Vac. Sci. & Technol. A **5**(4), 2591.

Lucas, T.E., 1965, Vacuum **15**, 222.

Lucatorto, T.B., et al., 1979, Capillary array-window with guard vacuum. Appl. Opt. **18**(14), 2505.

Luches, A., and M.R. Perrone, 1976, Use of perfluoropolyether fluids in particle accelerator vacuum systems. J. Vac. Sci. & Technol. **13**, 1097.

Luches, A., and A. Zecca, 1972, J. Vac. Sci. & Technol. **9**, 1237.

Lunelli, B., 1978, New cold trap. J. Vac. Sci. & Technol. **15**, 1192.

M

Madey, T.E., 1984, Early applications of vacuum, from Aristotle to Langmuir. J. Vac. Sci. & Technol. A **2**(2), 110.

Madey, T.E., and J.T. Yates, 1967, Suppl. Nuovo Cimento **5**, 483.

Maissel, L.I., and M.H. Francombe, 1973, An Introduction to Thin Films (Gordon and Breach, New York) 301pp.

Maissel, L.I., and R. Glang, 1970, Handbook of Thin Film Technology (McGraw-Hill, New York).

Malev, M.D., 1973a, Vacuum **23**, 43.

Malev, M.D., 1973b, Vacuum **23**, 359.

Malev, M.D., and E.M. Trachtenberg, 1973, Built-in getter-ion pumps. Vacuum **23**, 403.

Malev, M.D., and E.M. Trachtenberg, 1975, Vacuum **25**, 211.

Maliakal, J.C., P.J. Limon, E.E. Arden and R.G. Herb, 1964, J. Vac. Sci. & Technol. **1**, 54.

Maliakal, J.C., J.M. Crosby and C.B. Sibley, 1965, J. Vac. Sci. & Technol. **2**, 289.

Malinowski, M.E., 1978, Isotope effect in the pumping of hydrogen by titanium thin films. J. Vac. Sci. & Technol. **15**, 95.

Manes, M., and R.J. Grant, 1963, Calculation methods for the design of regenerative cryosorption pumping systems. Trans. 10th American Vacuum Symposium (Macmillan, New York) p. 122.

Mark, H., 1977, Space shuttle. J. Vac. Sci. & Technol. **14**, 1243.

Mark, J.T., and K. Dreyer, 1960, Ultra-high vacuum system developments for model-C stellarator. Trans. 6th American Vacuum Symposium (Pergamon Press, Oxford) p. 177.

Markley, F., R. Roman and P. Vosecek, 1962, Outgassing data for several epoxy resins and rubbers for the zero gradient synchroton. Trans. 2nd Int. Vacuum Congress (Pergamon Press, Oxford) p. 78.

Marks, R., 1957, Rev. Sci. Instrum. **28**, 381.

Marland, E.A., 1973, Herbert McLeod. Vacuum **23**, 171.

Marmy, P., 1983, Friction and wear in ultrahigh vacuum. Vakuum-Technik **32**(6), 163.

Marr, J.M., 1968, Leakage Testing Handbook, Report N68-20389 (General Electric, Schenectady).

Martin, J.H., 1948, Rev. Sci. Instrum. **19**, 404.

Martin, L.H., and R.D. Hill, 1949, A Manual of Vacuum Practice (University Press, Melbourne).

Marx, H., 1957, Glas und Keramik (Fachenbuch Verlag, Leipzig).

Mase, M., et al., 1988, Development of a new type of oil-free turbo vacuum pump. J. Vac. Sci. & Technol. A **6**(4), 2518.

Mason, B.F., and B.R. Williams, 1972, Rev. Sci. Instrum. **43**, 375.

Matheson, W.G., 1955, Vacuum systems and techniques in lamp and electronic tube industries. Trans. 1st American Vacuum Symposium (Pergamon Press, Oxford) p. 104.

Mathewson, A.G., 1974, The surface cleanliness of 316 L+N stainless steel, studied by SIMS and AES. Vacuum **24**, 505.

Mathewson, A., J. Kouptsidis and L. Hipp, 1977, In-situ glow discharge cleaning of the PETRA aluminium vacuum chambers. Proc. 7th Int. Vacuum Congress (Dobrozemsky, et al., Vienna) p. 235.

Mathewson, A.G., et al., 1987, Comparison of the synchrotron radiation induced gas desorption in aluminum vacuum chambers after chemical and argon glow discharge cleaning. J. Vac. Sci. & Technol. A **5**(4), 2512.

Matsunaga, M., and K. Hoshimoto, 1974, Frictional behavior of molybdenum disulfide in high vacuum. Proc. 6th Int. Vacuum Congress, Jpn. J. Appl. Phys., Suppl. 2, Pt. 1, p. 305.

Maurice, L., 1971, Pratique du Controle de l'Etanchéité a l'Helium (Soc. Franc. Ing. Techn. du Vide, Paris).

Maurice, L., 1974a, Dynamic seals. Proc. 6th Int. Vacuum Congress, Jpn. J. Appl. Phys., Suppl. 2, Pt. 1, p. 17.

Maurice, L., 1974b, A new molecular pump. Proc. 6th Int. Vacuum Congress, Jpn. J. Appl. Phys., Suppl. 2, Pt. 1, p. 21.

Maurice, L., et al., 1979, Oil backstreaming in turbomolecular and oil diffusion pumps. Proc. 25th American Vacuum Symposium. J. Vac. Sci. & Technol. **16**, 741.

McCracken, G.M., 1969, Vacuum **19**, 311.

McCracken, G.M., 1974, Interaction of light-ions with surfaces. Proc. 6th Int. Vacuum Congress, Jpn. J. Appl. Phys., Suppl. 2, Pt. 1, p. 269.

McCracken, G.M., and N.A Pashley, 1966, J. Vac. Sci. & Technol. **3**, 96.

McCulloh, K.E., 1983, Calibration of molecular drag vacuum gauges. J. Vac. Sci. & Technol. A 1(2/1), 168.

McCulloh, K.E., et al., 1986, Summary abstract: The NBS orifice flow primary high vacuum standard. J. Vac. Sci. & Technol. A 4(3), 362.

McIrvine, E.C., and R.C. Bradley, 1957, J. Chem Phys. **27**, 646.

McLeod, H.G., 1874, Philos. Mag. **48**, 110.

McRae, E.G., 1966, J. Chem. Phys. **45**, 3258.

Mehrhoff, T.K., and L.W. Barnes, 1984, Getter pumping speed measurements in the range 10^{-2} to 10^{-7} liter per second. J. Vac. Sci. & Technol. A **2**, 1210.

Meinke, C., and G. Reich, 1967, J. Vac. Sci. & Technol. **4**, 356.

Melfi, T.M., 1969, J. Vac. Sci. & Technol. **6**, 322.

Mennenga, H., 1980, Leak detection of small parts (with halogen, He turbo). Vakuum-Technik **29**(7), 195.

Menzel, D., and R. Gomer, 1964, J. Chem. Phys. **41**, 3311.

Merrill, R.P., and D.L. Smith, 1971, J. Vac. Sci. & Technol. **8**, 517.

Messer, G., 1977, Calibration of vacuum gauges in the range 10^{-7} Pa to 10 Pa with fundamental methods. Proc. 7th Int. Vacuum Congress (Dobrozemsky, et al., Vienna) p. 153.

Messer, G., and N. Treitz, 1977, Sensitive mass-selective outgassing rate measurements on baked stainless steel and copper samples. Proc. 7th Int. Vacuum Congress (Dobrozemsky, et al., Vienna) p. 223.

Messer, G., et al., 1987, High vacuum measured by the spinning rotor gauge. J. Vac. Sci. & Technol. A 5(4), 2440.

Metcalfe, R.A., and F.W. Trabert, 1962, Performance of a double-walled ultra-high vacuum chamber. Trans. 8th American Vacuum Symposium (Pergamon Press, Oxford) p. 1211.

Meyer, D.E., 1974, J. Vac. Sci. & Technol. **11**, 168.

Meyer, E.A., and R.G. Herb, 1967, J. Vac. Sci. & Technol. **4**, 63.

Mikhail, R.S., and S. Brunauer, 1975, Surface area measurements by nitrogen and argon adsorption. J. Colloid & Interface Sci. **52**, 572.

Miller, J.R., 1972, J. Vac. Sci. & Technol. **9**, 201.

Miller, H.C., 1973a, Gas desorption temperature of two molecular sieves. J. Vac. Sci. & Technol. **10**, 859.

Miller, J.R., 1973b, A thermally shielded atmospheric pressure standard leak. J. Vac. Sci. & Technol. **10**, 882.

Miller, J.E., 1982, Design of Tokamak vacuum vessels. J. Vac. Sci. & Technol. **20**(4), 1168.

Miller, C.F., and R.W. Shepard, 1961, Vacuum **11**, 58.

Miller, M.K., et al., 1976, A helium gas flow cryostat and specimen air-lock assembly for uhv use. J. Phys. E (Sci. Instrum.) **9**, 116.

Milleron, N., 1959, Some component designs permitting ultra-high vacuum with large oil diffusion pumps. Trans. 5th American Vacuum Symposium (Pergamon Press, Oxford) p. 140.

Milleron, N., 1965, Porous metal isolation traps and cryosorbents in vacuum technique. Trans. 3rd Int. Vacuum Congress, Vol. 2 (Pergamon Press, Oxford) p. 189.

Millner, T., F. Szalkay, et al., 1953, Vacuum Technology (in Hungarian), Vol. I-III (Nehzéipari Kiadó, Budapest).

Minter, C.C., 1958, Rev. Sci. Instrum. **29**, 793.

Minter, C.C., 1960, Rev. Sci. Instrum. **31**, 458.

Mirgel, K.H., 1972, J. Vac. Sci. & Technol. **9**, 408.

Miyahara, Y., 1986, Calculation of pressure distribution in an electron storage ring and comparison with monitored pressure and beam life time. J. Vac. Sci. & Technol. A **4**(1), 111.

Miyake, H., et al., 1988, Compatibility of spinning rotor gauge with tritium handling systems. J. Vac. Sci. & Technol. A **6**(1), 158.

Mizuno, H., and G. Horikoshi, 1977, Automatic modulation methods for Bayard–Alpert gauges. Proc. 7th Int. Vacuum Congress (Dobrozemsky, et al., Vienna) p. 129.

Moesta, H., and R. Renn, 1957, Vakuum-Technik **6**, 35.

Mog, D., 1976, Machinable glass-ceramic. Vacuum **26**, 25.

Moller, T.W., 1955, Factors influencing the calibration and application of standard leaks. Trans. 1st American Vacuum Symposium (Pergamon Press, Oxford) p. 121.

Mönch, G.C., 1937, Vakuumtechnik im Laboratorium (Wagner Sohn, Weimar).

Mönch, G.C., 1959, Neues and Bewährtes aus der Hochvakuumtechnik (VEB Knapp, Halle) 1000pp.

Mongodin, G., V.R. Piacentini and W. Sajnacki, 1974, Evaluation of a liquid-helium cryopumping system operating with an electron beam gun. Proc. 20th American Vacuum Symposium, J. Vac. Sci. & Technol. **11**, 340.

Moody, R.E., 1957, Versatile radio-frequency leak detector. Trans. 3rd American Vacuum Symposium (Pergamon Press, Oxford) p. 119.

Moore, W.J., 1960, Am. Sci. **48**, 109.

Moore, R.W., 1962, Cryopumping in the free-molecular flow regime. Trans. 2nd Int. Vacuum Congress (Pergamon Press, Oxford) p. 426.

Moore, B.C., 1972, J. Vac. Sci. & Technol. 9, 1090.

Moore, B.C., 1980, Method for degassing metals (electron bombardment). J. Vac. Sci. & Technol. 17(4), 836.

Moore, B.C., and R.G. Camarillo, 1973, J. Vac. Sci. & Technol. 10, 404.

Moore, H.E., and V.H. Frahm, 1974, Rev. Sci. Instrum. 45, 299.

Moore, G.E., and F.C. Unterwald, 1964, J. Chem. Phys. 40, 2639.

Moore, B.C., and J.A. Walker, 1979, Tracer slug leak detection. Proc. 25th American Vacuum Symposium, J. Vac. Sci. & Technol. 16, 698.

Morand, N., 1958, Traité Pratique de Technique du Vide (Gauthier Villars, Paris) 347pp.

Moraw, G., 1974, The influence of ionization gauges on gas flow measurements. Vacuum 24, 125.

Moraw, G., and R. Dobrozemsky, 1974, Attainment of outgassing rates below 10^{-13} Torr lit/sec cm^2 for Al and stainless steel after bakeout at moderate temperatures. Proc. 6th Int. Vacuum Congress, Jpn. J. Appl. Phys., Suppl. 2, Pt. 1, p. 261.

Moraw, M., and H. Prasol, 1978, Leak detection in large vessels. Vacuum 28, 63.

Morel, J., 1983, Pirani gauge for pressures 10^{-3} to 1000 mbar (in French). Le Vide 38(217), 301.

Morey, G.W., 1954, The Properties of Glass (Reinhold, New York).

Morimura, M., K. Nagakawa and Y. Nezu, 1974, Pressure measurement by Brownian motion. Proc. 6th Int. Vacuum Congress, Jpn. J. Appl. Phys., Suppl. 2, Pt. 1, p. 135.

Morrison, S.J., 1952, Engineer 113, 426.

Morrison, J., 1953, Rev. Sci. Instrum. 24, 546.

Moser, H., and H. Poltz, 1957, Z. für Instrkde 65, 43.

Moss, T.A., 1987, Conversion tables for pressure and leak rate. J. Vac. Sci. & Technol. A 5(5), 2962.

Mourad, W.G., T. Pauly and R.G. Herb, 1964, Rev. Sci. Instrum. 35, 661.

Muhlenhaupt, R.C., 1982, Comparison of cryo- and diffusion pump performance. J. Vac. Sci. & Technol. 20(4), 1005.

Mulder, B.J., 1976, Motion in uhv through the shape memory of Ni–Ti alloys. Vacuum 26, 31.

Mulder, B.J., 1977, Vacuum sealing of lithium fluoride windows with lead fluoride. J. Phys. E (Sci. Instrum.) 10, 591.

Muller, E., 1966, Cryogenics 6, 242.

Munchhausen, H., and F.J. Schittko, 1963, Vacuum 13, 549.

Munday, G.L., 1959, Nucl. Instr. & Methods 4, 367.

Munson, R.J., 1955, Rev. Sci. Instrum. **26**, 236.

Murakami, Y., 1973, Hydrogen pumping by a new catalytic pump. J. Vac. Sci. & Technol. **10**, 359.

Murakami, Y., 1983, Vacuum and surface technology in Tokamaks. Proc. 9th Int. Vacuum Congress (Madrid) p. 532.

Murakami, Y., and H. Ohtsuka, 1978, Pd alloy membrane pumps. Vacuum **28**, 235.

Murakami, Y., et al., 1974, Some considerations on large-sized catalytic pumps for nuclear fusion apparatus. Proc. 6th Int. Vacuum Congress, Jpn. J. Appl. Phys., Suppl. 2, Pt. 1, p. 89.

Murakami, Y., et al., 1987, Performance test of a ceramic rotor developed for turbomolecular pumps, for fusion use. J. Vac. Sci. & Technol. A **5**(4), 2599.

Murko-Jezovsek, M., et al., 1973, Cold welding in uhv. Vakuum-Technik **22**, 119, 153.

Myer, J.H., 1972, J. Vac. Sci. & Technol. **9**, 1106.

N

Naik, P.K., and R.G. Herb, 1968, J. Vac. Sci. & Technol. **5**, 42.

Naik, P.K., and S.L. Verma, 1974, A new modified orbitron pump. Proc. 6th Int. Vacuum Congress, Jpn. J. Appl. Phys., Suppl. 2, Pt. 1, p. 73.

Naik, P.K., and S.L. Verma, 1977, Performance of the modified orbitron pump. J. Vac. Sci. & Technol. **14**, 734.

Nair, C.V.G., and P. Vijendran, 1974, A new cryosorption pump for better ultimate vacua. Proc. 6th Int. Vacuum Congress, Jpn. J. Appl. Phys., Suppl. 2, Pt. 1, p. 93.

Nair, C.V.G., and P. Vijendran, 1977, Pumping speed of a sorption pump; determination of condensation coefficients. Vacuum **27**, 549.

Nakao, F., 1975, Determination of the ionization gauge sensitivity using the ionization cross-section. Vacuum **25**, 431.

Nakayama, K., and H. Hojo, 1974, Relative ion gauge sensitivities to various hydrocarbon gases. Proc. 6th Int. Vacuum Congress, Jpn. J. Appl. Phys., Suppl. 2, Pt. 1, p. 113.

Nakayama, K., Y. Akiyama and H. Hashimoto, 1968, Vacuum **18**, 65.

Nash, P.J., and T.J. Thompson, 1983, Vacuum gauge calibration by comparison. J. Vac. Sci. & Technol. A **1**(2/1), 172.

Nazarov, A.S., G.F. Iwanovsky and M.V. Kouznetsov, 1965, Getter-ion pumps with directly heated titanium evaporators. Trans. 3rd Int. Vacuum Congress (Pergamon Press, Oxford) p. 663.

Neal, R.B., 1965, J. Vac. Sci. & Technol. **2**, 149.

Nelson, K., 1980, Direct drive mechanical high vacuum pump. Proc. 26th American Vacuum Symposium, J. Vac. Sci. & Technol. **17**.

Nesseldreher, W., 1974, The effect of various parameters on the mass spectrograms of turbomolecular pumps (in German). Vakuum-Technik **23**, 163.

Nesseldreher, W., 1976a, Le Vide **31**, 74.

Nesseldreher, W., 1976b, The effect of parameters on the residual mass spectrograms of turbomolecular pumps. Vacuum **26**, 281.

Nester, R.G., 1956a, Rev. Sci. Instrum. **27**, 874.

Nester, R.G., 1956b, Rev. Sci. Instrum. **27**, 1080.

Neuhauser, R.G., 1979, Soft metal vacuum seals for glass and ceramics. Vacuum **29**, 231.

Neville, H.H., 1955, J. Sci. Instrum. **32**, 488.

Newman, F., 1925, Production and Measurement of Low Pressures (Benn, London).

Newstead, S.M., et al., 1984, The use of sliding metal electrical contacts in ultrahigh vacuum. J. Vac. Sci. & Technol. A **2**(4), 1603.

Nicolet, M., 1960, Les variations de la densité et du transport de chaleur par conduction dans l'atmosphére superieure. Space Research (North-Holland, Amsterdam) p. 64.

Nicollian, E.H., 1961, A laboratory leak detector using an Omegatron. Trans. 7th American Vacuum Symposium (Pergamon Press, Oxford) p. 80.

Nief, G.A., 1952, J. Chem. Phys. **49**, 149.

Nier, A.O., 1940, Rev. Sci. Instrum. **11**, 212.

Nier, A.O., 1947, J. Appl. Phys. **18**, 30; Rev. Sci. Instrum. **18**, 191.

Nöller, H.G., 1977, The significance of Knudsen numbers and laws of similitude in diffusion and ejector pumps. Vakuum-Technik **26**, 72.

Nöller, H.G., 1983, Production of clean vacuum with cryopumps. Proc. 9th Int. Vacuum Congress (Madrid) p. 217.

Nöller, H.G., G. Reich and W. Bächler, 1960, Diffusion pump and baffle systems of large suction speed for pressure lower than 10^{-8} Torr. Trans. 6th American Vacuum Symposium (Pergamon Press, Oxford) p. 72.

Nöller, H.G., H.D. Polaschegg and H. Schillalies, 1974, A high resolution electron spectrometer for Auger-electron spectroscopy (AES). Trans. 6th Int. Vacuum Congress (Kyoto) p. 343.

Normand, C.E., 1962, Use of a standard orifice in the calibration of vacuum gauges. Trans. 8th American Vacuum Symposium (Pergamon Press, Oxford) p. 534.

Norström, H., S. Berg and L.P. Andersson, 1977, Fast volume determination using a differential capacitance manometer. Vacuum **27**, 99.

Norton, F.J., 1957, J. Appl. Phys. **28**, 34.

Norton, F.J., 1962, Gas permeation through the vacuum envelope. Trans. 2nd Int. Vacuum Congress (Pergamon Press, Oxford) p. 8.

Nottingham, W.B., 1947, 7th Conf. Phys. Electron. (MIT, Cambridge).

Nottingham, W.B., and F.L. Torney, 1961, A detailed examination of the principles of ion gauge calibration. Trans. 7th American Vacuum Symposium (Pergamon Press, Oxford) p. 117.

Nuvolone, R., 1977, Degassing rates of 316 L stainless steel as a function of different heat treatments. Proc. 7th Int. Vacuum Congress (Dobrozemsky, et al., Vienna) p. 219.

O

Oatley, C.W., 1954, Br. J. Appl. Phys. **5**, 358.

Ochert, N., and W. Steckelmacher, 1951, Br. J. Appl. Phys. **2**, 332.

Ochert, N., and W. Steckelmacher, 1952, Vacuum **2**, 125.

Odaka, K., et al., 1987, Effect of baking temperature and air exposure on the outgassing rate of 316 L stainless steel. J. Vac. Sci. & Technol. A **5**(5), 2902.

Oguri, T., 1977, Ultra low noise Pirani gauge. Proc. 7th Int. Vacuum Congress (Dobrozemsky, et al., Vienna) p. 149.

O'Hanlon, J.F., 1979, Turbomolecular pumps for high gas flow applications. Proc. 25th American Vacuum Symposium, J. Vac. Sci. & Technol. **16**, 724.

O'Hanlon, J.F., 1980, A User's Guide to Vacuum Technology (Wiley, New York).

O'Hanlon, J.F., 1984, Vacuum pump fluids. J. Vac. Sci. & Technol. A **2**(2), 174.

O'Hanlon, J.F., 1987, An analysis of equations for flow in thin, rectangular channels. J. Vac. Sci. & Technol. A **5**(1), 98.

O'Hanlon, J.F., and D.B. Fraser, 1988, AVS recommended practices for pumping hazardous gases. J. Vac. Sci. & Technol. A **6**(3), 1226.

Ohsako, N., 1982, A Bayard–Alpert gauge from uhv to 10^{-1} Torr. J. Vac. Sci. & Technol. **20**(4), 1153.

Ohta, S., et al., 1983, Resistor network simulation of vacuum system flow. J. Vac. Sci. & Technol. A **1**(1), 84.

O'Kane, D.F., and K.L. Mittal, 1974, Plasma cleaning of metal surfaces. J. Vac. Sci. & Technol. **11**, 567.

Okano, T., K. Imura and G. Tominaga, 1977, A tantalum evaporation pump. Proc. 7th Int. Vacuum Congress (Dobrozemsky, et al., Vienna) p. 81.

Okano, T., et al., 1984, A Zr–Al composite-cathode sputter-ion pump. J. Vac. Sci. & Technol. A **2**(2), 191.

Omelka, L.V., 1970, J. Vac. Sci. & Technol. **7**, 257.

Ono, M., H. Shimizu and K. Nakayama, 1973, J. Vac. Sci. & Technol. **10**, 566.

Onusic, H., 1980, Practical way to calculate molecular conductance of pipes with constant annular cross section. J. Vac. Sci. & Technol. 17(2), 661.

Oran, W.A., and R.J. Naumann, 1977, Utilization of the vacuum developed in the wake zone of space vehicles. J. Vac. Sci. & Technol. 14, 1276.

Oran, W.A., and R.J. Naumann, 1978, Vacuum 28, 73.

Osterstrom, G.E., 1977, A brief review of turbomolecular pumps. Proc. 7th Int. Vacuum Congress (Dobrozemsky, et al., Vienna) p. 41.

Osterstrom, G.E., and T. Knecht, 1979, Grease lubrication of turbomolecular vacuum pump bearings. Proc. 25th American Vacuum Symposium, J. Vac. Sci. & Technol. 16, 746.

Osterstrom, G.E., and A.H. Shapiro, 1972, J. Vac. Sci. & Technol. 9, 405.

Ota, M., and N. Hirayama, 1987, Performance characteristics of single-stage centrifugal vacuum pumps at rough vacuum. J. Vac. Sci. & Technol. A 5(4), 2603.

Outlaw, R.A., and F.J. Brock, 1977, Orbiting molecular-beam laboratory. J. Vac. Sci. & Technol. 14, 1269.

Outlaw, R.A., F.J. Brock and J.P. Wightman, 1974, J. Vac. Sci. & Technol. 11, 446.

Owens, C.L., 1965, J. Vac. Sci. & Technol. 2, 104.

P

Pang, S.J., et al., 1987, Thermal desorption measurements for Al alloy 6063 and stainless steel 304. J. Vac. Sci. & Technol. A 5(4), 2516.

Panitz, J.K.G., et al., 1988, The tribological properties of MoS_2 coatings in vacuum, low relative humidity and high relative humidity environments. J. Vac. Sci. & Technol. A 6(3), 1166.

Pararas, A., et al., 1982, An O-ring sealed rotary feed-through for ultrahigh vacuum. J. Vac. Sci. & Technol. 21(4), 1031.

Parkash, S., and P. Vijendran, 1983, Sorption of active gases by non-evaporable getters. Vacuum 33(5), 295.

Parker, W.B., and J.T. Mark, 1961, A large ultra-high vacuum valve. Trans. 7th American Vacuum Symposium (Pergamon Press, Oxford) p. 21.

Parr, L.M., and C.A. Hendley, 1956, Laboratory Glass Blowing (Newnes, London).

Parris, M., et al., 1984, Vacuum measurement in Joint European Torus Project (in German). Vakuum-Technik 33(2), 52.

Partridge, J.H., 1949, Glass-to-Metal Seals (The Society of Glass Technology, Elmfield, Sheffield).

Pathak, H.A., 1982, Electromagnetic isolation valve. Vacuum **32**(6), 375.

Patrick, T.J., 1973, Outgassing and the choice of materials for space instrumentation. Vacuum **23**, 411.

Patrick, T.J., 1981, Vacuum properties of space materials. Vacuum **31**(8/9), 351.

Paty, L., 1968, Physics of Low Pressures (in Czech) (Academia, Praha) 298pp.

Paty, L., and P. Schürer, 1957, Rev. Sci. Instrum. **28**, 654.

Paul, G., 1973a, Molecular flow in vacuum systems with great sorbing surfaces (in German). Vakuum-Technik **22**, 201.

Paul, G., 1973b, Pumping speed of sputter-ion pumps of rotational symmetry (in German). Vakuum-Technik **22**, 243.

Paul, W., and H. Steinwendel, 1953, Z. Naturforsch. **8a**, 448.

Paul, W., H.P. Reinhard and U. von Zahn, 1958, Z. Phys. **152**, 143.

Pauly, T., R.D. Welton and R.G. Herb, 1960, Getter-ion pumps using cartridge evaporators. Trans. 7th American Vacuum Symposium (Pergamon Press, Oxford) p. 51.

Payne, A.R., and J.R. Scott, 1960, Engineering Design with Rubber (Interscience Publishers, New York).

Peacock, R.N., 1980, Practical vacuum seal selection. Proc. 26th American Vacuum Symposium. J. Vac. Sci. & Technol. **17**, 330.

Peacock, N.T., and R.N. Peacock, 1988, Some characteristics of an inverted magnetron cold cathode ionization gauge with dual feedthrough. J. Vac. Sci. & Technol. A **6**(3), 1141.

Pearce, S.J., and S.L. Barker, 1977, Fully screened, high-voltage vacuum feedthrough. J. Phys. E (Sci. Instrum.) **10**, 1231.

Peggs, G.N., 1976, The measurement of gas throughput. Vacuum **26**, 321.

Penning, F.M., 1937, Physica **4**, 71.

Penning, F.M., and K. Nienhuis, 1949, Philips Tech. Rev. **11**, 116.

Perkins, W.G., 1973, J. Vac. Sci. & Technol. **10**, 543.

Perkins, W.G., and D.R. Begeal, 1971, J. Chem. Phys. **54**, 1683.

Peters, J., and H. Pingel, 1977, A gate valve avoiding rf damage. Proc. 7th Int. Vacuum Congress (Dobrozemsky, et al., Vienna) p. 243.

Petit, B., and M.L. Feidt, 1983, Theoretical determination of ionization efficiency for an orbitron device. J. Vac. Sci. & Technol. A **1**(2/1), 163.

Petley, B.W., and K. Morris, 1968, J. Sci. Instrum. (Ser. 2) **1**, 417.

Pierini, M., 1984, Use of discharge intensity for evaluation of pumping characteristics of a sputter ion pump. J. Vac. Sci. & Technol. A **2**(2), 195.

Pierini, M., and L. Dolcino, 1983, A new sputter-ion pump. J. Vac. Sci. & Technol. A **1**(2/1), 140.

Pinsker, Z.G., 1953, Electron Diffraction (Butterworth, London).

Pirani, M., 1906, Verh. Dent. Phys. Ges. **8**, 686.

Pirani, M., and J. Yarwood, 1961, Principles of Vacuum Engineering (Chapman and Hall, London) 578pp.

Pittaway, L.G., 1970, Le Vide **25**, 146.

Pittaway, L.G., 1974, The design and operation of a new extractor gauge. Vacuum **24**, 301.

Polaschegg, H.D., 1978/1979, Surface analysis of thin films. Vakuum-Technik **27**, 238; **28**, 12.

Polaschegg, H.D., and E. Schirk, 1975, A vacuum lock for introducing solid samples into an uhv vessel (in German). Vakuum-Technik **24**, 136.

Polizzotti, R.S., and J.A. Schwarz, 1980, Versatile sample transfer mechanism. J. Vac. Sci. & Technol. **17**(2), 655.

Porteous, P., 1963, J. Sci. Instrum. **40**, 47.

Porter, J.R., 1988, Techniques for testing cryopump capacity. J. Vac. Sci. & Technol. A **6**(3), 1214.

Potter, W.G., 1971, Epoxide Resins (Springer, Berlin).

Poulsen, R.G., 1977, Plasma etching. Proc. 23rd American Vacuum Symposium, J. Vac. Sci. & Technol. **14**, 266.

Poulter, K.F., 1973, Calibration of a mass spectrometer in the uhv region. Vacuum **23**, 131.

Poulter, K.F., 1977, The calibration of vacuum gauges. J. Phys. E (Sci. Instrum) **10**, 112.

Poulter, K.F., 1978, Vacuum gauge calibration by the orifice flow method in the pressure range 10^{-4}–10 Pa. Vacuum **28**, 135.

Poulter, K.F., 1984, Effect of gas composition on vacuum measurement. J. Vac. Sci. & Technol. A **2**(2), 150.

Poulter, K.F., and C.M. Sutton, 1981, Long term behavior of ionization gauges. Vacuum **31**(3), 147.

Poulter, K.F., et al., 1980a, Comparison of vacuum standards. J. Vac. Sci. & Technol. **17**(3), 679.

Poulter, K.F., et al., 1980b, Reproducibility of performance of Pirani gauges. J. Vac. Sci. & Technol. **17**(2), 638.

Poulter, K.F., et al., 1983, Thermal transpiration correction in capacitance manometers. Vacuum **33**(6), 311.

Povh, B., and F. Lah, 1967, An analysis of the changes in Pirani gauge characteristics. Trans. 3rd Int. Vacuum Congress (Pergamon Press, Oxford) p. 287.

Powell, J.R., and D. McMullan, 1978, Increasing the sensitivity of mass spectrometer leak detection by selective gas accumulation. Vacuum **28**, 287.

Power, B.D., 1966, High Vacuum Pumping Equipment (Chapman and Hall, London) 412pp.

Power, B.D., and D.J. Crawley, 1954, Vacuum **4**, 415.

Power, B.D., and D.J. Crawley, 1960, Problems arising in the attainment of low pressure by fractionating vapour pumps in large demountable systems. Proc. 1st Int. Vacuum Congress (Pergamon Press, Oxford) p. 206.

Power, B.D., and F.C. Robson, 1962, Experiences with demountable ultra-high vacuum systems. Trans. 2nd Int. Vacuum Congress (Pergamon Press, Oxford) p. 1175.

Power, B.D., et al., 1974a, Single structure vapour pumping group. Vacuum **24**, 117.

Power, B.D., et al., 1974b, Characteristics of single structure vapour pumping groups. Proc. 6th Int. Vacuum Congress, Jpn. J. Appl. Phys., Suppl. 2, Pt. 1, p. 33.

Power, B.D., et al., 1977, An industrial vapour vacuum pump employing a porous element flash boiler. Proc. 7th Int. Vacuum Congress (Dobrozemsky, et al., Vienna) p. 13.

Powers, R.J., and R.M. Chambers, 1971, J. Vac. Sci. & Technol. **8**, 319.

Powle, U.S., and S. Kar, 1983, Pumping speed of liquid ring vacuum pumps. Vacuum **33**(5), 255.

Pressey, D.C., 1953, J. Sci. Instrum. **30**, 20.

Prévot, F., 1974, Vacuum problems in plasma physics and controlled nuclear fusion. Proc. 6th Int. Vacuum Congress, Jpn. J. Appl. Phys., Suppl. 2, Pt. 1, p. 225.

Prevot, F., and Z. Sledziewski, 1964, Le Vide **19**, 342.

Prevot, F., and Z. Sledziewski, 1968, Le Vide **23**, 1.

Prevot, F., and Z. Sledziewski, 1971, Le Vide **26**, 91.

Prevot, F., and Z. Sledziewski, 1972, J. Vac. Sci. & Technol. **9**, 49.

Prost, F.M., and T.C. Piper, 1967, J. Vac. Sci. & Technol. **4**, 53.

Prutton, M., 1971, Met. & Mater. **5**, 57.

Pupp, W., 1962, Vakuumtechnik, I. Grundlagen (1962) 109pp, II. Anwendungen (1964) 312pp (Verlag Thiemig, Munchen).

Pustovoit, Yu.M., 1974, The vacuum technique of nuclear-fusion installations. Proc. 6th Int. Vacuum Congress, Jpn. J. Appl. Phys., Suppl. 2, Pt. 1, p. 233.

R

Raible, F., 1955, Automatic leak testing. Trans. 1st American Vacuum Symposium (Pergamon Press, Oxford) p. 11.

Raj, K., and M.A. Grayson, 1981, Material evolution from magnetic liquid seals. Vacuum **31**(3), 151.

Ramananda Rao, B., 1975, Methods and techniques of gas analysis (in German). Fortschritte der Verfahrenstechnik **13**, 501.

Ramesh, V., and D.J. Marsden, 1973, Measurement of rotational and translational accommodation coefficients. Vacuum **23**, 365.

Ramesh, V., and D.J. Marsden, 1974, Prediction of translational accommodation coefficient. Vacuum **24**, 235.

Ratcliffe, R.T., 1964, Br. J. Appl. Phys. **15**, 79.

Rathbun, F.O., 1963, Design criteria for zero leakage connectors for launch vehicles. Report NASA N63-18159.

Rava, E., et al., 1987, Recent developments in turbomolecular pump rotor balancing techniques. J. Vac. Sci. & Technol. A **5**(4), 2530.

Reale, C., 1976, Fisica delle Pellicole Sottili (The Physics of Thin Films) (Tamburin, Milano, Italy).

Reddan, W.G., 1982, Vacuum vessel for Tokamak. J. Vac. Sci. & Technol. **20**(4), 1173.

Redhead, P.A., 1958, Can. J. Phys. **36**, 255.

Redhead, P.A., 1959, The production and measurement of ultra-high vacuum $(10^{-8}-10^{-13}$ mm Hg). Trans. 5th American Vacuum Symposium (Pergamon Press, Oxford) p. 148.

Redhead, P.A., 1960, Rev. Sci. Instrum. **1**, 343.

Redhead, P.A., 1961a, Errors in the measurement of pressure with ionization gauges. Trans. 7th American Vacuum Symposium (Pergamon Press, Oxford) p. 108.

Redhead, P.A., 1961b, Trans. Faraday Soc. **57**, 641.

Redhead, P.A., 1966, J. Vac. Sci. & Technol. **3**, 173.

Redhead, P.A., 1970, J. Vac. Sci. & Technol. **7**, 182.

Redhead, P.A., 1976, Uhv applied to physics. J. Vac. Sci. & Technol. **13**, 5.

Redhead, P.A., 1984, The measurement of vacuum pressures. J. Vac. Sci. & Technol. A **2**(2), 132.

Redhead, P.A., 1987, Ultrahigh vacuum pressure measurements: Limiting processes. J. Vac. Sci. & Technol. A **5**(5), 3215.

Redhead, P.A., and J.P. Hobson, 1965, Br. J. Appl. Phys. **16**, 1555.

Redhead, P.A., J.P. Hobson and E.V. Kornelsen, 1968, The Physical Basis of Ultra-high Vacuum (Chapman and Hall, London) 498pp.

Redman, I.D., 1963, Vacuum systems, techniques and material studies. Report ORNL 3472.

Rees, J.R., 1977, The positron–electron project PEP. IEEE Trans. Nucl. Sci. **NS-24**, 1836.

Reiber, L.M., and J. Lantaires, 1973, Le Vide **28**, 242.

Reich, G., 1961, The Farvitron – A new partial pressure indicator without a magnetic field. Trans. 7th American Vacuum Symposium (Pergamon Press, Oxford) p. 396.

Reich, G., 1982a, Spinning rotor gauge. Vakuum-Technik **31**(6), 172.

Reich, G., 1982b, Spinning rotor viscosity gauge. J. Vac. Sci. & Technol. **20**(4), 1148.

Reich, G, 1983, Accuracy in vacuum measurement. Proc. 9th Int. Vacuum Congress (Madrid) p. 195.

Reich, G., 1987, The principle of He enrichment in a counterflow leak detector with a turbomolecular pump with two inlets. J. Vac. Sci. & Technol. A **5**(4), 2641.

Reid, R.J., 1978, Pressure measurement in the SRS. Vacuum **28**, 499.

Reid, R.J., and B.A. Trickett, 1977, Optimisation of distributed ion pumps for the Daresbury synchrotron radiation source. Proc. 7th Int. Vacuum Congress (Dobrozemsky, et al., Vienna) p. 89.

Reimann, A.L., 1952, Vacuum Technique (Chapman and Hall, London).

Reinhard, H.P., 1983, The vacuum system of LEP. Proc. 9th Int. Vacuum Congress (Madrid) p. 273.

Reiter, F., and J. Composilvan, 1982, Thermal outgassing properties of Inconel 600. Vacuum **32**(5), 227.

Rettinghaus, G., and W.K. Huber, 1974, Backstreaming in diffusion pump systems. Proc. 6th Int. Vacuum Congress (p. A.853), Vacuum **24**, 249.

Rettinghaus, G., and W.K. Huber, 1977, Vacuum requirements for a satellite-borne mass spectrometer. Proc. 7th Int. Vacuum Congress (Dobrozemsky, et al., Vienna) p. 189.

Reynolds, F.L., 1955, Vakuum-Technik **4**, 181.

Reynolds, J.H., 1956, Rev. Sci. Instrum. **27**, 928.

Rhodin, T.N., and L.H. Rovner, 1961, Gas–metal reactions in oxygen at low pressures. Trans. 7th American Vacuum Symposium (Pergamon Press, Oxford) p. 228.

Riddiford, L., 1951, J. Sci. Instrum. **28**, 375.

Rigby, L.J., and C.R. Wright, 1968, Vacuum **18**, 274.

Riggs, F.B., 1962, Rev. Sci. Instrum. **33**, 1114.

Rijke, J.E. de, 1978, Performance of a cryo-ion pump system. Proc. 24th American Vacuum Symposium. J. Vac. Sci. & Technol. **15**, 765.

Ritehouse, J.B., and J.B. Singletary, 1968, Space Materials Handbook. Report AFML-TR-68-205; AD-692353.

Rivera, M., and R. le Riche, 1960, A differentially pumped ultra-high vacuum system. Trans. 6th American Vacuum Symposium (Pergamon Press, Oxford) p.56.

Riviére, J.C., and J.B. Thompson, 1965, Vacuum **15**, 353.

Robens, E., 1980, Cryopump for vacuum microbalance. Proc. 26th American Vacuum Symposium. J. Vac. Sci. & Technol. **17**, 277.

Roberts, J.A., 1957, Precision leaks for standardizing leak detection equipment. Trans. 4th American Vacuum Symposium (Pergamon Press, Oxford) p. 124.

Roberts, R.W., and E.L. Bahm, 1965, J. Vac. Sci. & Technol. **2**, 89.

Roberts, R.W., and T.A. Vanderslice, 1963, Ultrahigh Vacuum and its Applications (Prentice Hall, Englewood Cliffs, NJ) 199pp.

Robertson, A.J.B., 1957, Laboratory Glass Working (Butterworth, London).

Robertson, D.D., 1968, Evaluation of a high-rate titanium getter pump. Proc. 4th Int. Vacuum Congress (Institute of Physics, London) p. 373.

Robins, J.L., and J.B. Swan, 1960, Proc. Phys. Soc. (London) **76**, 857.

Robinson, N.W., 1968, The Physical Principles of Ultra-high Vacuum Systems and Equipment (Chapman and Hall, London) 270pp.

Robinson, C.F., and L.G. Hall, 1956, Rev. Sci. Instrum. **27**, 504.

Roehrig, J.R., and J.C. Simons, 1962, Accurate calibration of vacuum gauges to 10^{-9} Torr. Trans. 8th American Vacuum Symposium (Pergamon Press, Oxford) p. 511.

Roehrig, J.R., and G.F. Vanderschmidt, 1959, Advances in the design of vacuum gauges using radioactive materials. Trans. 6th American Vacuum Symposium (Pergamon Press, Oxford) p. 82.

Rogers, C., 1956, J. Techn. Assoc. Pulp & Paper Ind. **39**, 737.

Rogers, W.A., R.S. Buritz and D. Alpert, 1954, J. Appl. Phys. **25**, 868.

Rohring, H.D., et al., 1980, Investigation of hydrogen gettering. J. Vac. Sci. & Technol. **17**(1), 120.

Roller, K.G., 1988, Lubricating of mechanisms for vacuum service. J. Vac. Sci. & Technol. A **6**(3), 1161.

Romann, M.P., 1948, J. Sci. Instrum. **3**, 522.

Rommel, G., 1982, A cryogenic pump of 20000 liter/second. Vacuum **32**(2), 95.

Rosai, L., B. Ferrario and P. della Porta, 1978, Behavior of "Sorb-ac" wafer pumps in plasma machines. Proc. 24th American Vacuum Symposium, J. Vac. Sci. & Technol. **15**, 746.

Rose, W., 1950, Rev. Sci. Instrum. **21**, 772.

Rosebury, F., 1965, Handbook of Electron Tube and Vacuum Techniques (Addison-Wesley, Reading, USA) 597pp.

Rosenblum, S.S., 1986, Vacuum outgassing rates of plastics and composites for electrical insulators. J. Vac. Sci. & Technol. A **4**(1), 107.

Rosenthal, Y., et al., 1987, A simple sensitive helium leak detector using a quadrupole mass spectrometer. J. Vac. Sci. & Technol. A **5**(3), 389.

Roth, A., 1955, Vacuum Technology (in Rumanian) (Ed. Energetica, Bucharest) 646pp.

Roth, A., 1966, Vacuum Sealing Techniques (Pergamon Press, Oxford) 845pp.

Roth, A., 1967, Sealometry and Sealography. Proc. 3rd Int. Conf. on Fluid Sealing (British Hydromechanical Research Association, Cranfield, England) paper C2.

Roth, A., 1970a, Vacuum 20, 431.

Roth, A., 1970b, Nomographic evaluation of the gas load in vacuum systems. Report IA-1220 (Israel Atomic Energy Comm.).

Roth, A., 1971, The interface-contact vacuum sealing processes. Trans. 5th Int. Vacuum Congress (American Vacuum Society, New York) p. 14 [J. Vac. Sci. & Technol. 9, 14].

Roth, A., 1972, Vacuum 22, 219.

Roth, A., 1976, Vacuum Technology, 1st Ed. (North-Holland, Amsterdam).

Roth, A., 1982, Vacuum Technology, 2nd Ed. (North-Holland, Amsterdam).

Roth, A., 1983, Sealing mechanism in bakeable vacuum seals. J. Vac. Sci. & Technol. A 1(2), 211.

Roth, A., and A. Amilani, 1965, Sealing factors, their measurement and use in the design of vacuum gasket seals. Trans. 3rd Int. Vacuum Congress (Pergamon Press, Oxford) p. 181.

Roth, A., and A. Inbar, 1967a, ASTM J. Mater. 2, 567.

Roth, A., and A. Inbar, 1967b, Vacuum 17, 5.

Roth, A., and A. Inbar, 1968, Vacuum 18, 309.

Roussel, J., J.J. Thibault and A. Nanoboff, 1965, Le Vide 20, 249.

Rubet, L., 1966, Le Vide 21, 227.

Rubet, L., 1983, Calibration of He leaks. Le Vide 38(215), 21.

Rusch, T.W., and R.L. Erickson, 1976, Energy dependence of scattering ion yields in ISS. Proc. 22nd American Vacuum Symposium, J. Vac. Sci. & Technol. 13, 374.

Ruthberg, S., 1972, J. Vac. Sci. & Technol. 9, 186.

S

Sadler, P., 1973, Vacuum 23, 333.

Saksaganskii, G.L., 1988, Molecular Flow in Complex Vacuum Systems (Gordon and Breach, New York).

Sakuma, K., and M. Nagayama, 1974, Characteristics of a new diffusion pump fluid. Proc. 6th Int. Vacuum Congress, Jpn. J. Appl. Phys., Suppl. 2, Pt. 1, p. 37.

Salle, M.H., 1970, Le Vide 25, 203.

Salomon, G., and R. Houwink, 1967, Adhesion and Adhesives (Elsevier, New York).

Samuel, R.L., 1970, Vacuum **20**, 295.

Sands, A., and S.M. Dick, 1966, Vacuum **16**, 691.

Sänger, G., and A.K. Franz, 1984, Leak testing of spacecrafts (in German). Vakuum-Technik **33**(2), 42.

Sänger, G., *et al.*, 1982, Experiments with 20K and liquid He cryopumps in space simulation chambers. Vakuum-Technik **31**(3), 71.

Santeler, D.J., 1958, Outgassing characteristics of various materials. Trans. 5th American Vacuum Symposium (Pergamon Press, Oxford) p. 1.

Santeler, D.J., 1963, Vacuum **13**, 102.

Santeler, D.J., 1971, J. Vac. Sci. & Technol. **8**, 299.

Santeler, D., 1984, Leak detection – common problems and their solutions. J. Vac. Sci. & Technol. A **2**, 1149.

Santeler, D.J., 1986a, New concepts in molecular gas flow. J. Vac. Sci. & Technol. A **4**(3), 338.

Santeler, D.J., 1986b, Exit loss in viscous tube flow. J. Vac. Sci. & Technol. A **4**(3), 348.

Santeler, D.J., 1987a, Computer design and analysis of vacuum systems. J. Vac. Sci. & Technol. A **5**(4), 2472.

Santeler, D.J., and M.D. Boeckmann, 1987b, Combining transmission probabilities of different diameter tubes. J. Vac. Sci. & Technol. A **5**(4), 2493.

Santhanaman, S.J., and P. Vijendran, 1979, The effect of ambient and other pretreatment on the outgassing of certain materials in a vacuum. Vacuum **29**, 237.

Sar-El, H.Z., and E. Pellach, 1976, Extension of rate of rise method for calibrating vacuum gauges. Vacuum **26**, 137; Comment of W. Steckelmacher, Vacuum **26**, 547.

Saulgeot, C., 1977, A new range of Alcatel molecular pumps. Proc. 7th Int. Vacuum Congress (Dobrozemsky, *et al.*, Vienna) p. 37.

Sawada, T., 1974, Mass separation in a turbomolecular pump. Proc. 6th Int. Vacuum Congress, Jpn. J. Appl. Phys., Suppl. 2, Pt. 1, p. 1.

Schäfer, G., 1978, Cryopumps in science and technology. Vacuum **28**, 399.

Schäfer, G., and H.U. Häfner, 1987, Cryopumps for evacuating space simulation chambers. J. Vac. Sci. & Technol. A **5**(4), 2359.

Schäfer, G., and M. Schinkmann, 1973, Thermal load of liquid pool cryopumps (in German). Vakuum-Technik **22**, 10.

Schalla, C.A., 1975, Outgassing coefficient development for improved leak-tightness sensitivity in vacuum drift testing. Proc. 21st American Vacuum Symposium, J. Vac. Sci. & Technol. **12**, 430.

Schalla, C.A., 1980, Area/volume configuration influence on porous matter outgassing. J. Vac. Sci. & Technol. **17**(3), 705.

Schauer, F., 1977, Very low temperature vacuum joints. Vakuum-Technik **26**, 172.

Scheer, J.J., and J. Visser, 1982, Use of cryopumps in industry. Vakuum-Technik **31**(2), 34.

Scheffield, J.C., 1965, Rev. Sci. Instrum. **36**, 1269.

Scheflan, L., and M.B. Jacobs, 1953, The Handbook of Solvents (Van Nostrand, New York).

Schiller, S., U. Heisig and K. Goedicke, 1976, Ion plating (in German). Vakuum-Technik **25**, 65, 113.

Schittko, F.J., 1963, Vacuum **13**, 525.

Schittko, F.J., and C. Schmidt, 1975, Improvement of density ratio of light molecules by gas-ballast in molecular pumps (in German). Vakuum-Technik **24**, 110.

Schittko, F.J., F. Fünfer and C. Schmidt, 1977, Improved density ratio and pumping speed for light molecules in molecular pump systems. Proc. 7th Int. Vacuum Congress (Dobrozemsky, et al., Vienna) p. 33.

Schlier, R., 1961, J. Appl. Phys. **29**, 1162.

Schlier, A.R., 1982, Sample introduction system. J. Vac. Sci. & Technol. **20**(1), 100.

Schmidlin, F.W., 1962, Some investigations of cryotrapping. Trans. 9th American Vacuum Symposium (Macmillan, New York) p. 197.

Schrade, J., 1957, Les Resins Epoxy (Dunod, Paris).

Schrader, J.E., and H.M. Ryder, 1919, Phys. Rev. **13**, 321.

Schrag, G.M., and S.O. Colgate, 1987, A novel ultrahigh vacuum foreline valve. J. Vac. Sci. & Technol. A **5**(3), 388.

Schram, A., 1963, Le Vide **18**, 55.

Schrieffer, J.R., 1972, J. Vac. Sci. & Technol. **9**, 561.

Schuchman, J.C., 1983, Design of NSLS vacuum system. J. Vac. Sci. & Technol. A **1**(2/1), 196.

Schuchman, J.C., et al., 1979, Vacuum system for the national synchrotron light source. Proc. 25th American Vacuum Symposium. J. Vac. Sci. & Technol. **16**, 720.

Schuemann, W.C., 1962, A photo-current suppressor gauge for the measurement of very low pressures. Trans. 9th American Vacuum Symposium (Macmillan, New York) p. 428.

Schuetze, H.J., and F. Stork, 1962, Bayard–Alpert gauge with reduced X-ray limit. Trans. 9th American Vacuum Symposium (Macmillan, New York) p. 431.

Schuhmann, S., 1962, Study of Knudsen's methods of pressure division as a means of calibrating vacuum gauges. Trans. 9th American Vacuum Symposium (Macmillan, New York) p. 463.

Schulz, J., 1957, Rev. Sci. Instrum. **28**, 1051; J. Appl. Phys. **28**, 1149.

Schulz, G.J., 1962, Phys. Rev. **125**, 229.

Schulz, G.J., and A.V. Phelps, 1957, Rev. Sci. Instrum. **28**, 1051.

Schulz, W.D., and E. Usselmann, 1978, Modern turbomolecular pumps (in German). Vakuum-Technik **27**, 9.

Schulze, H., 1967, Vakuum-Technik **16**, 102.

Schütze, H.J., and E.W. Ehlbeck, 1962, The X-ray limit of ionization gauges. Trans. 2nd Int. Vacuum Congress (Pergamon Press, Oxford) p. 451.

Schwarz, H., 1960/1961, Arch. Tech. Mess. Dec. V. 1341-6, Feb. V. 1341-7; March V. 1341-8.

Schwarz, H., 1972, J. Vac. Sci. & Technol. **9**, 373.

Schwarz, H., 1977, Geometry and properties of the plasma in a quadrupole ion pump. J. Vac. Sci. & Technol. **14**, 731.

Schwarz, W., 1987, Thermal limitations of the pumping speed of refrigerator cooled cryosurfaces above $-90°C$. J. Vac. Sci. & Technol. A **5**(4), 2568.

Schwarz, H., and H.A. Tourtellotte, 1969, J. Vac. Sci. & Technol. **6**, 260.

Scuotto, F., 1986, A new gross leak calibration for leak detectors. J. Vac. Sci. & Technol. A **4**(5), 2287.

Sedgley, D.W., et al., 1987, Characterization of charcoals for He cryopumping in fusion devices. J. Vac. Sci. & Technol. A **5**(4), 2572.

Sedgley, D.W., et al., 1988, Helium cryopumping for fusion applications. J. Vac. Sci. & Technol. A **6**(3), 1209.

Segovia, J.L., and C.S. Martin, 1967, Behaviour of different ionization gauges at very low pressures. Trans. 3rd Int. Vacuum Congress (Pergamon Press, Oxford) p. 245.

Seibert, G., 1977, Spacelab experiments. J. Vac. Sci. & Technol. **14**, 1252.

Selhofer, H., and R. Wagner, 1977, Computer operated fully automatic vacuum plant for quality control. Proc. 7th Int. Vacuum Congress (Dobrozemsky, et al., Vienna) p. 211.

Sellenger, F.R., 1968, Vacuum **18**, 645.

Setina, J., et al., 1987, Vacuum tightness down to 10^{-15} mbar l/s range, measured with a spinning rotor viscosity gauge. J. Vac. Sci. & Technol. A **5**(4), 2650.

Sewell, P.B., and M. Cohen, 1965, Appl. Phys. Lett. **7**, 32.

Shand, E.B., 1958, Glass Engineering Handbook (McGraw-Hill, New York).

Sharma, J.K.N., and P. Mohan, 1976, A sensitive oil manometer, for the range $10-10^{-2}$ Torr. J. Phys. E (Sci. Instrum.) **9**, 618.

Sharma, J.K.N., and P. Mohan, 1988a, Use of porous plugs in the transition flow range for calibration of vacuum gauges. J. Vac. Sci. & Technol. A **6**(3), 1217.

Sharma, J.K.N., and D.R. Sharma, 1980, McLeod gauge, capillary depression and mercury vapor drag effect. J. Vac. Sci. & Technol. **17**(4), 820.

Sharma, J.K.N., and D.R. Sharma, 1982a, Comparison of two pumping speed measuring methods of oil diffusion pumps. Vacuum **32**(5), 253.

Sharma, J.K.N., and D.R. Sharma, 1982b, Study of molecular flow inside a test dome. Vacuum **32**(5), 283.

Sharma, J.K.N., and D.R. Sharma, 1988b, Measurement of the effective pressure distribution in axial direction in a dynamic vacuum system. J. Vac. Sci. & Technol. A **6**(4), 2508.

Shelby, J.E., 1972, Phys. Chem. Glasses **13**, 167.

Shen, G.L., 1987, The pumping of methane by an ionization assisted Zr/Al getter pump. J. Vac. Sci. & Technol. A **5**(4), 2580.

Shioyama, T., T. Takiguchi and S. Ogawa, 1978, Simple vacuum gauge using TaNi thin films in the pressure range 10^5–10^{-3} Pa. Proc. 24th American Vacuum Symposium, J. Vac. Sci. & Technol. **15**, 761.

Sickafus, E.N., 1974, J. Vac. Sci. & Technol. **11**, 299.

Siebert, J.F., and M. Omori, 1977, Anomalous behavior of Meissner coils in high-vacuum application. J. Vac. Sci. & Technol. **14**, 1307.

Siegbahn, K., ed., 1965, Alpha-, beta- and gamma-ray Spectroscopy (North-Holland, Amsterdam).

Sigmond, R.S., 1979, Vakuum Teknikk (in Norwegian) (Tapir Publishers, Trondheim, Norway).

Simon, H., 1924, Z. Tech. Phys. **5**, 221.

Simonds, H.R., 1959/1961, Sourcebook of the new Plastics (Reinhold, New York).

Simons, J.C., 1963, On uncertainties in calibration of vacuum gauges and the problem of traceability. Trans. 10th American Vacuum Symposium (Macmillan, New York) p. 246.

Simons, J.C., and R.E. King, 1967, Evaluation of uncertainties in vacuum gauge calibration. Trans. 3rd Int. Vacuum Congress (Pergamon Press, Oxford) p. 263.

Singleton, J.H., 1966, J. Vac. Sci. & Technol. **3**, 354.

Singleton, J.H., 1971, J. Vac. Sci. & Technol. **8**, 275.

Singleton, J.H., 1973, J. Phys. E **6**, 685.

Singleton, J.H., 1984, The development of valves, connectors and traps for vacuum systems during the 20th century. J. Vac. Sci. & Technol. A **2**(2), 126.

Singleton, J.H., and L.N. Yannopoulos, 1975, Yttrium and Scandium for radioactive electron emitters. Proc. 21st American Vacuum Symposium, J. Vac. Sci. & Technol. **12**, 414.

Sinharoy, S., and W.J. Lange, 1982, Behavior of small leaks in the presence of liquid or gaseous He at 4.2K. J. Vac. Sci. & Technol. **20**(4), 978.

Siu, M.C.I., 1973, Equations for thermal transpiration. J. Vac. Sci. & Technol. **10**, 368.

Skeits, J., 1958, Epoxy Resins (Reinhold, New York).

Skeits, J., 1962, Handbook of Adhesives (Chapman and Hall, London).

Sledziewski, Z., et al., 1974, The vacuum chamber of a Tokamak Plasma Experiment (TRF). Proc. 6th Int. Vacuum Congress, Jpn. J. Appl. Phys., Suppl. 2, Pt. 1, p. 217.

Smetana, F.O., and C.T. Carley, 1966, J. Vac. Sci. & Technol. 3, 49.

Smith, C.G., and G. Lewin, 1966, Free molecular conductance of a cylindrical tube with wall sorption. J. Vac. Sci. & Technol. 3, 92.

Smither, R.K., 1956, Rev. Sci. Instrum. 27, 964.

Snoek, C., and J. Kistemaker, 1965, Adv. Electron. & Electron Phys. 21, 67.

Snouse, T., 1971, J. Vac. Sci. & Technol. 8, 283.

Snyder, J.A., and E.C. Steel, 1962, Rev. Sci. Instrum. 33, 945.

Solbrig, C.W., and W.E. Jamison, 1965, J. Vac. Sci. & Technol. 2, 228.

Solomon, G.M., 1984, Air mass spectrometer leak detection using the special air leak test (SALT) chart. J. Vac. Sci. & Technol. A 2, 1157.

Solomon, G.M., 1986, Standardization and temperature correction of calibrated leaks. J. Vac. Sci. & Technol. A 4, 327.

Sommer, H., H.A. Thomas and J.A. Hipple, 1951, Phys. Rev. 82, 697.

Sørensen, G., and J.L. Whitton, 1977, Vacuum 27, 155.

Souchet, R., 1972, Le Vide 27, 125.

Sowell, R.R., et al., 1974, Surface cleaning by uv radiation. J. Vac. Sci. & Technol. 11, 474.

Spalvins, T., 1987, A review of recent advances in solid film lubrication. J. Vac. Sci. & Technol. A 5(2), 212.

Spinks, S.W., 1963, Vacuum Technology (Chapman and Hall, London) 125pp.

Springer, L., 1950, Laboratoriumsbuch für die Glasindustrie (Knapp, Halle).

Stansfield, B.L., et al., 1986, A simple ionization gauge for use in a magnetic field. J. Vac. Sci. & Technol. A 4(5), 2284.

Stanworth, J.E., 1950, Physical Properties of Glass (Clarendon Press, Oxford).

Stayt, J.W., 1982, A fail-safe automatic vent control for Pfeiffer turbopumps. J. Vac. Sci. & Technol. 21(4), 1028.

Steckelmacher, W., 1965, J. Sci. Instrum. 42, 63.

Steckelmacher, W., 1968, Performance specification of high vacuum pumps. Proc. 4th Int. Vacuum Congress (Institute of Physics, London) p. 67.

Steckelmacher, W., 1973a, Vacuum 23, 165.

Steckelmacher, W., 1973b, Vacuum 23, 307.

Steckelmacher, W., 1974, The flow of rarefied gases in vacuum systems and problems of standardization of measuring techniques. Trans. 6th Int. Vacuum Congress (Kyoto) p. 117.

Steckelmacher, W., 1976, personal communication.

Steckelmacher, W., 1978, The effect of cross-sectional shape on the molecular flow in long tubes. Vacuum **28**, 269.

Steckelmacher, W., and B. Fletcher, 1972, J. Vac. Sci. & Technol. **9**, 128.

Steckelmacher, W., and H. Henning, 1979, Comments on the "approximate calculation of transmission probabilities". Vacuum **29**, 31.

Steckelmacher, W., and D.M. Tinsley, 1962, Vacuum **12**, 153.

Steinherz, H.A., 1963, Handbook of High Vacuum Engineering (Reinhold, New York) 358pp.

Stern, S.A., and S.F. DiPaolo, 1969, J. Vac. Sci. & Technol. **6**, 941.

Stern, S.A., J.T. Mullhaupt, R.A. Hemstreet and F.S DiPaolo, 1965, J. Vac. Sci. & Technol. **2**, 165.

Stern, S.A., C.A. Hemstreet and D.M. Ruttenbur, 1966, J. Vac. Sci. & Technol. **3**, 99.

Stevens, C.M., 1953, Rev. Sci. Instrum. **24**, 148.

Steyskal, H., 1955, Arbeitsverfahren und Stoffkunde der Hochvakuumtechnik (Physik Verlag, Mosbach).

Stone, R.R., and R. Scully, 1979, Metal-seal valve for high radiation use. J. Vac. Sci. & Technol. **16**, 1049.

Störi, H., 1983, An in situ glow discharge cleaning method for LEP. Vacuum **33**(3), 171.

Stowers, I.F., 1978, Advances in cleaning metal and glass surfaces to micron-level cleanliness. Proc. 24th American Vacuum Symposium, J. Vac. Sci. & Technol. **15**, 751.

Strausser, Y.E., 1968, The effect of surface treatment on the outgassing rates of 304 stainless steel. Proc. 4th Int. Vacuum Congress (Institute of Physics, London) p. 469.

Strong, J., 1938, Procedures in Experimental Physics (Prentice-Hall, New York) p. 93–187.

Strubin, P., 1980, Study of a new method to control precisely the evaporation rate of Ti sublimation pumps. J. Vac. Sci. & Technol. **17**(5), 1216.

Stuart, R.V., 1983, Vacuum Technology, Thin Films and Sputtering; An Introduction (Academic Press, New York).

Suchannek, R.G., and J.R. Sheridan, 1974, Ionization gauge for relatively high pressures. Rev. Sci. Instrum. **45**, 19.

Suemitsu, M., et al., 1987, Aluminum alloy ultrahigh vacuum chamber for molecular beam epitaxy. J. Vac. Sci. & Technol. A **5**(1), 37.

Sullivan, J.J., 1966, Research/Development **17**(9), 64.

Sutherland, W., 1897, Philos. Mag. **43**, 83.

Sutton, C.M., and K.F. Poulter, 1982, A new reference ionization gauge (10^{-5} to 1 Pa). Vacuum **32**(5), 247.

Swartz, J.C., 1956, Evapor-ion pump characteristics. Trans. 3rd American Vacuum Symposium (Pergamon Press, Oxford) p. 83.

Sweetman, D.R., 1961, Nucl. Instrum. & Methods **13**, 317.

Swingler, D.L., 1968, Vacuum **18**, 669.

Szwemin, P.J., and S. Pytowski, 1977, On some improvements of ionization gauge stability. Proc. 7th Int. Vacuum Congress (Dobrozemsky, et al., Vienna) p. 141.

T

Tang, Z.Q., et al., 1987, The design and operation of a new extractor gauge. J. Vac. Sci. & Technol. A **5**(4), 2384.

Tasman, H.A., A.J. Boerboom and J. Kistemaker, 1963, Vacuum **13**, 33.

Tate, J.T., and P.T. Smith, 1932, Phys. Rev. **39**, 270.

Taylor, N.J., 1969, Vacuum **19**, 575.

Templemeyer, K.E., R. Dawdarn and R.I. Young, 1970, J. Vac. Sci. & Technol. **8**, 575.

Teutenberg, K., 1972, Uhv all-metal valves. Vakuum-Technik **21**, 169.

Thibault, J.J., J. Roussel and H. Jourdan, 1967, Le Vide **22**, 309.

Thieme, G., 1963, Vacuum **13**, 137.

Thomas, L.B., and R.E. Brown, 1950, J. Chem. Phys. **18**, 1367.

Thomas, E., and R. Leyniers, 1974, The McLeod gauge – Its birth, development and actuality. Trans. 6th Int. Vacuum Congress (Kyoto) p. 147.

Thomas, L.B., and E.B. Schofield, 1955, J. Chem. Phys. **23**, 861.

Thomas, L.B., C.L. Krueger and R.E. Harris, 1977, Mercury pumping pressure measurements in common gases. Proc. 7th Int. Vacuum Congress (Dobrozemsky, et al., Vienna) p. 161.

Thompson, W., and S. Hanrahan, 1977, Characteristics of a cryogenic extreme high-vacuum chamber. Proc. 23rd American Vacuum Symposium, J. Vac. Sci. & Technol. **14**, 643.

Thomson, S.L., and W.R. Owens, 1975, A survey of flow at low pressures. Vacuum **25**, 151.

Thornberg, S.M., 1988, Stepped linear piston displacement fundamental leak calibration system. J. Vac. Sci. & Technol. A **6**(4), 2522.

Thorpe, D.J., 1968, Vacuum **18**, 441.

Tilford, C.R., 1983, Reliability of high vacuum measurements. J. Vac. Sci. & Technol. A **1**(2/1), 152.

Tilford, C.R., 1985, Sensitivity of hot cathode ionization gauges. J. Vac. Sci. & Technol. A **3**(2), 546.

Tilford, C.R., and K.E. McCulloh, 1982, Performance characteristics of a broad range ionization gauge. J. Vac. Sci. & Technol. **20**(4), 1140.

Timoshenko, S., 1936, Theory of Elastic Stability (McGraw-Hill, New York).

Tjagunov, G.A., 1948, Foundations for Calculation of Vacuum Systems (in Russian) (Gossenergoizdat, Moscow).

Tobin, A.G., et al., 1987, Evaluation of charcoal sorbents for helium cryopumping in fusion reactors. J. Vac. Sci. & Technol. A 5(1), 101.

Toby, S., and K.O. Kutsche, 1957, Rev. Sci. Instrum. 28, 470.

Todd, C.J., 1973, Vacuum 23, 195.

Tom, T., 1972, Use of a metallic ion-source in cold cathode sputter-ion pumps. Proc. 5th Int. Vacuum Congress, J. Vac. Sci. & Technol. 9, 383.

Tom, T., and B.D. James, 1966, J. Vac. Sci. & Technol. 3, 300.

Toth, G., 1968, Speed and compressibility computations by a new diffusion-pumping theory based on the kinetic theory of non-uniform gases. Proc. 4th Int. Vacuum Congress (Institute of Physics, London) p. 300.

Toth, G., 1977, The relation of the phenomenological and kinetical theories of diffusion pumps. Proc. 7th Int. Vacuum Congress (Dobrozemsky, et al., Vienna) p. 17.

Trendelenburg, E.A., 1963, Ultrahochvakuum (Verlag Braun, Karlsruhe) 196pp.

Tretner, W., 1960, Vacuum 10, 31.

Tretner, W., 1960, Z. Angew. Phys. 11, 395.

Trickett, B.A., 1977, Vacuum systems for the Daresbury Synchrotron radiation source. Proc. 7th Int. Vacuum Congress (Dobrozemsky, et al., Vienna) p. 347.

Trickett, B.A., 1978, Vacuum 28, 471.

Truffier, J.L., and P.S. Choumoff, 1974, Accuracy of $(\Delta P/P) < 10^{-7}$ obtained on a differential interferometric manometer MDI used with recorder. Proc. 6th Int. Vacuum Congress, Jpn. J. Appl. Phys., Suppl. 2, Pt. 1, p. 139.

Tu, J.Y., et al., 1988, A further exploration of an important factor affecting the pumping performance of turbomolecular pumps. J. Vac. Sci. & Technol. A 6(4), 2535.

Turnbull, A.H., 1965, Vacuum 15, 3.

Turnbull, J.C., 1977, Barium, strontium and calcium as getters in electron tubes. Proc. 23rd American Vacuum Symposium, J. Vac. Sci. & Technol. 14, 636.

Turnbull, A.H., R.S. Barton and J.C. Riviere, 1962, An introduction to Vacuum Technique (Newnes, London) 190pp.

Turner, F.T., and M. Feinleib, 1962, Performance criteria for sorption pumps. Trans. 2nd Int. Vacuum Congress (Pergamon Press, Oxford) p. 300.

Turner, F.T., and W.H. Hogan, 1966, J. Vac. Sci. & Technol. 3, 252.

Turner, I.A., R.M. Pickard and C.R. Hoffman, 1962, J. Sci. Instrum. 39, 26.

Tuzi, Y., and M. Kobayashi, 1983, Directional detectors for measurement of gases released from surfaces. Proc. 9th Int. Vacuum Congress (Madrid) p. 203.

U

Ullman, I.R., 1962, Commercial seals as seats in a bakeable valve. Trans. 8th Int. Vacuum Congress (Pergamon Press, Oxford) p. 1323.

Unterlerchner, W., 1977, The influence of the mechanical properties of gasket materials on the reliability of bakeable uhv flanges. Proc. 7th Int. Vacuum Congress (Dobrozemsky, et al., Vienna) p. 247.

Unterlerchner, W., 1987, Some improvement work on ConFlat joints and their limit of reliability in a lirge-size ultrahigh vacuum system. J. Vac. Sci. & Technol. A 5(4), 2540.

V

Vacca, R.H., 1957, Recent advances in the Alphatron vacuum gauge. Trans. 4th American Vacuum Symposium (Pergamon Press, Oxford) p. 93.

Van Atta, C.M., 1965, Vacuum Science and Engineering (McGraw-Hill, New York) 459pp.

Van Essen, D., and W.C. Heerens, 1976, On the transmission probability for molecular gas flow through a tube. J. Vac. Sci. & Technol. 13, 1183.

Van Heerden, P.J., 1955, Rev. Sci. Instrum. 26, 1131.

Van Herwaarden, A.W., 1987, Thermal vacuum sensors; Doctor Thesis (Technische Universiteit, Delft).

Van Herwaarden, A.W., and P.M. Sarro, 1987, Double-beam integrated thermal vacuum sensor. J. Vac. Sci. & Technol. A 5(4), 2454.

Van Oostrom, A., 1962, A Bayard–Alpert type ionization gauge with a low X-ray limit. Trans. 8th American Vacuum Symposium (Pergamon Press, Oxford) p. 443.

Van Oostrom, A., 1972, Vacuum 22, 15.

Van Zyl, 1976, The linearity of a capacitance diaphragm manometer. Rev. Sci. Instrum. 47, 1214.

Varadi, P.F., and L.G. Sebestyen, 1956, J. Sci. Instrum. 33, 392.

Varga, J.E., and W.A. Bailey, 1973, Solid State Technol. 16(12), 79.

Varicak, M., 1956, Rev. Sci. Instrum. 27, 655.

Vaumoron, J., 1970, Le Vide 25, 26.

Venema, A., 1973/1974, The flow of highly rarefied gases. Philips Tech. Rev. 33, 43.

Venema, A., and M. Brandriga, 1958, Philips Tech. Rev. 20, 145.

Vijendran, P., and C.V.G. Nair, 1971, Vacuum 21, 159.

Visser, J., and J.J. Scheer, 1979, Twenty-Kelvin cryopumping in magnetron sputtering systems. Proc. 25th American Vacuum Symposium, J. Vac. Sci. & Technol. 16, 734.

Visser, J., B. Symersky and A.J.M. Geraerts, 1977, A versatile cryopump. Vacuum 27, 175.

Voege, W., 1906, Z. Phys. 7, 498.

Vollbrecht, H., 1974, Closed cylinder and sphere under external pressure (in German). Chemie-Technik 3, 331.

Von Ardenne, M., 1962, Tabellen zur angewandten Physik – Vakuumphysik (Verlag Wissenschaften, Berlin) 757pp.

Von Guericke, O., 1968, Neue Magdeburger Versuche über den Leeren Raum (Translation into German, of the Latin book published in 1672) (VDI Verlag, Dusseldorf) 291pp.

Von Zahn, U., 1963, Rev. Sci. Instrum. 34, 1.

W

Waclawski, B.J., 1983, A ultrahigh vacuum gasket removal tool. J. Vac. Sci. & Technol. A 1(1), 99.

Waelbroek, A., et al., 1984, Cleaning and conditioning of the walls of plasma devices by glow discharges in hydrogen. J. Vac. Sci. & Technol. A 2(4), 1521.

Wagener, S., and C.B. Johnson, 1951, J. Sci. Instrum. 28, 278.

Wagner, G., 1950, Erzeugung und Messung von Hochvakuum (Deuticke, Wien) p. 73.

Waldschmidt, E., 1954, Metallurgy 8, 749.

Waldschmidt, M., 1973, Vacuum 23, 435.

Walters, W.P., 1982, Mean free path of emitted gas from spacecraft. J. Vac. Sci. & Technol. 20(2), 255.

Ward, L., and J.P. Bunn, 1967, Introduction to the Theory and Practice of High Vacuum Technology (Butterworths, London) 216pp.

Warshawski, I., 1972, J. Vac. Sci. & Technol. 9, 196.

Warshawsky, I., 1982, Comparison of ion gauge calibration by several standard laboratories. J. Vac. Sci. & Technol. 20(1), 75.

Watanabe, F., 1987a, Point collector ionization gauge with spherical grid for measuring pressures below 10^{-11} Pa. J. Vac. Sci. & Technol. A 5(2), 242.

Watanabe, K., 1987b, Sensitivity of Bayard–Alpert gauges (for hydrogen isotopes). J. Vac. Sci. & Technol. A 5(2), 237.

Watanabe, F., et al., 1983, Modulating ion current pressure gauge. Vacuum 33(5), 271.

Watson, J.S., and P.W. Fisher, 1977, Cryosorption vacuum pumping under fusion reactor conditions. Proc. 7th Int. Vacuum Congress (Dobrozemsky, et al., Vienna) p. 363.

Watt, P.R., 1963, Molecular Stills (Reinhold, New York).

Webb, D.B., 1974, Oil lubrication/sealing on mechanical vacuum pumps. Trans. 6th Int. Vacuum Congress (Kyoto) p. 853.

Webber, R.J., and C.T. Lane, 1946, Rev. Sci. Instrum. 17, 308.

Weber, F., 1968, Dictionary of High Vacuum (German, English, French, Spanish, Italian, Russian) (Elsevier, Amsterdam).

Weber, W.H., and M.B. Webb, 1969, Phys. Rev. 177, 1103.

Wehner, G.K., 1955, Adv. Electron. & Electron Physics 7, 239.

Wehner, G.K., 1974, Sputtering in thin film analysis methods. Proc. 6th Int. Vacuum Congress, Jpn. J. Appl. Phys., Suppl. 2, Pt. 1, p. 459.

Weitzel, D.H., 1960, Rev. Sci. Instrum. 31, 1350.

Welch, K.M., 1973, The pressure profile in a long outgassing vacuum tube. Vacuum 23, 271.

Welch, K.M., 1976, New developments in sputter-ion pump configurations. Proc. 22nd American Vacuum Symposium, J. Vac. Sci. & Technol. 13, 498.

Westgaard, H., 1970, Vacuum 20, 143.

Weston, G.F., 1975, Materials for uhv. Vacuum 25, 469.

Weston, G.F., 1978, Pumps for uhv. Vacuum 28, 209.

Weston, G.F., 1979, Measurements of uhv, Vacuum 29, 277.

Weston, G.F., 1980, Partial pressure measurement in ultrahigh vacuum. Vacuum 30(2), 49.

Weston, G.F., 1984, Ultrahigh vacuum line components. Vacuum 34, 619.

Weston, G.F., 1985, Ultrahigh Vacuum Practice (Butterworth, London).

Wheeler, E.L., 1958, Scientific Glass Blowing (Interscience, New York).

Wheeler, W.R., 1963, Theory and application of metal gasket seals. Trans. 10th American Vacuum Symposium (Macmillan, New York) p. 159.

Wheeler, W.R., 1971, J. Vac. Sci. & Technol. 8, 337.

Wheeler, W., 1974, J. Vac. Sci. & Technol. 11, 332.

Wheeler, W.R., 1976, Recent developments in metal-sealed gate valves. Proc. 22nd American Vacuum Symposium, J. Vac. Sci. & Technol. 13, 503.

Wheeler, W.R., 1977, The design of large vacuum seals for nuclear fusion machines. Proc. 7th Int. Vacuum Congress (Dobrozemsky, et al., Vienna) p. 251.

Wheeler, W.R., and M.A. Carlson, 1962, Ultra-high vacuum flanges. Trans. 2nd Int. Vacuum Congress (Pergamon Press, Oxford) p. 1309.

White, W.C., and J.S. Hickey, 1948, Electronics 21, 100.

Whitman, C.B., 1987, Reclamation of vacuum pump fluids. J. Vac. Sci. & Technol. A 5(2), 255.

Wikberg, T., 1971, Vacuum 21, 337.

Wikberg, T., and A. Poncet, 1987, Large ultrahigh vacuum joints. J. Vac. Sci. & Technol. A **5**(4), 2624.

Williams, B.J., et al., 1968, Practical design aspects of a continuous air-to-air vacuum processing machine. Proc. 4th Int. Vacuum Congress (Institute of Physics, London) p. 753.

Willis, R.F., and B. Filton, 1972, J. Vac. Sci. & Technol. **9**, 651.

Wilson, R.R., 1941, Rev. Sci. Instrum. **12**, 91.

Wilson, D.B., 1969, Vacuum **19**, 323.

Wilson, N.G., 1975, Leak testing large vacuum systems. Proc. 21st American Vacuum Symposium, J. Vac. Sci. & Technol. **12**, 436.

Wilson, N.G., 1985, Management of vacuum leak-detection processes, calibration and standards. J. Vac. Sci. & Technol. A **3**(2), 533.

Wilson, S.R., 1987, Numerical modeling of vacuum systems using electronic circuit analysis tools. J. Vac. Sci. & Technol. A **5**(4), 2479.

Wilson, N.G., and L.C. Beavis, 1976, Handbook of Leak Detection, ed. W.R. Bottoms (American Vacuum Society, New York).

Wilson, N.G., and K.N. Watts, 1980, Design and operational characteristics of the cryopump high vacuum system for the 2500 liter target chamber of the helios laser facility. Proc. 26th American Vacuum Symposium, J. Vac. Sci. & Technol. **17**, 270.

Windsor, E.E., 1970, Vacuum **20**, 7.

Winkel, T., and J.L. Hemmerich, 1987, Helium leak detection in JET in the presence of high deuterium partial pressures. J. Vac. Sci. & Technol. A **5**(4), 2637.

Winkelman, C.R., and H.G. Davidson, 1975, Ultrasensitive leak detection. Proc. 21st American Vacuum Symposium, J. Vac. Sci. & Technol. **12**, 435.

Winkelman, C.R., and H.G. Davidson, 1979, Vacuum **29**, 361.

Winkler, A., 1987, Absolute calibration of small gas amounts. J. Vac. Sci. & Technol. A **5**(4), 2430.

Winkler, O., and R. Bakish, 1971, Vacuum Metallurgy (Elsevier, Amsterdam) 890pp.

Winter, H., 1975, The present state of controlled nuclear fusion research. Vacuum **25**, 497.

Winterbottom, W.L., 1973, Mass spectrometer calibration. J. Vac. Sci. & Technol. **10**, 871.

Winters, H.F., D.R. Denison and D.G. Bills, 1962, Rev. Sci. Instrum. **33**, 520.

Wishart, I., and G.H. Bancroft, 1961, A new design for ultra-high vacuum valves. Trans. 7th American Vacuum Symposium (Pergamon Press, Oxford) p. 13.

Wolgast, C.R., 1975, Design of the cryopumping vacuum system for ESCAR. IEEE Trans. Nucl. Sci. **NS-22**, 1496.

Wolsky, S.P., and A.W. Czanderna, 1975, Methods of Surface Analysis (Elsevier, New York).

Wolsky, A.W. Czanderna and S.P., eds., 1980, Microweighing in Vacuum and Controlled Environments (Elsevier, Amsterdam).

Wong, W., et al., 1988, An evaluation of the composition of the residual atmosphere above a commercial dry pump. J. Vac. Sci. & Technol. A 6(3), 1183.

Wood, S.D., and C.R. Tilford, 1985, Long term stability of two types of hot cathode ionization gauges. J. Vac. Sci. & Technol. A 3(2), 542.

Woods, R.O., 1973, J. Vac. Sci. & Technol. 10, 433.

Woods, R.O., and T.K. Devlin, 1974, A cryopumping system for balloon-borne mass spectrometers. Rev. Sci. Instrum. 45, 136.

Wössner, H., 1972, A 20K cryopump (in German). Vakuum-Technik 21, 101.

Wu, Y., and B. Dutt, 1972, J. Vac. Sci. & Technol. 9, 1248.

Wutz, M., 1965, Theorie und Praxis der Vakuumtechnik (Vieweg & Sohn, Braunschweig) 439pp.

Wutz, M., 1968, Vakuumpumpen nach dem Kreiskolbenprinzip. Proc. 4th Int. Vacuum Congress (Institute of Physics, London) p. 283.

Wutz, M., 1969, Vacuum 19, 1.

Wutz, M., 1979, Interpretation of the diffusion pump mechanism by gas-dynamic methods (in German). Vakuum-Technik 28, 136, 172.

Wutz, M., 1982, Establishing the characteristics of vapour ejector pumps. Vakuum-Technik 31(5), 146.

Wutz, M., H. Adam and W. Walcher, 1982, Theorie und Praxis der Vakuumtechnik (Vieweg, Braunschweig).

Wycliffe, H., 1987, Mechanical high vacuum pumps with an oil-free swept volume. J. Vac. Sci. & Technol. A 5(4), 2608.

Wyllie, H.A., 1957, J. Sci. Instrum. 34, 410.

X

Xu, Tingwei, and Wang Kaiping, 1982, The relation between the conductance of an elbow and the angle between tubes. Vacuum 32(10/11), 655.

Y

Yang, N., et al., 1987, Study of the structure and performance of a new type of molecular pump. J. Vac. Sci. & Technol. A 5(4), 2594.

Yarwood, J., 1943, High Vacuum Technique, 1st Ed. (Chapman and Hall, London).

Yarwood, J., 1967, High Vacuum Technique, 4th Ed. (Chapman and Hall, London) 274pp.

Yarwood, J., 1977, International vacuum standardization. Vacuum **27**, 87.

Yarwood, J., 1979, The production of vacuum: past, present, future. Vacuum **29**, 313.

Yatsu, K., et al., 1988, GAMMA 10 vacuum system and study of wall conditions. J. Vac. Sci. & Technol. A **6**(4), 2546.

Yoshimura, N., 1985, A differential pressure-rise method for measuring the net outgassing rates. J. Vac. Sci. & Technol. A **3**(6), 2177.

Young, J.R., 1961, Rev. Sci. Instrum. **32**, 85.

Young, J.F., 1966, Materials and Processes (Wiley, New York).

Young, J.R., 1966, J. Vac. Sci. & Technol. **3**, 345.

Young, J.R., 1970, J. Vac. Sci. & Technol. **7**, 210.

Young, D.M., and A.D. Crowell, 1962, Physical Adsorption of Gases (Butterworths, London).

Young, J.R., and F.P. Hession, 1963, A cold cathode discharge gauge for ultrahigh vacuum. Trans. 10th American Vacuum Symposium (Macmillan, New York) p. 234.

Young, J.R., and N.R. Whetten, 1962, Techniques for the admission of high purity gases to a vacuum system. Trans. 2nd Int. Vacuum Congress (Pergamon Press, Oxford) p. 625.

Yu-guo, F., 1981, Approximate calculation for molecular conductance. Vacuum **31**(7), 319.

Z

Zabielsky, M.F., and P.R. Blaszuk, 1976, Vacuum evaluation of a new perfluoroelastomer. J. Vac. Sci. & Technol. **13**, 644.

Zdanuk, E.J., R. Bierig and L.G. Rubin, 1960, Vacuum **10**, 382.

Zettler, J., and R. Sud, 1988, Extension of thermocouple gauge sensitivity to atmospheric pressure. J. Vac. Sci. & Technol. A **6**(3), 1153.

Zhilnin, V.S., and L.P. Zhilnina, 1968, Gas emission from materials. Proc. 4th Int. Vacuum Congress (Institute of Physics, London) p. 801.

Zincke, A., 1961, Technologie der Glasverschmelzungen (Akademischen Verlag, Leipzig).

Zobac, L., 1954, Foundations of Vacuum Technique (in Czech) (Statni Nakladatelstvi Techn., Lit., Praha) 313pp.

Subject Index